GW01373135

50p

Langmuir–Blodgett
Films

Langmuir–Blodgett Films

Edited by

GARETH ROBERTS

Thorn EMI plc
Hayes, Middlesex, United Kingdom
and University of Oxford
Oxford, United Kingdom

Plenum Press • New York and London

Library of Congress Cataloging-in-Publication Data

Langmuir-Blodgett films / edited by Gareth Roberts.
    p.   cm.
   Includes bibliographical references.
   ISBN 0-306-43316-8
   1. Thin films, Multilayered. 2. Monomolecular films.
I. Roberts, G. G. (Gareth Gwyn), 1940-
QC176.9.M84L36 1990
621.381'52--dc20

89-72111
CIP

© 1990 Plenum Press, New York
A Division of Plenum Publishing Corporation
233 Spring Street, New York, N.Y. 10013

Printed in the United States of America

Contributors

W. A. Barlow, Electronics Group, Imperial Chemical Industries plc, Runcorn, Cheshire WA7 4QE, England

S. D. Forrester, Department of Pure and Applied Chemistry, University of Strathclyde, Glasgow, Scotland

†C. H. Giles, Department of Pure and Applied Chemistry, University of Strathclyde, Glasgow, Scotland

R. A. Hann, ICI Imagedata, Brantham, Manningtree, Essex CO11 1NL, England

D. Möbius, Max Planck Institute for Biophysical Chemistry (Karl-Freidrich-Bonhoeffer Institute), D-3400 Göttingen-Nikolausberg, Federal Republic of Germany

M. C. Petty, Molecular Electronics Research Group, School of Engineering and Applied Science, University of Durham, Durham DH1 3LE, England

G. G. Roberts, Department of Engineering Science, University of Oxford, Oxford OX1 3PJ, England, and Thorn EMI plc, Hayes, Middlesex UB3 1HH, England

R. M. Swart, Corporate Colloid Science Laboratory, Imperial Chemical Industries plc, Runcorn, Cheshire WA7 4QE, England

†Deceased.

Preface

Monomolecular assemblies on substrates, now termed Langmuir–Blodgett (LB) films, have been studied for over half a century. Their development can be viewed in three stages. Following the pioneering work of Irving Langmuir and Katharine Blodgett in the late 1930s there was a brief flurry of activity just before and just after the Second World War. Many years later Hans Kuhn published his stimulating work on energy transfer. This German contribution to the field, made in the mid-1960s, can be regarded as laying the foundation for studies of artificial systems of cooperating molecules on solid substrates. However, the resurgence of activity in academic and industrial laboratories, which has resulted in four large international conferences, would not have occurred but for British and French groups highlighting the possible applications of LB films in the field of electronics.

Many academic and industrial establishments involved in high technology are now active in or maintaining a watching brief on the field. Nevertheless this important area of solid state science is still perhaps largely unfamiliar to many involved in materials or electronic device research. The richness of the variety of organic molecular materials suitable for LB film deposition offers enormous scope for those interested in their basic properties or their practical applications.

LB films are now an integral part of the field of molecular electronics. It seems inevitable that they will play some role in replacing inorganic materials in certain areas of application. Enthusiasts believe that a further stage will be reached with supermolecular assemblies in which entirely novel effects are discovered, and these findings in turn will lead to the fabrication of novel devices operating at the nanometer scale for the processing of information and its transmission and storage.

This book, which consists of contributions from leading research workers, is intended to impart to this broad audience and students new to the field a firm grounding in the subject. All aspects of this truly interdisciplinary and multi-disciplinary subject are covered, from a historical introduction to a concluding chapter describing the potential applications of LB films. Each chapter is more or less independent of the others and each author has been encouraged to write in his own style. Therefore omissions and variations in the level of treatment are inevitable. Unfortunately, the pace of research in the field will have made the references in some chapters representative rather than exhaustive. We apologize to those whose recent work is not cited and would direct readers to the proceedings of the 1989 International Conference.

In the course of writing, many research colleagues have provided assistance. The authors wish to acknowledge this support and also wish to thank Mrs. Frances Greenwood and Mrs. Pauline Morrell for their assistance with the preparation of the manuscript.

Gareth Roberts

Hayes and Oxford

Contents

Chapter 1

Historical Introduction

    C. H. Giles, S. D. Forrester, and G. G. Roberts

1.1. Introduction 1
1.2. Early Contributors to the Field of Monolayer Science 2
    1.2.1. Franklin (1706–1790) 3
    1.2.2. Shields (1822–1890) 4
    1.2.3. Aitken (1839–1919) 5
    1.2.4. Rayleigh (1842–1919) 6
    1.2.5. Pockels (1862–1935) 8
    1.2.6. Langmuir (1881–1957) 9
    1.2.7. Blodgett (1898–1979) 11
1.3. Conclusion 13
References 14

Chapter 2

Molecular Structure and Monolayer Properties

    R. A. Hann

2.1. Introduction 17
2.2. Formation and Stability of Monolayers 18
    2.2.1. Surface Pressure/Area Isotherms 19
    2.2.2. Modeling Molecular Structure 24
    2.2.3. Other Properties of Langmuir Films 25
    2.2.4. Mixed Monolayers 26

2.3. Langmuir–Blodgett Deposition   27
2.4. Types of Molecules Known to Form Monolayers and Langmuir–
      Blodgett Films   32
      2.4.1. Fatty Acids and Their Derivatives   32
              2.4.1.1. Chemical Structure of Fatty Acids   32
              2.4.1.2. Incorporation of Metal Ions   33
              2.4.1.3. Fatty Acids with Modified Hydrophobic Tailgroups . .   35
              2.4.1.4. Modifications to the Hydrophilic Headgroup   41
      2.4.2. Molecules Containing Five- or Six-Membered Rings   45
              2.4.2.1. Derivatives of Benzene   45
              2.4.2.2. Derivatives of Polycyclic Aromatic Hydrocarbons . . .   46
              2.4.2.3. Heterocyclic Compounds and Dyes   53
      2.4.3. Porphyrins and Phthalocyanines   59
      2.4.4. Mixed Monolayers of Unsubstituted Hydrophobic Materials . . .   67
2.5. Polymers and Polymerizable Materials   68
      2.5.1. Preformed Polymers   68
      2.5.2. Polymers Formed *in situ* by Addition of Carbon–Carbon
              Double Bonds   73
      2.5.3. Diacetylenes   76
      2.5.4. Oxirans   78
      2.5.5. Condensation Polymers   78
      2.5.6. Opportunities for New Materials   78
2.6. Exceptions   79
      2.6.1. Non-Langmuir–Blodgett Multilayers   79
      2.6.2. Nonaqueous Subphases   81
2.7. Prospects   82
2.8. Postscript   83
References   83

Chapter 3

Film Deposition

   M. C. Petty and W. A. Barlow

3.1. Deposition Principles   93
3.2. Subphase and Containment   98
3.3. Compression Mechanisms   100
      3.3.1. Single Movable Barrier   100
      3.3.2. Circular Trough   101
      3.3.3. Constant Perimeter Barrier   102
3.4. Ancillary Equipment   105
      3.4.1. Surface Pressure Measurement   105
      3.4.2. Surface Area Measurement   108
      3.4.3. Deposition Equipment   109

      3.4.4. Control Systems 110
3.5. Trough Environment 112
3.6. Experimental Techniques 113
      3.6.1. Surface Cleaning 113
      3.6.2. Monolayer Spreading 114
      3.6.3. Substrate Preparation 115
      3.6.4. Monolayer Transfer 118
3.7. Other Systems 120
      3.7.1. Continuous Fabrication 120
      3.7.2. Alternate-Layer Troughs 122
References 123

Chapter 4

Characterization and Properties

   M. C. Petty

4.1. Introduction 133
4.2. Evaluation of Film Thickness 134
      4.2.1. Interference Techniques 134
      4.2.2. Ellipsometry 136
      4.2.3. X-Ray Diffraction 137
      4.2.4. Neutron Diffraction and Reflection 140
      4.2.5. Alternative Methods 141
4.3. Film Structure 141
      4.3.1. Electron Microscopy 142
      4.3.2. Optical Microscopy 148
      4.3.3. IR and Visible Spectroscopy 149
      4.3.4. Raman Scattering 154
      4.3.5. Surface Analytical Techniques 157
      4.3.6. Other Characterization Methods 160
4.4. Optical Properties 162
      4.4.1. Refractive Index 162
      4.4.2. Nonlinear Effects 165
4.5. Electrical Characterization 167
      4.5.1. Specimen Preparation 167
      4.5.2. Quantum Mechanical Tunneling 169
      4.5.3. Direct Current Conduction through Multilayers 175
      4.5.4. Conductance 181
      4.5.5. Permittivity 187
      4.5.6. Permanent Polarization 190
      4.5.7. Dielectric Breakdown 192
4.6. Other Phenomena 193
      4.6.1. Mechanical Properties 193

4.6.2. Permeability Studies 195
References 196

Chapter 5

Spectroscopy of Complex Monolayers

D. Möbius

5.1. Introduction 223
5.2. Transmission, Absorption, and Reflection of Light 224
    5.2.1. Transmission 225
    5.2.2. Reflection 234
    5.2.3. Infrared Spectroscopy 249
5.3. Emission 251
    5.3.1. Steady State Fluorescence 252
    5.3.2. Fluorescence Decay 259
    5.3.3. Phosphorescence 263
5.4. Indirect Optical Methods 263
    5.4.1. Holographic Techniques 264
    5.4.2. Photoacoustic Spectroscopy 265
    5.4.3. Coupling to Surface Plasmons 266
References 267

Chapter 6

Monolayers and Multilayers of Biomolecules

R. M. Swart

6.1. Introduction 273
6.2. Langmuir Monolayers of Biological Molecules 276
    6.2.1. Phospholipids and Sterols 277
    6.2.2. Proteins 282
    6.2.3. Pigments 284
    6.2.4. Fatty Acids and Polymerizable Phospholipids 285
6.3. Preparation of Supported Multilayers 287
    6.3.1. Dipping Phospholipids onto a Vertical Support 288
    6.3.2. Dipping Phospholipids onto a Horizontal Support 289
    6.3.3. Dipping Phospholipids: General Considerations 290
    6.3.4. Dipping Polymerizable Derivatives of Phospholipids 291
    6.3.5. Oriented Multilayers by Other Methods 292
    6.3.6. Proteins and Pigments 293
6.4. Studies Employing Supported Molecular Layers 295
    6.4.1. Spectroscopic Studies 295
    6.4.2. Membrane Properties 297
    6.4.3. Proteins 299

6.4.4. Pigments and Photosynthesis 301
6.4.5. Electrical Properties. 301
6.4.6. Polymeric Surfaces 302
6.5. Concluding Remarks 303
6.6. Appendix 304
References 307

Chapter 7

Potential Applications of Langmuir–Blodgett Films

G. G. Roberts

7.1. Introduction 317
    7.1.1. Molecular Electronics 317
    7.1.2. Supermolecular Electronics 318
    7.1.3. Possible Niches for Langmuir–Blodgett Films 320
7.2. Patents Filed by Langmuir and Blodgett 321
    7.2.1. Image Reproduction 321
    7.2.2. Optical Devices 322
    7.2.3. Step Gauge 323
    7.2.4. Mechanical Filters 324
    7.2.5. Sensors 324
    7.2.6. Summary 324
7.3. Model Systems in Basic Research 325
    7.3.1. Energy Transfer in Complex Monolayers 325
    7.3.2. Magnetic Monolayers 326
    7.3.3. Determination of Electron Energy Ranges 328
    7.3.4. Biological and Permeable Membranes 329
7.4. Passive Thin-Film Applications 332
    7.4.1. Electron Beam Microlithography 332
    7.4.2. Lubrication 334
    7.4.3. Enhanced Device Processing 336
        7.4.3.1. Surface Acoustic Waves 336
        7.4.3.2. Liquid Crystal Alignment 337
7.5. Piezoelectric and Pyroelectric Organic Films 339
    7.5.1. Piezoelectric Properties 342
    7.5.2. Pyroelectric Properties 344
        7.5.2.1. Theoretical Consideration in a Thermal Imager 345
        7.5.2.2. Pyroelectric Langmuir–Blodgett Films 347
7.6. Applied Optics Using Langmuir–Blodgett Films 351
    7.6.1. Waveguiding in Langmuir–Blodgett Films 352
    7.6.2. Nonlinear Optics: Relevant Theory 353
        7.6.2.1. Second-Order Effects 354
        7.6.2.2. Third-Order Effects 356
    7.6.3. Optoelectronic Applications of Organic Films 357

7.6.4. Nonlinear Optics: Langmuir–Blodgett Films 358
    7.6.4.1. Second-Order Effects 358
    7.6.4.2. Electrooptic Effects 362
    7.6.4.3. Third-Harmonic Effects 362
    7.6.4.4. Summary 365
7.6.5. Optical Information Storage 365
7.7. Electrical Properties of Monolayers and Multilayers 368
    7.7.1. Electrical Conductivity in Langmuir–Blodgett Films 368
    7.7.2. Langmuir–Blodgett Films as Insulators 370
    7.7.3. Photoelectronic Behavior of Langmuir–Blodgett Films
        on Metals 371
    7.7.4. Molecular Rectification and Cooperation 373
        7.7.4.1. Molecular Pumping 373
        7.7.4.2. Superconductivity 374
        7.7.4.3. Organic Rectification 374
7.8. Langmuir–Blodgett Films on Semiconductors 375
    7.8.1. Metal–Insulator–Semiconductor Diodes 376
    7.8.2. Schottky Barrier Modification 380
        7.8.2.1. Photovoltaic Cells 380
        7.8.2.2. Electroluminescent Diodes 384
    7.8.3. Field-Effect Transistors 384
        7.8.3.1. Group III–V Semiconductors 386
        7.8.3.2. Silicon 388
7.9. Chemical/Biological Sensors and Transducers 389
    7.9.1. Conductivity Devices 390
    7.9.2. Field-Effect Devices 390
    7.9.3. Optical Sensors 392
    7.9.4. Acoustoelectric Sensors 394
        7.9.4.1. Quartz Oscillator 394
        7.9.4.2. Surface Acoustic Wave Oscillator 395
    7.9.5. Biosensors 397
7.10. Summary 400
References 401

Index 413

Chapter 1

Historical Introduction

C. H. GILES, S. D. FORRESTER, and G. G. ROBERTS

1.1. INTRODUCTION

When Lord Rayleigh (John William Strutt, 3rd Baron Rayleigh, 1842–1919) received a registered letter* from Brunswick in Germany, on the morning of Monday, January 12, 1891, he little thought that this was the opening of a new phase in his physical research and that it would lay the foundation for an entirely new and revolutionary branch of physical chemistry—the study of monolayers.

Many concepts fundamental to science are universally accepted and have been absorbed into popular thinking. Some which spring to mind at random include the reality of gases and the vacuum, atoms, gravity, electromagnetic radiation, sound waves, and a host of others. The monolayer concept, however, though equally widely applicable, is not popularly familiar. Yet wherever a gas or a liquid is in contact with another liquid or a solid surface there is a layer of molecules, often regularly oriented, at the interface, and which, if so, differs in this respect from those in the bulk of the two phases. This is the oriented monolayer concept which, in its development, includes one of the most interesting stories in the history of

*For a reproduction of the envelope and opening page of the letter from Agnes Pockels, see the work of Forrester and Giles.[1] Her contributions to monolayer research are discussed in Section 1.2.5.

C. H. GILES • Late of Department of Pure and Applied Chemistry, University of Strathclyde, Glasgow, Scotland.   S. D. FORRESTER • Department of Pure and Applied Chemistry, University of Strathclyde, Glasgow, Scotland.   G. G. ROBERTS • Department of Engineering Science, University of Oxford, Oxford, England and Thorn EMI plc, Hayes, Middlesex, England.   Dr. Giles died shortly after completing this chapter. He will long be remembered for his well researched studies in the early history of surface chemistry.

science and provides the foundation stone of modern colloid and interface science. It has universal importance in Nature, in biology, and in very many industrial and also domestic operations. We usually regard Langmuir as its discoverer, knowing that he received the Nobel Prize in 1932 for his work in surface science, but the practical effects of the oriented monolayer had been observed from time immemorial mainly in the calming effect of oil on waves.

Mankind must have noticed early in history that when oil is spilled on water it spreads out to form a thin film, which has the property of dampening surface waves and ripples. In a recent, very detailed survey of the literature, from the classical times of Plutarch, Aristotle, and Pliny to the present day, Scott[2] concludes that "knowledge regarding the causes of this effect has progressed only very slowly. . . . Real advances in the theory of the subject have occurred most irregularly and unevenly, and even today there is neither an adequate explanation nor even a widespread appreciation among scientists of what has and has not been established."

According to Tabor,[3] the earliest "written" record of observations of the spreading of oil on water is in cuneiform on clay tablets, dating from Hammurabi's period (18th century BC) in Babylonia. "The Babylonians were apparently extremely superstitious . . . (and) . . . one of the lesser known forms of divination involved pouring oil on water (or water on oil) and observing the types of spreading behaviour that occurred." Tabor provides an illustration of a portion of a clay tablet, recording the omens dealing with the spreading of oil on water and a translation of their messages. "The Greeks who learned of the practice a thousand years later, had a word for it: they called it lecanomancy, from lekani = bowl, manteia = divination, the whole word implying divination by examining a liquid in a bowl."

Fulford[4] cites two recorded cases of pouring oil on stormy seas in AD 429 and *ca* 651. In both cases the calming property seems to have been attributed more to the *holiness* of the oil than to its *oiliness!* Little more appears to have been published on the subject until the much repeated and translated 1774 paper of Benjamin Franklin,[5] the theory until then, according to Scott,[2] being that attributed to Aristotle by Plutarch, namely, that "the oil produces calm by smoothing the water surface so that the wind can slip over it without making an impression."

The earliest technical application of organic monolayer films is believed to be the Japanese printing art called *sumi-nagashi*. The dye comprising a suspension of submicron carbon particles and protein molecules is first spread on the surface of water; the application of gelatin to the uniform layer converts the film into a patchwork of colorless and dark domains.[6] These distinctive patterns can then be transferred by lowering a sheet of paper onto the water surface.

1.2. EARLY CONTRIBUTORS TO THE FIELD OF MONOLAYER SCIENCE

In this section we have endeavored to trace the development of thought on the subject of monolayers by outlining the contributions of key individuals. Langmuir acknowledged his debt to his forerunners, especially Rayleigh. Here we briefly review their work and attempt to sort out the threads which make up the story of the discovery of the monolayer. The evidence on which statements below are based is given in detail in earlier papers.[7−9]

1.2.1. Franklin (1706–1790)

The first attempt to place the subject of monolayers on a scientific basis was made by Benjamin Franklin, the versatile American statesman in the eighteenth century AD! It was during his frequent visits to Europe as the principal representative of the American States in their critical discussions on sovereignty with the French and the British that he carried out his original experiments on the spreading of oil on various stretches of English water. Franklin's often-quoted and picturesque account to the Royal Society[5] was used many times in the 19th century as a basis for research. It reads thus:

> In 1757, being at sea in a fleet of 96 sail bound against Louisburg, I observed the wakes of two of the ships to be remarkably smooth, while all the others were ruffled by the wind, which blew fresh. Being puzzled with the differing appearance, I at last pointed it out to our captain, and asked him the meaning of it? 'The cooks', says he, 'have, I suppose, been just emptying greasy water through the scuppers, which has greased the sides of those ships a little', and this answer he gave me with an air of some little contempt as to a person ignorant of what everybody else knew. In my own mind I at first slighted his solution, though I was not able to think of another. But recollecting what I had formerly read in Pliny, I resolved to make some experiment of the effect of oil on water, when I should have opportunity. . . .
>
> At length being at Clapham where there is, on the common, a large pond, which I observed to be one day very rough with the wind, I fetched out a cruet of oil, and dropped a little of it on the water. I saw it spread itself with surprising swiftness upon the surface. . . . I then went to the windward side, where [the waves] began to form; and there the oil, though not more than a teaspoonful, produced an instant calm over a space several yards square, which spread amazingly, and extended itself gradually till it reached the lee side, making all that quarter of the pond, perhaps half an acre, as smooth as a looking glass. After this, I contrived to take with me, whenever I went into the country, a little oil in the upper hollow joint of my bamboo cane, with which I might repeat the experiment as opportunity should offer; and I found it constantly to succeed.
>
> In these experiments, one circumstance struck me with particular surprise. This was the sudden, wide and forcible spreading of a drop of oil on the face of the water, which I do not know that anybody has hitherto considered. If a drop of oil is put on a polished marble table, or on a looking-glass that lies horizontally; the drop remains in its place spreading very little. But when put on water it spreads instantly many feet around, becoming so thin as to produce the prismatic colours, for a considerable space, and beyond them so much thinner as to be invisible, except in its effect of smoothing the waves at a much greater distance. It seems as if a mutual repulsion between its particles took place as soon as it touched the water, and a repulsion so strong as to act on other bodies swimming on the surface, as straws, leaves, ships, etc., forcing them to recede every way from the drop, as from a center, leaving a large clear space. The quality of this force and the distance to which it will operate, I have not yet ascertained but I think it a curious enquiry, and I wish to understand whence it arises. . . .

It appears that it was only later, in 1762, that this original observation was brought back to Franklin's mind, and developed into the above publication, which had so many later repercussions. In that year Franklin returned from London to Philadelphia for a period, and on the voyage he happened to notice some curious effects of vibration at the interface between oil and water in a lamp hanging in his cabin. After writing about this, in a letter to a scientific friend, he made his Clapham experiment, probably some time in 1770 or 1771. The actual location of the Clapham pond has been identified and, thanks to information given to us by the Yale University Library, the location and owner of the house where he was staying at the time are now known.[7]

Franklin's account of his experiments, which applied the principles of observation, experiment, and theoretical deduction in a new area, stimulated further investigations mainly in Germany, as recorded by Scott.[2]

1.2.2. Shields (1822–1890)

During the second half of the last century, public interest was being shown in Britain regarding the question of legislating that all ships should carry supplies of oil for pouring on stormy seas. From 1860 onward *Chambers' Journal* had been publishing accounts of the successful use of the operation and editorials had been encouraging its more widespread use.[10] One Victorian lady, Miss C. F. Gordon Cumming, who had traveled widely in Europe, India, and the Far East, made a plea that all ships should carry tanks of oil for slow discharge over the sides in dangerous seas. She also suggested that there should be more extensive use of "oil bags" hung over the sides of ships and having a slow leak of oil. She even requested that one should be provided on each life buoy.[11]

Shortly afterward, much public interest in wave damping by oil was aroused by the large-scale experiments at Peterhead and Aberdeen harbors carried out by John Shields, the proprietor of a linen mill, in Perth, Scotland.[7] He had noticed that, when some oil had been spilled on the surface of a pond at his works on a day when a high wind was blowing, the oil spread in all directions and calmed the surface. On another day of similar weather conditions, he had a length of rubber tubing laid along the bottom of the pond toward the center—"keeping one end on land. We then took a flask of oil, and poured it into the tube until it was fully charged, it then began to ascend in beautiful beads to the surface, and spread with lightning rapidity, stilling the whole pond almost instantly, and not more than a gill of oil was used. I was then convinced of the great utility that this could be turned into, and to my mind then and is still only a mechanical problem, how to get the oil *where* it is wanted and when it is wanted. I continued the experiments on a small scale for about a couple of years, and every time was more and more convinced that it would be a practical power for saving life and property at sea at no distant date. I then resolved at my own expense to give it a more practical test. . . ." On August 30, 1879, he lodged a provisional British patent specification (sealed on February 20, 1880)[12a] for a simple device for spreading oil from valves in undersea pipes, and later another from buoys at the entrances to harbors.[12b] Patents were also taken out in the USA and France. In the first full-scale experiment, conducted at the mouth of Peterhead harbor, bottles of oil were dropped overboard from a steam tug on a very stormy day "and the effect was something extraordinary."

Shields's trials were not reported in the scientific press but were widely reported in the daily press. They created so much public interest that they formed the subject of discussion in Parliament.[8,13,14] Despite receiving no financial help from the Government to continue his investigation on the large scale, Shields repeated his experiment at Aberdeen Harbour on December 4, 1882 in the presence of, among others, an observer from the Board of Trade,[15,16] but no further trials seem to have been made. The records make it clear that numerous practical difficulties were encountered, including damage to the piping system by underwater currents and the need for constant replenishment of the oil. Aitken[17] reports that the effect lasted for one hour after the pumping ceased and the oil used cost no less than £10 (a large sum in 1882).

Shields's work created considerable interest in the United States. A report compiled by the US Life-Saving Service about that time[18] gave prominence to the Scottish harbor experiments but concluded that "The nature of the phenomena presented by a rough sea, the relative influence of the different agencies concerned in its action, and just how far this action can be controlled, can only be determined by carefully compared experiments which would require an expenditure of time and funds not at present at our disposal: therefore, the subject is herein treated only as a matter of practical observation and not of exhaustive scientific enquiry."

1.2.3. Aitken (1839–1919)

The description of Shields's work in the Scottish press seems to have rekindled scientific interest in the subject. It is believed that the first scientist to take up the matter was John Aitken, a graduate in engineering of the University of Glasgow. Aitken was a bachelor of independent means, who did not take up paid employment because of ill-health. He fitted out part of his home in Falkirk (Scotland) as a workshop and laboratory and made classical observations of natural phenomena with apparatus he designed and made himself. In one such investigation he devised and constructed apparatus to test theories of the calming action of oil.[17] In a circular vessel he arranged that a jet of air should blow over the surface and cause it to take up a rotary motion. This motion was measured by hanging a completely submerged horizontal paddle in the middle of the vessel by means of a fine platinum wire attached to a torsion head. A needle, rigidly attached to the paddle, measured on a circular scale the amount of torsion produced by the moving water. With suitable precautions for ensuring uniformity of conditions, Aitken measured the amount of torsion of different currents of air on clear water surfaces and on those having a thin layer of oil, and found that[17] there was no decrease in the amount of deflection after the oil was added to the water, ". . . therefore oil does not reduce the bite, grip or friction of the air on the surface," thus contradicting the theory of wave-damping by oil that had long been held. Aitken then devised experiments to explain the phenomenon: these are discussed elsewhere.[8]

1.2.4. Rayleigh (1842–1919)

There was thus a more or less continuous thread of interest in oil films on water from Franklin's time to the 1880s.

While all these earlier experiments had been proceeding Rayleigh had begun to take an interest in all forms of wave motion. His interest in sound waves led him to make observations on the behavior of colliding water jets and water drops, and the effect on them of static electricity.[19,20] In 1879 he published a paper[21] on "The influence of electricity on colliding water drops," in which he examined experimentally the conditions in which vertical jets break into drops which rebound from each other. This effect he considered "results from the charge they carry, though they coalesce if subjected to a mild electric charge. Two jets issuing side-by-side horizontally rebound, but coalesce when charged with static electricity." But he also commented: "that coalescence of the jets would sometimes occur in a capricious manner, without the action of electricity or other apparent cause. I have reason to believe that some, at any rate, of these irregularities depended upon a want of cleanness in the water. The addition to the water of a very small quantity of soap makes the rebound of the jets impossible. The last observation led me to examine the behavior of a fine vertical jet of slightly soapy water; and I found, as I had expected, that no scattering took place." In a later paper, in 1882,[22] he writes: "It has been already shown that the normal scattering of a nearly vertical jet is due to the rebound of the drops when they come into collision. If by any means, the drops can be caused to amalgamate at collision the appearance of the jet is completely transformed. This result occurs if a feebly electrified body be held near the place of resolution into drops, and it was also observed to follow the addition of a small quantity of soap to the water of which the jet was composed."

Further experiments seemed to prove that the real agent was not soluble soap at all and that the amalgamation was due to a monolayer of free fatty acid in the soap. Rayleigh had thus, among the many subjects exercising his active mind, an interest in water surfaces and a few years later he published five papers dealing in various ways with water surfaces.[23–27]

In the second of these[24] he discussed the effect of superficial layers of olive oil upon the surface tension of water. We can surmise from a comment in this paper that he was beginning to suspect that surface films are of molecular thickness, because he mentions "the great interest which attaches to the determination of molecular magnitudes." He noted that the surface tension of water can be lowered by "contamination" with a surface film of insoluble grease or oil, and then permanent changes in tension accompany changes in area of the surface. He estimated from measurements with films of olive oil on a water surface, just able to prevent the movement of camphor, that such films are between 10 and 20 Å thick.

Rayleigh had been convinced of the reality of molecules since 1865, and

Figure 1.1. Top: Benjamin Franklin (courtesy of Emmet Collection, New York, Public Library). Bottom left: Lord Rayleigh. Bottom right: Agnes Pockles.

thought that the oil films on water ultimately extended until they were one molecule thick. He believed that if the thickness of such an extended layer could be determined it would give the first direct measurement of the size of an organic molecule, but he had not found a method of making an exact measurement.

1.2.5. Pockels (1862–1935)

For a direct measurement of molecular sizes, Rayleigh was indebted to Agnes Pockels whose simple apparatus later became the model for what is now termed a Langmuir trough. Giles and Forrester[8,9] have described her domestic background and how she worked on the problem for about 10 years, since the age of 18, literally on the kitchen table. Yet in her letter in 1891 to Rayleigh she described the methods which have remained to this day the essentials of monolayer research.

She described her use of a rectangular tin trough, 70 cm × 5 cm × 2 cm, filled with water to the brim, with a 1.5 cm wide strip of tin laid across it, just in contact with the water. Thus by moving the strip she could vary the area of and also completely clean an enclosed water surface. By this means she examined the variation in the surface tension of an oil-contaminated water surface, using a balance which measured the force required just to lift from it a small disk (a button!). She found that it was essential first to clean the water surface by drawing the tin strip across it. This was a critical technique on which all subsequent work has depended.

On reading her letter Rayleigh at once realized the importance of the simple apparatus and method she had described. Then, consistent with his character of generosity toward those he considered deserving of recognition, after satisfying himself on certain points* he sent the letter to the leading British scientific journal, with a covering note recommending its publication. The editor published it in full,[28] together with Rayleigh's note. We suspect he inserted this note to anticipate any criticism for publishing the contribution from an unknown and apparently nonprofessional authoress. The first pressure–area isotherms, so familiar now in monolayer research, were published by Pockels. An example is given in Figure 1.2. It is truly amazing to note that translating her data to modern-day units yields a molecular area for stearic acid of 2.2 nm.[29,30] Rayleigh shortly afterward published two further papers dealing with surface films,[31,32] but he did not proceed to develop Miss Pockels's method until a few years later. His time was apparently largely occupied with other work, especially the discovery of the rare gases. Only in 1899[33] did he return to a quantitative study of oil films on water, commenting "The tension of slightly contaminated surfaces was made the subject of special experiments by Miss Pockels, who concluded that a water-surface can 'exist in two sharply contrasted conditions; the normal condition, in which the displacement of the partition (altering the density of the contamination) makes no impression upon the tension, and the anomalous condition, in which every increase or decrease alters

*In a second letter she comments "With regard to your curiosity about my personal status, I am indeed a lady!"

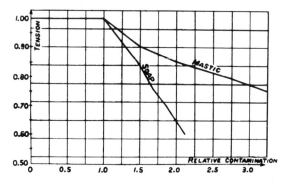

Figure 1.2. The first pressure–area diagram published by Agnes Pockels.[30]

the tension.' It is only since I have myself made experiments upon the same lines that I have appreciated the full significance of Miss Pockels' statement. The conclusion that, judged by surface-tension, the effect of contamination comes on suddenly, seems to be of considerable importance, and I propose to illustrate it further by actual curves embodying results recently obtained." In this paper he described his use of the Pockels method and described the size of an olive oil molecule in the monolayer on water as "about 1 nm." It is interesting that Franklin's observation that a teaspoonful (approximately 2.5 ml) of olive oil spread over half an acre of water gives a similar thickness of oil film.[7]

1.2.6. Langmuir (1881–1957)

Rayleigh's proposal that it was possible to experiment with films only a single molecule in thickness increased the level of activity in the field even though many were reluctant to accept his viewpoint. For example, Devaux[34,35] using additional elegant methods measured the thickness of various films and showed that they represented molecular diameters. In particular, he was the first to spread polymers as films using proteins and cellulose. He also reported that egg albumen spreads on water to give a remarkably elastic monolayer. A little later Hardy[36,37] discovered that oils which do not contain polar functional groups cannot be spread on a water surface in the same manner as animal and vegetable oils. He was the first to postulate the orientation of polar molecules on surfaces and to consider their function in the field of lubrication. On the other hand, he was wrong in his assertion that the cohesive forces between molecules were long range in character. It was Langmuir who surmised correctly that the forces were short range and acted only between molecules in contact. Stimulated by the experimental work of Marcelin[38] and those already quoted, he was responsible for laying the scientific foundation of monomolecular films.

Irving Langmuir was born on January 31, 1881 in Brooklyn, New York. In 1903 he received the degree of Metallurgic Engineering from the Columbia University School of Mines. He received his doctorate under the supervision of Nernst in Gottingen, Germany, for a dissertation on incandescent lamp bulbs. Following three unproductive years as a teacher at the Stevens Institute of Technology in New Jersey he joined the General Electric Research Laboratory in Schenectady, New York. During his 41 years on the staff he published 229 papers and 63 patents.[39] His discovery in 1912, that the addition of nitrogen and argon in a light bulb protected the tungsten wires from deterioration, led to the gas-filled lamp with its higher intrinsic efficiency. He also made many other notable contributions in the fields of thermionic emission and electrical discharges in gases. However, these significant and classic works were overshadowed in basic scientific value by his research in surface chemistry, which accounted for approximately 25% of his prolific output and for which he was awarded the Nobel Prize in Chemistry in 1932. In later life he became more concerned with scientific education and the philosophy of science but remained as a consultant with the General Electric Company until his death in 1957.

His famous theory,[40] first fully described in 1915, enabled adsorption to be explained simply, clearly, and quantitatively. In attributing adsorption primarily to the existence of unsaturated valency forces on the surfaces of solids and liquids he was also laying the basis of his later theories of catalytic reactions. His theory has all the hallmarks of a great scientific generalization: the ability to explain and coordinate a mass of different phenomena, together with the capacity to withstand the growth of a superstructure of later theoretical development, and to act as a constant stimulus to ever-widening research. It has stood the test of time, and in its simplicity and elegance it has an appeal to the artistic sense possessed by the true scientist. Yet like many other great theories, it did little more than bring together in a new relation facts or hypotheses already well known: the surface nature of adsorption, the kinetic theory of gases, and the range of intermolecular attractive forces. Langmuir's great contribution, however, was to give these facts unity by relating them[41] all to an additional hypothesis, one which had lain unnoticed for 16 years in the "Proceedings of the Royal Society." This was Rayleigh's suggestion that extended layers of polar oils on water are one molecule thick.[33] Langmuir gave generous acknowledgement to Pockels and Rayleigh and commented on the apparent neglect of their method and conclusions during the first two decades of this century. It seems strange that the active workers in the field had ignored this vital clue to an understanding of their data. Even Ramsay, who had been associated with Rayleigh in the discovery of the rare gases in the 1890s and who must have had frequent contact with him, apparently failed to see the significance of Rayleigh's hypothesis; otherwise he would surely have conveyed it to his colleague Miss Homfray, who published many experimental results of adsorption of gases by solids,[42] and seems to have been unaware of it. As we state above, Langmuir himself remarked on the lack of recognition given to this particular conclusion of Rayleigh, and was careful to acknowledge his indebtedness to it in his own formulation of the basic hypothesis of oriented monolayers on both liquid and solid surfaces. Nevertheless, Langmuir's work was largely carried out independently of the earlier investigators.

For his measurements of the spreading pressures of thin films, Langmuir developed a number of new techniques including the surface film balance with which his name is now associated. In this device, a movable float separates a clean water surface from the area covered with a film; the deflection of the float then provides a direct measure of the forces involved. Langmuir confirmed that his films had the thickness of a single molecular layer and also concluded that the molecules were orientated at the water surface, with the polar functional group immersed in the water and the long nonpolar chain directed almost vertically from the surface. His experiments provided strong support for the existence of short-range forces and explained clearly the basis on which certain molecules did or did not form good monolayer films. A schematic diagram of Langmuir's original film balance[41] is shown in Figure 1.3; it is salutory to remind oneself that this elegant technique provided information on the sizes and shapes of organic molecules before X-ray diffraction was used for this purpose.

1.2.7. Blodgett (1898–1979)

Miss Katharine Blodgett joined the research staff of the General Electric Company in Schenectady in 1919. Her MS degree had been obtained at the University of Chicago. She holds the distinction of being the first woman scientist to join the GE research staff and the first to obtain a doctorate from the Cavendish Laboratory in Cambridge, England.

By 1919 Blodgett, under Langmuir's guidance, had already been able to transfer fatty acid monolayers from water surfaces to solid supports such as glass slides. In fact, the final sentence of a paper[43] read by Langmuir at that time to the Faraday Society reads, "The writer is much indebted to Miss Katharine Blodgett for carrying out much of the experimental work." In this paper he had emphasized the importance of single monolayers on the wettability of surfaces. Built-up monolayer assemblies are now referred to as Langmuir–Blodgett (LB) films as distinct from Langmuir films, a term reserved for a floating monolayer. The first formal report

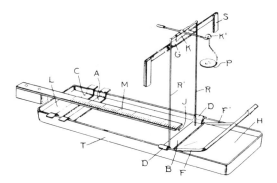

Figure 1.3. Langmuir's original film balance.[41]

Figure 1.4. Top left: Irving Langmuir. Top right: Katharine Blodgett. Bottom: Langmuir and Blodgett in conversation with Dr. I. Schaefer.

describing the preparation of LB films did not appear until 1935[44] (with a received date of January 24, 1934). The long delay in its appearance, as Gaines[45] explains in detail, was due to the fact that both Langmuir and Blodgett were actively engaged in other work of importance to their Company. Langmuir himself was deeply involved in the period 1919–1934 in several areas of research involving the adsorption of gases and the behavior of filaments in vacuum—topics of relevance in the development of radio. Blodgett during this period had been studying the mean free path of electrons in ionized mercury vapors at the University of Cambridge. However, in November 1933, Langmuir's attention was drawn to the problem of lubricating the jewelled bearings of meters and this seems to have been an important factor in restimulating his interest in oil films, especially on solid surfaces. Gaines[45] has provided a list of the thirty-one relevant publications by Langmuir and Blodgett, either separately or jointly, between 1934 and 1952. A great deal of their pioneering work is also summarized in the admirable book by Gaines,[46] together with the contributions of other early investigators including Adam and Rideal.

Many of their important contributions, such as research on skeletonized films, speed of dipping, monolayer overturning, etc., are referred to in later chapters. Blodgett concentrated a great deal of her effort on the optical properties of multilayers[47] while Langmuir[48] turned his attention increasingly to protein monolayers.

1.3. CONCLUSION

As we have described, Rayleigh first developed an interest in water surfaces about 1879 and Agnes Pockels carried out her first experiments on oil films on water in 1882. Langmuir is rightly regarded as the father figure in the field because of his momentous contributions to surface science. Multilayer films were a later development which did not follow naturally from the original monolayer theory and practice. Following a brief flurry of activity just before and just after the Second World War, the field of Langmuir–Blodgett films lay relatively dormant. In the mid-sixties Hans Kuhn began his stimulating experiments on monolayer organization which are described in detail in Chapter 5 of this book. However, the upsurge of interest in Langmuir–Blodgett films, which has resulted in four large International Conferences (the first in Durham, England, the second in Schenectady, New York, the third in Göttingen, West Germay, and the fourth in Tsukuba, Japan) would not have occurred but for the commercial potential of thin films deposited using the Langmuir trough. These practical possibilities are mentioned in the final chapter.

Science too often is related as a sequence of inventions or breakthroughs that occur spontaneously followed by rapid progress toward a final omniscient state. By relating the historical development of the subject of monomolecular assemblies and mentioning some of the key contributors over the past two centuries, it should be evident that this is not the case with Langmuir–Blodgett films. Despite the large volume of research, much of the subject continues to have an empirical basis and there is still no genuine application for such layers except perhaps as model systems in fundamental research.

REFERENCES

1. S. D. Forrester and C. H. Giles, *Chem. Ind. (London)*, 318 (1972).
2. J. C. Scott, in: *History of Technology* (A. R. Hall and N. Smith, eds.), Vol. 3, p. 163, Mansell, London (1978).
3. D. Tabor, *J. Colloid Interface Sci.*, **75**, 240 (1980).
4. G. D. Fulford, *Isis*, **59**, 198 (1968).
5. B. Franklin, *Philos. Trans. R. Soc. London*, **64**, 445 (1774).
6. T. Terada, R. Yamamoto, and T. Watanabe, *Sci. Pap. Inst. Phys. Chem. Res. (Tokyo)*, **23**, 173 (1984).
7. C. H. Giles, *Chem. Ind. (London)*, 1616 (1969).
8. C. H. Giles and S. D. Forrester, *Chem. Ind. (London)*, 80 (1970).
9. C. H. Giles and S. D. Forrester, *Chem. Ind. (London)*, 43, (1971).
10. *Chambers' Journal*, 25 (1860); 511, 811 (1878); 497 (1879); 140 (1883); quoted from Scott.[2]
11. C. F. Gordon Cumming, *From the Hebrides to the Himalayas* (2 Vols.), Vol. 1, p. 347, Sampson Low, Marston, Searl and Rivington, London (1876).
12. a. J. Shields, British Patent 3490 (1879); cf. U.S. Patent 289720 (1883) and U.S. Patent 334295 (1886). b. British Patent 1112 (1882); cf. French Patent 148160 (1882).
13. *Hansard's Parliamentary Debates*, 3rd Series, Vol. 273, cols. 6–15, Cornelius Buck, London (1882).
14. C. F. Gordon Cumming, *The Nineteenth Century*, **11**, 572 (1882).
15. *The Daily Free Press*, Aberdeen, 5 December 1882. See Giles.[7]
16. Official report on the use of oil at sea, for modifying the effect of breaking waves. *Board of Trade Journal*, **1**, 211 (1886). Quoted from Scott.[2]
17. J. Aitken, *Proc. R. Soc. Edinburgh* **12**, 56 (1882–4).
18. B. C. Sparrow, *Annual Report of the United States Life-Saving Service*, Chapter II, Reports of Committees, Section 1, p. 427 (1883).
19. S. D. Forrester and C. H. Giles, *Chem. Ind. (London)*, 469 (1979).
20. Lord Rayleigh, *Proc. Lond. Math. Soc.*, **10**, 4 (1879).
21. Lord Rayleigh, *Proc. R. Soc. London*, **28**, 406 (1879).
22. Lord Rayleigh, *Proc. R. Soc. London*, **34**, 130 (1882).
23. Lord Rayleigh, *Proc. R. Soc. London*, **47**, 281 (1890).
24. Lord Rayleigh, *Proc. R. Soc. London*, **47**, 364 (1890).
25. Lord Rayleigh, *Proc. R. Inst. London*, **13**, 85 (1890).
26. Lord Rayleigh, *Proc. R. Soc. London*, **48**, 127 (1890).
27. Lord Rayleigh, *Philos. Mag.*, **30**, 386 (1890).
28. A. Pockels, *Nature (London)*, **43**, 437 (1891).
29. A. Pockels, *Nature (London)*, **46**, 418 (1892).
30. A. Pockels, *Nature (London)*, **48**, 152 (1893).
31. Lord Rayleigh, *Philos. Mag.*, **33**, 363 (1892).
32. Lord Rayleigh, *Philos. Mag.*, **33**, 468 (1892).
33. Lord Rayleigh, *Philos. Mag.*, **48**, 321 (1899).
34. H. Devaux, *Kolloid-Z.*, **58**, 260 (1932).
35. H. Devaux, *Annu. Rep. Smithsonian Inst.*, 261 (1913).
36. W. B. Hardy, *Proc. R. Soc. London, Ser. A* **86**, 610 (1912); **88**, 303 (1913).
37. W. B. Hardy, *Collected Scientific Papers*, **508**, 550, Cambridge University Press (1939).
38. A. Marcelin, *Ann. Phys.*, **1**, 19 (1914).

39. I. Langmuir, in: *The Collected Works of Irving Langmuir,* Volumes 1–12 (C. G. Suits and H. E. Way, eds.), Pergamon Press, London (1961).
40. I. Langmuir, *J. Am. Chem. Soc.,* **37,** 1139 (1915); **38,** 2221 (1916); **40,** 1361 (1918).
41. I. Langmuir, *J. Am Chem. Soc.,* **39,** 1848 (1917).
42. I. F. Homfray, *Proc. R. Soc. London,* Ser. A, **84,** 99 (1910).
43. I. Langmuir, *Trans. Faraday Soc.,* **15,** 62 (1920).
44. K. B. Blodgett, *J. Am. Chem. Soc.,* **57,** 1007 (1935).
45. G. L. Gaines Jr., Proc. 1st International Conference on Langmuir–Blodgett Films (G. G. Roberts and C. W. Putt, eds.), *Thin Solid Films,* **99** (1983).
46. G. L. Gaines Jr., *Insoluble Monolayers at Liquid Gas Interfaces,* Wiley, New York (1966).
47. K. B. Blodgett, *Phys. Rev.,* **51,** 964 (1937).
48. I. Langmuir and V. J. Schaefer, *Chem. Rev.,* **24,** 181 (1939).

Molecular Structure and Monolayer Properties

R. A. HANN

2.1. INTRODUCTION

There is something about the unique combination of properties of a water surface that makes water by far the most favored subphase for Langmuir film formation. This chapter will be concerned with the nature of the interactions between a water surface (more properly an air–water interface) and an insoluble monolayer. The chemical features of molecules that form stable monolayers will be illustrated, together with the factors that lead to their successful deposition onto substrates as solid multilayers. There has been much recent interest in the study of polymerizable monolayers and an extension of the earlier work on monolayers of preformed polymers. As polymers have especial importance in connection with potential applications, they are dealt with in a separate section. The systematic chemical names of many of the materials discussed in this chapter are long and complex. No attempt has therefore been made to make consistent use of formal nomenclature; common names and abbreviations are used in the text and alternatives are sometimes given in the captions to the figures.

R. A. HANN • ICI Imagedata, Brantham, Manningtree, Essex CO11 1NL, England.

2.2. FORMATION AND STABILITY OF MONOLAYERS

The surface of a liquid always has excess free energy; this is due to the difference in environment between the surface molecules and those in the bulk. In particular, hydrogen bonding forces in water tend to set up loosely defined networks that will inevitably be modified near the surface. The thermodynamics of liquid surfaces has been reviewed by Gaines,[1] and only a few of the most relevant results will be noted here. The *surface tension* (γ) of a plane interface is given by the partial differential:

$$\gamma = (\partial G/\partial s)_{T,P,n_i} \tag{2.1}$$

where G is the Gibbs free energy of the system, s is the surface area, and the temperature T, pressure P, and composition n_i are held constant. The surface tension of water is 73 mNm^{-1} (= $dyncm^{-1}$) at 20 °C and atmospheric pressure. This is an exceptionally high value compared to most other liquids and goes some way toward explaining water's preeminence as a subphase.

A classic monolayer-forming material such as stearic acid (Figure 2.1) has two distinct regions in the molecule: a *hydrophilic* ("water-loving") headgroup ($-CO_2H$), which is easily soluble in water, and a long alkyl chain ($C_{17}H_{35}-$), which provides a hydrophobic ("water-hating") or *oleophilic* ("oil-loving") tail. When a

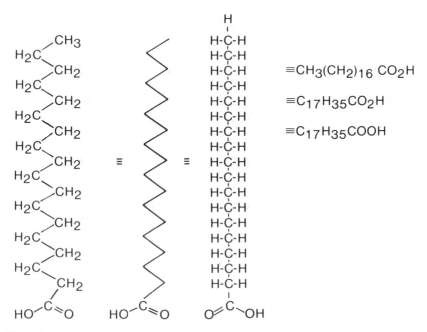

Figure 2.1. Various representations of the chemical formula of stearic acid (octadecanoic acid).

solution of stearic acid in a water-immiscible solvent such as chloroform is placed on a water surface, the solution spreads rapidly to cover the available area. As the solvent evaporates, a monolayer is formed as dictated by the *amphiphilic* (amphi = either) nature of the stearic acid molecules; the headgroups are immersed in the water surface and the tailgroups remain outside. This is illustrated schematically in Figure 2.2a.

When the distance between stearic acid molecules is large, their interactions are small, and they can be regarded as forming a two-dimensional gas. Under these conditions the surface monolayer has relatively little effect on the water's surface tension. If a barrier system (as described in Chapter 3) is used to reduce the area of surface available to the monolayer, the molecules exert a repulsive effect on each other. This two-dimensional analog of a pressure is normally called the *surface pressure* and denoted by a capital Greek pi (Π). For a plane surface at equilibrium, the relationship

$$\Pi - \gamma - \gamma_0 \qquad (2.2)$$

holds, where γ is the surface tension in the absence of a monolayer, and γ_0 the value with the monolayer present. It follows that the maximum possible surface pressure for a monolayer on a water surface at 20 °C is 73 mNm^{-1}, and normally encountered values are much lower.

2.2.1. Surface Pressure/Area Isotherms

The single most important indicator of the monolayer properties of a material is given by a plot of surface pressure as a function of the area of water surface available to each molecule. This is carried out at constant temperature and is accordingly known as a surface pressure/area isotherm, and is often abbreviated to "isotherm." Equilibrium values can be measured on a point-to-point basis, but it is more common to record a pseudoequilibrium isotherm by compressing the film at a constant rate while continuously monitoring the surface pressure (see Chapter 3). Depending on the material being investigated, repeated compressions and expansions may be necessary to achieve a reproducible trace. A typical surface pressure/area isotherm for stearic acid is shown in Figure 2.3.

A number of distinct regions are immediately apparent on examining the isotherm. As the surface area is reduced from its initial high value, there is a gradual onset of surface pressure until an approximately horizontal region is reached (see

               a   b   c

Figure 2.2. Monolayer of stearic acid on a water surface: (a) expanded, (b) partly compressed, (c) close packed.

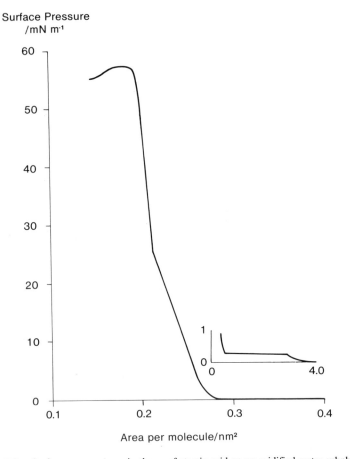

Surface Pressure
/mN m⁻¹

Area per molecule/nm²

Figure 2.3. Surface pressure/area isotherm of stearic acid on an acidified water subphase.

inset, Figure 2.3). In the horizontal region the hydrophobic chains, which were originally distributed near the water surface, are being lifted away (Figure 2.2b). The surface pressure at which this occurs is usually very small (<1 mN m^{-1}) owing to the weakness of interaction between water and the tailgroups; thus this portion of the isotherm is often not resolved by the apparatus. There follows a second abrupt transition to a steeply sloping linear region, where the compressibility (Figure 2.2c) defined by the equation

$$c = -1/A \ (\partial A/\partial\Pi)_{T,P,n_i} \tag{2.3}$$

is approximately constant.

At a surface area of just over 0.20 nm²molecule⁻¹ there is an abrupt increase of slope. This is clearly also due to a phase change and represents a transition to an ordered solidlike arrangement of the two-dimensional array of molecules. The compressibility in this region is also constant, but is lower by about a factor of 10. If this second linear portion of the isotherm is extrapolated to zero surface pressure, the intercept gives the area per stearic acid molecule that would be expected for the hypothetical state of an uncompressed close-packed layer. This value of 0.22 nm²molecule⁻¹ is close to that occupied by stearic acid molecules in single crystals, thus confirming the interpretation of a compact film as a two-dimensional solid.

At smaller surface areas the phenomenon of *collapse* occurs and the compressibility approaches infinity. The onset of collapse depends greatly on such factors as the past history of the film and the rate at which the film is being compressed. In this type of collapse it is believed that molecular layers are riding on top of each other and disordered multilayers are being formed (Figure 2.4); it can readily be imagined that a process of this kind will be intrinsically variable and will exhibit induction periods and dependence on history.

The question arises as to the circumstances under which an insoluble monolayer is thermodynamically stable. Two basic requirements are that the subphase and the surrounding atmosphere must be saturated with the material concerned. These requirements can often be ignored in practice because of very low equilibrium concentrations or because of slow kinetics, but they are important in setting some of the limits on the range of materials that are stable as a monolayer.

Conditions for stability against collapse can in principle be established by measuring the *equilibrium spreading pressure* (ESP) and corresponding surface area of the material. This is the surface pressure that is spontaneously generated when a crystalline sample of the solid material is placed in contact with a water surface. Provided that sufficient time is allowed for equilibration to occur one can, in principle, be sure that the monolayer which has been formed by molecules detaching themselves from the crystal surface and spreading over the subphase is in equilibrium with the crystals themselves. At any surface pressure higher than this there should therefore be a tendency for the monolayer to aggregate into crystals;

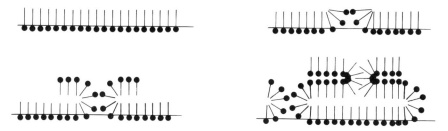

Figure 2.4. Collapse of a monolayer.

the process is similar to the formation of an equilibrium vapor over a crystalline solid—if the vapor pressure is exceeded, molecules will redeposit onto the solid. Unfortunately, the situation is complicated by the fact that equilibrium is often approached very slowly; this means that the ESP may not be attained in the course of an experiment. The practical *benefit* of slow equilibrium is the resultant ability to handle films at considerably more than their ESP without discernible collapse. Variations in ESP can also be observed depending on the size of crystals or even the crystal face exposed to the subphase. It would be expected from all this that amphiphilic *liquids* should have much better-defined ESP and collapse behavior; this is observed in practice.

Gaines[1] has reviewed early work in this area, but the mechanism of collapse remains of interest. A recent study[2] of arachidic acid (closely related to stearic acid—see later) has found that at surface areas between 0.204 and 0.233 nm^{2}molecule^{-1} the collapse shows the characteristics of a nucleation and growth process, as would be expected if crystalline material were being generated. At areas of less than 0.204 nm^{2}molecule^{-1} the collapse appears to be due to the sliding formation of multilayers as discussed earlier, and no collapse is observed at areas greater than 0.233 nm^{2}molecule^{-1}, which corresponds to the equilibrium spreading pressure. It must be remembered that although collapse is measurable in this type of lengthy experiment, monolayers of arachidic acid can normally be manipulated at a surface pressure of 30 mNm^{-1} or more without visible collapse on a time scale of hours.

Accepting then that large portions of the isotherm refer to nonequilibrium conditions, it is still worthwhile trying to describe the state of the molecules that make up these films. The identification of the low compressibility region with a two-dimensional solid still appears to be attractive, as does the two-dimensional gas model for a highly expanded monolayer. It is therefore tempting to describe the higher compressibility linear region as liquid and this term has been frequently used after its suggestion by Harkins[3] that these films should be described as *liquid-condensed*. A completely different type of isotherm is sometimes observed, in which there are no clear phase transitions and the area per molecule remains at values much higher than those required for close packing. This type of film is known as *liquid-expanded* in Harkins's classification. This type of isotherm is most frequently observed when studying molecules in which some distruption of the hydrophobic chain causes a difficulty in packing. Typical examples are provided by oleic acid and 2-ethylpalmitic acid whose isotherms are shown in Figure 2.5; these materials and the disruption of chain packing will be discussed in more detail later, but at present it should be noted how highly expanded these films are relative to the stearic acid monolayer. The appearance of these different regions of the isotherm can be described in purely thermodynamic terms,[4] by taking account of the activity of the subphase surface.

The above classification is in fact simplistic, as the shape of the isotherm depends greatly on temperature and a number of different phase transitions can be observed if the temperature is varied. A wide investigation[5] of a range of homo-

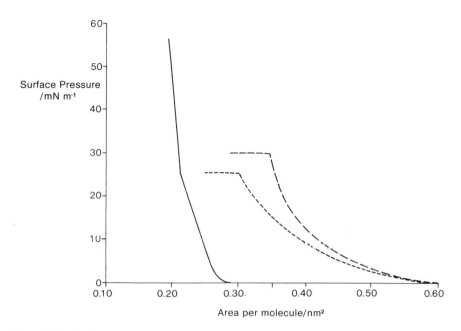

Figure 2.5. Surface pressure/area isotherm of 2-ethylpalmitic acid (dashed line) and oleic acid (dotted line), with stearic acid (solid line) as a comparison. (After Stenhagen[47] and Adam and Dyer.[45])

logs of stearic acid has shown that a reduction in chain length can, to some extent, be traded for an increase in temperature. In this way it is possible to construct[5] a model sequence of isotherms that covers a wider temperature range than is practicable with a single material; Figure 2.6 is a simplified version of this. There is insufficient space here to discuss the complex combination of interactions which are involved in the interpretation of these data.

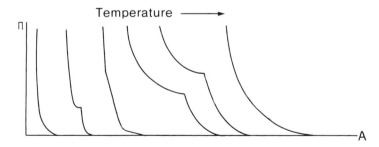

Figure 2.6. Change of shape of surface pressure/area isotherm with increasing temperature. (Diagrammatic; after Baret *et al.*[5])

2.2.2. Modeling Molecular Structure

Although the crystal structure of stearic acid is well established, most of the materials discussed later in this chapter and most new materials will be of unknown structure. How then do we determine whether the area per molecule observed in any experiment is reasonable, and how do we decide on the most likely molecular arrangement on the water surface? If there is only one hydrophilic group in the molecule, then there is little doubt that this will be on the water surface, and if the molecule is very similar to stearic acid, the area per molecule should also be similar. For more complex cases more powerful methods are needed; fortunately these are available and one at least is simple and inexpensive.

Chemists have for years been concerned with the spatial or steric arrangement of atoms within molecules and have developed ways of representing molecules as semicoalesced "hard spheres" to give a very accurate representation of their packing requirements. The stearic acid formula (2.1) is essentially a connection diagram and gives little idea of the steric requirements of the molecule. In contrast, Figure

Figure 2.7.   Space-filling representation of stearic acid molecule.

2.7 shows up features such as the zigzag nature of the hydrocarbon $(CH_2)_n$ chain as well as the existence of a "sphere of influence" around each atom. This figure was computer generated by one of the available systems that allows the construction, rotation, measurement, and juxtaposition of molecules on a VDU. At present the cost of such a system is high (although decreasing rapidly) and it is fortunate that there is an alternative system in the form of "CPK" models. These are simply assembled from the component "atoms" by plugging them together to give a "molecule" that can be picked up, twisted, measured, and aligned with other "molecules." Even if a computer system is available, the building of these models remains an essential way of getting a "feel" of what the molecules are like.

2.2.3. Other Properties of Langmuir Films

Surface potential measurements have been used as an additional source of information on the arrangement of molecules on the water surface. Interpretation of these measurements is complicated by the unknown effects that the monolayer is having on the water structure. The observed potential changes relative to a pure water surface are therefore due to a composite of monolayer properties and changes induced in the water. In spite of this, surface potential measurements can be very informative particularly where comparisons between similar materials are being made (see Gaines[1] for a further discussion and Tredgold and Smith[6] for surface potential measurements on deposited layers).

The mechanical properties of monolayers are also of significance for the propagation of surface waves and because of their effect on deposition. Three different surface viscosities can be defined: the surface dilational viscosity and the in-plane and out of-plane surface shear viscosities. All of these can be non-Newtonian and influence the dynamic behavior of films. The in-plane shear viscosity is probably the most frequently measured of these properties, and is important in determining flow behavior; laser light scattering promises to be a technique suitable for a wider study of viscoelastic properties.[7]

In a recent attempt[8] to elucidate the difference between "solid" and "liquid" regions of the isotherm, use was made of a viscometer which recorded the relaxation after a torsional perturbation and could detect very small residual stresses. A simple liquid, no matter how viscous, should relax completely leaving no residual torque. A solid would be expected to show an elastic deformation and leave a residual torque. Stearyl alcohol $(C_{18}H_{37}OH)$ shows the expected behavior giving a residual torque above the phase transition, and no residual torque below it. The higher homologs show a range of unexpected effects; for example, arachidyl alcohol $(C_{20}H_{41}OH)$ behaves as a solid in the "liquid" region, and after the phase transition to a "solid" it behaves as a liquid. This reinforces the caveats given earlier that the nature of the phase transitions may be complex. The behavior of lipid layers (see Chapter 6) is often interpreted in terms of liquid-crystal and gel phases[9] and this type of model may also be appropriate to the simpler molecules discussed here.

2.2.4. Mixed Monolayers

Mixed monolayers have been studied extensively. Where the mixture consists of a major component "doped" with a low concentration of a second material (as in many studies on film-forming dyes) there is usually little doubt that the minor component is evenly dispersed in the major one, provided that both components are amphiphilic. It is also possible to spread monolayers of mixtures of a film former, such as stearic acid, with materials that are either totally hydrophobic or largely hydrophilic; in this type of "mixed monolayer," which is discussed in more detail later, the second species is banished to the tail or head region.

Where there are two (ore more) *major* components the question of miscibility arises more urgently, in the same sort of way as with bulk liquids and solids. It is usually possible to determine whether two components are miscible by studying a range of properties such as viscosity and surface potential, both for the individual components and for the mixture. If the materials are immiscible the components will exist as domains (Figure 2.8), and provided that these are large enough to be representative of bulk materials (say 0.1 μm in diameter[1]) and small enough not to be resolved by the measurement (say 1 mm diameter for surface potential measurements), then the above properties will simply be a weighted mean of the individual values. If, by contrast, the materials are miscible, the properties are free to take some other value. It is therefore easier to establish miscibility in monolayers, as several measurements must be taken on apparently immiscible films to ensure that the measured property is not fortuitously the weighted mean. Some detailed suggestions have been made regarding the diagnosis of mixing in the region of the low pressure ("gas–liquid") phase transition.[10]

The collapse pressure is often a useful guide to miscibility. If two immiscible components have well-defined and different collapse pressures (determined on sepa-

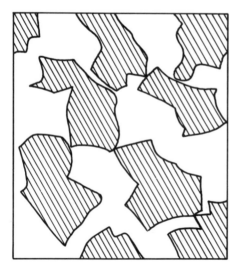

Figure 2.8. Domains in a monolayer formed from two immiscible components (diagrammatic).

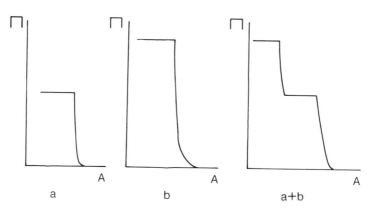

Figure 2.9. Surface pressure/area isotherm of two immiscible monolayer materials of different collapse pressures and the resultant dual collapse of the mixed monolayer.

rate monolayers) the resultant monolayer should start to collapse at the lower value; if compression is continued until all the domains of the first material have collapsed, the surface pressure should rise until the higher collapse pressure is reached (Figure 2.9). A true mixture will give only a single collapse, probably at a pressure different from that of either component.

2.3. LANGMUIR–BLODGETT DEPOSITION

A wealth of useful information about molecular sizes and intermolecular forces can be obtained from studies of monolayers on the water surface, but the great resurgence of interest in this area of science has been largely due to the fact that films can be transferred from the water surface onto a solid substrate using what has become universally known as the Langmuir–Blodgett (LB) technique.

In the most commonly used method, the substrate (e.g., a glass slide) is first lowered through the monolayer so that it dips into the subphase (e.g., water) and then withdrawn; in order to maintain constant conditions during this process the surface pressure is kept constant using one of the methods described in Chapter 3. The value of surface pressure which gives best results depends on the nature of the monolayer and is established empirically. It is, however, possible to place limits on the search, as materials can seldom be successfully deposited at surface pressures of less than 10 mNm^{-1}, and at surface pressures above 40 mNm^{-1} collapse and film rigidity often pose problems. It is traditional to carry out LB deposition using films in the "solid" phase but this does not appear to be essential as long as the film is not expanded. If the slide used has a hydrophilic surface, deposition follows the sequence of events shown in Figure 2.10. The water wets the slide's surface and the meniscus turns up, and there is no mechanism for deposition at this stage. As the slide is withdrawn (Figure 2.10b) the meniscus (which is still turned up) is wiped

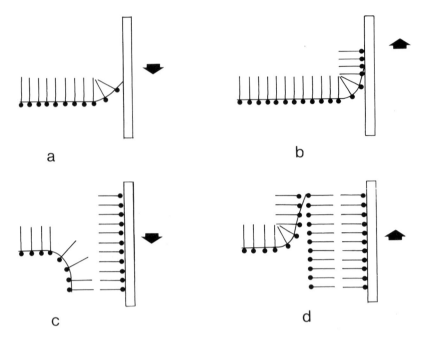

Figure 2.10. Deposition of multilayers by the Langmuir–Blodgett technique: (a) first immersion, (b) first withdrawal, (c) second immersion, (d) second withdrawal.

over the slide's surface and leaves behind a monolayer in which the hydrophilic groups are turned toward the hydrophilic surface of the slide. It will be apparent that there must initially be a liquid film between the slide and the deposited monolayer and that bonding of the monolayer to the slide will only be complete after the intervening layer of water has drained away or evaporated. For this reason, particular care is usually taken to ensure complete drying of the deposited film before further treatment. The water film present during withdrawal will tend to fill any microirregularities on the slide's surface, so that the monolayer bridges gaps between asperities. When the water evaporates, the monolayer is left unsupported and may collapse and form defects,[11] or if it is sufficiently robust (e.g., a polymerized film) it can remain[12] as an integral film over gaps of up to 500 μm.

The rate at which a slide can be withdrawn from the water is typically about 1 mms^{-1} and depends partly on the rate at which the liquid film drains from the monolayer/slide interface and partly on the dynamic properties of the monolayer on the water surface. It can be imagined that a highly viscous monolayer will be unable to adjust itself so as to maintain a homogeneous film in the neighborhood of a rapidly moving slide. Particular difficulty arises where a monolayer has been polymerized on the water surface to give an essentially rigid solid film. Even with monomeric materials, unusually high monolayer viscosity can lead to unsatisfactory deposition; this has recently been quantified by Daniel and co-workers[13] and by Malcolm.[14]

It is not uncommon to see the Wilhelmy plate displaced laterally during compression of a monolayer. This will clearly lead to errors in surface pressure measurements and is an undesirable consequence of monolayer viscosity. The motion of the water-supported monolayer during a process of LB deposition will clearly depend greatly on the viscosity of the layer and also on the dipping and barrier geometry. The time-honored technique of following movements of the surface layer by first sprinkling on fine particles (e.g., sulfur) has been recently modified by Daniel and Hart,[15] who have photographed the displacement of a regular array of thin floating PTFE disks during the course of an LB deposition experiment.

The flow patterns observed are represented in Figure 2.11. From these it can be seen that the "history" of the deposited film depends very much upon the initial alignment of the substrate to the moving barrier, the streamlines at the substrate's surface being different for arrangements a and b. This is confirmed by observed effects on the morphology of the deposited LB film. The influence on surface flow of plateau-type transitions in the isotherm has also been investigated.[14]

The opposite effect occurs where a monolayer has exceptional mobility. Peterson and co-workers[16,17] have shown that some materials can be deposited very rapidly (1 cms^{-1}) provided that the monolayer has not been allowed to age for too long on the water surface. The aging process may be a type of crystallization or may be due to solvent loss, but it is certainly[18] affected by the presence of minute concentrations of counterions, as is the speed of drainage. If this observation can be applied to a wider range of materials it will be of enormous practical importance.

The second trip into the subphase (Figure 2.10c) differs from the first in that the slide is now hydrophobic; the meniscus turns down and a second monolayer is deposited with its tailgroups in contact with the exposed tailgroups on the slide. The second trip out of the subphase exactly resembles the first, except that the new monolayer is now being deposited onto the hydrophilic headgroups of the mono-

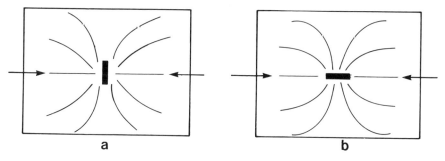

Figure 2.11. Lines of motion of a monolayer during deposition of LB films onto a substrate aligned (a) parallel and (b) perpendicular to the direction of barrier motion. (After Daniel and Hart.[15])

layer already present. This type of deposition, in which layers are laid down each time the substrate moves across the phase boundary, is known as Y type.[19] It is also possible to have deposition occur only when the substrate enters the subphase (X type) or leaves it (Z type), depending on the nature of the monolayer, substrate, subphase, and the surface pressure.

If the substrate is initially hydrophobic, deposition will normally start on the first immersion into the subphase. If the deposition is Y type there will therefore be an even number of layers deposited at the end of each complete cycle, in contrast to the odd number of layers deposited onto hydrophilic substrates. Common polymers are normally hydrophobic and show this behavior. It is possible to make glass hydrophobic by rubbing it with ferric stearate[20] or by silylating the surface. This treatment may also be used[21] to pretreat the surface of a wide range of compound semiconductors and enhance the quality of LB film deposition. Metals which oxidize readily tend to have hydrophilic surfaces because of the nature of the oxide films. Noble metals, such as gold, which do not readily oxidize, are often regarded as hydrophobic, although it has recently been argued[21] that a *clean* gold surface is hydrophilic but still picks up a monolayer on its first immersion because of the strong oleophilic interaction between the surface and the hydrocarbon chains.

If a hydrophilic substrate is being coated, and if the monolayer material shows poor adhesion to the substrate, then the second immersion in the subphase can simply lead to the monolayer peeling off the slide and respreading on the water surface. Under these circumstances, repeated dipping always leads to a single deposited monolayer on the slide. This process can also occur if inadequate time is allowed for drying and drainage of the first layer. To reduce the risk of problems of this kind, it is often desirable to deposit one or more monolayers of stearic acid onto a slide in order to define a good surface before attempting to deposit multilayers of other materials which may be inherently less good film formers. A slide rendered hydrophobic in this way can be coated with highly viscous monolayers by lowering it horizontally. This type of deposition has been extensively studied by Fukuda and co-workers, who have used it for the preparation of multilayers from rigid films of aromatic compounds.[22,23] (see Section 2.4.2) and polymers[24] (see Section 2.5.1).

The question arises as to whether the manner of deposition determines the final structure of the film, or whether some reorganization takes place after the multilayer has been dipped. At first sight one would expect multilayers deposited in X,Y modes to have the structure shown in Figure 2.12. However, it seems that multilayers of stearic acid always have Y-type structure irrespective of whether they have been deposited in an X or Y type of sequence,[25,26] so that simple molecules of this type are able to invert at some stage during the dipping process. Larger, more rigid, molecules impart greater conformational stability to the monolayer, and it is possible to carry out quite complex manipulations without disturbing molecular orientation as in the elegant studies on dyes by Kuhn and co-workers.[27]

The structural details of LB films will be dealt with elsewhere (Chapter 4). It should be noted at this stage that the first monolayer usually consists of two-

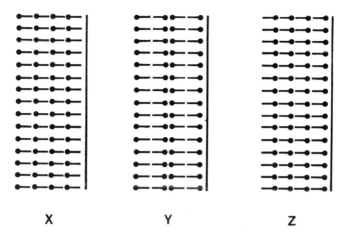

Figure 2.12. Structures of X, Y, and Z multilayers. X and Z type depositions do not normally guarantee that the multilayers will have the corresponding structure.

dimensional crystalline domains which are approximately aligned in the direction of dipping. Subsequent layers may either be epitaxially deposited onto these initial domains, or they may have an independent domain structure, depending on the state of the monolayer on the water surface (see, for example, Peterson.[28]

The *transfer ratio* is often used as a measure of the quality of deposition. It is defined as the ratio of the area of monolayer removed from the water surface to the area of substrate coated by the monolayer. The area removed from the water surface is easily measured by the mechanism used to maintain a constant surface pressure and there is then a direct electrical readout of surface area. Under most circumstances a transfer ratio of unity is taken as a criterion for good deposition, and one would then expect the orientation of molecules on the slide to be very similar to their orientation on the water. Occasionally, there is a large but consistent deviation from a value of unity; this points to a situation in which the molecular orientation is changing during transfer. Variable transfer ratios are almost always a sign of unsatisfactory film deposition.

The energy balance during film deposition has been studied in some detail by Saint Pierre and Dupeyrat[29] by measuring the vertical force on the slide during a dipping procedure. After subtraction of Archimedes flotation effects the work of immersion and emersion (removal) of the slide can be determined by integration. Formal analysis shows that deposition during immersion will only occur if the energy of interaction of the tailgroups per unit area (W_{tt}) satisfies the inequality $-W_{tt} > \gamma$.

This implies that there is a maximum value of surface tension (γ) beyond which deposition will not occur; indeed there is ample experimental evidence to the effect that a minimum surface pressure ($\Pi = \gamma - \gamma_0$) is a necessary prerequisite for deposition. From their measurements of the integrated work of immersion the authors conclude[29] that the interaction energy between pairs of alkyl chains is 1.35×10^{-20} J.

The emersion process is more complicated because dehydration of the interlayer requires an energy input ($2W_{deh}$). Accordingly, the condition for deposition requires a high value for the head–head interaction energy (W_{hh}): $-W_{hh} > 2W_{deh} + \gamma$.

"Normal" deposition of Y type layers depends on the fact that head–head interactions are usually stronger than tail–tail interactions, although they are also greatly dependent on the presence of cations and pH variations in the subphase. The work of deposition will additionally depend on the extent of dehydration during emersion. An extension of this work by another group[30] has led to the suggestion[31] that the above interpretation may be simplistic. The technique, however, also provides an excellent way of monitoring deposition and gives a valuable indication of irregularities in transfer.

2.4. TYPES OF MOLECULES KNOWN TO FORM MONOLAYERS AND LANGMUIR–BLODGETT FILMS

2.4.1. Fatty Acids and Their Derivatives

The model material discussed so far in this chapter has been stearic acid (Figure 2.1) $C_{17}H_{35}CO_2H$, and the detailed composition of the aqueous subphase has been ignored. This section will be concerned with materials related to stearic acid and will also show how minor constituents of the subphase can drastically alter the monolayer properties.

2.4.1.1. Chemical Structure of Fatty Acids

A simple fatty acid such as stearic acid consists of a linear, saturated, alkyl chain (C_nH_{2n+1}) terminated by a carboxylic acid group. Under conditions of fixed subphase composition a carboxylic acid group ($-CO_2H$) can be regarded as providing a constant amount of hydrophilicity to a molecule. Alkyl groups are known to be hydrophobic, and it would therefore be expected that overall hydrophilicity would increase with the length of the alkyl group (i.e., the value of n). This is borne out in practice: provided that the subphase pH is sufficiently low (see later), acids containing more than 13 carbon atoms (i.e., $n > 12$) can be spread to form a monolayer, but the stability varies greatly with chain length. For example, myristic acid ($n = 13$) shows[32] a loss of 0.1% min^{-1} from the surface area of a monolayer at 20 °C and 10 mNm^{-1}. This loss is not due to collapse, but is caused by material dissolving into the subphase; it is commonly observed in materials having an inadequate hydrophobic/hydrophilic balance. In contrast, stearic acid ($n = 17$) was shown by Gaines[1] to lose less than 0.001% min^{-1} under similar conditions. Because of their stability, stearic and arachidic ($n = 19$), and to a lesser extent behenic ($n = 21$) acids have been the workhorses of monolayer and LB studies, although monolayers have been spread[33] from acids with very long chains ($n = 35$).

As their name implies, these materials are all acidic; in fact they are weak acids and when the subphase has been acidified with a mineral acid to pH 4, their ionization is completely suppressed so that they behave as neutral molecules. At higher pH (i.e., less acidic subphases) ionization occurs to form hydrogen ions in the subphase and carboxylate ions ($C_nH_{2n+1}CO_2{}^-$) in the film. Although pure water has a pH of 7, carbon dioxide from the atmosphere dissolves in the subphase and slightly acidifies it to pH 5.8. The pH at which half the molecules of an acid are ionized is known as the pK_A of the acid; the value expected for the pK_A of stearic acid in solution is 4.8, but on the water surface it is shifted[34] to 5.6, although the situation is further complicated in the presence of metal cations in the aqueous phase.

2.4.1.2. Incorporation of Metal Ions

At pH < 7 there must be cations other than H^+ present in the subphase. If these cations are singly charged (e.g., N^+ or K^+) the solubility of the acid increases, and micelles can be formed unless measurements are taken very rapidly. Soap is the sodium salt of a mixture of fatty acids, and films of soap and other detergent molecules have been studied extensively using an oil/water trough in which a layer of oil floating on top of the water stabilizes the monolayer at the interface. If the cations are doubly charged (e.g., Ca^{2+}, Ba^{2+}, and Cd^{2+}) the resultant "soaps" are insoluble in water and the monolayers are stable at an air–water interface. The isotherm of stearic acid on a subphase containing 10^{-4} M Cd^{2+} ions is shown in Figure 2.13. The extension of the solid region relative to that observed in the absence of Cd^{2+} is due to the crosslinking action of Cd^{2+} ions, which require two stearate ions ($C_{17}H_{35}CO_2^-$) for electroneutrality. Incorporation of Cd^{2+} into the structure also accounts for the small change in surface area per molecule. Triply charged cations (e.g., Fe^{3+}, Al^{3+}) in the subphase tend to give extremely rigid films with a high shear modulus.[35]

The extent of incorporation of ions into the film will depend on the pH as will the state of the incorporated ion. The collapse properties depend on the pH even in the absence of added multivalent ions[36] and studies of these properties have led to an estimate[37] of 8.2 for the pK_A of a stearic acid monolayer; this differs from the figure of 5.6 noted earlier and shows the problems of interpretation that can arise in monolayer systems. The collapse properties are further complicated in the presence of multiply charged ions.

The rigid films obtained when there are triply charged ions in the subphase cannot normally be deposited by the LB technique[19] and even concentrations as low as 10^{-5} M Al^{3+} in the subphase can prevent deposition. It is remarkable that the same concentration (10^{-5} M) of Cu^{2+} in the subphase prevents deposition, but that a lower concentration (2×10^{-6} M) was found by Blodgett and Langmuir[19] to aid deposition of very large numbers of layers.

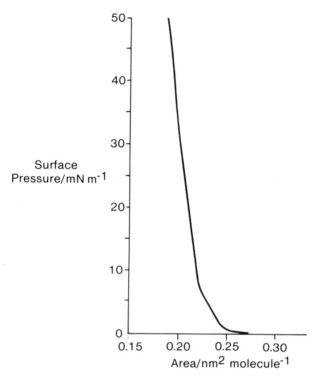

Figure 2.13. Surface pressure/area isotherm of stearic acid on a subphase containing 2.5×10^{-4} M cadmium chloride (temperature 20 °C, pH 5.0).

Metal ions are incorporated into the multilayer to an extent that depends on pH. For example, four regimes have been identified[38] in the stearic acid/Ca^{2+} system. Below pH 4.2 stearic acid is deposited; between pH 4.2 and 6.4 a homogeneous mixture of stearic acid and calcium stearate is deposited, with the calcium concentration rising to 60% at pH 6.4. Between pH 6.4 and pH 8, the calcium stearate begins to cluster into two-dimensional aggregates termed surface micelles by the authors.[38] Above pH 8 virtually all the material is present as a mosaic of surface micelles. Segregation has also been observed in mixed monolayers of barium stearate[39] and attention drawn to the rigidity of pure salt relative to acid/salt mixtures.[40] The ions of Ca^{2+}, Sr^{2+}, Ba^{2+} are unusual among divalent ions in being able to exist over a wide pH range without forming a precipitate by reacting with hydroxide ions. Pb^{2+}, for example, reacts[40] as does Cd^{2+}, which is one of the most commonly used additives to the subphase. The presence of Cd^{2+} in a film of arachidic acid has been shown to change the structure from monoclinic with chains at 25° to the normal[41] to orthorhombic with near-vertical chains.[42] In neither case did there appear to be any preferred orientation induced by the dipping process.

Another ion that has been deliberately incorporated into multilayers is Mn^{2+}, because of its magnetic properties.[43] A trivalent ion, nominally Cr^{3+}, has been successfully dipped by raising the pH to 6.7, at which it is substantially all present as $Cr^+(OH)_2$ in the film.[44] The free acids and their salts differ greatly in their solubility in organic solvents. This differential solubility allows LB multilayers to be modified by a process termed *skeletonization* by Blodgett and Langmuir,[19] in which the acid component of a mixed acid/salt monolayer is dissolved away to leave a film of slightly reduced thickness but greatly reduced density and refractive index. The optical properties of these skeletonized films will be discussed in Chapter 7.

Other, more complex manipulations can be carried out with multilayer assemblies. For example, a layer of water-soluble polymer such as PVA can be spread from solution onto an LB film. If the polymer layer is now peeled away, it can be shown that in favorable circumstances the top monolayer, and only the top monolayer, is removed. This layer can be transported elsewhere on the polymeric carrier, which is then dissolved away.[27]

2.4.1.3. Fatty Acids with Modified Hydrophobic Tailgroups

The alkyl group (C_nH_{2n-1}) of fatty acids may be replaced by chains containing one or more double bonds. These unsaturated groups are centers of chemical reactivity and are particularly important as polymerizable groups. A double bond upsets the ordering of the chain by introducing a kink to the overall structure. This effect can be clearly seen in its effect on the melting points of corresponding compounds; for example, stearic acid (with no double bonds) melts at 70 °C while oleic acid (Figure 2.14) $C_{17}H_{33}CO_2H$, with the same number of carbon atoms and one double bond, melts at 14 °C. This disruption is reflected in the monolayer behavior, as can be seen by comparison of the isotherms[45] in Figure 2.5, although the packing is still sufficiently good to allow multilayer deposition.[46] The effect of introducing a branch into a saturated alkyl chain without altering the number of carbon atoms in the molecule is shown[47] by the isotherm of 2-ethylpalmitic acid (Figure 2.15), also shown in Figure 2.5. The melting point (38 °C) also reflects the disruption introduced in the packing.

The position of the perturbing group within the chain and its geometry can also influence the packing and hence the monolayer properties. For example, oleic acid has a *cis* double bond with both parts of the chain on the same side of the bond while elaidic acid (Figure 2.16) has a *trans* double bond, which gives a straighter structure and a melting point of 52 °C. As expected, this is intermediate in properties between stearic and oleic acids and has also been deposited as multilayers.[46] The disruption is maximized if there is a methylene sidegroup, where both parts of the chain are attached to one end of the double bond, as in 2-methylene octadecanoic acid (Figure 2.17). Although this has an extra carbon atom, it still forms expanded monolayers. If the double bond is terminal (i.e., at the end of the chain), disruption is greatly reduced and useful LB materials can be obtained, particularly if the chain is length-

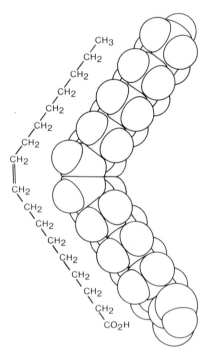

Figure 2.14.   Skeleton and space-filling formulas of oleic acid (Z-octadeca-9-enoic acid).

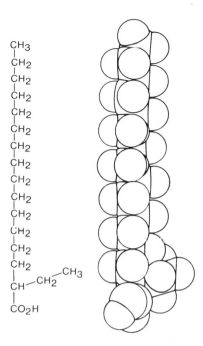

CH3
|
CH2
|
CH2
|
CH2
|
CH2
|
CH2
|
CH2
|
CH2
|
CH2
|
CH2
|
CH2
|
CH2
|
CH2
|
CH2
| CH3
| CH2╱
CH╱
|
CO2H

Figure 2.15. Skeleton and space-filling formulas of 2-ethylpalmitic acid (2-ethylhexadecanoic acid).

ened. The best-studied example is omega-tricosenoic acid (Figure 2.18) which Barraud's group has investigated extensively because of its polymerizability (see later) and suitability for LB deposition.

The effect of adding a second weakly polar group into the chain can be seen in the case of oleic epoxide (Figure 2.19). This has a relatively high melting point (56 °C) compared to oleic acid, probably because of the interactions between dipoles in adjacent molecules. The isotherm (Figure 2.20) expectedly does not show a well-defined solid phase, but a more remarkable feature is the residual surface pressure at very high areas per molecule.[48] This is most readily interpreted in terms of the hydrophilic character of the epoxide group (oxiran ring) causing the molecules to arch over and occupy a relatively large surface area. When the surface pressure is increased sufficiently, the second point of contact is detached and the area per molecule thereby reduced, the whole process being an exaggerated version of the usual displacement of hydrocarbon chains from the surface at very low pressures. It is not possible to build up multilayers of this molecule using the LB technique. The longer chain terminal epoxide (Figure 2.21) gives a more conventional isotherm and can readily be built up as multilayers[48] as can a variety of longer chain epoxides with the epoxy ring in the middle of the molecule.[49] The diacetylenic acids containing the group ($-C{\equiv}C-C{\equiv}C-$) have been studied extensively because of

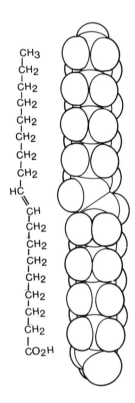

Figure 2.16. Skeleton and space-filling formulas of elaidic acid (*E*-octadeca-9-enoic acid).

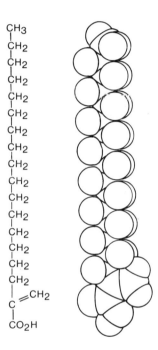

Figure 2.17. Skeleton and space-filling formulas of 2-methyleneoctadecanoic acid.

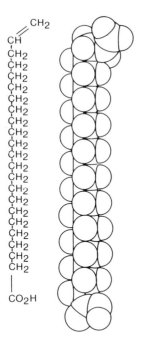

CH₂
CH
CH₂
CH₂
CH₂
CH₂
CH₂
CH₂
CH₂
CH₂
CH₂
CH₂
CH₂
CH₂
CH₂
CH₂
CH₂
CH₂
CH₂
CH₂
CH₂
CO₂H

Figure 2.18.   Skeleton and space-filling formulas of omega-tricosenoic acid (22-tricosenoic acid).

their polymerizability both in monolayers and in LB multilayers, which will be discussed later. The triple bond differs from a double bond by virtue of its linearity, which is reflected in the melting point (58 °C) and compact isotherm of the seventeen carbon acid (Figure 2.22).

It is also possible to replace the hydrogen atoms in the alkyl chain by fluorine atoms, to give a partially or totally fluorinated material. A fluorocarbon chain $(CF_2)_n$ differs from a hydrocarbon chain $(CH_2)_n$ in three principal ways: it is slightly more hydrophobic, it is much more rigid, and it has a higher cross-sectional area. A carboxylic acid headgroup attached to a fluorinated carbon atom is much more acidic than a normal carboxylic acid. The totally fluorinated molecule $C_{10}F_{21}CO_2H$ formed monolayers, although these were unstable above 15 mNm^{-1}. In contrast, the partially fluorinated species $C_8F_{17}(CH_2)_nCO_2H$ where $n = 2,4,6$ (see Figure 2.23) all formed stable monolayers and could be deposited as multilayers, the second two giving very well ordered films.[50] This work complements an earlier study[51] in which the stability of monolayers of partially fluorinated fatty acids increased progressively as the length of the fluorinated chain was increased from 1 to 7 carbon atoms. The stability of these materials is consistent with the more hydrophobic nature of the CF₂ chain; the instability of the totally fluorinated material is probably due to the high acidity of the fluorocarbon acid which leads to a

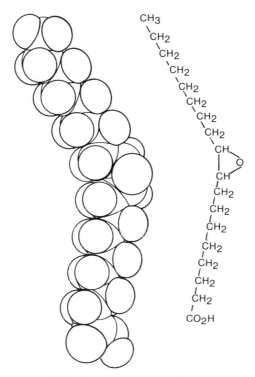

```
                                    CH3
                                    \
                                     CH2
                                      \
                                       CH2
                                        \
                                         CH2
                                          \
                                           CH2
                                            \
                                             CH2
                                              \
                                               CH2
                                                \
                                                 CH2
                                                  \
                                                   CH
                                                   |    O
                                                   CH
                                                   |
                                                   CH2
                                                   |
                                                   CH2
                                                   |
                                                   CH2
                                                   |
                                                   CH2
                                                   |
                                                   CH2
                                                   |
                                                   CH2
                                                   |
                                                   CH2
                                                   |
                                                   CO2H
```

Figure 2.19. Skeleton and space-filling formulas of oleic epoxide (*cis*-9,10-oxiranoctadecanoic acid).

highly ionized film. The measured area per molecule of 0.32 nm² is consistent with data on solid PTFE. Addition of Al^{3+} to the subphase has a beneficial effect on deposition in contrast to its effect on unfluorinated fatty acids.

2.4.1.4. Modifications to the Hydrophilic Headgroup

So far, all the modifications described in this section have involved the tail group, while leaving the hydrophilic headgroup untouched. The carboxylic acid group can, however, be modified, for example, by conversion to an ester ($-CO_2R$) or an amide[45] ($-CO_2NH_2$). Ethyl stearate ($C_{17}H_{35}CO_2C_2H_5$) is an example of an ester that forms[52] compact, low compressibility monolayers (Figure 2.24), while vinyl stearate ($C_{17}H_{35}CO_2C_2H_3$) differs only in the fact that the ester headgroup is unsaturated and can be polymerized both as monolayers and as LB multilayers. Dibasic esters such as $CH_3O_2C(CH_2)_nCO_2CH_3$, in which there is a weakly hydrophilic group at either end of the chain, have also been studied.[53] At large areas per

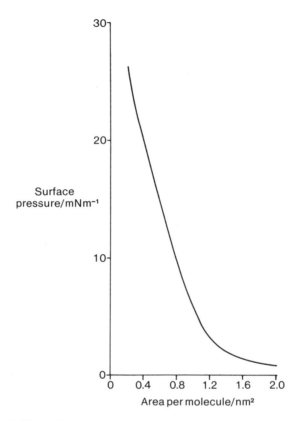

Figure 2.20.   Surface pressure/area isotherm of oleic epoxide (19 °C, pH 3.5).

molecule they lie flat on a water surface, but on compression they stand upright, as evidenced by surface-potential measurements.

Lipids are a particularly important class of esters of fatty acids because they form the active matrix of biological membranes. The general rules developed above with regard to chain length, unsaturation, etc., also apply to these materials, but the situation is often complicated by the presence of more than one hydrophobic tail in the molecule. These materials are reviewed extensively in Chapter 6.

Fatty alcohols have the general formula $C_nH_{2n+1}OH$; their monolayer properties follow the same trends as the acids, except that they are not subject to dissociation so that the subphase pH is largely immaterial. As with the acids, monolayers are formed for members of the series with thirteen carbon atoms or more, and stability increases with chain length. A significant difference lies in the mechanism of loss from the film; the carboxylic acids lose material mainly by dissolution into

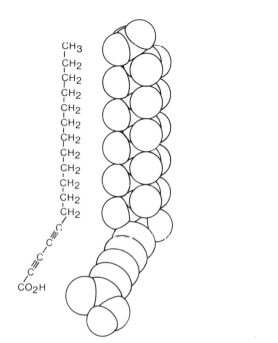

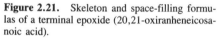

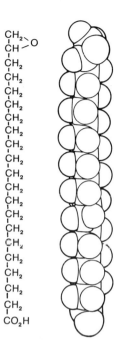

Figure 2.21. Skeleton and space-filling formulas of a terminal epoxide (20,21-oxiranheneicosanoic acid).

Figure 2.22. Skeleton and space-filling formulas of a diacetylene acid (heptadeca-2,4-diynoic acid).

the subphase while the alcohols lose material mainly by evaporation. Another difference lies in the difficulty of forming LB multilayers from alcohols.[54] Other oxygen-containing headgroups such as ketone ($-COCH_3$) have been studied[53] on the water surface.

The amines have the ($-NH_2$) headgroup. They have been less extensively studied, but are of great interest because they are bases, i.e., they will give cations in the presence of acids. Their behavior is accordingly to some extent a mirror image of the behavior of acids. In particular, they are most stable on subphases with a high pH value and form easily dissolved expanded monolayers on acidic subphases. As with acids they can be stabilized by a multiply charged counterion such as sulfate (SO_4^{2-}) or hydrogen phosphate (HPO_4^{2-}), and octadecylamine ($C_{18}H_{37}NH_2$) has been deposited as LB multilayers with the latter counterion[55] although the pH range for deposition was restricted. More recently, Gaines[56] has shown that the longer chain amine docosylamine ($C_{22}H_{45}NH_2$) deposits much more easily. He has also highlighted the fact that amines react reversibly with atmospheric carbon dioxide and that this has complicated the interpretation of earlier work with these materials.

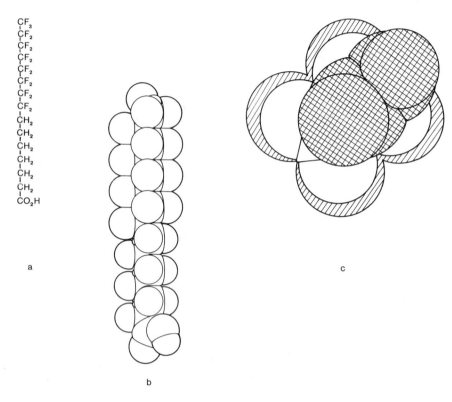

CF$_3$
CF$_2$
CF$_2$
CF$_2$
CF$_2$
CF$_2$
CF$_2$
CF$_2$
CH$_2$
CH$_2$
CH$_2$
CH$_2$
CH$_2$
CH$_2$
CO$_2$H

a

c

b

Figure 2.23. Skeleton (a) and space-filling (b and c) formulas of a fluorinated fatty acid [heptad-ecafluoroheptadecanoic acid, C$_8$F$_{17}$(CH$_2$)$_6$CO$_2$H]. Formula (b) shows how the greater size of the fluorine atoms would cause them to interfere when the chain is in this conformation. The view from the carboxylic acid end (c) also shows the greater bulk of the fluorine atoms.

Other headgroups that have been studied include phosphine oxide,[57] various phosphoric acid derivatives,[58,59] amine oxide,[60] and nitrile.[61] More recently most of the various possible sulfur-containing headgroups have been studied[62]: alkyl chains attached to the headgroups SCH$_3$, SOCH$_3$, and SO$_2$CH all give monolayers on the water surface, while the more polar group S$^+$(CH$_3$)$_2$ gives films that are soluble at high surface pressure. It is well known that sulfonic acids (the sulfur equivalent of carboxylic acids) are more soluble than the corresponding carboxylic acids, and sulfate esters tend to form soluble films except in the presence of dissolved salts in the subphase.[63] With the exception of lipids and polymerizable materials there has been relatively little recent work on materials related to fatty acids and the reader is referred to Gaines's[1] and Adam's[64] textbooks for an account of the earlier work.

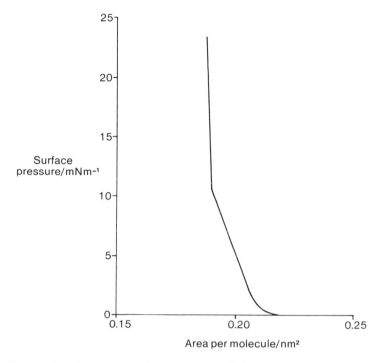

Figure 2.24. Surface pressure/area isotherm of ethyl stearate (ethyl octadecanoate).

2.4.2. Molecules Containing Five- or Six-Membered Rings

The molecules considered in this section are all covered by the chemical description aromatic, that is, they contain at least one five- or six-membered ring that is fully unsaturated, i.e., it contains three double bonds (if a six-membered ring) or two double bonds and a lone pair of electrons from an N,O, or S atom (if five-membered). The archetypal aromatic compound is benzene (Figure 2.25a) with a single six-membered ring; several rings can be fused together as in anthracene (Figure 2.25b) or aliphatic chains can replace the hydrogen atoms. The aromatic rings are hydrophobic (although not as hydrophobic as alkyl chains containing the same number of carbon atoms—see later) and require the attachment of hydrophilic groups in order to spread on the water surface.

2.4.2.1. Derivatives of Benzene

Much of the early work on simple derivatives of benzene has been confined to monolayer studies such as Adam's investigation of oxygenated benzenes.[65] Studies[66] of a group of azo compounds and stilbenes are of particular interest because they have followed by optical absorption measurements the transition from an expanded monolayer where the molecules are lying flat on the water surface, to a compressed monolayer where the molecules (and hence the absorption axis) become normal to the surface. LB films have, however, been deposited from hexadecylphenol (Figure 2.26), from tetradecylbenzoic acid (Figure 2.27) as part of a study of the influence of the aromatic ring on molecular packing,[67] and an epoxidized benzoic acid (Figure 2.28) has been deposited for polymerization purposes.[49] Monoclinic multilayers of p-octadecyloxyaniline (Figure 2.29) have been deposited.[68]

Several liquid-crystal-forming materials have been studied,[69,70] the most stable monolayers being formed from a material known as T15 (Figure 2.30), which can also be deposited with difficulty as Z-type LB multilayers. Other liquid crystalline materials, both monomeric and polymeric, have been studied only as monolayers,[71,72] but the azo compound (Figure 2.31) has been deposited as multilayers[73] which are structurally Y type irrespective of whether deposition is Y or Z type. Mutlilayers of this last material form crystalline domains (up to 1 mm across) on heating. In contrast to the rodlike materials so far mentioned, a discotic compound (Figure 2.32) has also been examined[74] and found to have a high ESP (10 mNm^{-1}) although it forms multilayers on the water surface even at this pressure.

2.4.2.2. Derivatives of Polycyclic Aromatic Hydrocarbons

The derivatives of anthracene (Figure 2.25b) have been more extensively investigated than any other group of materials except the fatty acid derivatives. The original investigation was by Stewart,[75] who obtained stable films from anthracene derivatives (Figure 2.33) with hydrophilic X groups and long chains in the R position. The more recent work by the ICI group[76–78] has extended the range of film-forming materials to shorter R groups and a wide range of hydrophilic X groups. The slight solubility of the materials with a short R group can be compensated for by repeatedly compressing and respreading monolayers until stability is obtained.

The effect of chain length on monolayer formation (after this recompression process) can be seen in Figure 2.34 for a series of anthracenepropanoic acids (R = C_nH_{2n+1}; X = —CH$_2$CH$_2$CO$_2$H) at a subphase pH of 6. As expected, the longest chains give the most stable films, the material with $n = 4$ (9-butyl-anthr-10-ylpropan-3-oic acid, henceforth abbreviated C4) being too soluble to give compact films at this pH. When the subphase pH is reduced, the stability of C4 films is increased and, at pH 4 stable, compact films are obtained (Figure 2.35). In contrast, the derivative with $n = 12$ gives isotherms which are independent of pH over the range investigated.

It is interesting that C4 has a total of 21 carbon atoms in the molecule and has a melting point of 127 °C. By the criteria applied to simple fatty acids one would expect it to have some margin of stability over a wide pH range and even suppose

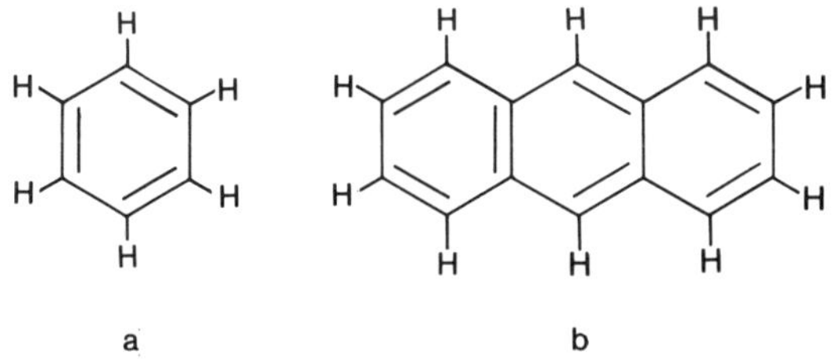

Figure 2.25. Unsubstituted aromatic compounds: (a) benzene, (b) anthracene.

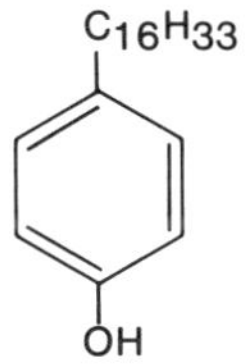

C16H33

OH

Figure 2.26. 4-Hexadecylphenol.

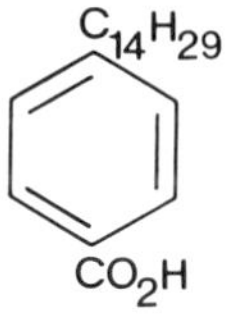

C14H29

CO2H

Figure 2.27. 4-Decylbenzoic acid.

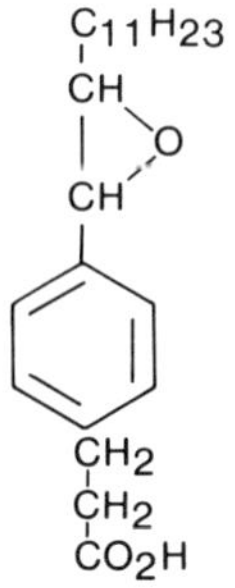

C11H23
CH
O
CH

CH2
CH2
CO2H

Figure 2.28. 3-(1,2-Oxirantridecaphenyl) pro-panoic acid.

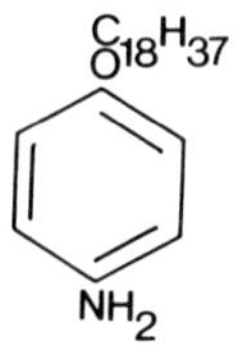

C18H37
O

NH2

Figure 2.29. 4-Octadecyloxyaniline.

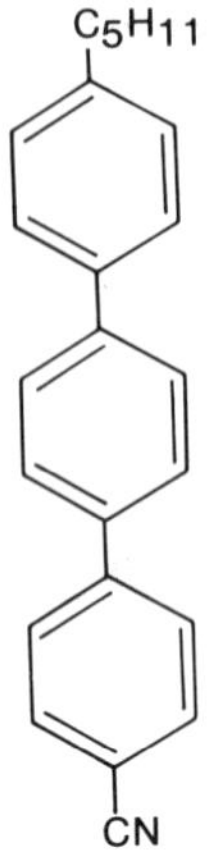

Figure 2.30. T15 liquid crystalline terphenyl derivative.

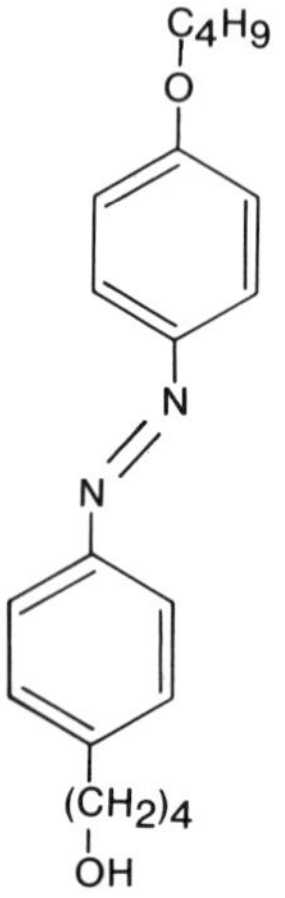

Figure 2.31. Liquid crystalline azo dye.

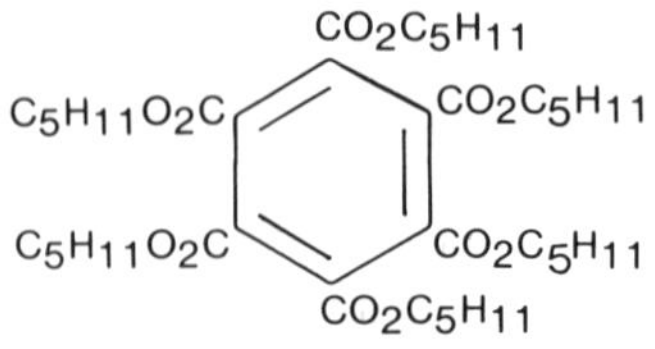

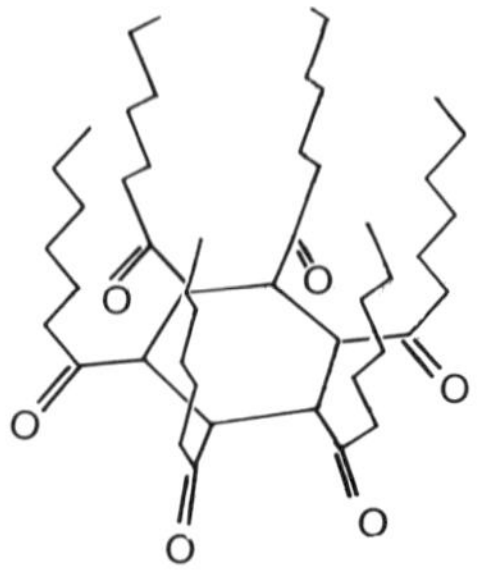

Figure 2.32. Discotic liquid crystalline ester of mellitic acid.

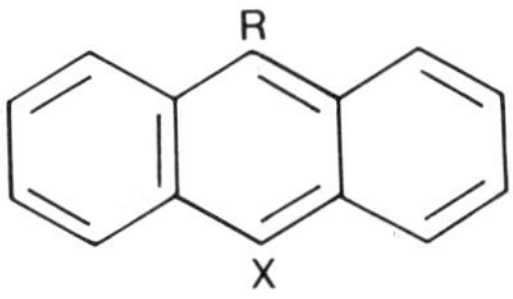

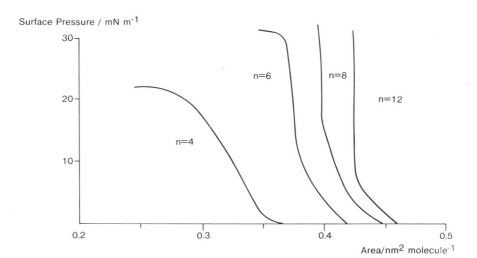

Figure 2.33. Monolayer-forming anthracene derivatives (see text for explanation of R and X).

Figure 2.34. Surface pressure/area isotherm of anthracene derivatives (Figure 2.33; $R = C_nH_{2n+1}$, $X = CH_2CH_2CO_2H$), with varying values of n (temperature 20 °C, subphase pH 6.0).

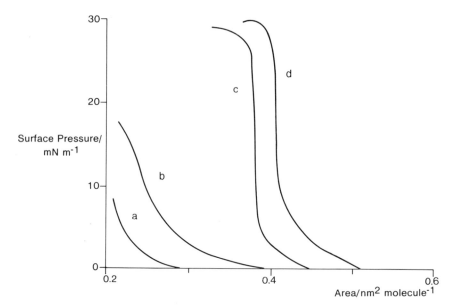

Figure 2.35. Surface pressure/area isotherms of anthracene derivatives, showing the influence of subphase pH on C4 (Figure 2.33; R = C_4H_9, X = $CH_2CH_2CO_2H$). Curve (a) was recorded at pH 8, curve (b) at pH 6, and curve (c) at pH 4. The isotherm (d) of a long-chain homolog (Figure 2.33; R = $C_{12}H_{25}$, X = $CH_2CH_2CO_2H$) is unchanged over the same pH range (temperature 20 °C).

that the lower homologs would have reasonable stability. However, it will be clear from Figure 2.36 that a reduction in the chain length of the X group to X = —CH_2CO_2H or X = CO_2H produces unstable films even at a subphase pH of 4; the same is true if the R group is shortened to C_3H_7. We can therefore conclude that although the insertion of an aromatic ring system such as anthracene enhances the molecular packing (as evidenced by the high melting point), the ring system must be less hydrophobic than an alkyl chain of corresponding length.

The effectiveness of a wide range of different headgroups is also shown in Figure 2.36. Some of the trends are closely analogous to those observed in the alkyl series discussed earlier. With R held constant at C_4H_9, introduction of a double bond (X = —$CH{=}CH{\cdot}CO_2H$) reduces monolayer stability, as does introduction of a second acid group [X = —$CH_2CH(CO_2H)_2$]. If the carboxylic acid group is replaced by an alcohol group, for instance in the material with X = —CH_2OH, the monolayer stability is much improved. However, as with the aliphatic alcohols deposition as LB films is unsatisfactory.

In contrast, multilayers (up to 500 layers) of excellent quality have been built up from C_4 and their X-ray structure has been investigated.[78] The multilayers have the expected head-to-head structure, but the $C_4H_9^-$ tail groups are interleaved with those of the next layer so that the molecular spacing is reduced (Figure 2.37).

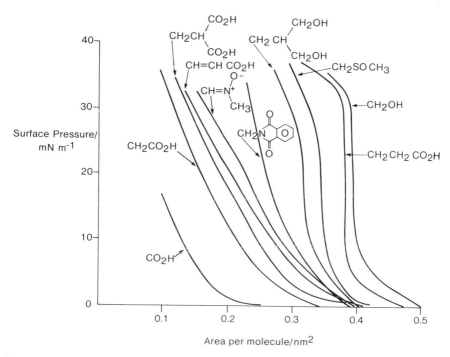

Figure 2.36. Influence of headgroup on the surface pressure/area isotherm of anthracene derivatives (Figure 2.33; R = C$_4$H$_9$, X as indicated) at a temperature of 20 °C and subphase pH 4.

Interleaving of this kind has also been proposed in a different aromatic system.[79] It must, however, be remembered that it is difficult to distinguish between interleaving and a situation where the alkyl groups have an unusually large angle of tilt. More recently, there have been more detailed studies of C4 multilayers by electron diffraction[80] and these have allowed the tilt angle of the anthracene ring to be fixed at 60° ± 1°.

Monolayer studies have also been carried out on a number of derivatives of polycyclic hydrocarbons with more than three rings. As the number of fused rings is increased, the length of alkyl chain necessary to stabilize the monolayer is decreased. For example,[76] perylenebutyric acid (Figure 2.38) forms very stable monolayers, although the film is too rigid to permit deposition.[81] Pyrene derivatives have also been investigated: the short chain compound [Figure 2.39, X = (CH$_2$)$_3$CO$_2$H] was too soluble to form stable monolayers,[76] but the longer chain acid [Figure 2.39, X = (CH$_2$)$_{15}$CO$_2$H] and more recently a series of pyrene substituted lipids have been deposited in order to study their fluoresence properties.[82,83] Undoubtedly, pyrene derivatives of intermediate chain length could form stable monolayers.

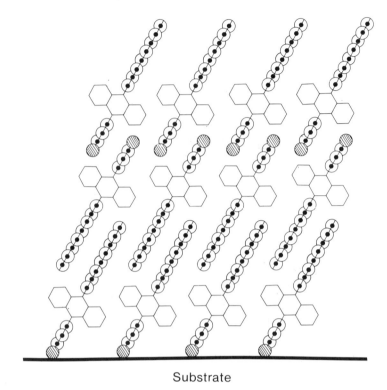

Substrate

Figure 2.37. Proposed structure of multilayers of an anthracene derivative (Figure 2.33; R = C_8H_{17}, X = $CH_2CH_2CO_2H$), showing interleaving of the alkyl chains. (After Belbeoch *et al.*[79])

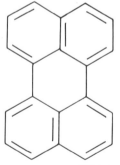

$(CH_2)_3CO_2H$ **Figure 2.38.** Perylenebutyric acid.

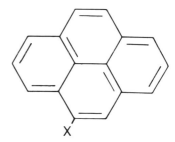

Figure 2.39.   Pyrene derivatives (see text for explanation of X).

2.4.2.3. Heterocyclic Compounds and Dyes

Rings containing atoms other than carbon are termed heterocyclic. Replacement of one of the carbon atoms in a benzene ring by a nitrogen atom renders the molecule much more hydrophilic, and if the nitrogen is quaternized by addition of another group it becomes charged and can act as a very effective polar headgroup. N-docosylpyridinium TCNQ (Figure 2.40) has recently been prepared as LB films which become highly conducting on doping with iodine.[84] The bispyridinium salt (Figure 2.41) has been deposited as mixed multilayers with stearic acid.[82] This material has two positive charges and two hydrophobic chains on the surface active species and has been used in studies of photoinduced electron transfer processes.

The cyanine dyes are a particularly interesting class of materials that have been widely studied by Kuhn, Möbius, and co-workers, as well as many other groups (for reviews see elsewhere).[27,85−87] A typical cyanine dye has two nitrogen-containing rings separated by a chain of carbon–carbon double bonds. In the structural formula (for example, in the compounds illustrated in Figure 2.42) only one of these carries a formal positive charge and the other one is neutral. In fact the positive charge is delocalized between the two nitrogen atoms, and the molecules as a consequence have strong optical absorption and are often fluorescent. Many of the investigations carried out on these materials relate to their optical properties, and will be dealt with in more detail in Chapter 5. Most often they are substituted with two long hydrophobic chains and are deposited as mixed monolayers with a diluent such as stearic acid.

A result which could have wide-ranging consequences is the epitaxial deposition[88] of one of these (Figure 2.42c) onto a single crystal of gypsum (calcium sulfate). A novel squarylium cyanine (Figure 2.43) has been deposited as normal LB

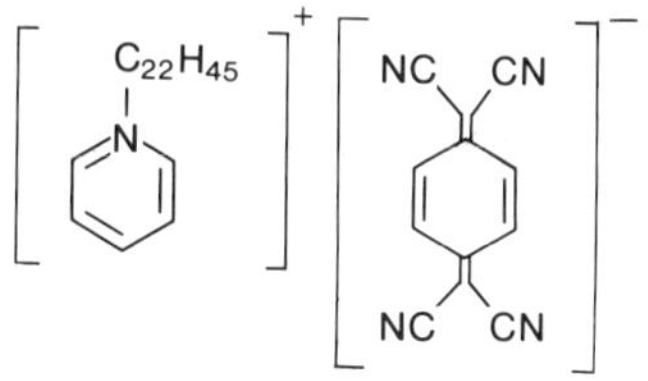

Figure 2.40.   N-Docosylpyridinium TCNQ salt.

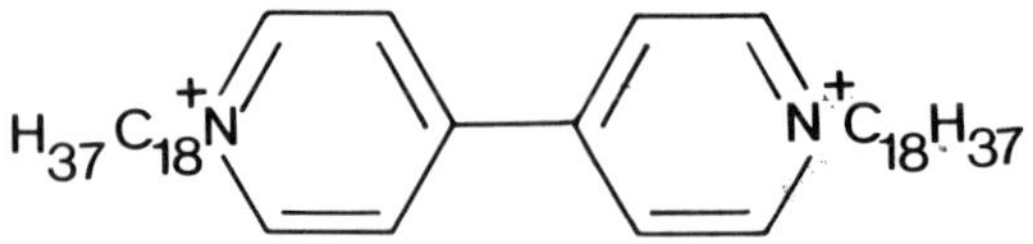

Figure 2.41. Bis (*N*-octadecylpyridinium) dication.

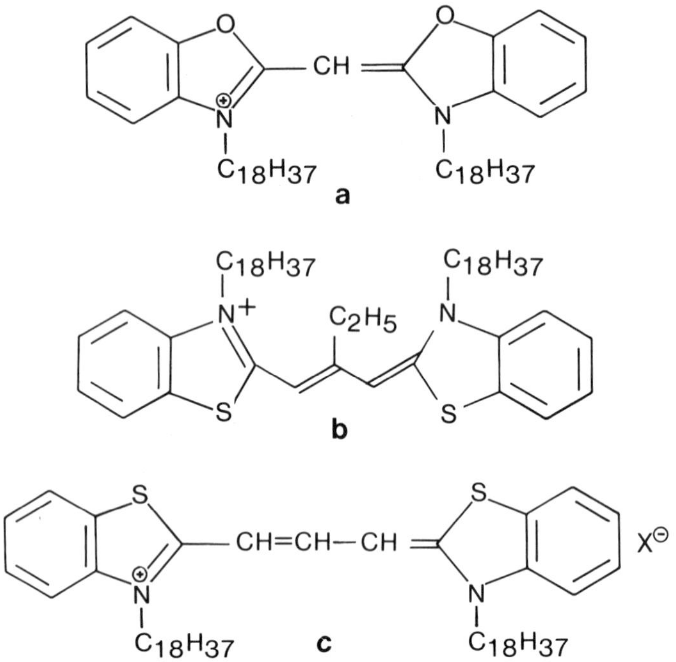

Figure 2.42. Typical cyanine dyes.

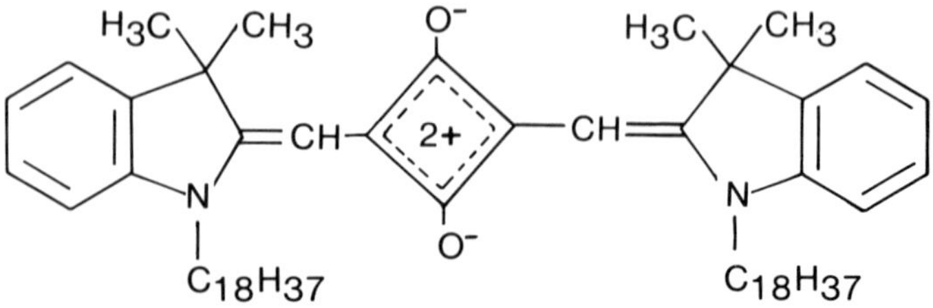

Figure 2.43. Squarylium cyanine derivative.

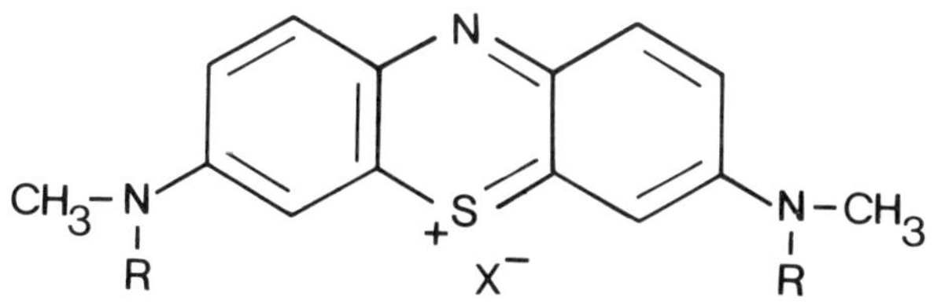

Figure 2.44. Methylene Blue dye and derivatives.

films, but surprisingly it could also be deposited as Y type bilayers at higher surface pressures, where the molecules form bilayers on the water surface.[89] Bilayers of other positively charged species have been deposited from aqueous micelles.[90]

Giles has studied dye films on a water surface, and has shown that even quite soluble dyes such as methylene blue (Figure 2.44, R = CH$_3$) will form *multilayer aggregates* on the water surface.[91] The technique also provides a model for studying the mechanism of textile dyeing.[92] Stable monolayers have been formed from a long chain derivative (Figure 2.44, R = C$_{18}$H$_{37}$), and multilayers have been built up on glass.[93] Similar substitution of anthraquinone dyes has given[22,23] a series of monolayer-forming materials, which can be built up on glass by the horizontal deposition technique, although the monolayers are too rigid for normal LB dipping. A long chain derivative of thioindigo (Figure 2.45) is particularly interesting because of its ability to undergo photochemical isomerization.[94] This property is shared by the spiropyrans which undergo easily visible changes of color on isomerization (see, for example, Ando *et al.*[95]) The electron transfer properties of LB films incorporating a triphenylmethane derivative (Figure 2.46) have been studied.[96]

The merocyanine dyes are similar to the cyanine dyes in that a single molecular formula does not adequately describe the charge distribution within the molecule. A typical merocyanine dye (such as the one illustrated in Figure 2.47) has two aromatic rings separated by a π-bonded system; one ring has electron-donating properties and the other has electron-withdrawing groups. There is some donation of electrons across the molecule, and the electronic structure is best represented as a mixture of the neutral and polar structures. Much of the early work on LB films of these

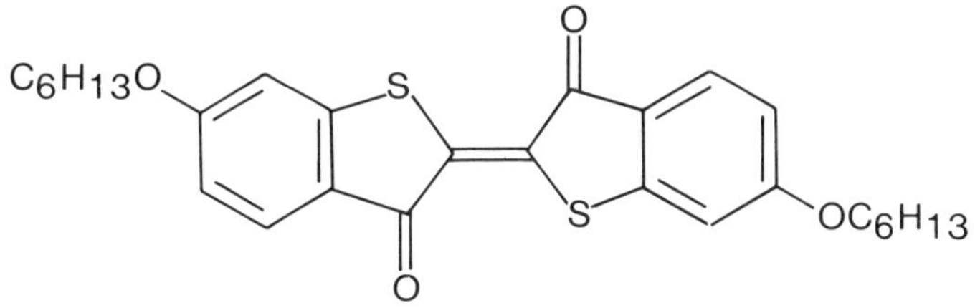

Figure 2.45. Alkyl substituted thioindigo.

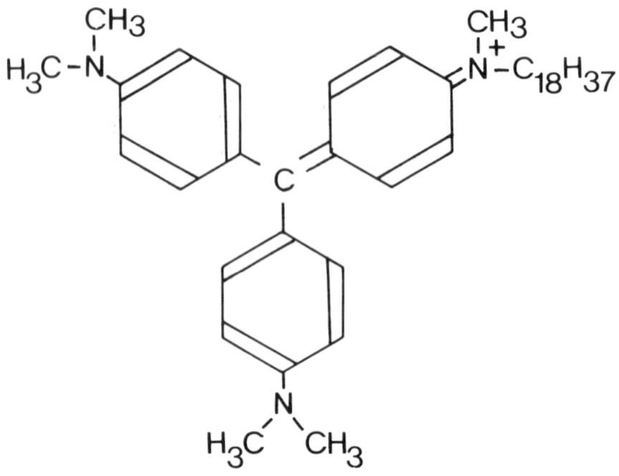

Figure 2.46. Triphenylmethane dye.

materials was carried out by Sugi and co-workers,[97–99] although they have been examined much more widely in recent years. One of the reasons for interest in the merocyanine structure is because of the potential for large second-order nonlinear optical effects (e.g., frequency doubling and the electrooptic effect) in dyes where electron donor and acceptor groups are spaced at either end of the chromophore. A wide range of dyes has been examined for this purpose and many schemes have been devised for the efficient deposition of multilayers in which all the active molecules are arranged with their dipoles aligned in the same direction (a necessary requirement for second-order nonlinear effects) either by using Y or Z type deposition or by alternating the monolayer material (see Chapter 3).

Chapter 7 discusses fully the potential of LB films in the field of electrooptics. However, it is worth noting at this stage that the first observation[100] of frequency doubling in an LB film was in a monolayer of an azo dye (Figure 2.48) with electron donor and acceptor groups at either end of the chromophore. Vectorial effects were later demonstrated[101] for another azo dye (Figure 2.49) with a series of unusual headgroups, and electrooptic measurements subsequently reported.[102] The mero-

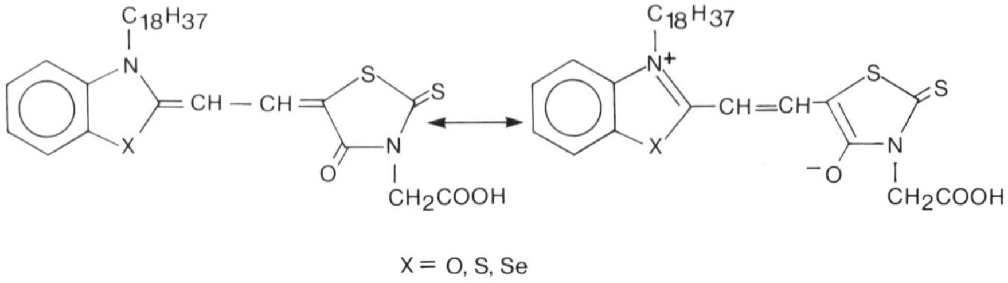

X = O, S, Se

Figure 2.47. Typical merocyanine dye showing the dipolar nature of the molecule.

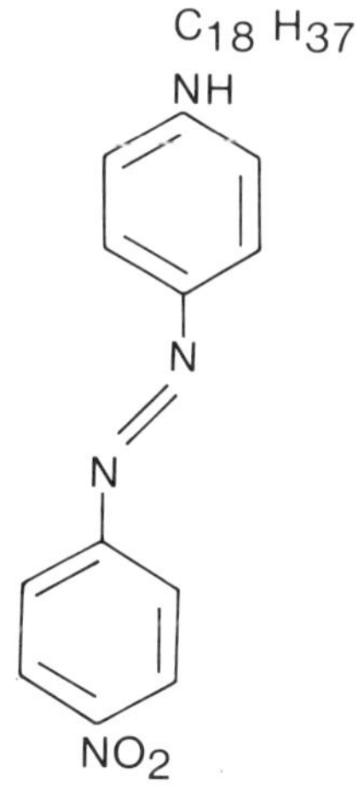

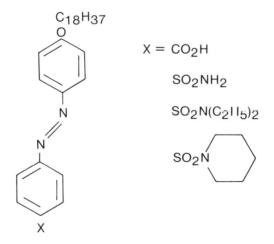

$C_{18}H_{37}$

NH

N

N

NO_2

Figure 2.48. 4-Butoxy-4'-(4-hydroxybutyl)azobenzene. This was the first material to show second harmonic generation by a monolayer.

cyanine dye "Gedye" (Figure 2.50) described by Gaines[103] has become much used for the preparation of vectorial films,[104–107] the last reference being the first reported observation of second harmonic generation by LB multilayers. A long chain derivative (Figure 2.51) of the classic nonlinear material MNA has been deposited as mixed multilayers in an alternating structure, but these films did not initially show nonlinear properties.[108] Most recently, second harmonic generation has been demonstrated by an Anglo-French group[109] in multilayers of an azo dye (Figure 2.52).

$C_{18}H_{37}$

O

N

N

X

$X = CO_2H$

SO_2NH_2

$SO_2N(C_2H_5)_2$

SO_2N

Figure 2.49. Azo dye with a range of unusual headgroups.

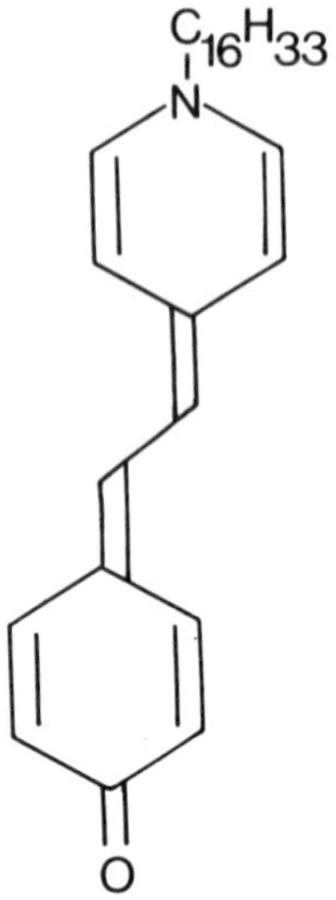

Figure 2.50. "Gedye"; many groups have deposited this material as AB type films.

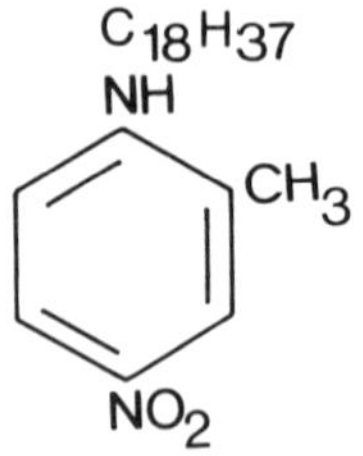

Figure 2.51. Long chain derivative of MNA.

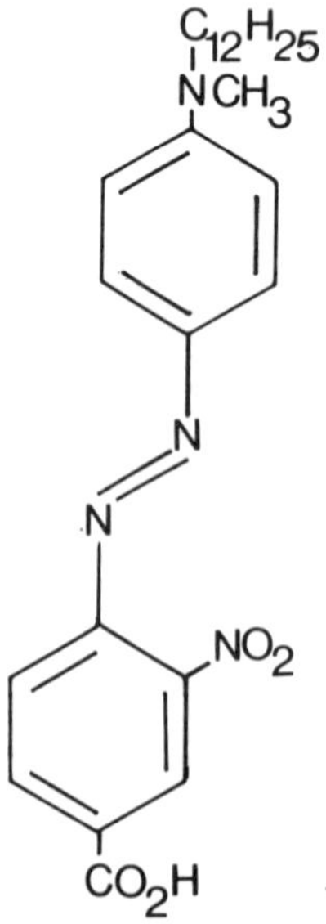

Figure 2.52. DPNA (4-(4-(N-n-dodecyl-N-methylamino)phenylazo)-3-ni-trobenzoic acid.

2.4.3. Porphyrins and Phthalocyanines

Porphyrin (Figure 2.53) is a macrocyclic substance whose highly colored derivatives are almost ubiquitous in nature. The derivatives differ in having a wide number of possible groups attached to the periphery of the ring system; the central hydrogen atoms are most often replaced by a metal ion as, for example, in chlorophyll-*a* (Figure 2.54). This material is particularly important because it is actively involved in the conversion of sunlight into chemical energy by the photosynthetic membranes of green plants. Because of this its LB properties have received much attention[110,111] and model systems have been built up from monolayers of chlorophyll and mixtures with stearic acid and electron acceptors. As it is extremely unstable chemically, the ambient conditions have to be carefully controlled and much of the earlier work is suspect.

The simplest porphyrins to have been investigated[112] are the tetraphenylporphyrin derivatives (Figure 2.55); these do not occur in nature but are relatively easily synthesized. The simpler examples deviate from the normal requirement of having well-defined head and tail groups and this is reflected in their monolayer properties. The parent material (Figure 2.55; $R_1 = R_2 = H$, $M = 2H$) and its derivatives with $M = Co$ or Zn give isotherms with an extrapolated area of only $0.13–0.17$ nm^2 per molecule. This contrasts with the expected value of 1.6 nm^2 per molecule to be expected if the molecules lay flat on the surface and 0.7 for a vertical array of molecules. The material must therefore form multilayer stacks in which the structure and stability of the film (collapse pressure 40 mNm^{-1}) is governed mainly by molecular packing rather than by hydrophilic/hydrophobic interactions.

In contrast,[113] the material with $M = Mg$ has an extrapolated surface area of 0.66 nm^2 per molecule, which is nearly consistent with a monolayer of vertical molecules; the collapse pressure is, however, low (15 mNm^{-1}). Replacement of phenyl groups of pyridyl groups (which are more hydrophilic) gave larger areas (0.8 nm^2 per molecule), but substitution with long alkyl chains (Figure 2.55; $R_1 = C_{18}H_{37}, R_2 = H$) did not improve stability. Multilayers have been prepared from many of these materials but the structure is unknown.

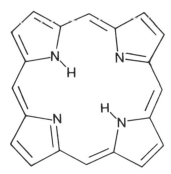

Figure 2.53. Porphyrin.

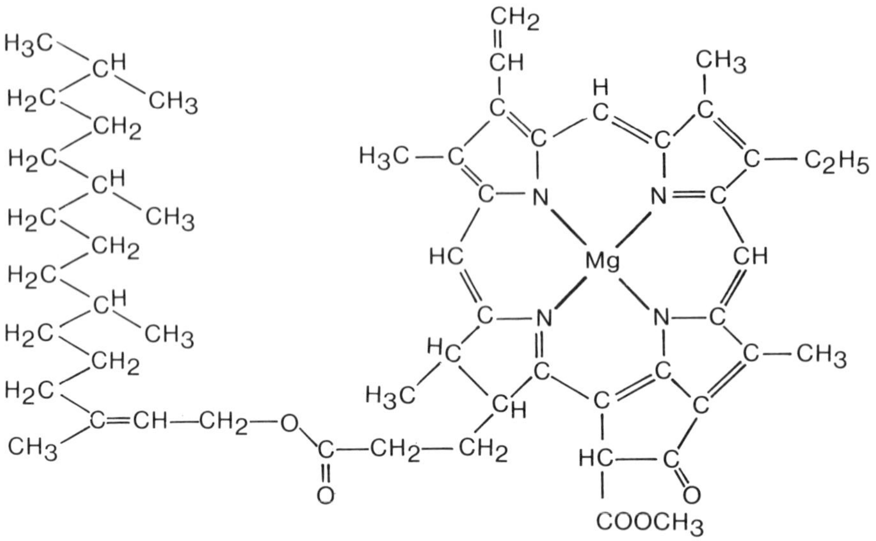

Figure 2.54. Chlorophyll-*a*, the green pigment of plants.

Tetraphenylporphyrin derivatives with appended alkoxy groups (Figure 2.55; $R_1 = C_n H_{2n+1}O$) have also been studied[114] as mixed monolayers with behenic acid ($C_{21}H_{43}CO_2H$). The macrocycle will spread from a mixed solution and LB films of the mixture have been deposited, but it is excluded from the main part of the fatty acid film. Depending on the alkyl chain length the flat ring system is accommodated at either the hydrophilic or hydrophobic faces of the film.

Whitten's group has investigated[115] the insertion of metal ions from the subphase into the central position of the porphyrin. In the series of synthetic tetraphenyl-porphyrin derivatives (Figure 2.55; $R_1 = NHCOX$, $R_2 = H$ for $X = CH_3$, $C_{11}H_{23}$, $C_{15}H_{31}$), the last member incorporated Cu^{2+} most rapidly into monolayers and multilayers, although in homogeneous solution the first member reacts fastest. Two other hydrophobically substituted porhyrins, mesoporphyrin 1X dioctadecyl ester and *meso*-tetra-(-4-carboxyphenyl) porphyrin, incorporate Cu^{2+} very slowly both in solution and as LB films. The porphyrin ring apparently lies in an accessible hydrophilic site in the first material and progressively more remote hydrophobic sites in the others. There are important differences in behavior of different ions with the first porhyrin; many ions react rapidly both in solution and in the monolayer, but Mg^{2+} reacts rapidly in the monolayer but not in solution, while Ni^{2+} and Fe^{2+} react rapidly in solution but not in the monolayer.

The group has also studied extensively the oxidation–reduction chemistry and photochemistry of porphyrins, for example, photooxidation reactions[116] and reductive addition.[117] Their investigation of the structural quality of magnesium and

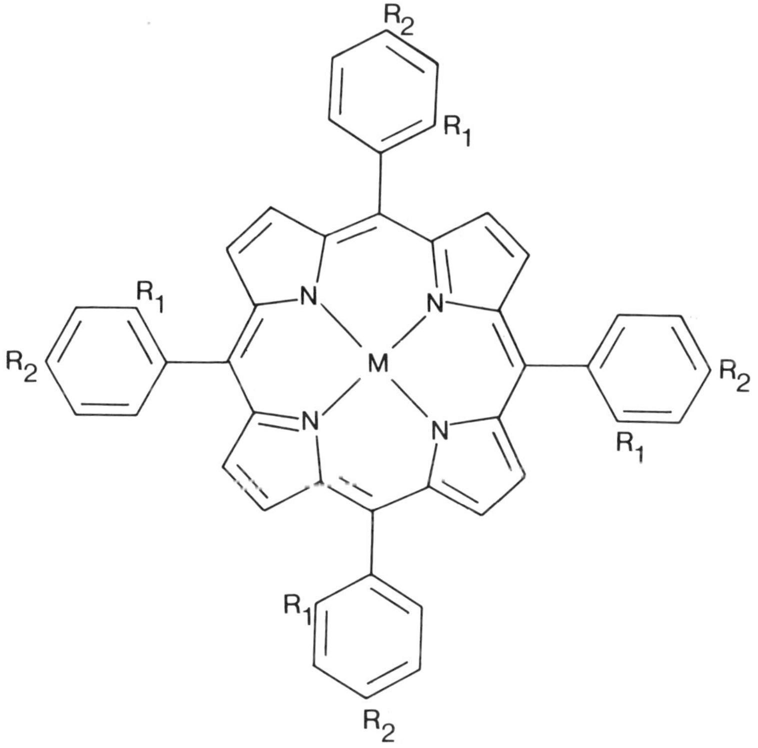

Figure 2.55. Tetraphenylporphyrin; see text for explanation of R_1, R_2, and M.

copper porphyrin films provides a warning that simple criteria like optical clarity and smooth deposition can be deceptive.[118]

Tredgold's group has investigated[119] two derivatives of naturally occurring porphyrins, protoporhyrin IX dimethyl ester and mesoporphyrin IX dimethyl ester, and observed limiting areas of 0.55 nm^2 per molecule, which correspond to a nearly vertical packing of the molecular planes. Very good quality multilayers were built up by Z deposition in contrast to the Y deposition observed for chlorophyll. Barraud's group[120] has substituted the periphery of tetraphenyl porphyrin with hydrophilic groups with long alkyl chains attached, for example [Figure 2.55; R_1 = H; R_2 = —O—CH(CO$_2$H)C$_{18}$H$_{37}$]. This has the effect of ensuring that the molecules lie flat on the water surface, where they occupy limiting areas of 1.60–2.30 nm^2 per molecule. The materials have been deposited as Y type films on solid substrates, as have mixed layers of behenic acid and alkyl substituted derivatives.[121] In the latter films the macrocycle is tilted at an angle of about 67° to the substrate. If each phenyl group in TPP is replaced by a quaternized (positively charged) pyridinium group with attached alkyl chains, the resultant material (Figure 2.56) forms[122] mono-

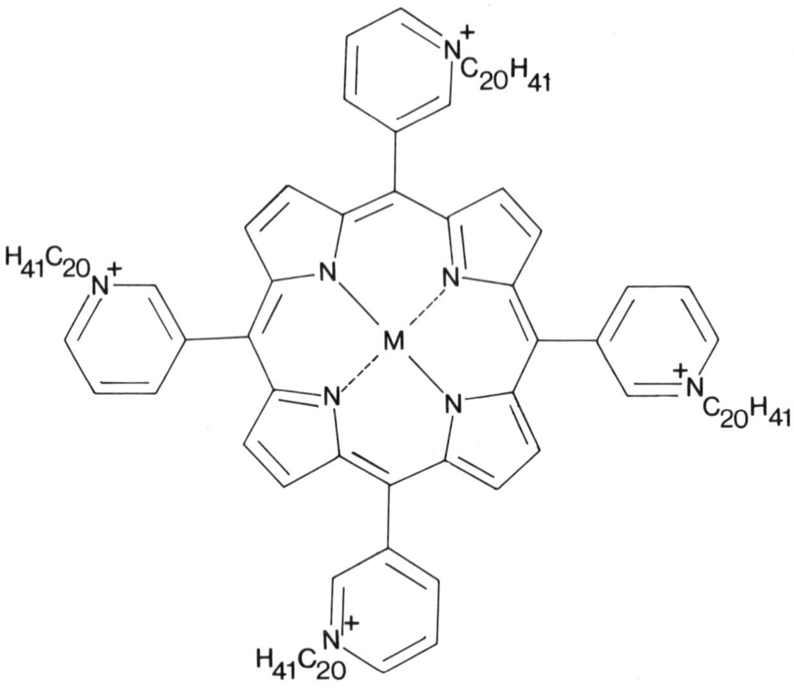

Figure 2.56. Tetrapyridiniumporphyrin—the positive charge ensures high hydrophilicity of the pyridinium rings.

layers flat on the water surface, with an area of 2.2 nm^2 per molecule. If the porphyrin nucleus is further substituted with electron-withdrawing cyano groups, the area per molecule is reduced to 1.5 nm^2. Both materials can be deposited as LB films. The central metal atom can be replaced by cobalt and the resulting material deposited as LB films; in the presence of suitable additional ligands and spacer layers these films can reversibly bind oxygen gas.[123]

The phthalocyanines are a related group of highly colored macrocycles which have been widely studied as evaporated films,[124] but until recently they have not been studied on the trough. The parent member of the series (Figure 2.57; $R_1 = R_2$ = H, M = 2H) and most simple derivatives are insoluble in organic solvents. However, Roberts and co-workers[125] showed that the soluble dilithium salt (Figure 2.57; $R_1 = R_2$ = H, M = 2Li) could be spread from a solvent onto an aqueous subphase where it was rapidly converted to the metal-free parent molecule. They also showed that films could be spread from solutions of tetra-t-butylphthalocyanine (Figure 2.58; M = 2H); this material is unusual because the bulky t=butyl groups greatly increase its solubility and therefore its metal derivatives have been studied intensively. For convenience, the macrocycle will be abbreviated to TBP. The films could be reproducibly deposited as multilayers, although the low surface areas

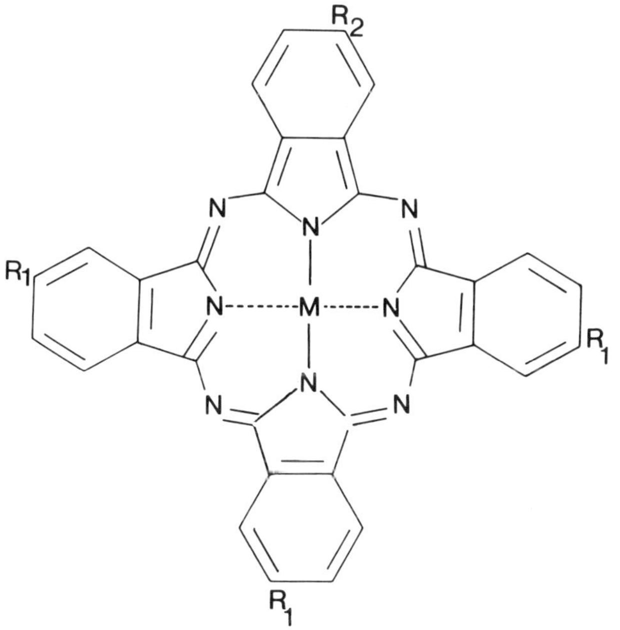

Figure 2.57. Phthalocyanine. See text for explanation of R_1, R_2, and M.

observed (0.05–0.12 nm² per molecule for the unsubstituted material and 0.27 nm² per molecule for TBP) indicate that the films must consist of multilayers in a similar manner to many of the porphyrin derivatives discussed above. When the central hydrogen atoms are replaced with copper or zinc [Figure 2.58; R_1 = R = —C(CH₃)₃, M = Cu or Zn] Hann and coworkers have shown[126] that films with much larger surface areas (0.96 and 1.2 nm² per molecule, respectively) are formed. These areas are consistent with the film on the water surface being present as a true monolayer in which the very large spatial requirement of the *t*-butyl groups dominates the packing. There has been a subsequent report[127] of monolayer formation by metal-free TBP, although the very high compressibility of the film leaves some doubt as to its nature. Other metal derivatives of TBP (Figure 2.58; M = Zn,VO, Mn) have been investigated recently[128]; the film properties appear to be improved by the addition of a few percent of ethanol to the subphase. Even so, the quality of the films prepared from the above phthalocyanines is significantly lower than that of conventional LB materials although the films themselves are unusually robust (see below). The nature of the central metal ion plays an important part in determining film quality as well as area per molecule, although the mechanism of the effect is not as yet understood. Recent work on dicholorosilicon TBP[129] (Figure 2.58; M = Cl—Si—Cl) and lead TBP[130] (Figure 2.58; M = Pb) has shown that these materials give much higher quality films than other TBP derivatives.

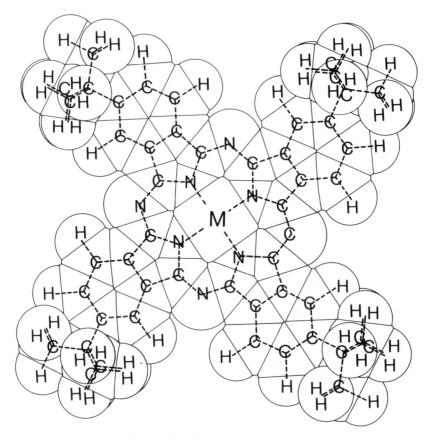

Figure 2.58. Tetra-*tert*-butylporphyrin.

These molecules show little evidence of conventional hydrophilic/hydrophobic character in their chemical constitution. It appears that their bulk and rigidity are the dominating factors in determining their packing on the water surface and after build up into LB multilayers. These films are nonetheless extremely robust; they show very great temperature and chemical stability and exceptional adhesion to their substrates. They are far better in this respect than any other monomeric materials so far investigated, although some of their properties are rivaled by those of polymers (see later). The stability of these materials is highlighted by a recent electron micrograph obtained[131,132] from a single monolayer of copper TBP (Figure 2.58; M = Cu).

Other soluble phthalocyanines have been investigated. These include[133] one with mixed substitution (Figure 2.57; R_1 = $CH_2NH \cdot C_3H_7$, R_2 = H,M = Cu), which gives rather more hydrophilic character to the molecule, and enables a good LB multilayer to be deposited. There are also symmetrically substituted phthalocyanines with hydrophilic/hydrophobic groups. One of these is the compound (Figure 2.57; $R_1 = R_2 = C_{12}H_{25}OCH_2$, M = Cu) which apparently lies flat on the water surface[134] with a surface area of 1.8 nm^2 per molecule; multilayers have been built up from this material. In contrast, another range of compounds (Figure 2.57; $R_1 = R_2 = C_6H_5O$, $C_6H_5 \cdot CMe_2$ C_6H_4O, $C_{18}H_{31}O$ Me_3CCH_2O, M = Fe,Co,Ni,Pd,Pt,Cu,Zn,Pb, (where Me = CH_3) gives multilayers[135,136] on the water surface, with areas of 0.4–0.6 nm^2 per molecule. Problems were experienced with deposition of these materials unless they were prepared as mixed monolayers with octadecanol ($C_{18}H_{37}OH$); this material presumably occupies an element of free volume that would otherwise cause disorder in the packing.

Apart from the stability and adhesion properties of the phthalocyanines, their particular interest lies in their electrical conduction and especially the modulation of this in the presence of reactive gases. The electrical measurements on phthalocyanines and their application to gas detection are reviewed in Chapters 4 and 7. Electrochromic effects have been observed in multilayers of phthalocyanine.[137] These and other applications of phthalocyanine LB films are fully discussed in Chapter 7.

The metal derivatives of porphyrins and phthalocyanines are examples of *complexes* or *coordination compounds*. Several other complexes have been deposited as LB films, and although they are not macrocyclic there are similarities to the main subjects of this section because the presence of the metal ion at the center of the molecule holds the organic *ligands* in a rigidly controlled geometry. Examples are the series of complexes which are formed[138] by spreading the ligand (Figure 2.59; R = $C_6H_5CH_2$, C_6H_{11}, $C_{12}H_{23}$, $C_{12}H_{25}OCOCH_2$) on as containing copper ions, and the ruthenium complex (Figure 2.60) at one time thought to cleave water on illumination.[139]

A metal complex of particular interest is the "sandwich" structure ferrocene (Figure 2.61). This has recently been spread as a monolayer by attaching[140] an amphiphilic group ($C_{18}H_{37}CON$ or $C_{17}H_{35}CO_2$). The extrapolated molecular area is 0.25 nm^2 per molecule indicating that the molecules are close packed or even multilayered on the water surface. The optically measured film thickness of built up films is 0.24 nm^2 per layer. Multilayers have also been built up[141] from ferrocene in which an amphiphilic group ($C_{18}H_{37}CONH$) is attached to each ring. This work is of particular interest, because this material has been built up as an ordered

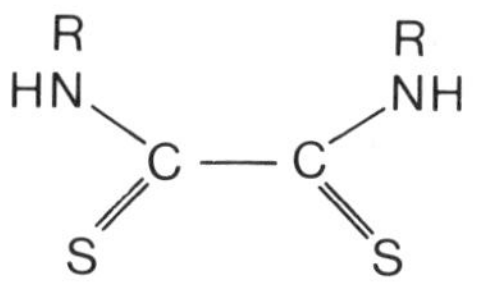

Figure 2.59. Alkyldithiocarbamate ligand.

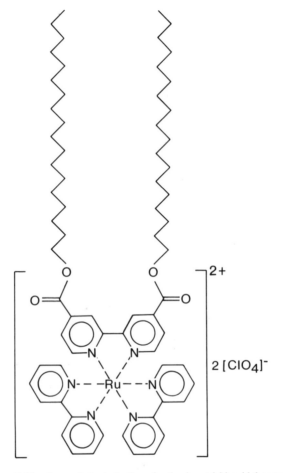

Figure 2.60.   Long chain derivative of ruthenium trisbipyridyl complex.

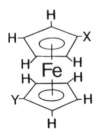

Figure 2.61.   Ferrocene derivatives. See text for explanation of X and Y.

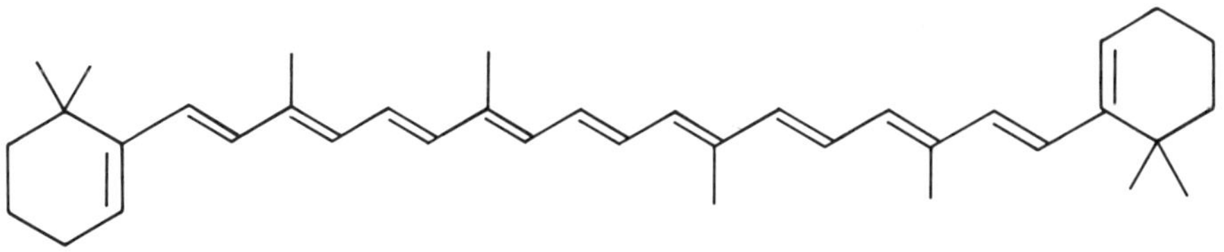

Figure 2.62.   Beta-carotene.

structure with other electronically active materials and vectorial electrical properties demonstrated.

2.4.4. Mixed Monolayers of Unsubstituted Hydrophobic Materials

A range of rigid hydrophobic molecules has been shown to form mixed mono-layers with fatty acids, even though the materials themselves cannot be spread on a water surface. The highly unsaturated long-chain material beta-carotene (Figure 2.62) forms mixed monolayers with arachidic acid.[142] It is not clear where the carotene is accommodated in the monolayers or built up films, as the molecules are much longer than arachidic acid. Quinquethienyl (Figure 2.63) has been shown to form mixed monolayers in which the molecular axis is aligned with the host arach-idic acid.[143] The polycyclic aromatic hydrocarbons anthracene (Figure 2.25b), perylene, benzperylene (Figure 2.64), and chrysene form mixed monolayers with cadmium arachidate, whereby the aromatic molecules migrate to the hydrophobic interface.[144] These layers can be built up in the same way as monolayers and provide a very attractive means of obtaining LB films of complex molecules without having to embark on a program of chemical synthesis to obtain amphiphilic deriva-tives. Although it appears that the hydrocarbons are molecularly dispersed at very high dilutions (1 : 100 hydrocarbon : acid) there is spectroscopic evidence for ag-gregation at higher concentrations.[144] Later work[145] has shown that, at any rate in the case of benzperylene (Figure 2.64) the aggregates are three-dimensional microcrystals.

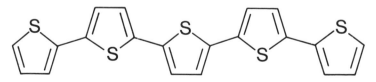

Figure 2.63.   Quinquethienyl.

Figure 2.64. 1,12-Benzperylene.

2.5. POLYMERS AND POLYMERIZABLE MATERIALS

2.5.1. Preformed Polymers

Polymers can be spread from solutions in the same way as the monomeric materials described in the previous sections. If there are sufficient hydrophilic groups to allow spreading but insufficient to cause solubility, then stable monolayers can be obtained. Structural requirements appear to be rather less stringent for polymers, and a wide range of materials has been examined, including polyacrylates,[146,147] polymethacrylates,[148] poly(vinylbutyral),[149] poly(vinylmethylether),[150] poly(vinylacetate),[151] poly(vinylfluoride) and poly(vinylidenefluoride),[152] and silicone copolymer.[153] Most of the work has been concerned with studies of the physicochemical properties of the materials and the relationship between two- and three-dimensional solutions of polymers. Polymers show the same general types of behavior as monomers; for example, they form expanded and condensed films but do not show sharp phase transitions. The compressibility of condensed films varies markedly with temperature and can be extremely low at the theta temperature.[146] Mixed monolayers of compatible polymers can be formed, as well as mixed monolayers of polymers and simple molecules, such as poly(vinylstearate) and stearic acid.[154]

It may seem strange that it was not until recently that preformed polymers were deposited by the LB technique. Part of the explanation must lie in the fact that many polymer films are so rigid that any attempt to immerse a substrate simply leads to a hole in the monolayer. However, Hodge, Tredgold, and co-workers[155] have reported that the monolayer properties of a series of copolymers which can successfully be deposited as LB multilayers. These materials are formed by a copolymerization of maleic anhydride (Figure 2.65a) with a range of compounds containing a terminal double bond (Figure 2.65b). The resulting polymer can be further transformed to give several variations of the hydrophilic grouping (Figure 2.65).

The structural requirements for polymers to give stable films and good deposition have been extensively investigated by this group.[156,157] It is important to have a main polymer chain with hydrophilic groups regularly distributed at short intervals along it. If there are long sections without such groups, then these will tend to form

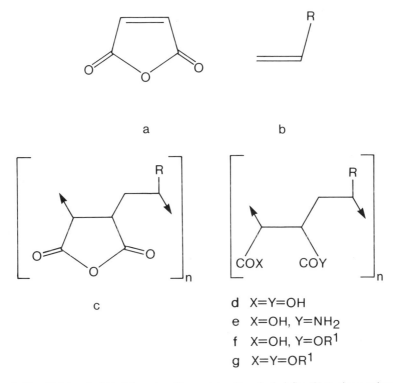

a

b

c

d X=Y=OH
e X=OH, Y=NH$_2$
f X=OH, Y=OR^1
g X=Y=OR^1

Figure 2.65. Maleic anhydride (a) reacts with a variety of terminal olefins (b) to give a polymer (c), which can be further converted to more hydrophilic derivatives (d), (e), (f), and (g).

ill-defined loops clear of the water surface (Figure 2.66). The hydrophobic tail-groups of the polymer structure have a detailed influence on the monolayer proper-ties, but in general there is much greater toleration of structural variation than with monomer systems; for example, because of the intrinsically lower solubility of polymers quite short tailgroups will give insoluble monolayers as evidenced by many of the preformed polymers mentioned in the first paragraph of this section.

The maleic anhydride copolymers are particularly useful due to their regular distribution of hydrophilic headgroups and the range of structural variation that can

Figure 2.66. A polymer with widely spaced hydrophilic groups will form irregular loops over the water surface.

be built in. The anhydride headgroup (Figure 2.65c) gives large-area films which slowly hydrolyze to the diacid (Figure 2.65d). The diesters (Figure 2.65e) collapse, presumably because the headgroup is insufficiently hydrophilic, but the acid amide (Figure 2.65f) and acid ester (Figure 2.65g) both give very stable compact films. The acid ester is generally preferred, as the acid amide is less soluble in the spreading solvent (normally ethyl acetate or ethyl formate with these materials). Of the wide range of tailgroups investigated, perhaps the most interesting[158] are the compounds containing liquid-crystal type groups (Figure 2.67). These materials have been deposited as LB films of up to 400 layers, the first two materials giving Y-type deposition and the last material Z-type deposition. It is remarkable that the relatively small structural change between these polymers (Figures 2.67b and 2.67c) should cause a major change in both the deposition mode and the isotherm. All these polymers gave very good quality films as evidenced by their ability to support optical waveguided modes. X-ray studies[159] have indicated that the packing in polymers with aromatic X groups is quite close while aliphatic X groups tend to give films with much free volume. In both cases there is a well-defined layer structure.

Some of the polypeptides also undergo LB deposition, although this was not

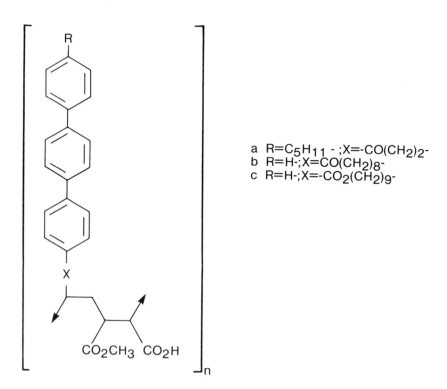

a $R=C_5H_{11}-;X=-CO(CH_2)_2-$
b $R=H-;X=CO(CH_2)_8-$
c $R=H-;X=-CO_2(CH_2)_9-$

Figure 2.67. Liquid crystal polymers.

investigated in earlier studies.[160] Poly(γ-methyl-L-glutamate) (Figure 2.68) gives a two-region isotherm (Figure 2.69) which is a fairly common feature of polymer isotherms; there is a lower pressure region corresponding to a monolayer and a higher pressure region corresponding to a bilayer. When LB deposition is carried out in these two regions, 1 and 2 layers, respectively, are deposited on each stroke.[161] Other polypeptides appear[162] to give well-ordered LB films only when mixed with stearic acid.

Some cellulose derivatives have also been deposited onto substrates.[24] These form extremely rigid films and deposition was by means of a horizontal lifting technique. Although acrylates and methacrylates have been reported[156] as inherently less suitable for deposition than the maleic anhydride copolymers discussed earlier, multilayers of poly(octadecylmethacrylate) (Figure 2.70) have been built up.[163] A quaternary ammonium salt terminated oligobutadiene has been built up into X-type multilayers,[164] although it was necessary to crosslink each layer with ultraviolet light before depositing the next one.

Ringsdorf and co-workers have employed a different approach to the problem of ensuring sufficient flexibility to allow LB deposition of polymer films.[165] They have constructed lipid type molecules in which the polymerizable group is attached to the headgroup by a flexible hydrophilic chain, a typical grouping being R—$(CH_2CH_2O)_4$—OC—$C(CH_3):CH_2$. The flexibility of the $(CH_2CH_2O)_n$ group allows the reactive double bond to polymerize to completion on the water surface or in multilayers. The work has recently been extended to include fluorocarbon groups in the hydrophobic part of the molecule.[166]

A novel approach[167] to preparing a monolayer of a polymer involves spreading a monolayer of a monomeric precursor molecule (Figure 2.71a) which has long chains attached to reactive groups on either side of a benzene ring. The subphase is a dilute acidic solution of an aromatic diamine (Figure 2.71b), which slowly reacts with the monolayer to give a polymeric film of PPTA (Figure 2.71c). This polymer cannot otherwise be spread on the water surface; the long chains served to stabilize the monolayer and were eliminated as amines during the reaction and dissolved in the acidic subphase. Multilayers have been deposited but not extensively investigated.

Very recently, a polyionic material (Figure 2.72a) has been spread on the water surface[168] to give films in which the molecules are apparently lying flat with an

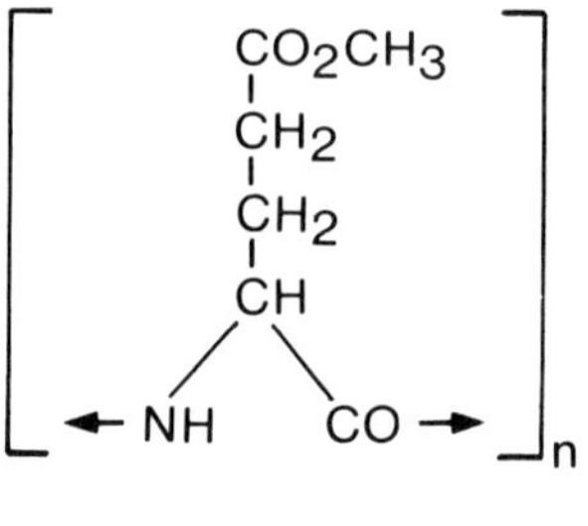

Figure 2.68. Poly(γ-methyl-L-glutamate).

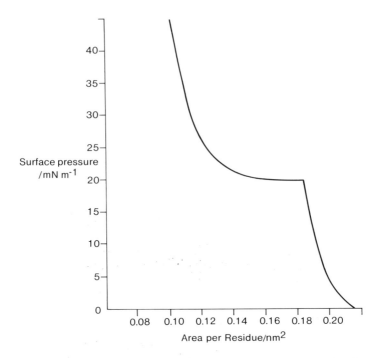

Figure 2.69. Surface pressure/area isotherm of poly(γ-methyl-L-glutamate) showing the monolayer and bilayer regions. (After Winter and Tredgold.[161])

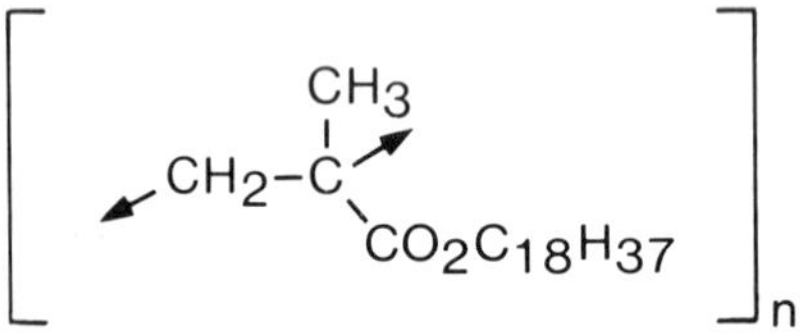

$$\left[\begin{array}{c} \underset{\displaystyle CH_2-C}{CH_2-C}\overset{\displaystyle CH_3}{\underset{\displaystyle CO_2C_{18}H_{37}}{}} \end{array}\right]_n$$

Figure 2.70. Poly(octadecylmethacrylate).

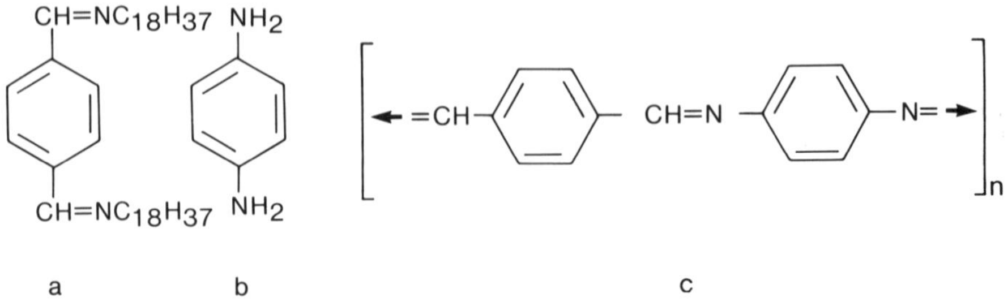

a   b   c

Figure 2.71. The diimine (a) monolayer reacts with phenylenediamine (b) in the subphase to give a polymeric film of PPTA (c), which no longer has long chain substitution.

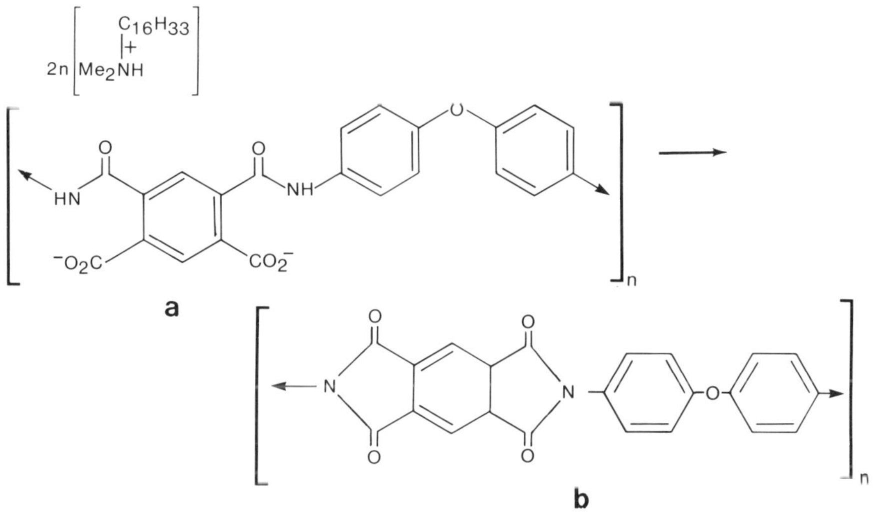

Figure 2.72. The polymeric salt (a) is deposited as LB film which can then be converted to ultrathin layers of polyimide (b).

area of 1.4 nm^2 per repeat unit. These have been deposited[169] as LB films (up to 200 Z-type layers) of apparently good quality. It is particularly exciting that these LB films can be converted quantitively to the corresponding polyimide (Figure 2.72b) by subsequent chemical treatment. It has not yet been established if these LB polyimides have all the chemical stability normally associated with this type of polymer, but it has been shown by x-ray diffraction that the layer structure is maintained, and both the diffraction and stylus measurements lead to the conclusion that each monolayer is only 0.4 nm thick. This value is unusually small, but is entirely consistent with expectation.

2.5.2. Polymers Formed *in situ* by Addition of Carbon–Carbon Double Bonds

One of the first materials to be investigated[170] for its polymerization on the water surface and as LB multilayers[171] was vinyl stearate ($C_{17}H_{35}CO_2CH{=}CH_2$). This will polymerize on UV or γ irradiation in the absence of oxygen. The multilayers formed are of X type, and this structural arrangement is retained after polymerization, although minor reorganization takes place. Polymerization on the water surface gives highly rigid films that cannot be deposited by the LB technique.

Most of the polymerization reactions described in this chapter are of the same general type as the vinyl stearate reaction (Figure 2.73) in which the polymer is formed by the double bonds on adjacent molecules being converted to links in a chain, from which the other groups dangle. This type of polymerization usually occurs by a free radical process (i.e., the intermediates are species with an odd

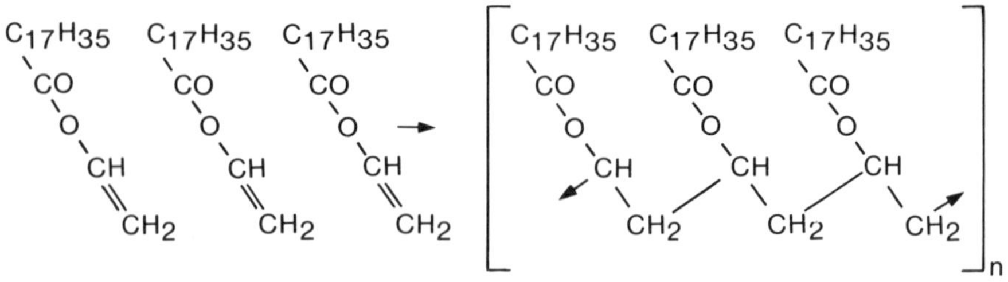

$$C_{17}H_{35} \quad C_{17}H_{35} \quad C_{17}H_{35} \qquad \left[C_{17}H_{35} \quad C_{17}H_{35} \quad C_{17}H_{35} \right]$$

Figure 2.73. Polymerization of vinyl stearate.

electron on the growing chain). Such processes are readily quenched by impurities, particularly oxygen, and it is therefore usually necessary to carry out the polymerization in an inert atmosphere. Another feature of this type of addition polymerization is observed when a monolayer is polymerized at constant pressure on the water surface: a reduction in surface area by several percent takes place. In multilayers this can cause problems, such as cracking, if the change cannot be accommodated by reorientation of the molecules.

Similar polymerizations have been carried out on multilayers of cadmium octadecylfumarate (Figure 2.74). When the LB film was built up of alternate layers of stearate and octadecylfumearate, it was found that polymerization occurred at the same rate as in the absence of the intervening layers. Complete sheets of polymer were therefore being formed with no need for interaction between the reacting headgroups in adjacent layers. A styrene derivative (Figure 2.75) can only be deposited as mixed films with stearic acid, but can readily be polymerized by irradiation.[173]

Ringsdorf's group[174] has investigated the polymerization of a series of dienes $(C_{13}H_{27}CH{=}CH{-}CH{=}CHX; \; X \; = \; CO_2H, \; CHO, \; CH_2OH)$ as monolayers on water. Although none of these will polymerize in the crystalline form, in the melt, or in solution, they all polymerize on the water surface. This shows the importance of orientation in determining the possibility of polymerization, and is an encouragement for looking at materials that might not be expected to polymerize at an appreciable rate under normal conditions. The group has extended its studies[175] to model lipid materials, in which two of these polymerizable chains are attached to a hydrophilic headgroup (Figure 2.76). These materials can be polymerized as monolayers, but the main interest lies in the generation of model biological membranes in the form of liposomes (spherical bilayers). More recently[165] the molecules incor-

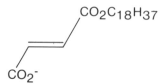

Figure 2.74. Octadecylfumarate anion.

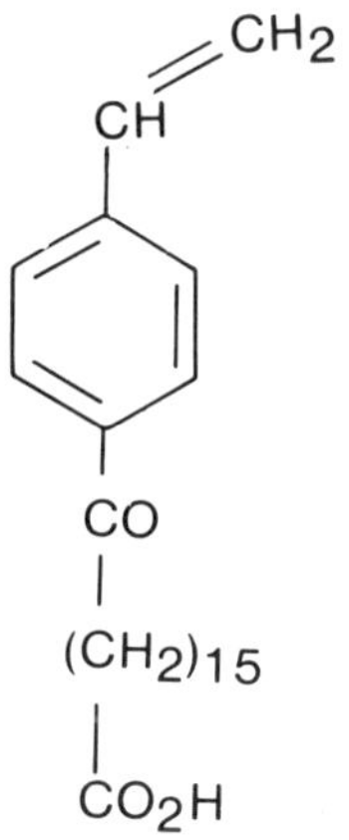

$$CH_2$$
$$CH$$

CO

$(CH_2)_{15}$

Figure 2.75.   LB film-forming styrene derivative. CO_2H

porating flexible linkages described in the previous section have been polymerized to unusually high degrees of polymerization both on the water surface and in LB films.

Multilayers of 2-octadecylacrylic acid, $C_{18}H_{37}C(\!=\!CH_2)CO_2H$, have been polymerized by UV or electron beam irradiation.[176] Higher doses of electron beam irradiation will depolymerize the material, thus allowing it to be used as a positive or negative lithographic resist. Two other naturally occurring fatty acids containing double bonds have been evaluated as electron beam resists; these are elaidic acid[177] and *trans*-13-docosenoic acid.[178] This application of LB films will be dealt with in more detail in Chapter 7.

By far the best investigated[179] material for use as a resist is omega-tricosenoic acid (Figure 2.19). The material forms stable monolayers with a collapse pressure of

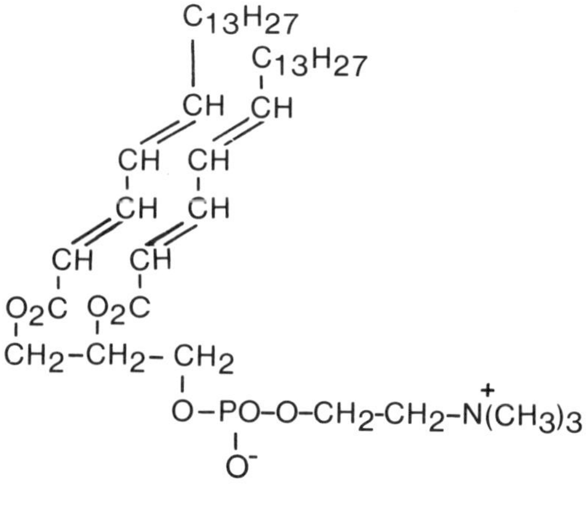

$C_{13}H_{27}$
$C_{13}H_{27}$
CH   CH
CH   CH
CH   CH
CH   CH
O_2C O_2C
CH_2–CH_2– CH_2
O–PO–O–CH_2–CH_2–$\overset{+}{N}(CH_3)_3$
O^-

Figure 2.76.   Lipid analog with two polymerizable groups.

more than 35mN m^{-1} and an extended region of low compressibility with an extrapolated surface area of 0.27 nm^2 per molecule. This value is higher than the 0.22 nm^2 typical of a $(CH_2)_n$ chain and shows the perturbing influence of the double bond. It may, however, be beneficial in allowing an unusually great freedom of movement, perhaps accounting for the very fast dipping rates achievable with this material[16] in the absence of metal ions in the subphase.[180] The multilayers are very readily polymerized, but because of the reorganization caused by bond length changes the length of the polymer chains is limited. This limitation probably results in a very high uniformity in the chain length of the polymer molecules (i.e., they are monodisperse), and this factor together with the close packing of molecules within a monolayer allows this material to be used as a very high contrast electron beam resist (see Chapter 7).

2.5.3. Diacetylenes

Of all the classes of material investigated for polymerization as monolayers and LB films, by far the most attention has been paid to the diacetylenes. They have the general formula R—C≡C—C≡C—X and can be readily polymerized by γ or UV radiation (Figure 2.77). Although the effect of the reaction is similar to the

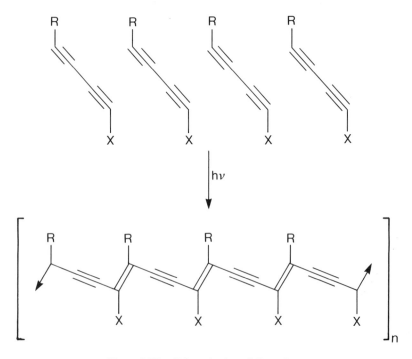

Figure 2.77. Polymerization of diacetylenes.

double bond additions in Section 2.5.2., there are important differences. The intermediate is a carbene, with two nonbonded electrons on a carbon atom, and the reaction proceeds so rapidly that oxygen quenching is not a problem. Polymerization is known to occur in single crystals of diacetylenes provided that the sizes and shapes of groups R and X are within limits which have been well defined.[181] There is, in effect, a limited tolerance of distortion of the unit cell during the polymerization process, and the diacetylene groups must be in a suitable orientation for mutual reaction. The polymer formed is of great interest because it retains the crystal structure of the monomer and because of the highly conjugated backbone, which has however proved not to be as highly conductive as was at first hoped due to the difficulty of introducing dopants.

Long chain diacetylenes with a hydrophilic group on one end have been prepared[182] for LB studies, such as[183] the materials $C_{12}H_{25}$—C≡C—C≡C—X where X = CO_2H, CH_2OH, both of which give good quality Y type layers[183]; the second material is an exception to the general rule about poor deposition of alcohols. The LB films polymerize readily to give colored (usually yellow) films which are unusually tough and solvent-resistant. The toughness can be judged from an experiment in which an electron microscope grid with a 0.5 mm square hole in it was coated on either side with a polydiacetylene acid, C_9H_{19}C≡C—C≡C—$(CH_2)_8CO_2H$, to give a bilayer that was stable in air or water for weeks.[12] The ability of diacetylenes to polymerize in LB films is governed by a similar set of rules to those determined for bulk polymerization,[184,185] however the rigidity of the materials tends to lead to stresses and cracking in the polymer films.[186]

Monolayers and multilayers of diacetylenes and their polymers are not single crystals, but consist of an array of two-dimensional domains. These can be observed from the electron diffraction pattern[183] or from optical microscopy via crossed polarizers.[12] It is remarkable that a single monolayer of polymer can be clearly examined by this optical technique, and this facility has allowed detailed study of the factors influencing domain size. The domains appear to be formed during the spreading and compression of the monolayer and remain constant during dipping and polymerization, unless an excessive rate of dipping is used.[187] It appears that vibration due to the trough mechanism may cause some readjustment of diacetylene monolayers and can lead to varying extents of area change during polymerization on the water surface.[188]

The domain structure is a factor limiting the usefulness of diacetylenes in electron beam resists; polymerization tends to continue to the edge of a domain so that it would be desirable to obtain domain sizes substantially smaller than the 0.5–300 μm range normally observed. Integrated optics applications,[189] however, require as large a domain size as possible, and it has been claimed[190] that a horizontal electric field of only 10^4 $V\,m^{-1}$ above the water surface can increase the domain size to the order of 1 mm.

Diacetylenes have also been incorporated into lipidlike molecules and polymerized as model membranes.[191,192] Relative to the diene systems mentioned above, they suffer from excessive rigidity, which limits transport and other active processes, although they are extremely robust. The subject has been reviewed elsewhere.[193]

2.5.4. Oxirans

An oxiran is a three-membered ring including an oxygen atom. It is more commonly called an epoxide and this grouping is well known for entering into polymerization reactions when suitably catalyzed. There is relatively little change in volume during oxiran polymerization (Figure 2.78), as the opening of the oxiran ring leads to an increase in volume which nearly compensates for the volume loss due to joining the molecules together. Langmuir–Blodgett films of a long chain oxiran (Figure 2.79) have been polymerized by electron beam irradiation[194] and depolymerized by higher doses. Although the low volume change offers the promise of defect-free structures, the electron beam sensitivity of these compounds at present appears to be too low for resist applications.

2.5.5. Condensation Polymers

In a condensation polymerization reaction the constituent molecules join together while eliminating another molecule (often water); the reacting molecules are usually (but not invariably) of two different kinds. Although not many examples of this type of process have been studied in monolayer systems, the condensation[195] of a monolayer of octadecylurea (Figure 2.80a) with formaldehyde (Figure 2.80b) in the subphase has been shown to give a film of the polymer (Figure 2.80c). A similar reaction has been used[196] to graft a monolayer of octadecanal ($C_{17}H_{35}CHO$) onto a preformed polymer (poly-1-lysine) which is initially dissolved in the subphase.

2.5.6. Opportunities for New Materials

There can be little doubt that the polymeric and polymerizable materials which have been investigated to date represent only a small fraction of the possible types of

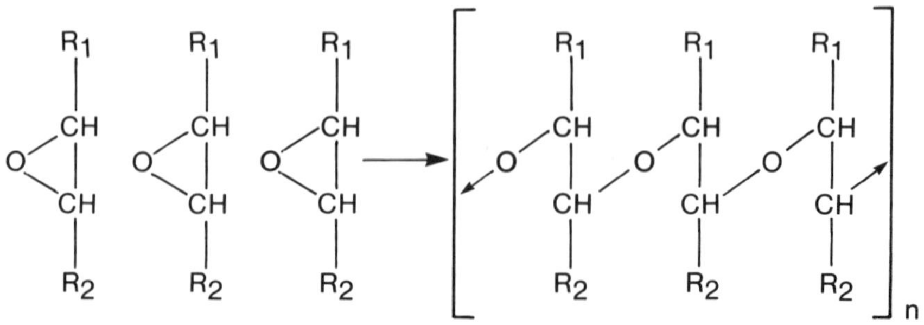

Figure 2.78. Polymerization of the oxiran ring.

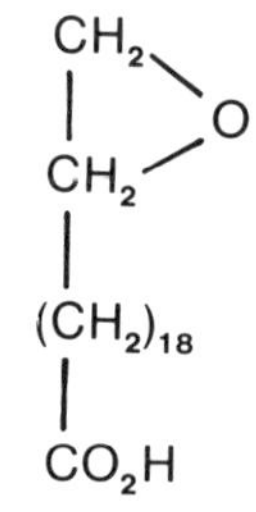

$$CH_2 \diagdown$$
$$\mid \quad \diagup O$$
$$CH_2 \diagup$$
$$\mid$$
$$(CH_2)_{18}$$
$$\mid$$

Figure 2.79. Oxiran used for electron beam polymerization. CO_2H

polymeric LB film. In view of the much greater structural stability to be expected from polymer films relative to monomer films and because of their potential applications as resists, we can be sure that the scope of this area will be greatly extended in the next few years.

2.6. EXCEPTIONS

2.6.1. Non-Langmuir–Blodgett Multilayers

Why do we need a trough at all? A novel approach due to Sagiv and co-workers[197–199] has been to build up multilayers by successive adsorption and reaction of appropriate molecules; the type of materials used and structures prepared are shown in Figure 2.81.

The molecules adsorb to form a monolayer on the surface; the trichlorosilane headgroup reacts with the substrate to give a permanent chemical attachment, and each subsequent layer is chemically attached to the one before in a very similar way to that used in systems for supported synthesis of proteins. One problem that is widely recognized with the protein synthesis system is the difficulty of obtaining 100% reaction at each stage, and this desirable situation is normally approached by ensuring that the reactant molecules cover the support very sparsely. This is of course not possible in built up layer systems.

The consequence of these problems is that the structure tends to deteriorate very rapidly as the number of layers is increased and a recent X-ray study[200] of a nominally three-layer sample indicated that it contained a substantial proportion of a

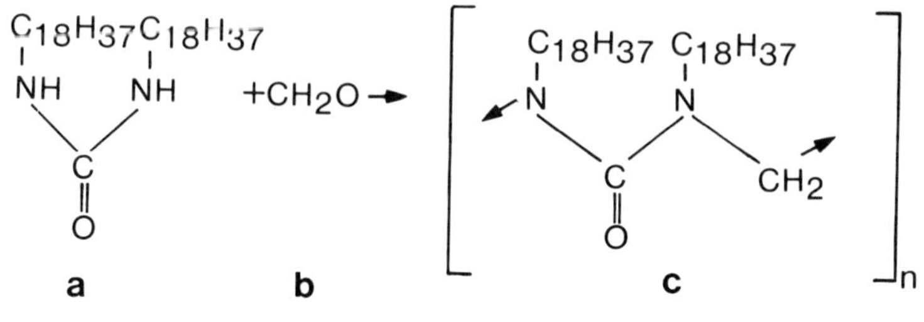

Figure 2.80. Bisdodecylurea (a), formaldehyde (b), and the resultant polymer (c).

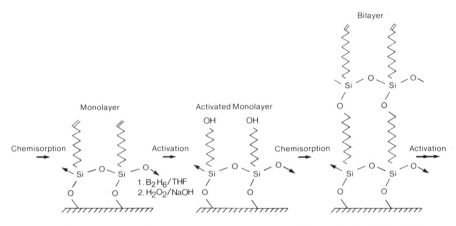

Figure 2.81. Preparation of chemically attached polymeric multilayers by the Sagiv technique.

two-layer film. One of the other apparent problems with the technique is that the reactive headgroup is significantly larger than a carbon chain, so that the packing is less close than in an LB film. Recent work has increased the bulk of the tailgroup either by replacing[201] some of the CH_2 groups with the bulkier CF_2 or by incorporating rings[202] into its structure. In spite of this, the covalently bonded monolayers have been shown to have good barrier properties to both aqueous and organic solutions of permanganate.[203]

The arriving molecules in an adsorption experiment are constrained to react with specific groups on the substrate, and once reacted they cannot move. One might expect that this would lead to a different structure from that given by deposition of an LB monolayer, where the individual molecules are free to adjust their positions to improve the overall packing. Recent work by Peterson and Russell[17] has demonstrated the difference in the abilities of the two types of film to support epitaxial growth of subsequent LB multilayers. A stearic acid LB monolayer has large regions with long-range order that can be visualized by the epitaxial deposition technique; a chemically attached stearic acid layer has no such long-range order.

A similar method of using adsorption to build up multilayers has recently been described by Tredgold's group.[204] A substrate is alternately immersed into an aqueous solution of a quaternary ammonium salt (Figure 2.82a) and a hexadecane solution of a fluorinated acid (Figure 2.82b). The plot of reciprocal capacitance against number of layers (up to 7) shows some scatter, but indicates that the quality of multilayers is better than that of Sagiv-type multilayers. As the method lends itself to the preparation of alternating films it warrants further investigation.

The problem of using adsorption for monolayer preparation is highlighted by a study[205,206] of fatty acid adsorption onto alumina. The acids form close-packed monolayers only after many hours of equilibration, although contact angles reach limiting values much more quickly. It seems that there is a dynamic process of

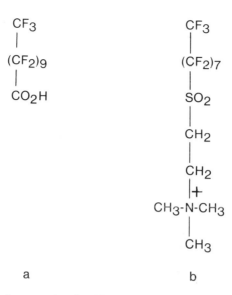

Figure 2.82. Materials for preparation of multilayers by sequential adsorption. The substrate is dipped alternately into a nonaqueous solution of perfluorodecanoic acid (a) and an aqueous solution of a fluorinated quaternary ammonium salt (b).

adsorption, desorption, and readsorption, which ultimately leads to the desired close-packed monolayer; but this clearly cannot occur in a system where the adsorbant molecules become chemically bound to the substrate.

For the time being we must conclude that a trough is necessary for the preparation of deposited films of more than about three layers and if the ordering and quality of the films is important. Further developments may change the picture, and there is already a good case for using the adsorption–reaction technique for putting down a single layer whose firm attachment to the substrate is a key criterion.

2.6.2. Nonaqueous Subphases

At the beginning of this chapter the special virtues of water as a subphase were extolled. There are however exceptions to the general rule; one category "cheats" a little by using mixtures of water with other solvents, such as glycerol,[207] ethanol,[128] butanol, or hexanol.[208] Pure glycerol has recently been used as a subphase with acetonitrile as the spreading solvent,[209] and behenic acid and a charge-transfer salt have been successfully deposited under these conditions, but owing to the high viscosity of glycerol compression speeds must be kept very low. Mercury has a much higher surface tension than water, and the problem of solution of organic materials does not arise, so that in addition to standard materials like stearic acid, water-soluble materials such as benzoic acid ($C_6H_5CO_2H$) form stable films.[210]

Even straight-chain hydrocarbons such as octacosane ($C_{18}H_{38}$) will form mono-layers in which the chains are horizontal, although they collapse before the chains become vertical. The subject of monolayers on mercury has been reviewed by Smith.[211]

The type of subphase that departs furthest from the rules would be nonpolar liquids, in particular hydrocarbons; in fact mineral oil, hexadecane ($C_{16}H_{34}$), and the ester tricresyl phosphate have all been used as subphases for spreading mono-layers of a wide range of materials.[212,213] These include a fluorinated quaternary ammonium salt, namely $C_7F_{15}CONH(CH_2)_3N^+(CH_3)_3 I^-$, and also silicate esters and a number of polymers ranging from silicones to proteins. The maximum surface pressures are very low (less than $1 \ mNm^{-1}$) and there is a tendency to form multilayers even at lower surface pressures than this.

These are, however, the exceptions that prove the rule. Water is cheap, plen-tiful, and relatively earily purified on a large scale. It is unlikely that we shall ever see its displacement as the major subphase for monolayer studies.

2.7. PROSPECTS

If we look back at the range of materials discussed in this chapter, it can be seen that a wide variety of hydrophilic groups has proved suitable for Langmuir film formation, and a smaller number for LB deposition. There appears to be little restriction on the tailgroup composition, as long as sufficient hydrophobic character is built in, and if a molecule is marginally stable on the water surface it can usually be converted to a stable molecule by increasing the length of the alkyl chain substitution. Even so, it seems that at present we have only begun to explore the possible variety of head and tail groups and their combinations.

The phthalocyanines provide an example of materials that do not fit in with the classical pattern of an amphiphilic molecule. The LB films produced from these are nevertheless exceptionally tough, stable, and adherent. It would not be surprising if other materials are found that correspond to the phthalocyanine model or yet another way of spreading films on the water surface.

Another area that is in its infancy is the study of deposition of preformed polymers. Tredgold's group has pioneered this work and are establishing the general rules of the area. Undoubtedly there will be much more active study both of the maleic anhydride polymers and other polymers designed for dipping. The search for applications will bring a requirement for more stable LB films, and this almost inevitably will mean polymeric films. The alternative route to formation of poly-mers is via deposition of monomers and subsequent polymerization. This has been more widely studied, but with a rather narrow range of polymerizable groups. Almost certainly this range will be expanded in the next few years.

Many potential applications of organic thin films require a noncentrosym-metric structure. This can be generated by the LB technique if different materials are deposited on the upstroke and the downstroke, and a number of special trough modifications have been designed for this purpose (see Chapter 3). The molecules of greatest interest are dye-type molecules and the next few years will see intense activity on the synthesis of these materials, suitably substituted for LB deposition. There will also be a major effort to incorporate these molecules into polymeric LB films and to study optimization of the interaction between head groups on the alternate layers.

2.8. POSTSCRIPT

Since the completion of this chapter there have been numerous advances in the LB field. There have been two international conferences, one in Göttingen, Germany (1987) and one in Tsukuba, Japan (1989). The proceedings of both conferences are published[214,215] and provide a good idea of recent trends.

This period has seen much greater industrial research commitment to LB films, particularly by Japanese companies. Accordingly, it is now essential to scan the Japanese patent literature in order to keep abreast of new developments, especially in the field of polymers.

REFERENCES

1. G. L. Gaines, Jr., *Insoluble Monolayers at Liquid–Gas Interfaces*, Wiley–Interscience, New York, London, Sydney (1966).
2. G. Gabrielli, G. G. T. Guarni, and E. Ferroni, On the mechanism of collapse of arachidic acid films at the water/air interface, *J. Colloid Interface Sci.*, **54**, 424–429 (1976).
3. W. D. Harkins, T. F. Young and E. Boyd, The thermodynamics of films: Energy and entropy of extension and spreading of insoluble monolayers, *J. Chem. Phys.*, **8**, 954–965 (1940).
4. G. L. Gaines, Jr., The thermodynamic equation of state for insoluble monolayers. 1. Uncharged films, *J. Chem. Phys.*, **69**, 924–930 (1978).
5. J. F. Baret, H. Hasmonay, J. L. Firpo, J. J. Dupin, and M. Dupeyrat, The different types of isotherm exhibited by insoluble fatty acid monolayers. A theoretical interpretation of phase transitions in the condensed state, *Chem. Phys. Lipids*, **30**, 177–187 (1982).
6. R. H. Tredgold and G. W. Smith, Surface potential studies on Langmuir–Blodgett multilayers and adsorbed monolayers, *Thin Solid Films*, **99**, 215–220 (1983).
7. J. C. Earnshaw, Laser light scattering from Langmuir films, *Thin Solid Films*, **99**, 189–195 (1983).
8. K. Miyano, B. M. Abraham, J. B. Ketterson, and S. Q. Xu, The phases of insoluble monolayers: Comparison between the surface pressure–molecular area (Π–A) diagram and shear modulus measurements, *J. Chem. Phys.*, **78**, 4776–4777 (1983).
9. A. Georgallas and D. A. Pink, A new theory of the liquid condensed–liquid expanded phase transition in lipid monolayers, *Can. J. Phys.*, **60**, 1678–1681 (1982).
10. N. Furasaki, Comments on miscibility in monolayers, *J. Colloid Interface Sci.*, **90**, 551–553 (1982).
11. J. J. Bikerman, Formation and structure of multilayers, *Proc. R. Soc. London, Ser. A*, **170**, 130–144 (1939).
12. D. Day and H. Ringsdorf, Polymerization of di-acetylene carbonic acid monolayers at the gas–water interface, *J. Polym. Sci., Polym. Lett. Ed.*, **16**, 205–210 (1978).
13. M. R. Buhaenko, J. W. Goodwin, R. M. Richardson, and M. F. Daniel, The influence of shear

viscosity of spread monolayers on the Langmuir–Blodgett process, *Thin Solid Films*, **134**, 217–226 (1985).

14. B. R. Malcolm, Studies of the flow of molecular monolayers during compression and the effect of a plateau in the pressure area curve, *Thin Solid Films*, **134**, 201–208 (1985).

15. M. F. Daniel and J. T. T. Hart, The effect of surface flow on the morphology of Langmuir–Blodgett films, *J. Mol. Electronics* (in press).

16. I. R. Peterson, G. J. Russell, and G. G. Roberts, A new model for the deposition of ω-Tricosenoic acid Langmuir–Blodgett film layers, *Thin Solid Films*, **109**, 371–378 (1983).

17. I. R. Peterson and G. J. Russell, The deposition and structure of Langmuir–Blodgett films of long chain acids, *Thin Solid Films*, **134**, 143–152 (1985).

18. G. Veale and I. R. Peterson, Novel effects of counterions on Langmuir films of 22-tricosenoic acid, *J. Colloid Interface Sci.*, **103**, 178–189 (1984).

19. K. B. Blodgett, Films built by depositing successive monomolecular layers on a solid surface, *J. Am. Chem. Soc.*, **57**, 1007–1022 (1935).

20. K. B. Blodgett and I. Langmuir, Built up films of barium stearate and their optical properties, *Phys. Rev.*, **51**, 964–982 (1937).

21. I. R. Peterson, G. Veale, and C. M. Montgomery, The preparation of oleophilic surfaces for Langmuir–Blodgett deposition, *J. Colloid Interface Sci.*, **109**, 527–530 (1986).

22. K. Fukuda, H. Nakahara, and T. Kato, Monolayers and multilayers of anthraquinone derivatives containing long alkyl chains, *J. Colloid Interface Sci.*, **54**, 430–438 (1976).

23. H. Nakahara and K. Fukuda, Orientation control of chromophores in monolayer assemblies of long chain dyes and the effects on some physical properties, *Thin Solid Films*, **99**, 45–52 (1983).

24. T. Kawaguchi, N. Nakahara, and K. Fukuda, Mono- and multilayers of amphiphilic cellulose esters and some novel comblike polymers, *J. Colloid Interface Sci.*, **104**, 290–293 (1985).

25. I. Fankuchen, The structure of built up films on metals, *Phys. Rev.*, **53**, 909 (1938).

26. S. Bernstein, Comparison of X-ray photographs taken with X and Y built-up films, *J. Am. Chem. Soc.*, **60**, 1511 (1938).

27. H. Kuhn and D. Möbius, Systems of monomolecular layers—Assembling and physico-chemical behaviour, *Angew Chem., Int. Ed. Engl.*, **10**, 620 (1971).

28. I. R. Peterson, Optical observation of monomer Langmuir–Blodgett film structure, *Thin Solid Films*, **116**, 357–366 (1984).

29. M. Saint-Pierre and M. Dupeyrat, Measurement and meaning of the transfer process energy in the building up of Langmuir–Blodgett multilayers, *Thin Solid Films*, **99**, 205–213 (1983).

30. J. B. Peng, B. M. Abraham, P. Dutta, and J. B. Ketterson, The contact angle of lead stearate covered water on mica during the deposition of Langmuir–Blodgett assemblies, *Thin Solid Films*, **134**, 187–193 (1985).

31. P. Dutta, Paper presented at the Second International Conference on Langmuir–Blodgett Films, Schenectady, NY, U.S.A. (July 1–4, 1985).

32. L. Ter Minassian Saraga, Adsorption and desorption at liquid surfaces. III. Desorption of unimolecular layers, *J. Chem. Phys.*, **52**, 181–200 (1955).

33. H. E. Ries, Jr. and W. A. Kimball, Monolayer structure as revealed by electron microscopy, *J. Phys. Chem.*, **59**, 94–99 (1955).

34. J. J. Betts and B. A. Pethica, The ionization characteristics of monolayers of weak acids and bases, *Trans. Faraday Soc.*, **52**, 1581–1589 (1956).

35. B. M. Abraham, J. B. Ketterson, K. Miyano, and A. Kueny, Shear rigidity of spread stearic acid monolayers on water, *J. Chem. Phys.*, **75**, 3137–3141 (1981).

36. S. Q. Xu, K. Miyano, and B. Abraham, The effect of pH on monolayer stability, *J. Colloid Interface Sci.*, **89**, 581–583 (1982).

37. P. Joos, Effect of the pH on the collapse pressure of fatty acid monolayers. Evaluation of the surface dissociation constant, *Bull. Soc. Chim. Belg.*, **80**, 277–281 (1971).

38. R. D. Newman, Calcium binding in stearic acid monomolecular films, *J. Colloid Interface Sci.*, **53**, 161–171 (1975).

39. Yu A. Bobrov and L. D. Saginov, Infrared Fourier spectroscopic study of the micro-heterogeneity of mixed Langmuir barium stearate–stearic acid multilayers, *Zh. Fiz. Khim.*, **57**, 972–976 (1983).
40. C. Vogel, J. Corset, F. Billouchet, M. Vincent, and M. Dupeyrat, Properties of monomolecular layers of stearic acid in presence of lead (2+) or barium salts, related to the composition and structure of Langmuir–Blodgett multilayers, *J. Chim. Phys.Phys.-Chim. Biol.*, **77**, 947–951 (1980).
41. P. A. Chollet, Determination by infrared absorption of the orientation of molecules in mono-molecular layers, *Thin Solid Films*, **52**, 343–360 (1978).
42. J. F. Rabolt, F. C. Burns, W. E. Schlotter, and J. D. Swalen, Anisotropic orientation in molecular monolayers by infrared spectroscopy *J. Chem. Phys.*, **78**, 946–952 (1983).
43. M. Pomerantz, F. H. Dacol, and A. Segmüller, Preparation of literally two-dimensional magnets, *Phys. Rev. Lett.*, **40**, 246–249 (1978).
44. M. Simovic and L. Dobrilovic, Study of radioactive labeled chromium(III) ion adsorption on stearic acid monomolecular layer, *J. Radioanal Chem.*, **44**, 345–354 (1978).
45. N. K. Adam and J. W. W. Dyer, The molecular structure of thin films, *Proc. R. Soc. London, Ser. A*, **106**, 694–709 (1924).
46. A. E. Alexander, Built-up films of unsubstituted and substituted long-chain compounds, *J. Chem. Soc.*, 777–781 (1939).
47. E. Stenhagen, Monolayers of compounds with branched hydrocarbon chains. I. Di-substituted acetic acids, *Trans. Faraday Soc.*, **36**, 597–606 (1940).
48. P. A. Delaney, R. A. W. Johnstone, B. L. Eyres, R. A. Hann, I. McGrath, and A. Ledwith, Formation of stable Langmuir–Blodgett multilayers from oxiran carboxylic acids, *Thin Solid Films*, **123**, 353–360 (1985).
49. R. A. W. Johnstone, R. A. Hann, P. A. Delaney, and S. K. Gupta, unpublished work.
50. S. Mijata and H. Nakahara, Langmuir–Blodgett film formation of fluorinated carboxylic acids and film properties, Paper presented at the Second International Conference on Langmuir–Blodgett Films, Schenectady, NY, U.S.A. (July 1–4, 1985).
51. M. Bernett and W. A. Zisman, The behaviour of monolayers of progressively fluorinated fatty acids adsorbed on water, *J. Phys. Chem.*, **67**, 1534–1540 (1963).
52. A. E. Alexander and J. H. Schuman, Orientation in films of long-chain esters, *Proc. R. Soc. London, Ser. A*, **161**, 115–127 (1937).
53. N. K. Adam, J. F. Danielli, and J. B. Harding, The structure of surface films XXI—surface potentials of dibasic esters, alcohols, aldoximes and ketones, *Proc. R. Soc. London, Ser. A*, **147**, 491–499 (1934).
54. E. P. Honig, J. H. Th. Hengst, and D. den Engelsen, Langmuir–Blodgett deposition ratios, *J. Colloid Interface Sci.*, **45**, 92–102 (1973).
55. J. G. Petrov, H. Kuhn, and D. Möbius, Three-phase contact line motion in the deposition of spread monolayers, *J. Colloid Interface Sci.*, **73**, 66–75 (1979).
56. G. L. Gaines, Jr., Langmuir–Blodgett films of long chain amines, *Nature*, **298**, 544–545 (1982).
57. W. Rettig, C. Koth, and H. D. Doerfler, Characterization of miscibility properties of binary monolayers by measurement of equilibrium spreading pressures, *Colloid Polym. Sci.*, **260**, 345–348 (1982).
58. E. C. Hunt, The interacton of alkyl phosphate monolayers with metal ions, *J. Colloid Interface Sci.*, **29**, 105–110 (1969).
59. R. A. Uphaus, G. F. Vandegri, and E. P. Horwitz, Monolayer characterization of several short alkyl chain phosphoric acid extractants, *J. Colloid Interface Sci.*, **90**, 380–389 (1982).
60. E. D. Goddard and H. C. Kung, Monolayer properties of a long chain amine oxide, *J. Colloid Interface Sci.*, **43**, 511–520 (1973).
61. N. K. Adam, The properties and molecular structure of thin films II: condensed films, *Proc. R. Soc. London, Ser. A*, **101**, 452–472 (1922).
62. T. J. Lewis, D. M. Taylor, J. P. Llewellyn, S. Salvagno, and C. J. M. Stirling, Langmuir films composed of sulphur-containing molecules, *Thin Solid Films*, **133**, 243–252 (1985).

63. J. Mingins and B. A. Pethica, Properties of ionised monolayers. Part 5—surface potentials and pressures of insoluble monolayers of sodium octadecyl sulphate, *Trans. Faraday Soc.*, **59**, 1892–1905 (1963).

64. N. K. Adam, *The Physics and Chemistry of Surfaces*, Clarendon Press, Oxford (1938).

65. N. K. Adam, The structure of thin films. Part X1—oxygenated derivatives of benzene, *Proc. R. Soc. London, Ser. A*, **119**, 628–644 (1928).

66. J. Heeseman, Studies on monolayers. 1. surface tension and absorption spectroscopic measurements of monolayers of surface-active azo and stilbene dyes, *J. Am. Chem. Soc.*, **102**, 2166–2176 (1980).

67. C. Naselli, J. P. Rabe, J. F. Rabolt, and J. D. Swalen, Thermally induced order–disorder transitions in Langmuir–Blodgett films, *Thin Solid Films*, **134**, 173–178 (1985).

68. V. I. Troitskii, Study of the structures of Langmuir films, *Deposited Doc. USSR, VINITI 3278*, 60–64 (1981); *Chem. Abstr.*, **97**, 47776a (1982).

69. M. F. Daniel, O. C. Lettington, and S. M. Small, Investigations into the Langmuir–Blodgett films formation ability of amphiphiles with cyano head groups, *Thin Solid Films*, **99**, 61–69 (1983).

70. M. F. Daniel, O. C. Lettington, and S. M. Small, Langmuir–Blodgett films of amphiphiles with cyano headgroups *Mol. Cryst. Liq. Cryst.*, **96**, 373–385 (1983).

71. H. Diep-Quang and K. Uberreiter, Monolayers of some liquid-crystal forming compounds, *Colloid Polym. Sci.*, **258**, 1055–1061 (1980).

72. K. A. Suresh, A. Blumstein, and F. Rondelez, Langmuir monolayers of mesomorphic monomers and polymers, *J. Phys. (Orsay. Fr.)*, **46**, 453–460 (1985).

73. R. Jones, R. H. Tredgold, A. Hoorfar, R. A. Allen, and P. Hodge, Crystal formation and growth in Langmuir–Blodgett multilayers of azobenzene derivatives. Optical and structural studies, *Thin Solid Films*, **134**, 57–66 (1985).

74. F. Rondelez, D. Koppel, and B. K. Sadashiva, Two dimensional films of discotic molecules at an air–water interface, *J. Phys. (Paris)*, **43**, 1361–1377 (1982).

75. F. H. C. Stewart, Unimolecular films from certain anthracene derivatives, *Aust. J. Chem.*, **14**, 57–63 (1961).

76. J. H. Steven, R. A. Hann, W. A. Barlow, and T. Laird, Influence of chemical structure on the monolayer properties of polycyclic aromatic molecules, *Thin Solid Films*, **99**, 71–79 (1983).

77. P. S. Vincett, W. A. Barlow, F. T. Boyle, J. A. Finney, and G. G. Roberts, Preparation of Langmuir–Blodgett built-up multilayer films of a lightly substituted model aromatic, anthracene, *Thin Solid Films*, **60**, 265–277 (1979).

78. P. S. Vincett and W. A. Barlow, Highly organized aromatic molecular systems using Langmuir–Blodgett films: Structure, optical properties and probable epitaxy of anthracene-derivative multilayers, *Thin Solid Films*, **71**, 305–326 (1980).

79. B. Belbeoch, M. Roulliay, and M. Tournarie, Evidence of chain interdigitation in Langmuir–Blodgett films, Paper presented at the Second International Conference on Langmuir–Blodgett Films, Schenectady, NY, U.S.A. (July 1–4, 1985).

80. I. R. Peterson, G. J. Russell, D. B. Neal, M. C. Petty, G. G. Roberts, T. Ginnai, and R. A. Hann, An electron diffraction study of LB films from a lightly substituted anthracene derivative, *Philos. Mag.*, **B54**, 71–79 (1986).

81. M. F. Daniel, Personal communication.

82. D. Möbius, Designed monolayer assemblies, *Ber. Bunsenges. Phys. Chem.*, **82**, 848–858 (1978).

83. P. K. J. Kinnunen, J. A. Virtanen, A. P. Tulkki, R. C. Ahuja, and D. Möbius, Pyrene-fatty acid containing phospholipid analogs: Characterization of monolayers and Langmuir–Blodgett assemblies, *Thin Solid Films*, **132**, 193–203 (1985).

84. A. Ruaudel-Teixier, M. Vandevyver, and A. Barraud, Novel conducting Langmuir–Blodgett films, *Mol. Cryst. Liq. Cryst.*, **120**, 319–322 (1985).

85. H. Kuhn, Electron transfer in monolayer assemblies, *Pure Appl. Chem.*, **51**, 341–352 (1979).

86. H. Kuhn, Information, electron and energy transfer in surface layers, *Pure Appl. Chem.*, **53**, 2105–2122 (1981).

87. D. Möbius, Molecular cooperation in monolayer organizates, *Acc. Chem. Res.*, **14**, 63–68 (1981).
88. G. R. Bird, G. Debuch, and D. Möbius, Preparation of a totally ordered monolayer of a chromophore by rapid epitaxial attachment, *J. Phys. Chem.*, **81**, 2657–2663 (1977).
89. Y. Kawabata, T. Sekiguchi, M. Tanaka, T. Nakamura, H. Komizu, M. Matsumoto, E. Manda, M. Saïto, M. Sugi, and S. Iizima, Formation and deposition of super monomolecular layers by means of surface pressure control, *Thin Solid Films*, **133**, 175–180 (1985).
90. M. Schimomura and T. Kunitake, Chromophore orientation in the Langmuir–Blodgett films of bilayer-forming amphiphiles, Paper presented at the Second International Conference on Langmuir–Blodgett Films, Schenectady, NY, U.S.A. (July 1–4, 1985).
91. C. H. Giles, V. G. Agribotri, and N. McIver, Measurement of aggregation number of monionic dyes with the Langmuir film balance, *J. Colloid Interface Sci.*, **50**, 24–31 (1975).
92. C. H. Giles, Dyeing in two dimensions. A review of the use of the monolayer method in the study of dye–fiber reactions, *J. Soc. Dyers Colour*, **94**, 4–12 (1978).
93. S. J. Valenty, Monolayer films of surfactant derivatives of methylene blue, *J. Colloid Interface Sci.*, **68**, 486–491 (1979).
94. D. G. Whitten, Photochemical reactions in organized monolayer assemblies. I. *Cis–trans* isomerization of thioindigo dyes, *J. Am. Chem. Soc.*, **96**, 594–596 (1974).
95. E. Ando, J. Miyazaki, K. Morimoto, H. Nakahara, and K. Fukuda, J-aggregation of photochromic spiropyran in Langmuir–Blodgett films, *Thin Solid Films*, **133**, 21–28 (1985).
96. K. Sakai, M. Saito, M. Sugi, and S. Iizima, Molecular p–n junction photodiodes of Langmuir multilayer semiconductors, *Jpn. J. Appl. Phys.*, **24**, 865–869 (1985).
97. M. Sugi, M. Saito, T. Fukui, and S. Iizima, Effect of dye concentration in Langmuir multilayer photoconductors, *Thin Solid Films*, **99**, 17–20 (1983).
98. T. Fukui, M. Saito, M. Sugi, and S. Iizima, Thermochromic behaviour of merocyanine Langmuir–Blodgett films, *Thin Solid Films*, **109**, 247–254 (1983).
99. S. Kuroda, M. Sugi, and S. Iizima, Origin of stable spin species in Langmuir–Blodgett films of merocyanine dyes studied by ESR and ENDOR, *Thin Solid Films*, **133**, 189–196 (1985).
100. O. A. Aktipetrov, N. N. Akhimediev, E. D. Mishina, and V. R. Novak, Second-harmonic generation on reflection from a monomolecular Langmuir layer, *JETP Lett.*, **40**, 207–209 (1983).
101. L. M. Blinov, N. V. Dubinin, L. V. Mikhnev, and S. G. Yudin, Polar Langmuir–Blodgett films, *Thin Solid Films*, **120**, 161–170 (1984).
102. L. M. Blinov, N. V. Dubinin, and S. G. Yudin, Linear Stark effect in optically anisotropic multimolecular structures, *Opt. Spektrosk.*, **56**, 280–286 (1984).
103. G. L. Gaines, Jr., Solvatochromic compound as an acid indicator in nonaqueous media, *Anal. Chem.*, **48**, 450–451 (1976).
104. M. F. Daniel and G. W. Smith, Preparation of non-centrosymmetric Langmuir–Blodgett films with alternating merocyanine and stearylamine layers, *Mol. Cryst. Liq. Cryst.*, **102**, 183–198 (1984).
105. G. W. Smith, M. F. Daniel, J. W. Barton, and W. Ratcliffe, Pyroelectric activity in non-centrosymmetric Langmuir–Blodgett multilayer films, *Thin Solid Films*, **132**, 125–134 (1985).
106. I. R. Girling, N. A. Cade, P. V. Kolinsky, and C. M. Montgomery, Observation of a second-harmonic generation from a Langmuir–Blodgett monolayer of a merocyanine dye, *Electron Lett.*, **21**, 169–170 (1985).
107. I. R. Girling, N. A. Cade, P. V. Kolinksky, J. D. Earls, G. H. Cross, and I. R. Peterson, Observation of second harmonic generation from Langmuir–Blodgett multilayers of a hemicyanine dye, *Thin Solid Films*, **132**, 101–112 (1985).
108. H. Nakanishi, S. Okada, H. Matsuda, M. Kato, M. Sugi, M. Saito, and S. Iizima, Fabrication of polar structures by use of Langmuir–Blodgett techniques, Paper presented at the Second International Conference on Langmuir–Blodgett Films, Schenectady, NY, U.S.A. (July 1–4, 1985).
109. I. Ledoux, D. Josse, P. Vidakovic, J. Zyss, R. A. Hann, P. F. Gordon, B. D. Bothwell, S. K. Gupta, S. Allen, P. Robin, E. Chastaing, and J-C. Dubois, Second harmonic generation by Langmuir–Blodgett multilayers of an organic azo dye, *Europhys. Lett.*, **3**, 803–809 (1987).

110. S. M. de B. Costa and G. Porter, Model systems for photosynthesis. IV. Photosensitization by chlorophyll a monolayer at a lipid-water interface, *Proc. Roy. Soc.* (London) **341**, 167–176 (1974).

111. R. Jones, R. H. Tredgold, and J. E. O'Mullane, Photoconductivity and photovoltaic effects in Langmuir–Blodgett films of chlorophyll-*a*, *Photochem. Photobiol.*, **32**, 223–232 (1980).

112. R. A. Bull and J. E. Bulkowski, Tetraphenylporphyrin monolayers: formation at the air water interface and characterization on glass supports by absorption and fluorescence spectroscopy, *J. Colloid Interface Sci.*, **92**, 1–12 (1983).

113. F. R. Hopf, D. Möbius, and D. G. Whitten, Environmental control of reactions; enhancement of μ-oxo dimer formation from iron(III) porphyrins in organised monolayer assemblies as a model for membrane catalysis, *J. Am. Chem. Soc.*, **98**, 1584–1586 (1976).

114. M. Vandevyver, A. Barraud, Ruaudel-Teixier, P. Maillard, and C. Gianotti, Structure of porphyrin multilayers obtained by the Langmuir Blodgett technique, *J. Colloid Interface Sci.*, **85**, 571–585 (1982).

115. R. H. Schmehl, G. L. Shaw, and D. G. Whitten, Modification of interfacial reactivity by the microenvironment. Metalation of porphyrins in monolayer films and assemblies, *Chem. Phys. Lett.*, **58**, 549–592 (1978).

116. B. E. Horsey, F. R. Hopf, R. H. Schmehl, and D. G. Whitten, Photochemistry of free base and metalloporphyrin complexes in monolayers, *Porphyrin Chem. Adv.*, 17–28 (1979).

117. J. A. Mercer-Smith and D. G. Whitten, Photochemical reactivity in organised assemblies. 15. Photoreactions of metalloporphyrins in supported monolayer assemblies and at assembly–solution interfaces. Reductive addition of palladium complexes with surfactant and water soluble di-alkylanilines, *J. Am. Chem. Soc.*, **101**, 6620–6625 (1979).

118. G. Adler, Inhomogeneities in the structure of built-up monolayers of two porphyrin compounds, *J. Colloid Interface Sci.*, **72**, 164–169 (1979).

119. R. Jones, R. H. Tredgold, and P. Hodge, Langmuir–Blodgett films of simple esterified porphyrins, *Thin Solid Films*, **99**, 25–32 (1983).

120. A. Ruaudel-Teixier, A. Barraud, B. Belbeoch, and M. Roulliay, Langmuir–Blodgett films of pure porphyrins, *Thin Solid Films*, **99**, 33–40 (1983).

121. M. Vandevyver, A. Barraud, and A. Ruaudel-Teixier, Structure of porphyrin multilayers obtained by the Langmuir–Blodgett technique, *Mol. Cryst. Liq. Cryst.*, **96**, 361 (1983).

122. A. Barraud, C. M. Ardle, and A. Ruaudel-Teixier, Synthesis and properties of copper cyanopor-phyrins in Langmuir–Blodgett films, *Thin Solid Films* (in press).

123. C. Lecomte, C. Baudin, F. Berleur, A. Ruaudel-Teixier, A. Barraud, and M. Momenteau, An example of molecular building: Alternate Langmuir–Blodgett films of cobalto-*meso*-porphyrins designed to bind dioxygen, *Thin Solid Films*, **133**, 103–112 (1985).

124. F. H. Moser and A. L. Thomas, *The Phthalocyanines*, Vol. 1, CRC Press Inc., Boca Raton (1983).

125. S. Baker, M. C. Petty, G. G. Roberts, and M. V. Twigg, The preparation and properties of stable metal-free phythalocyanine Langmuir–Blodgett films, *Thin Solid Films*, **99**, 53–59 (1983).

126. R. A. Hann, W. A. Barlow, B. L. Eyres, M. V. Twigg, and G. G. Roberts, Langmuir–Blodgett films of substituted aromatic hydrocarbons and phthalocyanines, *Proc 2nd Int. Workshop on Molecular Electronic Devices*, Washington, U.S.A., 1983, Marcel Dekker, New York, 1987.

127. G. J. Kovacs, P. S. Vincett, and J. H. Sharp, Stable, tough, adherent Langmuir–Blodgett films: Preparation and structure of ordered, true monolayers of phthalocyanine, *Can J. Phys.*, **63**, 346–349 (1985).

128. G. G. Roberts, M. C. Petty, S. Baker, M. T. Fowler, and N. J. Thomas, Electron devices incorporating stable phthalocyanine Langmuir–Blodgett films, *Thin Solid Films*, **132**, 113–123 (1985).

129. Y. L. Hua, G. G. Roberts, M. M. Ahmad, M. C. Petty, M. Hanack, and M. Rein, Monolayer films of a substituted silicon phthalocyanine, *Philos. Mag.*, **B53**, 105–113 (1986).

130. R. A. Hann and S. K. Gupta (to appear).

131. J. R. Fryer, R. A. Hann, and B. L. Eyres, Single organic monolayer imaging by electron micros-copy, *Nature*, **313**, 382–384 (1985).

132. R. A. Hann, S. K. Gupta, J. R. Fryer, and B. L. Eyres, Electrical and structural studies on copper tetra-*tert*-butylphthalocyanine Langmuir–Blodgett films, *Thin Solid Films*, **134**, 35–42 (1985).

133. S. Baker, G. G. Roberts, and M. C. Petty, Phthalocyanine Langmuir–Blodgett film gas detector, *IEE Proc.*, **130**, 260–263 (1983).

134. D. W. Kalina and S. W. Crane, Langmuir–Blodgett films of soluble copper octa(dodecoxymethyl)phthalocyanine, *Thin Solid Films*, **134**, 109–119 (1985).

135. A. W. Snow and N. L. Jarvis, Molecular association and monolayer formation of soluble phthalocyanine compounds, *J. Am. Chem. Soc.*, **106**, 4706–4711 (1984).

136. W. R. Barger, A. W. Snow, H. Wohltjen, and N. L. Jarvis, Derivatives of phthalocyanines prepared for deposition as thin films by the Langmuir–Blodgett technique, *Thin Solid Films*, **133**, 197–206 (1985).

137. H. Yamamoto, T. Sugiyama, and M. Tanaka, Electrochromism of metal-free phthalocyanine Langmuir–Blodgett films, *Jpn. J. Appl. Phys.*, **24**, L305–L307 (1985).

138. A. Suzuki, K. Ohkawa, S. Kanda, M. Emoto, and S. Watari, Monolayers and built-up films of *N,N'*disubstituted dithiooxamide copper(II) coordination compounds, *Bull. Chem. Soc. Jpn.*, **48**, 2634–2638 (1975).

139. G. Sprintschnik, H. W. Sprintschnik, P. P. Kirsch, and D. G. Whitten, Photochemical cleavage of water: A system for solar energy conversion using monolayer-bound transition metal complexes, *J. Am. Chem. Soc.*, **98**, 2337–2338 (1976).

140. H. Nakahara, K. Fukuda, and M. Sato, Langmuir–Blodgett films of ferrocene derivatives with long alkyl chains, *Thin Solid Films*, **133**, 1–10 (1985).

141. M. Fujihara, K. Nishiyama, and H. Yamada, Photoelectrochemical responses of optically transparent electrodes modified with Langmuir–Blodgett films consisting of surfactant derivatives of electron donor, acceptor, and sensitizer molecules, *Thin Solid Films*, **132**, 77–82 (1985).

142. M. Pincus, S. Windreich, and I. R. Miller, Preparation of stable mixed monolayers of beta-carotene and their transfer to glass slides, *Biochem. Biophys. Acta*, **311**, 317–319 (1973).

143. U. Schoeler, K. H. Tews, and H. Kuhn, Potential model of dye molecule from measurements of the photocurrent in monolayer assemblies, *J. Chem. Phys.*, **61**, 5009–5016 (1974).

144. H. Schreiber, Farbstoff-Aggregate in monomolekularen Cd-Arachidatschichten und Photo-EMK dieser Mischschichten, Inaugural Dissertation, University of Marburg (1968).

145. J. R. Fryer, R. A. Hann, and B. L. Eyres, unpublished work.

146. A. Takahashi, A. Yoshida, and M. Kawaguchi, Test of scaling laws describing the concentration dependence of surface pressure of a polymer monolayer, *Macromolecules*, **15**, 1196–1198 (1982).

147. M. Kawaguchi, A. Yoshida, and A. Takahashi, Experimental determination of the temperature–concentration diagram of Daoud and Jannink in two-dimensional space by surface pressure measurements, *Macromolecules*, **16**, 956–961 (1983).

148. G. Gabrielli, M. Puggelli, and P. Baglioni, Orientation and compatibility in monolayers. II. Mixtures of polymers, *J. Colloid Interface Sci.*, **86**, 485–500 (1982).

149. M. Koyama, R. Tomioka, M. Ueno, and K. Meguro, Monolayers of poly(vinyl butyrals), *Colloid Polym. Sci.*, **252**, 372–376 (1974).

150. D. D. Eley, M. J. Hey, and J. Speight, Influence of salts on poly(vinyl methyl ether) at the air/aqueous-solution interface. Part I. Spread monolayers, *J. Chem. Soc.*, **79**, 755–763 (1983).

151. G. Gabrielli, P. Baglioni, and E. Ferroni, Bi- and tri-dimensional solutions of poly(vinyl acetate), *Colloid Polym. Sci.*, **257**, 121–127 (1979).

152. E. Ferroni, G. Gabrielli, and M. Puggelli, Mixed poly(vinylidene fluoride)–poly(vinyl fluoride) monolayers, *Chim. Ind. (Milan)*, **42**, 147–50 (1967).

153. G. L. Gaines, Jr., Monolayers of dimethyl-siloxane-containing block copolymers, *Adv. Chem. Ser.*, **144**, 338–346 (1975).

154. K. Fukuda, T. Kato, S. Machida, and Y. Shimizu, Binary mixed monolayers of poly(vinyl stearate) and simple long chain compounds at the air/water interface, *J. Colloid Interface Sci.*, **68**, 82–95 (1979).

155. R. H. Tredgold and C. S. Winter, Langmuir–Blodgett monolayers of preformed polymers, *J. Phys. D*, **15**, L55–58 (1982).

156. P. Hodge, E. Khoshdel, R. H. Tredgold, A. J. Vickers, and C. S. Winter, *Br. Polym. J.*, **17**, 368–370 (1985).

157. C. S. Winter, R. H. Tredgold, A. J. Vickers, E. Khoshdel, and P. Hodge, Langmuir–Blodgett films from preformed polymers: derivatives of Octadec-1-ene-maleic anhydride copolymers, *Thin Solid Films*, **134**, 49–55 (1985).

158. A. J. Vickers, R. H. Tredgold, E. Khoshdel, P. Hodge, and I. Girling, An investigation of liquid crystal side chain polymeric Langmuir–Blodgett films as optical waveguides, *Thin Solid Films*, **134**, 43–48 (1985).

159. R. H. Tredgold, A. J. Vickers, A. Hoorfar, P. Hodge, and E. Khoshdel, X-ray analysis of some porphyrin and polymer Langmuir–Blodgett films, *J. Phys. D*, **18**, 1139–1145 (1985).

160. B. R. Malcolm, Structure and properties of monolayer of synthetic polypeptides at the air–water interface, *Prog. Surf. Membr. Sci.*, **1**, 183–229 (1973).

161. C. S. Winter and R. H. Tredgold, Langmuir–Blodgett multilayers of polypeptides, *Thin Solid Films*, **123**, L1–L3 (1985).

162. T. Furuno, H. Sasabe, R. Nagata, and T. Akaike, Structure of Langmuir–Blodgett films of poly(1-benzyl-L-histidine)-stearic acid, *Thin Solid Films*, **133**, 141–152 (1985).

163. S. J. Mumby, J. F. Rabolt, and J. D. Swalen, Structural characterization of a polymer monolayer on a solid surface, *Thin Solid Films*, **133**, 161–164 (1985).

164. P. Christie, M. C. Petty, G. G. Roberts, D. H. Richards, D. Service, and M. J. Stewart, The preparation and dielectric properties of polybutadiene Langmuir–Blodgett films, *Thin Solid Films*, **134**, 75–82 (1985).

165. R. Elbert, A. Laschewsky, and H. Ringsdorf, Hydrophilic spacer groups in polymerizable lipids: formation of biomembrane models from bulk polymerised lipids, *J. Am. Chem. Soc.*, **107**, 4134–4141 (1985).

166. A. Laschewsky, H. Ringsdorf, and G. Schmidt, Polymerization of hydrocarbon and fluorocarbon amphiphiles in Langmuir–Blodgett multilayers, *Thin Solid Films*, **134**, 153–172 (1985).

167. A. K. Engel, T. Yoden, K. Sanui, and N. Ogata, Aliphatic chains as disposable aids to monolayer formation: poly(*p*-phenylene terephthalaldimine) oligomers formed at the air/water interface, Paper presented at the Second International Conference on Langmuir–Blodgett Films, Schenectady, NY, U.S.A. (July 1–4, 1985).

168. M. Kakimoto, M. Suzuki, T. Konishi, Y. Imai, M. Iwamoto, and T. Hino, Preparation of mono- and multilayer films of aromatic polyimides using Langmuir–Blodgett technique, *Chem. Lett.*, 823–826 (1986).

169. M. Suzuki, M. Kakimoto, T. Konishi, Y. Imai, M. Iwamoto, and T. Hino, Preparation of monolayer films of aromatic polyamic acid alkylamine salts at the air–water interface, *Chem. Lett.*, 395–398 (1986).

170. H. Z. Friedlander, U.S. Patent 3, 031, 721 (1962); *Chem. Abstr.*, **57**, 14008 (1962).

171. A. Cemel, T. Fort, Jr., and J. B. Lando, Polymerization of vinyl stearate multilayers, *J. Polym. Sci.*, **10**, 2061–2083 (1972).

172. D. Naegele, J. B. Lando, and H. Ringsdorf, Polymerization of cadmium octadecylfumarate in multilayers, *Macromolecules*, **10**, 1339–1344 (1977).

173. D. G. Whitten and P. R. Worsham, Light induced polymerisation of surfactant styrene derivatives in monolayer films and assemblies, *Org. Coat. Plast. Chem.*, **38**, 572–575 (1978).

174. H. Ringsdorf and H. Schupp, Polymerization of substituted butadienes at the gas–water interface, *J. Macromol. Sci., Chem.*, **A15**, 1015–1026 (1981).

175. B. Hupfer, H. Ringsdorf, and H. Schupp, Poly-reactions in oriented systems, 21a Polymeric phospholipid monolayers, *Makromol. Chem.*, **182**, 247–253 (1981).

176. G. Fariss, J. Lando, and S. Rickert, Electron beam resists produced from monomer–polymer Langmuir–Blodgett films, *Thin Solid Films*, **99**, 305–315 (1983).

177. K. Shih, S. E. Rickert, and J. B. Lando, Characterisation of thin films of elaidic acid prepared through Langmuir–Blodgett technique and its application as an electron resist, Paper presented at the Second International Conference on Langmuir–Blodgett Films, Schenectady, NY, U.S.A. (July 1–4, 1985).

178. J. Tan, S. E. Rickert, and J. B. Lando, Electron beam resists from Langmuir–Blodgett films, Paper

presented at the Second International Conference on Langmuir–Blodgett Films, Schenectady, NY, U.S.A. (July 1–4, 1985).

179. A. Barraud, C. Rosilio, and A. Ruaudel-Teixier, Solid-state electron induced polymerisation of ω-tricosenoic acid multilayers, *J. Colloid Interface Sci.,* **62,** 509–523 (1977).

180. I. R. Peterson and G. J. Russell, Deposition mechanisms in Langmuir–Blodgett films, *Br. Polym. J.,* **17,** 364–367 (1985).

181. R. H. Baughman, Solid state synthesis of large polymer single crystals, *J. Polym. Sci., Polym. Phys. Ed.,* **12,** 1511–1535 (1974).

182. B. Tieke, G. Wegner, D. Naegele, and H. Ringsdorf, Polymerization of tricosa-10, 12-diynoic acid in multilayers, *Angew. Chem., Int. Ed. Engl.,* **15,** 764–765 (1976).

183. B. Tieke, H. J. Graf, G. Wegner, B. Naegele, H. Ringsdorf, A. Banerjie, D. Day, and J. B. Lando, Polymerization of mono- and multilayer forming diacetylenes, *Colloid Polym. Sci.,* **255,** 521–531 (1977).

184. B. Tieke, G. Lieser, and K. Weiss, Parameters influencing the polymerisation and structure of long-chain diynoic acids in multilayers, *Thin Solid Films,* **99,** 95–102 (1983).

185. B. Tieke and G. Lieser, Influences of the structure of long chain diynoic acids on their polymerisation properties in Langmuir–Blodgett multilayers, *J. Colloid Interface Sci.,* **88,** 471–486 (1982).

186. F. Kajzar and J. Messier, Solid state polymerisation and optical properties of diacetylene Langmuir–Blodgett multilayers, *Thin Solid Films,* **99,** 109–116 (1983).

187. G. Fariss, J. Lando, and S. Rickert, Electron beam resists produced from monomer–polymer Langmuir–Blodgett films, *Thin Solid Films,* **99,** 305–315 (1983).

188. G. Veale, D. R. J. Milverton, and M. N. Wybourne, The influence of control mechanism feedback parameters on the photoreactions of diacetylene monolayers. *Thin Solid Films,* **136,** 141–145 (1986).

189. C. W. Pitt and L. M. Walpita, Lightguiding in Langmuir–Blodgett films, *Thin Solid Films,* **68,** 101–127 (1980).

190. F. Grunfeld and C. W. Pitt, Diacetylene Langmuir–Blodgett layers for integrated optics, *Thin Solid Films,* **99,** 249–255 (1983).

191. D. Day, H. H. Hub, and H. Ringsdorf, Polymerisation of mono- and bifunctional diacetylene derivatives in monolayers at the gas–water interface, *Isr. J. Chem.,* **18,** 325–329 (1979).

192. H. H. Hub, B. Hupfer, H. Koch, and H. Ringsdorf, Polymerization of lipid and lysolipid like diacetylenes in monolayers and liposomes, *J. Macromol. Sci., Chem.,* **A14,** 701–716 (1981).

193. L. Gros, H. Ringsdorf, and H. Schupp, Polymeric antitumor agents on a molecular and on a cellular level?, *Angew. Chem., Int. Ed. Engl.,* **20,** 305–325 (1981).

194. B. Boothroyd, P. A. Delaney, R. A. Hann, R. A. W. Johnstone, and A. Ledwith, Electron irradiation of polymerisable Langmuir–Blodgett multilayers as model resists for electron-beam lithography, *Br. Polym. J.,* **17,** 360–363 (1985).

195. A. Barraud, A. Ruaudel-Teixier, and C. Rosilio, Solid-state chemical reactions in organic mono-molecular layers, *Ann. Chim.,* **10,** 195–200 (1975).

196. S. J. Valenty, Amine-aldehyde polycondensation chemistry in monolayer films. An approach toward achieving a light energy transducing molecular organizate, *Macromolecules,* **11,** 1221–1228 (1978).

197. L. Netzer and J. Sagiv, A new approach to construction of artificial monolayer assemblies, *J. Am. Chem. Soc.,* **105,** 674–676 (1983).

198. L. Netzer, R. Iscovici, and J. Sagiv, Adsorbed monolayers versus Langmuir–Blodgett monolayers—Why and How? 1: from monolayer to multilayer, by adsorption, *Thin Solid Films,* **99,** 235–241 (1983).

199. L. Netzer, R. Iscovici, and J. Sagiv, Adsorbed monolayers versus Langmuir–Blodgett monolayers—Why and How? II: Characterization of built-up films constructed by stepwise adsorption of individual monolayers, *Thin Solid Films,* **100,** 67–76 (1983).

200. M. Pomerantz, A. Segmüller, L. Netzer, and J. Sagiv, Coverage of Si substrates by self-assembling monolayers and multilayers as measured by IR, wettability and X-ray diffraction, *Thin Solid Films,* **132,** 153–162 (1985).

201. J. Gun and J. Sagiv, Formation and structure of self-assembling hydrocarbon and fluorocarbon

monolayer films, Paper presented at the Second International Conference on Langmuir–Blodgett Films, Schenectady, NY, U.S.A. (July 1–4, 1985).

202. L. Netzer and J. Sagiv, Self assembling monolayers of some bifunctional surfactants and their chemical modification, Paper presented at the Second International Conference on Langmuir–Blodgett Films, Schenectady, NY, U.S.A. (July 1–4, 1985).

203. R. Maoz and J. Sagiv, Penetration controlled reactions in organised monolayer assemblies III. Organic permanganate interaction with self-assembling monolayers of long chain surfactants, *Thin Solid Films,* **132,** 135–151 (1985).

204. R. H. Tredgold, C. S. Winter, and Z. I. El-Badawy, Multiple monolayer adsorption—a new technique for the production of noncentrosymmetric films, *Electron. Lett.,* **21,** 554–555 (1985).

205. D. L. Allahara and R. G. Nuzzo, Spontaneously organised monolayer assemblies 1. Formation, dynamics and physical properties of *n*-alkanoic acids adsorbed from solution on an oxidised aluminium surface, *Langmuir,* **1,** 45–52 (1985).

206. D. L. Allara and R. G. Nuzzo, Spontaneously organised molecular assemblies 2. Quantitative infrared spectroscopic determination of equilibrium structures of solution-adsorbed *n*-alkanoic acids on an oxidised aluminium surface, *Langmuir,* **1,** 52–66 (1985).

207. D. A. Cadenhead and R. J. Demchak, Monolayers of elaidic acid on aqueous glycerol solutions, *J. Colloid Interface Sci.,* **24,** 484–490 (1967).

208. M. Nakagaki and H. Ichihashi, Intermolecular interaction in egg lecithin monolayers spread on aqueous alcohol solutions, *Fac. Pharm. Sci.,* **98,** 577–584 (1978).

209. A. Barraud, J. Leloup, and P. Lesieur, Monolayers on a glycerol subphase, *Thin Solid Films,* **133,** 113–116 (1985).

210. A. H. Ellison, Surface pressure–area properties of organic monolayers on mercury, *J. Phys. Chem.,* **66,** 1867–1872 (1962).

211. T. Smith, Monomolecular films on mercury, *Adv. Colloid Interface Sci.,* **3,** 161–221 (1972).

212. A. H. Ellison and W. A. Zisman, Force–area properties of films of polyfluoroquaternary ammonium compounds on hydrocarbon liquid substrates, *J. Phys. Chem.,* **59,** 1233 (1955).

213. A. H. Ellison and W. A. Zisman, Surface activity at the organic liquid/air interface, *J. Phys. Chem.,* **60,** 416–421 (1956).

214. *Thin Solid Films,* **159,160** (1988).

215. *Thin Solid Films,* **178,179,180** (1989).

Film Deposition

M. C. PETTY AND W. A. BARLOW

3.1. DEPOSITION PRINCIPLES

There are a number of different ways in which a floating monolayer may be trans-ferred to a solid plate (to be referred to as the substrate). This chapter is concerned exclusively with the method reported by Langmuir,[1] extensively applied by Blodgett,[2,3] and which is now firmly associated with the names of these workers. Other transfer mechanisms can result from different interactions between the mono-layer, subphase, and solid. Examples are the "touching" method of Schulman et al.[4] and the "lifting" technique suggested by Langmuir and Schaefer[5]; these are reviewed in the book by Gaines.[6] The lifting technique has found some application by Fukuda et al.,[7−10] who have used it to deposit monolayers of certain long chain aromatics, and also by workers on phospholipid membranes, in order to produce hydrophilic outer surfaces.[213] A similar horizontal deposition technique has been described by Lando et al.[11,12]; this would seem to be useful for the transfer of polymerized or highly rigid monolayers to solid supports. A further variation is discussed by Kossi and Leblanc,[13] who have reported on a combined lifting and touching technique to produce model membrane systems.

  Although the Langmuir–Blodgett (LB) method is one of the classical tech-niques of surface chemistry, the detailed mechanisms by which floating monolayers are transferred to solid substrates are still poorly understood.[14] The molecular

M. C. PETTY • Molecular Electronics Research Group, School of Engineering and Applied Science, University of Durham, Durham DH1 3LE, England.   W. A. BARLOW • Electronics Group, Imperial Chemical Industries plc, Runcorn, Cheshire WA7 4QE, England.

interactions involved in the deposition of the first layer may be quite different than those that are responsible for the transfer of subsequent layers. For example, in the former case a chemical reaction can take place between the monolayer and the substrate,[15] resulting in a strongly bound first layer. For some materials, film deposition also seems likely to be associated with a distinct phase change: from a two-dimensional liquid crystalline phase on the water surface, to a closer packed solid crystalline form on the substrate.[16-20] However, despite much experimental evidence and a number of theoretical treatments of LB film deposition (see, for example, Clint and Walker[21] and Stephens[22]), many of the phenomena noted by the original workers have never been completely explained; examples include the different modes of film transfer, and the variation of the speed at which different materials can be deposited. A few ideas will be outlined in this chapter; however, a full understanding of LB film formation must await further developments in both theory and experiment.

A schematic diagram illustrating the commonest form of LB film deposition is shown in Figure 3.1. In this example the substrate is hydrophilic and the first monolayer is transferred (like a carpet) as the substrate is raised through the sub-phase (Figure 3.1b)—the substrate may therefore be placed in the subphase before the monolayer is spread, or may be lowered into the subphase through the compressed monolayer. Subsequently a monolayer is deposited on each traversal of the surface (Figure 3.1c). As shown in Figure 3.1d, these stack in a head-to-head and tail-to-tail configuration; this deposition mode is referred to as Y type. In the particular example shown in Figure 3.1, a multilayer structure containing only an odd number of layers can be produced. However, if the solid substrate is hydrophobic a monolayer will be deposited as it is first lowered into the subphase; thus a Y-type film containing an even number of monolayers can be fabricated.

Although Y-type layers are the most easily produced multilayers, monolayers which deposit only as the substrate is being inserted into the subphase or only as the substrate is being removed have been reported. These deposition modes are referred to as X type and Z type, respectively; schematic diagrams for the deposition and the expected molecular arrangement for the two types of layers are shown in Figure 3.2. Mixed deposition modes have also been observed; for instance, XY deposition refers to complete transfer of the monolayer as the substrate is being lowered into the subphase, but only partial transfer as the substrate moves up through the monolayer air interface.[23-25] Film deposition may usefully be characterized by reference to a deposition (transfer) ratio, τ, given by Langmuir et al.[26]:

$$\tau = A_L/A_S \tag{3.1}$$

where A_L is the decrease in the area occupied by the monolayer on the water surface (held at a constant pressure), and A_S is the coated area of the solid substrate. Honig et al.[27] have suggested the use of another parameter, ϕ, in order to help quantify the various deposition modes; this may be calculated from

$$\phi = \tau_u/\tau_d \tag{3.2}$$

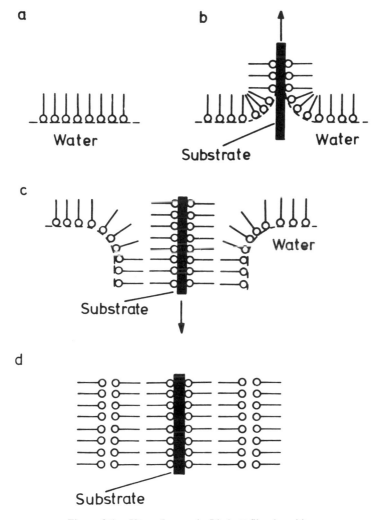

Figure 3.1. Y-type Langmuir–Blodgett film deposition.

where τ_u and τ_d are the deposition ratios on the upward and downward passages, respectively, of the substrate. Thus for pure Y-type deposition $\phi = 1$, for X type deposition $\phi = 0$, and for Z-type deposition ϕ is infinite. If asymmetric substrates are used (e.g., a glass slide metallized on just one surface), then some care must be exercised in interpreting the measured deposition ratio; it is unlikely that τ will be identical for the different surfaces. Neuman and Swanson[28] have also noted that the value of τ given by equation (3.1) should really be considered as an apparent deposition ratio, because it is not known whether the "transferred" film has completely deposited on the solid substrate. In addition to monolayer losses arising from desorption, evaporation, or collapse, some film molecules may dissolve in the

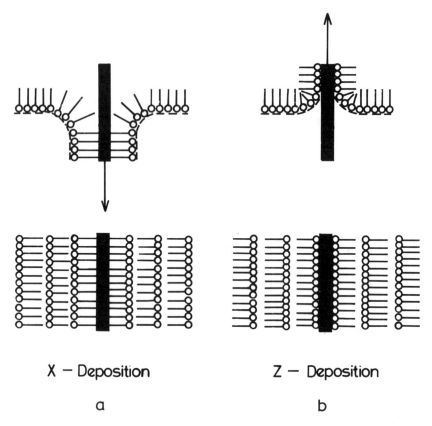

Figure 3.2.   (a) X-type deposition; (b) Z-type deposition.

subphase at the moment of transfer, or after deposition. Neuman and Swanson have employed radiometric techniques to investigate such possibilities; however, the local transfer ratio for calcium stearate monolayers was found to be unity during both immersion and withdrawal, except under conditions where calcium stearate surface micelles were believed to be present in the monolayer ($>$ pH 6.4).

Fatty acid monolayers normally deposit as Y-type layers. However, X-type deposition is possible with suitable changes in the dipping conditions; X-type deposition is usually favored by high pH values.[3,6] It is not customary for an X-type film to be formed on the first dipping cycle, but the character of the film becomes more and more X-type as the dipping proceeds.[3,23,27] Appropriate conditions also enable long chain esters to be deposited as either X- or Y-type layers[29–31]; in this case a Y-type deposit is favored by a high surface pressure, a low temperature, and a rapid withdrawal of the substrate from the subphase.[31] X-type deposition has also been reported for oleic acid,[32] phospholipids,[33] and for a quarternary ammonium salt-terminated polybutadiene material[34]; in the latter case a process involving ultraviolent irradiation to initiate crosslinking of the polymer chains was used after each monolayer had been transferred to the substrate.

It should be noted that the final molecular arrangements in an LB layer may not always be as shown in Figures 3.1 and 3.2. For fatty acids, X-ray diffraction measurements have revealed that the spacing between the hydrophilic head groups is nearly the same, and equal to twice the length of the hydrocarbon chain, whether they are deposited as X-type or as Y-type films.[35-38] Molecules in the X-type films must therefore rearrange, during or shortly after deposition, so that the structure finally produced is essentially identical to that of Y-type films. In contrast, some long chain esters (such as ethyl stearate or vinyl stearate) have been shown always to produce an X-type structure, irrespective of how the multilayers are built up[29-31] (however, Enkelmann and Lando[39] describe a method for building up Y-type vinyl stearate layers by polymerizing each bilayer after deposition, and thus freezing in the original tail-to-tail molecular arrangement). It has been suggested that, for a particular material, the final structure of the LB multilayer will be identical to that of the bulk crystal.[31] A number of studies of the transformation of X- to Y-type layers have been reported.[22,23,27,29,40-43] The theoretical approach of Stephens[22] is based on the principles of ionic equilibria; monolayer transfer is shown to depend on the pH of the subphase and the distribution of soluble cations and anions between the bulk of the subphase and the monolayer. The importance of these factors has also been noted by Gaines[44] from observations of contact angles during the deposition process. Both Langmuir[40] and Honig[23] have proposed models for fatty acid layers in which the overturning of the molecules occurs underwater, while Charles[41] and Stephens[22] suggest that this process takes place actually as the monolayer is being transferred to the solid substrate. However, as Honig[23] notes, the latter mechanism can result in films which are hydrophilic; this is not observed in practice.[40]

There are a number of reports of Z-type deposition. Most of these concern aromatic materials with relatively short or no carbon chains; examples include substituted anthracene derivatives,[45] porphyrins,[46] liquid crystal type molecules with cyano head groups,[47] phthalocyanines,[48-51] azobenzene derivatives,[52] polymers,[53,54] and certain dye materials.[214,215] Z-type deposition has also been observed for biological monolayer systems.[55-57] All such materials are very different from "classical" LB film molecules (e.g., long chain fatty acids); alternative processes may therefore be involved in their transfer to a substrate. It is not yet clear whether Z-type films rearrange once deposited, although Jones et al.[52] report X-ray data suggesting that Z-deposited layers of azobenzene derivatives in fact possessed a Y-type structure. An interesting series of long chain azo-dye materials has been studied by Blinov et al.[58,59] By raising and lowering a substrate through the floating monolayer, Y-type deposition was achieved. However, both X-type and Z-type structures could also be obtained by moving the substrate in one direction through the floating monolayer and in the opposite direction through the clean water surface. In some cases, Z-type deposition has been achieved by decompression of the monolayer during the downstroke.[74,216,217]

A number of techniques are now available to accurately measure contact angles and interaction energies during monolayer transfer[14,21,24,28,44,60–62,218]; the application of these may well lead to a much better insight into the various LB deposition modes.

3.2. SUBPHASE AND CONTAINMENT

Although workers have experimented with monomolecular films on a variety of different subphases (e.g., water,[1] mercury,[63] hydrocarbons,[64] glycerol[65]), almost all the work on transferred monolayers has concentrated on aqueous subphases; the discussion in this section will therefore be restricted to this. The quality of water used in LB film work is of the utmost importance.[219] In most monolayer experiments the quantity of monolayer-forming material that is used is in the μg range; the subphase volume can be many liters. Thus ppb (parts per 10^9) impurities in the subphase, if they are surface active, can be a problem. For example, it has been demonstrated that the presence of cations, in concentrations as low as 10^{-8} M, may affect fatty acid films on the water surface.[66] Low concentrations of certain metallic ions have also been shown to impair crystallinity, inhibit epitaxial deposition, and affect the maximum deposition speed of 22-tricosenoic acid.[67] The normally accepted practice is to use water that has been at least double-distilled in glass or quartz equipment.[6,27,68,69] Bücher et al.[70] describe a system for producing quadruple-distilled water; two of the distillation steps utilize potassium permanganate as an oxidizing agent to remove organic molecules.

Some workers use deionized water for the subphase; this can be conveniently produced "on-tap." Deionizing systems, using mixed anion and cation ion-exchange resins, have the capability of producing water which has a specific resistance > 18 MΩcm; this contrasts with the distillation process, by which specific resistances of only 0.5–1 MΩcm can be achieved.[71] Although the deionization process will reduce ionic contaminants to an acceptable level, organic surface active impurities can remain or even be introduced[72]; these can affect the behavior of monolayer films on the water surface. Gaines[73] has shown that isotherms for fatty acid films on deionized water exhibit larger extrapolated areas per molecule, higher compressibility, higher collapse pressure, and slightly more curvature and hysteresis than similar films on a redistilled water subphase. A reduction in organic contamination can be achieved by the use of an activated carbon filter; a further decrease may be made if the deionization system is preceded by a reverse osmosis unit. Whether the levels of such contaminants are reduced below an acceptable level will inevitably depend on the use to which the LB film is put. Criteria for the purity of water include its surface tension, ageing, and surface compression behavior.[219]

Once it has been produced, the handling of the water is also important. Freshly distilled water is sterile, but cannot remain so without very careful storage. On exposure to the atmosphere the water will also absorb carbon dioxide. Both Gaines-[73] and Albrecht[69] have reported significant shifts of isotherms obtained for fatty acid films on "used" as opposed to "fresh" water. If a deionizing and/or a reverse osmosis system is used, then some attention should be given to the distribution of the water to the trough; the use of soft plastic piping containing large quantities of plasticizer can easily reverse the work of a sophisticated (and expensive) water purification system.

Monolayer investigations and multilayer deposition usually require the addition of other materials to the subphase, e.g., to vary the pH or to produce salt formation in the monolayer. Some workers have reported on the addition of alcohol to aqueous subphases to initiate a chemical reaction in the floating monolayer[12] or to improve transfer of rigid monolayers.[74] In all cases care must be taken to ensure that substances added to the subphase are not a source of contamination by using only the highest purity chemicals available.[219]

The requirements of a trough material for monolayer studies or film deposition are relatively straightforward. The material must be inert, in particular it must not release impurities into the subphase; it should preferably be hydrophobic to enable easy cleaning; it should withstand organic solvents as well as inorganic acids; finally, to aid fabrication, it should be easily shaped. Most of the early work was carried out using metal or glass basins coated with paraffin wax to render them hydrophobic. Simple glass containers are favored by many workers; these can be easily fabricated and produced in a variety of sizes. A silanizing treatment may be used to render the glass hydrophobic.[75,76] However, glass containers are never completely inert and will react with the aqueous subphase, and, over a period of time, impurities may be released into the water. The interactions between glass and water are complex and will depend on the exact composition of the glass and the pH of the water; possible effects and techniques for cleaning glass have been discussed, in some detail, in the book by Holland.[77]

The use of certain (plasticizer-free) plastic materials is also popular. Most workers have favored troughs constructed from polytetrafluoroethylene (PTFE),[78–90] although other materials such as nylon,[91] polyethylene,[41,92] and perspex[42] have been reported; both polypropylene[68,93] and polyvinylchloride[94] have been used to fabricate components of Langmuir trough systems.

PTFE is hydrophobic, oleophobic, and resists almost all chemicals. Unfortunately the material does possess pores which may pick-up and slowly release surface active materials. Therefore meticulous cleaning is essential to avoid contamination; hot chromic acid has been used for this.[95] PTFE can be easily machined like other thermoplastics, but it cannot be formed by heating and shaping in the normal way. The difficulty in bonding PTFE to other materials and the fact that it exhibits cold-flow under relatively low loads can make the construction of intricate PTFE components difficult. To overcome this particular problem some workers have used PTFE tape[6,21,88,96] and spray-deposited PTFE to coat their Langmuir troughs and trough components. In the case of the latter, problems can arise from pin holes and from impurities.[69,97] Copolymers of tetrafluoroethylene (e.g., with hexafluoropropylene) can offer similar chemical resistance as PTFE (although with increased wettability[98]), but can be shaped more readily. These may prove useful materials in LB trough systems.[99]

3.3. COMPRESSION MECHANISMS

Traditionally, surface pressure/area isotherms at the liquid–air interface have been made in simple horizontal float balances. These consist of a shallow rectangular container in which a liquid subphase is added until a meniscus appears above the rim. The barrier for manipulation of the spread film then rests across the edge of the container. Numerous modifications of the early film balances have been reviewed in 1966, by Gaines.[6] Since this time, the increase in interest in Langmuir–Blodgett films (as opposed to floating monolayers) has led to greater attention being placed on trough design and better control systems. The three basic types of trough system currently in use for LB film deposition are summarized in the following sections.

3.3.1. Single Movable Barrier

In troughs of this type the container which holds the liquid subphase forms an integral part of boundary of the compression system; therefore a good seal is required between the edges of the trough and the movable barrier. The principle of operation is illustrated in Figure 3.3. The barrier may simply be moved via a suitable gearing system to an electric motor, or may be moved indirectly by magnetic means.[21,68] The technique is very much that first described by Langmuir[2] and developed by Langmuir and Schaefer,[66] using glass or metal basins. Most of the trough need not be very deep (typically a few mm). As shown in Figure 3.4, a "well" can allow for the deposition of monolayers onto relatively large substrates. Commercial versions of the single movable barrier trough are produced by KSV Chemicals (Finland), Lauda (West Germany), and Atemeta (France). Such systems usually have maximum monolayer areas in the range 1000–4000 cm^2 and total subphase volumes of 0.5–5 liters. It should be noted that one disadvantage of Langmuir troughs of this kind concerns the critical position of the subphase men-

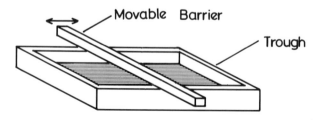

Figure 3.3. Simple Langmuir–Blodgett trough with a single movable barrier.

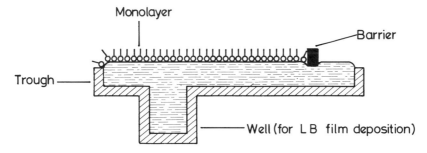

Figure 3.4. Cross section of trough showing a "well" to facilitate LB film deposition.

iscus. If the troughs are to be used over an extended period of time, then some means is required to compensate for evaporation and keep the liquid level constant. A control mechanism, capable of maintaining the subphase level constant to within 30 μm, has been described by den Engelsen et al.[100]: a float shifts the ferrite core of a differential transformer when it is displaced and this in turn opens or closes a magnetic valve in the tube connecting the trough to a liquid reservoir.

An alternative design of the single movable barrier trough may be used which effectively eliminates the problem of maintaining a constant water level. In this case a floating barrier maintains contact with the sides, rather than with the top, of the rectangular trough. Such systems have been described by Sher and Chanley,[101] by Schoeler,[102] and by Walpita.[103] In order to keep frictional forces between the barrier and trough sides to a minimum, while at the same time preventing the organic monolayer from escaping, the barrier design may be more complicated than that shown in Figure 3.3.[102,103]

3.3.2. Circular Trough

The single movable barrier trough may be used to compress a monolayer in a circular as well as in a rectangular geometry. In fact, a surface balance based on a cylindrical arrangement is also possible.[104] A circular trough has the advantage of a simple reliable motor drive for the barriers, where linear troughs require special guidance of the barrier and transmission of the circular motor motion into a linear one (although one drawback may be the gradient in the speed of the monolayer motion). Such a system was reported in 1963 by Sucker.[81] Smith[105] has also described the construction of a circular trough; this was used in the investigation of monolayers at the mercury–gas interface. A commercial trough, based on a circular geometry, is marketed by Nima Technology (UK).

In the multicompartment circular trough described by Fromherz[86] and manufactured by Mayer-Feintechnik (West Germany) there are two independent drive barriers enclosing the monomolecular film, thus enabling the monolayer to be transported as well as compressed on the liquid surface; Figure 3.5 shows a sim-

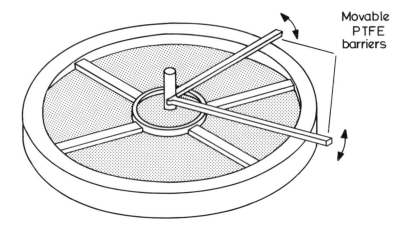

Figure 3.5. Simple version of Fromherz multicompartment circular trough. (After Fromherz.[86])

plified diagram of the system. The compartments (eight in Fromherz's design) are separated by walls which are somewhat lower than the edges of the trough. The body and the two barriers are made from PTFE; both barriers are mobile, either independently or clamped together. The entire trough (diameter 19 cm) is pressed into a base plate of anodized aluminum (diameter 35 cm) by means of two aluminum rings, so that the PTFE plate is able to expand if the temperature changes. The provision of a well in one of the compartments facilitates LB film deposition. However, the multicompartment design also provides the possibility to spread a monolayer on one subphase, shift it to a compartment with a different subphase, and then return it to the original subphase, as shown in Figure 3.6. This procedure may be applied for adsorption studies[106,107] or to study chemical reactions. After each reaction step the product may be characterized *in situ* or the monolayer may be removed from the surface using the Langmuir–Blodgett technique.

3.3.3. Constant Perimeter Barrier

Some of the difficulties encountered with compression systems of conventional design can be overcome by enclosing the monolayer within a continuous flexible plastic barrier. Zilversmit[108] describes a trough using a PTFE band (165 × 3.5 cm and 0.1 mm thick) stretched around four vertical glass spools and a rod which can be rotated to take up the slack as the area is changed. Other reports on the use of PTFE tape to contain the surface active molecules have been made by Clements,[109] by Somasundaran *et al.*,[99] and by Watkins.[83] Mendenhall *et al.*[110] have described a system in which part of the monolayer enclosure is made of a rubber dental dam

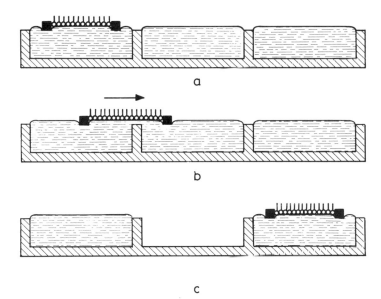

Figure 3.6. Monolayer manipulation using the multicompartment trough. a, b, and c demonstrate how the monolayer may be moved between the surface compartments.

whose expansion and contraction controls the confined area. The use of a flexible polyvinylchloride ring to contain a monolayer has also been discussed.[94]

The commercial version of the constant perimeter barrier type of trough, supplied by the Joyce-Loebl Company (UK) and based on research at ICI plc and the University of Durham, is similar to that described by Blight, Cumper, and Kyte.[93] Figure 3.7 shows a schematic diagram of this system: the compression barrier (a PTFE-coated fiberglass belt) is located by a system of six PTFE rollers and these in turn are secured by two mobile overarms which move symmetrically inward or outward thus maintaining the barrier taut at all times. The maximum and minimum areas that may be produced with this type of barrier are shown in Figure 3.8a. A possible alternative arrangement which only uses one movable overarm is illustrated in Figure 3.8b.[111] A constant perimeter barrier avoids leakage problems and the need for careful water-level adjustment; moreover, being independent of the basin containing the subphase, it can be easily removed, cleaned, and replaced.

A trough which has been designed for use with either a constant perimeter (PTFE) barrier or a solid (PTFE) movable barrier has been described by Tabak and Notter.[87] The equipment allowed these workers to directly compare dynamic isotherms obtained with the two different systems. As expected, problems of surface leakage were encountered with the single movable barrier. However, in the case of the constant perimeter barrier trough, distorted isotherms were obtained for dipalmitoyl lecithin at high surface pressures (> 50 mNm^{-1}). It was suggested that,

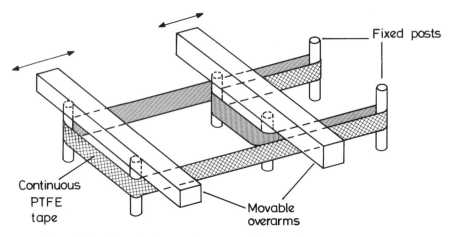

Figure 3.7. Schematic diagram of constant perimeter barrier compression system.

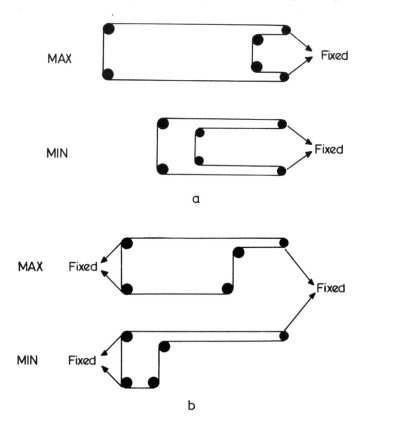

Figure 3.8 Plan showing the maximum and minimum areas obtainable with the constant perimeter barrier trough: a, system with two moving overarms (as illustrated in Figure 3.7); b, alternative arrangement with one moving overarm.

at these pressures, the equilibrium time between the film in the narrow channel and the bulk of the film outside becomes significant. This problem has also been noted by Somasundaran *et al.*[99] Thus, for viscous monolayers, small gradients in the monolayer surface pressure could exist in the constant perimeter system that are not present to the same extent with the single solid barrier trough. Such layers are difficult to deposit as LB films irrespective of the style of trough used to form the floating monolayer.

3.4. ANCILLARY EQUIPMENT

3.4.1. Surface Pressure Measurement

There are two fundamentally different approaches to the measurement of surface pressure: the Langmuir balance and the Wilhelmy plate. The former method is a differential technique with a sensitivity in the 10^{-3} mNm^{-1} range[112]; a clean portion of the subphase surface is separated from the monolayer-covered area by a partition and the force acting on this is measured. The partition usually consists of a movable float connected to a conventional balance with which the magnitude of the force is determined. With Langmuir's original film balance[113] jets of air, directed at the water surface between the float (a length of waxed paper) and the trough sides, were used to contain the monolayer. Figure 3.9 shows the approach that is used in more recent Langmuir balances: in order to avoid leakage a PTFE float is connected by thin PTFE foils to the edges of the trough. Torsion systems are used extensively for the measurement of forces in Langmuir film balances. In the technique illustrated in Figure 3.9, the force due to the film pressure displaces the float until the reaction force of a flat spring is equal. The displacement of the float is

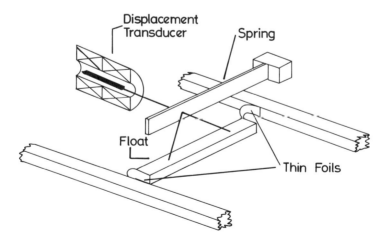

Figure 3.9. Langmuir balance system for surface pressure measurement. (After Albrecht.[69])

usually very small ($\sim$ 10 μm) and introduces an insignificant error in the determination of the area of the monolayer. It is conveniently measured using a displacement transducer, such as a linear variable differential transformer (LVDT).[69,80,114,115] A technique using a commercial electrobalance,[116] and systems involving a light source and a pair of phototransistors[112,117] have also been described.

Modifications to the Langmuir balance, which avoid surface leakage problems, have been reported by Malcolm and Davies,[118] and also by Albrecht and Sackmann.[88] In the former case a comparatively rigid nonfloating barrier, in the form of a thin metal strip, is used. The force exerted by the monolayer on one side of this is measured by observing the deflection of the strip with a microscope. Albrecht and Sackmann use a float consisting of a closed rectangular frame of PTFE-covered steel foil which floats on the subphase surface. The deformation of the frame is measured by four strain gauges fixed to the frame in a symmetric arrangement; the gauges form a bridge circuit which is activated by a lock-in amplifier. The resolution of this system was found to be better than 20 μNm^{-1}. However, one disadvantage of the arrangement is that it does not allow the monolayers to be compressed to zero area since about 15% of the surface is necessary to allow the film to flow around the measuring frame.

In the Wilhelmy method,[119] an absolute measurement is made by determining the force due to surface tension on a plate or other object suspended so that it is partially immersed in the subphase; this may be compared with a similar absolute measurement on a clean surface. The sensitivity limit of this technique is approximately 5×10^{-3} mN m^{-1}.[112] Figure 3.10 shows a diagram of the arrangement. The forces acting on the plate consist of gravity and surface tension downward, and buoyancy due to displaced water upward. For a rectangular plate of dimensions l, w,

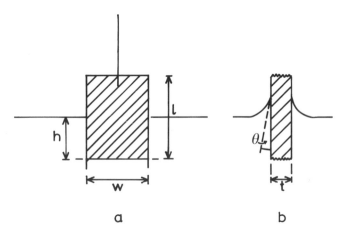

a   b

Figure 3.10.   Diagrams (not to scale) showing forces acting on a Wilhelmy plate: a, front view; b, side view.

and t (as shown in Figure 3.5), and of material of density ρ_P, immersed to a depth h in a liquid of density ρ_L the net downward force, F, is given by

$$F = \rho_p glwt + 2\gamma(t + w)\cos\theta - \rho_L gtwh \qquad (3.3)$$

where γ is the surface tension of the liquid. θ is the contact angle on the solid plate (see Figure 3.10), and g is the gravitational constant. The effects of tilting the Wilhelmy plate and the validity of measurements made with a roughened plate have been discussed by Jordan and Lane.[120]

The usual procedure for the use of a Wilhelmy balance involves maintaining the plate completely wetted by the liquid (although Takenaka[121] has reported the use of a hydrophobic plate) and measuring either the change in F for a stationary plate or the change in h for a constant applied force, when the surface tension is altered. The former is generally favored by most workers; in this case the change in force, ΔF, is related to the change in surface tension, $\Delta\gamma$, by the equation

$$\Delta\gamma = \Delta F/2(t + w) \qquad (3.4)$$

If the plate is thin enough so that $t \ll w$, then

$$\Delta\gamma = \Delta F/2w \qquad (3.5)$$

It should be noted that, in monolayer experiments, it is normal to refer to a measurement of surface pressure. This surface pressure, Π, is generally considered to be equal to the reduction of the pure liquid surface tension by the film, i.e.,

$$\Pi = -\Delta\gamma \qquad (3.6)$$

Various methods have been used for measuring the forces involved with a Wilhelmy plate; most of these have been discussed in some detail by Gaines.[6] One technique that is used extensively is to directly couple the plate to a sensitive electrobalance. Alternatively, the force exerted vertically by the surface tension may be transformed into a minute displacement by means of a spring. This displacement is measured by converting it into a voltage with a differential transformer,[69,122] as shown in Figure 3.11.

There are a number of disadvantages of the Wilhelmy balance. The main problem concerns the contact angle of the liquid on the plate. It is apparent from equation (3.3) that the evaluation of the surface pressure requires that the contact angle be known, and that it does not change during the experiment. It is fairly easy to ensure that the plate is completely wetted at the beginning of the experiment, when the freshly cleaned plate is immersed in the clean liquid surface. However, under increasing surface pressure, monolayers may be deposited onto the plate, changing the value of θ.

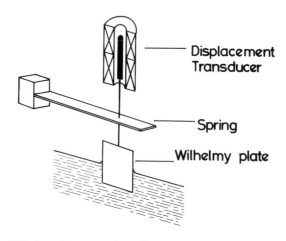

Figure 3.11. Wilhelmy plate system for surface pressure measurement. (After Albrecht.[69])

The wetting problem may be minimized by choosing appropriate materials for the plate: quartz, glass, mica, and platinum have all been used successfully.[6] One of the most effective materials is a clean filter paper[27,122–124]: an investigation of different grades of paper has been reported by Gaines.[124] A fresh plate can easily be used for each monolayer experiment, minimizing contamination problems from the previous monolayer. Contamination from the plate itself can be reduced by keeping the filter paper under the water surface after cutting it from a larger piece, and then cleaning the water surface before taking the plate out.[125]

The exact position of the Wilhelmy plate with respect to the moving barrier(s) in the LB trough may also affect the pressure measurement. For example, Malcolm[126,127] has demonstrated that the presence of the plate can perturb the monolayer flow patterns during compression. This may be particularly important when the monolayer is condensed and does not flow freely (as in the case of some polymers). However, such problems may be minimized by the use of a Wilhelmy plate centrally mounted with symmetrical compression from both ends of the trough.

3.4.2. Surface Area Measurement

The surface area of the monolayer contained within the barriers of any of the three types of compression system described above may be conveniently measured by using a simple potentiometer to monitor the position of the compression barrier (see, for example, Honig *et al.*[27]). Hybrid linear potentiometers (in which the wire windings are coated with a conducting plastic) possess extremely high resolutions and are preferred. A voltage derived from the potentiometer will need calibrating against the trough area. This is relatively straightforward since, in most of the compression systems in use, the barrier position is linearly related to the monolayer area.

3.4.3. Deposition Equipment

The LB technique requires the solid substrate to be raised and lowered through the air/monolayer interface. This can be accomplished using a simple mechanical device, either operated manually or coupled to a low-speed electric motor. In the commercial Joyce–Loebl trough an electric motor simply turns a micrometer via a suitable gearing.[128] Such a system requires careful design and construction in order that vibrations from the motor are not transmitted to the substrate during the deposition process. An alternative arrangement has been suggested by Nathoo[85]: a PTFE piston is pulled slowly upward in a PTFE cylinder by the suction from a pump, and then falls downward under gravity when the suction is reduced to an appropriate value. A similar system, based on a conventional glass syringe and using water rather than air as the working fluid, has been used by Agarwal et al.[129] The commercial trough produced by Atemeta (and based on research at the Centre d'Etudes Nucleaires de Saclay, France) utilizes an unusual dipping mechanism; this is constructed from a driving arm and a slave arm rotating on a common axis. The driving arm is controlled by a motor–sensor system, and the slave arm is guided by the driving arm but remains free for large manual displacements.

The method normally adopted to monitor the deposition process of an LB film (Y-type transfer) is shown in Figure 3.12a (see, for example, Honig et al.[27]); the film area is simply recorded as a function of time. However, this does suffer from the disadvantage that variations in the deposition ratio can be obscured by changes in the speed of dipping. Figure 3.12b shows a technique which overcomes this difficulty; in this case the film area is plotted as a function of dipping head (i.e., substrate) position, rather than time (a voltage corresponding to the position of the substrate may be derived from a linear potentiometer, in a similar way to that by

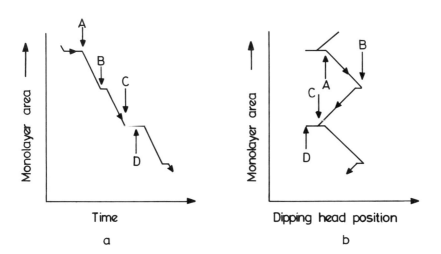

Figure 3.12. Techniques for measuring monolayer deposition: a, monitoring monolayer area versus time; b, monitoring monolayer area versus substrate (dipping head) position.

which the monolayer area is monitored). In both diagrams the transfer of the film to the substrate on the downward motion of the dipping head starts at point A. At B the direction of the substrate motion changes; the monolayer area temporarily remains constant while the meniscus changes from that shown in Figure 3.1c to that shown in Figure 3.1b. The deposition on the upward motion of the substrate terminates at C. At D the direction of the substrate motion again changes.

3.4.4. Control Systems

For the deposition of LB layers onto a solid support it is important to be able not only to measure the surface pressure of the monolayer, but also to control it. In this way a constant position on the isotherm can be maintained while material is being removed from the surface of the subphase. Control of the surface pressure may be accomplished using a simple feedback control loop, as shown in Figure 3.13; a Wilhelmy plate is the pressure sensor in this case. A voltage, derived either from an electrobalance or from a displacement transducer (e.g., an LVDT), is compared with a voltage corresponding to the desired surface pressure. The differential signal activates a motor which drives the compression barrier to either decrease or increase the monolayer area. An important consideration in this simple control system is the role of the monolayer itself; this forms an integral part of the control feedback loop. Thus the response speed and bandwidth of the control system will depend critically upon the properties of this layer. Using the same control unit it is possible to obtain excellent control action with one type of monolayer material, but to set the system into oscillations with another. To compensate for this it is therefore essential to introduce some additional (and adjustable) damping into the

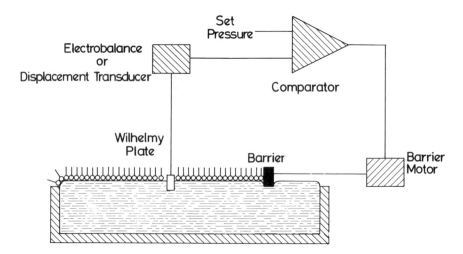

Figure 3.13.   Feedback loop to control monolayer surface pressure.

feedback loop. Commercial control systems for Langmuir troughs are capable of keeping the surface pressure constant to within approximately 0.1 mNm^{-1}.

A block diagram showing the experimental variables that are usually controlled in an LB trough system is given in Figure 3.14. The trough itself must be serviced so that it can readily be filled and emptied, and the surface of the subphase cleaned. Careful control of pH is required for the deposition of many types of monolayer; it is therefore common for a pH electrode to be permanently installed in the trough (utmost care is needed to ensure that the electrode does not become accidently contaminated with LB layers—this may seriously affect its performance). For a fixed surface pressure, the thermodynamic state of a particular monolayer material can be changed by a variation in the temperature.[6] This important fact is often ignored by workers.[219] Temperature control can be achieved by placing the entire deposition system in a constant-temperature enclosure, mounting the trough on a thermostated base, or by surrounding the subphase with circulating water. In all temperature-control arrangements the presence of convection currents in the subphase should be minimized. Finally, in a complete trough installation, some control of deposition parameters, such as the deposition speed, the distance over which the

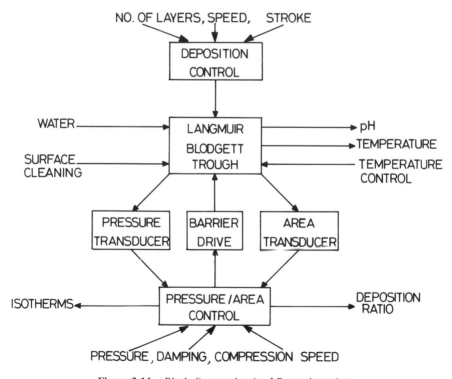

Figure 3.14. Block diagram showing LB trough services.

solid substrate is lowered into the subphase (stroke), and the number of layers deposited, is desirable.

As expected, more recent Langmuir troughs, including many of the commercial versions, have included microcomputer control.[69,90,130–132] This can be a significant advantage, as large amounts of data can be collected and manipulated. For example, the analysis of a family of isotherms and the display of derivatives of surface pressure–area curves can help avoid a great deal of tedious manual work. Moreover, computer-based systems can easily be adapted from one experiment to another by simply changing the software.

3.5. TROUGH ENVIRONMENT

There are a number of factors which may disturb the regular array of molecules in a compact floating monolayer. Any of these can result in a deterioration in the quality of a deposited LB film. Probably the most important parameters, other than temperature which has already been discussed, are vibration, airborne contamination, and ambient lighting.

The monomolecular layer on the subphase will be disturbed by ripples on the surface; the elimination of these is desirable if the monolayer is to be transferred to a solid substrate.[133] The degree of isolation required will obviously depend on the location of the trough. In many instances siting the trough on a rigid support and/or the use of solid rubber antivibration mounts may be adequate.[85] Pneumatic isolation systems, which are commonly used to support optical benches, have also found general application in LB facilities. The simplest variation is a rigid table top mounted on one or more inflated tire inner tubes.[43,134] Commercial antivibration tables based upon similar principles normally have four damped pneumatic springs supporting a rigid table top and are usually stated to possess resonant frequencies in the range 0.5 to 3 Hz. One disadvantage of using such tables to support a Langmuir trough is that accidental bumping of the top can set up large-amplitude vibrations in which the top rotates about an axis in the horizontal plane. Very large surface waves can be set up in the trough by this motion. Severn and Batchelder[135] have undertaken a comparison of different vibration isolation systems for use with a Langmuir trough. In fact, the best system investigated was found to be the least expensive: this consisted of a rigid table top supported by the polyethylene air bubbles of a commercial packing material (although under a heavy load, air inevitably leaks out of the bubbles and the material has to be replaced).

Air-bound contamination is difficult to avoid completely. Precautions such as simple covers or enclosures for the trough are desirable. A substantial reduction in the number of airborne particles in the vicinity of the subphase may be made by locating the trough in a microelectronics-type clean room.[30,41,136,137] However, such facilities are expensive and may be unnecessary for some experiments. The possibility of contamination originating from the trough operator should not be neglected (an average person simply standing or sitting still can "emit" 10^5 particles

> 0.3 μm in diameter per minute[71]!). Therefore it would seem sensible for operators to take precautions such as the use of coats, caps, gloves (made from synthetic fabrics such as polyester or nylon), and even face masks when working over the subphase (alternatively, techniques such as subphase surface cleaning and monolayer spreading could be fully automated). For certain monolayer materials it may also be desirable to site the trough in a container which can be flooded with an inert gas[89]; however, the flexibility of the experimental arrangement will be greatly reduced.

The control of ambient light may be necessary for particular monolayer materials.[138] For instance, amphiphilic diacetylenes will polymerize in the monolayer form if irradiated with ultraviolet light.[139] Although a relatively intense source is required for full polymerization, the use of simple filters to prevent UV radiation falling on the monolayer is a wise precaution. Alternatively, the installation of a UV lamp above the subphase surface will permit the controlled polymerization of the floating monolayer.[12,103,140]

3.6. EXPERIMENTAL TECHNIQUES

Meticulous attention to experimental detail is required for all monolayer and LB film work. A frequent and thorough cleaning procedure for all the equipment, particularly before a new monolayer material is to be used, is recommended for success. Furthermore, instruments such as pH meters and microbalances must be periodically calibrated according to the manufacturers instructions. Once the cleaning and calibration have been carried out, it is advisable always to check the reproducibility of an isotherm for a known monolayer material, e.g., a fatty acid. It should also be noted that an LB trough system (filled with water) presents a relatively hostile environment to the precision mechanical arrangements of the barrier and dipping mechanisms. Thus periodic attention should be paid to these to ensure trouble-free operation.

In his book, Gaines[6] recommends workers wishing to prepare LB layers to read, carefully, the original papers of Langmuir and Blodgett[3,141]; this is still excellent advice.

3.6.1. Surface Cleaning

Adequate cleaning of the surface of the subphase prior to monolayer spreading is an essential part of obtaining high-quality LB layers. With the single barrier troughs, this can be readily accomplished by sweeping the compression barrier (or a second barrier) over the surface; this was the technique used by Blodgett in her studies of film deposition.[3] The usual approach to cleaning any trough system is to first reduce the area to a minimum value and then to clean the surface of this area with a glass capillary tube (or a series of tubes) attached to a suitable pump. The capillary tube may be used by hand, in which case precautions should be taken to prevent contamination falling onto the surface, or it can be conveniently fixed in position at a small distance from the subphase surface. In the latter case it is important to ensure that the air gap between the capillary tube and the subphase surface is so adjusted that the water surface sucked up during the cleaning all leaves the capillary when the pump is turned off.[68] In the Atemeta trough the surface cleaning is accomplished with the aid of two arrays of mobile capillary tubes. Surface cleaning may be aided by spreading and removing a monolayer of an appropriate material,[142] or simply by "washing" the subphase surface with a solvent.[143] The success of the cleaning can be determined by monitoring the surface pressure as the area is reduced. The operation may well have to be repeated a number of times before this is below an acceptable value; the level of the subphase may also have to be adjusted.

3.6.2. Monolayer Spreading

Most monolayer-forming materials are applied to the subphase by first dissolving them in a solvent. The necessary properties for such a solvent have been discussed by Gaines[6]: it must be able to dissolve an adequate quantity of the monolayer material (concentrations of $0.1-1$ mg ml^{-1} are typical); it must not react chemically with the material or dissolve in the subphase; finally, the solvent must evaporate within a reasonable period of time so that no trace remains in the condensed monolayer. A solvent which evaporates too quickly may be a problem, as this could prevent the accurate determination of the solution concentrations. Although there have been a number of reports on the effects of spreading solvents on monolayer properties, some of these were probably the result of contamination, possibly originating from the interaction of the solvent and the paraffin wax coating of the trough that were used.[6] However, a detailed investigation by Mingins et al.[144] has revealed that, even when precautions were taken to minimize such contamination, different solvents can still have an influence on the pressure–area isotherms for particular monolayer materials.

Solvents which are commonly used for monolayer spreading include n-hexane (boiling point 69 °C), benzene (bp 80 °C), chloroform (bp 61 °C), and ethyl ether (bp 35 °C). For the spreading of materials which are not soluble in such nonpolar solvents, special techniques have to be devised. One approach is to use mixed solvents, which give sufficient solubility for the monolayer material but do not introduce serious water-solubility problems[6]; examples include hexane–ethanol, benzene–ethanol, and chloroform–methanol.[96,145–147] Chloroform–alcohol–water mixtures have been used for the spreading of biological materials.[107,148] (We note that the toxicity of the spreading solvents should not be ignored by the experimenter during monolayer and multilayer work, and a suitable extraction system should be included in the vicinity of the trough.)

Some solvents possessing much higher boiling points than those listed above have also been used. Möbius[149] reports on the use of hexadecane to help in obtaining a homogeneous distribution of dye molecules in an inert matrix. The hexadecane is thought to act as a sort of "molecular lubricant," which is slowly squeezed out under the applied surface pressure. Steele et al.[150] have reported that the use of hexadecane in addition to chloroform as a spreading solvent markedly affects the morphology of diacetylene LB films. Baker et al.[48] have discussed the use of mixed spreading solvents containing mesitylene (1,3,5-trimethyl benzene) to obtain reproducible isotherms of phthalocyanine, while Hann et al.[151] have used xylene to obtain monolayers of similar materials. The effect of different spreading solvents on monolayer formation of phthalocyanine materials has been discussed by Roberts et al.[152] Although solvents such as decane, hexadecane, and pentadecane have been shown to evaporate completely from some monolayer materials,[153,154] this might not be true for all such solvent/monolayer combinations. The possibility therefore exists that some of the spreading solvent may remain in the monolayer after deposition onto a substrate.

The application of the spreading solution to the subphase is generally accomplished by allowing drops to fall from a suitable delivery pipette (e.g., a micro-syringe) held a few millimeters away. The usual technique is to apply these drops to the center of the subphase surface contained within the barrier, allowing each drop to spread out and evaporate before the next is applied. In this way contamination is spread toward the extremities of the monolayer and away from the "dipping" area. However, under certain circumstances it may be preferable to distribute the drops over the surface to be covered; this would seem to be particularly appropriate in the case of the constant perimeter barrier troughs where, in order to obtain accurate isotherms, it is essential to get the monolayer material into the narrow channels.[155] The effect of varying the concentration of the spreading solution and also of varying the pressure of the monolayer immediately after spreading has been investigated by LaMer and Barnes.[156] It was found that, provided sufficient precautions were taken to prevent contamination of the monolayer, the spreading technique was not critical. Special techniques have been developed for spreading proteins on the surface of a subphase.[6,148,157]

Automated spreading is a feature of some computer-controlled LB troughs.[90] This is particularly useful if a large number of layers is required and the floating monolayer needs to be frequently replenished.

3.6.3. Substrate Preparation

Condensed monolayers can be transferred to a variety of solids. Some of these substrates are used simply as an inert support for the LB film; others are an integral part of the system under study (e.g., metal electrodes or semiconductors for conductivity measurements). For the production of multilayer systems the adhesion of the first layer to the underlying solid is particularly critical, and this step should be carefully controlled. There are many parameters associated with the substrate surface which may influence deposition. Its exact chemical composition, particularly in respect to native oxide layers, may affect ion exchange in the first layer deposited. This has been found to be the case for fatty acid materials deposited onto certain substrates.[15] Thus when a calcium stearate film is transferred onto an aluminum plate, a layer of aluminum stearate is obtained; a similar ion exchange occurs with a tin substrate.[158] On noble metal substrates, which hardly oxidize, no such chemical reaction occurs[15] and fatty acids adhere poorly.[159] Poor adhesion may result in recrystallization of the monolayer.[160]

The physical structure of the substrate surface may also be important in determining the quality of the deposited layer. By measuring deposition ratios for fatty acid films deposited onto surfaces that had been deliberately roughened, Bikerman[161] concluded that, at the moment of deposition, the monolayer will bridge over voids; monolayers could also be transferred to a fine wire gauze. However, such films were probably supported by a water layer which subsequently drained or dried, resulting in the film collapse.[6,161] More recently, Albrecht et al.[162] have shown that LB films of a diacetylene polymer would, in fact, bridge small (0.2×0.05 μm) holes in polypropylene supports if the support surface were smooth enough.

In spectroscopic studies, glass and quartz slides are commonly used as supports for monolayer systems. The surface of a glass slide can be cleaned by flaming or using a chemical treatment. In the latter case this can be achieved by acid treatment, ultrasonic methods, vapor degreasing, and washings.[77] Two techniques, which have proved equally satisfactory for the deposition of long chain fatty acids from a Cd^{2+}-containing subphase have been described by Möbius.[68] In the first, the silica or quartz slides are treated in hot chromic acid, rinsed in water, soaked in dilute sodium hydroxide, and finally rinsed again. The other technique involves putting the slides into an ultrasonic cleaning tank (frequency 50 kHz) containing an aqueous alkaline cleaning solution at 85 °C.

Surface preparation techniques for several solid metal plates (Cr, Pt, Cu, and Al) have been given by Gaines.[163] The effect of different surface treatments and of crystallographic orientation for single-crystal silver substrates has been discussed, in some detail, by Spink.[160] Vacuum deposited metal layers have also been used as solid supports for monolayer systems. An oxide layer is invariably present on evaporated layers of metals like Al and Pb; these surfaces tend to be hydrophilic and, as discussed above, can react chemically with head groups of particular monolayers. Noble metals, such as Au, tend to be hydrophobic. However, Smith[164] has suggested that this is not an intrinsic property of the metal surface, but is simply due to contamination. Au evaporated in an ultrahigh vacuum,[164] or in a high vacuum system that has been previously baked before use,[165] is shown to be water wettable. However, these surfaces become quickly hydrophobic when exposed to air in the laboratory.

There has been much work reported for LB film/semiconductor systems. The surfaces of semiconductors are extremely complicated. With the same semiconductor, surfaces with very different physical[166] and/or chemical[167] properties may be obtained. It is likely that a number of different surface treatments may be used to enable LB films to be deposited, in a reproducible fashion, onto a particular semiconductor surface. Whether the treatments lead to fundamentally different experimental results will very much depend on the type of investigation that the LB film/semiconductor system is to be used in. For instance, Sykes *et al.*[168] report that aqueous bromine or bromine/methanol solutions produce a relatively thick oxide film on the surface of InP, while etchants such as hydrofluoric acid and inorganic acid/hydrogen peroxide mixtures result in an almost oxide-free substrate. Differences in the surfaces produced by these two types of etchant were manifest in the electrical properties of metal/LB film/semiconductor structures fabricated on them. Apart from surface treatment it is also possible that the crystallographic orientation of single-crystal semiconductor substrates, the doping material, and the doping type (n or p type) may influence LB film deposition. Such effects have yet to be fully investigated.

Table 3.1 shows a range of semiconductor materials that have been used in LB film investigations and lists surface treatments which have been found to result in good LB film deposition. Of particular note is the use of dichlorodimethylsilane or

Table 3.1. Semiconductor Surface Preparation for LB Film Deposition

Semiconductor	Surface treatment	LB film material	References
Si single crystal	HF followed by refluxing in CCl_3CH_3 with small addition of $Si(CH_3)_2Cl_2$ *or* 5 min in $[(CH_3)_3 Si]_2$ NH	ω-Tricosenoic acid α-Octadecylacrylic acid	169—171
α-Si:H thin film	2.5 min in 40% HF:40% NH_4F (1 : 5 ratio by volume) follwed by 10 min in chlorine water	Cadmium stearate	172
GaAs epitaxial layer	1 min in Br_2/methanol (1 : 2000 by volume), 5 min in conc HCl followed by methanol rinse	ω-Tricosenoic acid	173
GaP single crystal	3 min in $H_2SO_4 : H_2O_2 : H_2O$ (4 : 1 : 1 by volume) followed by 1 min $H_2O_2 : H_2O$ (1 : 20 by volume) containing 2 g NaOH per 100 ml of solution	Preformed polymers Cadmium stearate Chlorophyll	174–176
GaP epitaxial film	30 s in 3 g $K_3Fe(CN)_6$, 0.24 g KOH and 10 cm^3 H_2O at 80 °C	Cadmium stearate ω-Tricosenoic acid Phthalocyanine	177
InP single crystal	$H_2SO_4 : H_2O_2 : H_2O$ (4 : 1 : 1 by volume) followed by *either* a 20% solution of Br_2/HBr/H_2O (1 : 17 : 35 by volume) *or* HCl : H_2O_2 : H_2O (1 : 1 : 10 by volume)	Cadmium stearate	167
InSb, CdTe, CdS, ZnS (single crystals); ZnSe (epitaxial films)	Bromine in methanol (0.5–1% by volume)	Cadmium stearate Substituted anthracene derivative (for CdTe)	178–182

hexamethyldisilazane to prepare the silicon surface.[169–171] These silanization treatments result in a very hydrophobic surface for the semiconductor; long-chain fatty acid type layers are therefore transferred to this surface on the first downstroke of the substrate through the monolayer (in contrast with the scheme illustrated in Figure 3.1). The silanization technique has also been used to prepare glass substrates for LB film deposition.[27,68,70]

3.6.4. Monolayer Transfer

One of the most important experimental variables during the film transfer process is the speed at which the substrate is moved through the monolayer/air interface. The floating monolayer may be transferred to the substrate quite rapidly as it is lowered into the subphase. However, on withdrawal through the monolayer it is important not to raise the substrate faster than the rate at which water drains from the solid. This drainage is not due to gravity but is a result of the adhesion between the monolayer being transferred and the material on the substrate which acts along the line of contact and so drives out the water film.[183] The rate at which films can be built up is thus limited by the rate at which the ascending substrate sheds water, and the "speed" refers to this limiting rate and not to the actual rate of building.[40,141] Langmuir[40] used the term "zipper angle" to refer to the angle formed by the water meniscus against the solid plate as it is withdrawn from the subphase; if the plate comes out wet, the zipper angle is zero, but a large zipper angle (50°–60°) is observed when the monolayer becomes tightly bound to the solid, expelling the water layer rapidly. A "fast" film is one which can be built in successive layers at a rate of 20 or more layers per minute on a glass substrate 2.5 cm long.[3,141]

It is normal to transfer the initial monolayer onto a prepared substrate relatively slowly; speeds of 10 μms^{-1} to mms^{-1} are typical.[184,185] However, much faster speeds are possible once the initial layer has been transferred. Peterson et $al.$[184] have described an investigation of the drainage speed of 22-tricosenoic acid; this was shown to be independent of the spreading solvent used, but exhibited a marked dependence on the time that elapsed between the spreading and dipping of the monolayer. Relatively fast dipping speeds ($\sim$ 1 cms^{-1}) were possible for monolayers that were deposited within about one hour of spreading, however much slower speeds were required for "aged" layers. It was suggested that the drainage speed was dependent upon the crystallite size in the floating monolayer, this factor increasing with the time that the monolayer remains on the subphase surface. The best ordered films were in fact produced by depositing the first layer from an "aged" monolayer and subsequent layers from freshly spread material. Evidence for the epitaxial deposition of such layers has been discussed by Peterson et $al.$[185,186] Such experiments correlate with the observations of Blodgett and Langmuir[141] who noted that . . . "Speed is commonly the greatest for the first layer which are deposited, but gradually slackens as more layers are added." This effect was dependent on the monolayer material: films of magnesium stearate were found to lose "speed" after about 50 layers, while films of barium–copper–stearate retained their "speed" for up to more than 1000 layers. The loss of speed was thought to be due to disorder in the molecular orientation of successive layers. Other studies by Peterson and coworkers[19,67] have suggested that the drainage speed of the 22-tricosenoic acid is also related to the presence of metal ions in the subphase; ions such as sodium and cadmium were shown to reduce the initial maximum drainage speed and to cause a variation of maximum drainage speed with time. Buhaenko *et al.*[187] have discussed the influence of monolayer viscosity on the LB film deposition speed and presented data showing a correlation between the maximum drainage speed and the monolayer viscosity. The surface viscosity must be below a certain value for successful deposition; too high a viscosity is symptomatic of a brittle monolayer, which is merely broken on insertion or withdrawal of the substrate.

The drainage speed (also referred to as the three-phase contact line motion) for a number of different monolayer materials has been reported by Petrov *et al.*[188,189] The maximum deposition rate was found to be determined by the reactivity between the monolayer head groups and the surface of the solid substrate. Fast deposition rates were found in the case of oppositely charged head groups in the monolayer and on the substrate surface (achieved by alternate deposition of a fatty acid and a fatty amine). The addition of divalent ions (e.g., Cd^{2+}) to the subphase was also found to increase the maximum deposition speed of arachidic acid. However, this effect was shown not to be associated with the decrease in thickness of the diffuse double layer at both interfaces. At a pH value of 2, relatively fast deposition rates were found for stearic acid, in the absence of divalent ions in the subphase. This was attributed to the formation of hydrogen bonds between the undissociated carboxyl groups at the surface of the solid substrate and the transferred monolayer, resulting in dimerization of the carboxylic acid molecules.

The influence of surface flow patterns on the morphology of LB films has been studied by Daniel and Hart,[190] and by Malcolm.[220] Changes in the flow pattern near the substrate were shown to have a marked effect on film morphology, as observed optically. In particular, convergent flow produced shearing forces which tended to both elongate and break up domains in the spread monolayer prior to deposition. Using a commercial constant perimeter barrier trough, Daniel and Hart showed that improved film morphology was obtained with the substrate perpendicular to the advancing barriers. Alternatively, Malcolm has proposed a new belt profile which can be used to deposit the monolayer with minimal distortion. The use of a diamond-shaped constant-perimeter barrier[130] is also claimed to produce acceptable monolayer flow patterns.

Electric fields have been shown to affect the motion of domains in lipid monolayers.[221] They might also be expected to influence the transfer of monomolecular layers to substrates.[191] In this respect Grunfeld and Pitt[192] have obtained encouraging results on the use of weak electric fields ($\sim 10^4$ Vm^{-1}) in the control of domain size in diacetylene polymer films. Magnetic fields could also influence the deposition of certain materials.

Once deposited, it is possible to modify the properties of monolayer and multilayer films in a number of ways.[6] Long-chain fatty acid type materials are invariably deposited as a mixture of the fatty acid and a salt. The exact composition of the deposited layer depends on the type of ions in the subphase and the pH of the latter.[41,193,194] The free acid may be removed from the film by soaking the LB layers in an appropriate solvent. This "skeletonization" process can produce marked changes in the optical properties of the LB film.[141] A hydrophobic soap multilayer can be rendered water-wettable with dilute aqueous solutions containing certain polyvalent cations[195]; this treatment is thought to involve the overturning of molecules in the LB film. Materials which have been deposited as monolayers may be polymerized by exposure to radiation of an appropriate wavelength.[12,103,139,147,196−201,222] Finally, care should be taken not to expose monolayer or multilayer films to a high vacuum for long periods, as this might result in desorption from the substrate[159]; LB films are probably best stored in a desiccator in an inert gas.

3.7. OTHER SYSTEMS

3.7.1. Continuous Fabrication

Most of the troughs that have been described in the literature (including those that are available commercially) are specifically designed for small-scale production of LB films: the number of layers that may be deposited from one monolayer depends on the areas of the subphase and substrate. When the amount of prepared film has been used up, the deposition has to be interrupted and a fresh monolayer spread. One possible way of achieving continuous deposition of LB layers has been described by Barraud and Vandevyver.[92] Figure 3.15 shows a schematic diagram of this type of trough. In essence it consists of three compartments, separated by rotating rollers which are partially immersed in the subphase. The operation depends on the ability of the rollers to transfer the amphiphilic molecules from one of the trough compartments to the next. The Langmuir sequence which is distributed in time (i.e., spreading the solution, waiting for the solvent to evaporate, compressing the film, and finally transferring the compressed monolayer to a solid plate) is replaced by a sequence that is distributed in space; all of the elementary operations take place simultaneously in the different compartments.

Two distinct working conditions can be obtained, depending on the roller material. In the first, the "hydrophilic" mode of operation, the molecules can be picked up by the roller at virtually zero pressure and compressed to a few mNm^{-1}; this mode is obtained with rollers made of most common metals (e.g., stainless steel, aluminum) or glass. The transfer rate is independent of the roller speed and varies only with the surface concentration of uncompressed molecules.

The other mode of operation is the so-called "hydrophobic" mode. In this mode, which is obtained with materials such as PTFE, molecules which have been precompressed to a few mNm^{-1} may be transferred to the other side of the roller and compressed to very high surface pressures; the molecular transfer is shown as

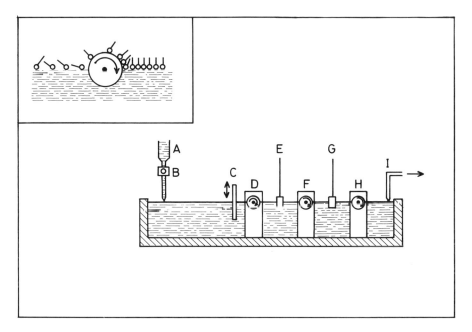

Figure 3.15. Trough for continuous deposition of monolayers (after Barraud and Vandevyver[(92)]). The inset shows a roller working in the "hydrophobic" mode to compress the monolayer.

the inset to Figure 3.15. The transfer rate is directly proportional to the rotation speed of the roller. Since the roller is hydrophobic, but is nevertheless covered with a thin water layer, its speed cannot be reduced below a minimum value. This is one disadvantage of the system since transfer cannot be completely stopped.

The continuous deposition trough works in the following manner. The first compartment (the farthest to the left in Figure 3.15) is the largest and is devoted to spreading and drying. The solution drips via a microelectrovalve (B) from a reservoir (A). Roller D, which separates the first two compartments, works in the hydrophilic mode. Close to roller D an aperture of variable width (C) is inserted between the first compartment and the roller; this controls the flow of molecules from the first compartment to the second. The second compartment therefore contains molecules compressed at a surface pressure of a few mNm^{-1}. Roller F, which separates the second compartment from the third, operates in the hydrophobic mode. Its speed is adjusted according to the amount of film removed from the third compartment, in which the compressed film is used. The surface pressures in the second and third compartments can be monitored via the Wilhelmy plates E and G. When material is not removed from the third compartment, some excess film accumulates there because of the residual speed of roller F. If this is the case, the surface pressure increases and an auxiliary hydrophobic roller is actuated to take the excess film to a fourth compartment, where it is removed via a capillary tube connected to a pump (I). The overall system is therefore a two-stage compression system in which molecules are injected according to the amount of film consumed. Although the prototype is still of laboratory size (total trough dimensions 80 cm × 15 cm, a few cm deep), it shows the practicability of the method for possible industrial requirements: continuous film production, fast compression, high efficiency, and large film areas.

3.7.2. Alternate-Layer Troughs

A number of the possible applications for LB films rely on the fabrication of supermolecular structures (see Chapter 7); in particular, alternate-layer structures of two different types of molecules are of great interest. The construction of such a molecular array would be very time consuming using a conventional Langmuir trough, as layers of the two different materials would have to be alternately spread onto, and removed from, the subphase surface.[202] A number of possible techniques to simplify the deposition of such alternate-layer structures have now been documented.

In 1964 Puterman et al.[43] described a double trough (fabricated from PTFE) for producing alternate-layer structures. A double-gated canal made possible the transfer of the substrate under water. A simpler approach is described by Daniel et al.[203]; the LB trough has two compartments where different monolayers may be kept separate but share a common liquid subphase. The fixed barrier separating the compartments incorporates a leak-proof gate, which permits the easy horizontal transfer of the immersed substance between the compartments. A variation on this has been described by Kato[223]; a double-gate arrangement is used in this instance. A single-gate alternate-layer system has been discussed by Barraud and Leloup in a 1983 patent.[204] Furthermore, a number of other ways of constructing alternate-layer LB films were also described; one of these has been realized as a fully working system.[205]

Other methods for the deposition of alternate-layer structures are based on the original suggestion of Bikerman and Schulman[206] and the observations of Bikerman[207] and Ries[208,209]: a small cylinder, when rotated in a stearic acid layer, held at constant pressure, was found to result in the transfer of the fatty acid from the surface of the subphase to the roller. In the technique reported by Girling and Milverton,[210] the roller (made from PTFE) is placed across a Langmuir trough so that it divides the surface into two separate areas. A different monolayer is spread over each surface area and then compressed to an appropriate surface pressure with a rigid barrier. Rotation of the roller about its long axis results in a continuous transfer of both monolayer materials, at constant surface pressure, from the surface to the roller. Fluorescence quenching was used to demonstrate that this system could be used to produce alternating layers of dye molecules.[210]

An alternate-layer trough based on a constant perimeter barrier has also been produced[211,212]; a schematic diagram is shown in Figure 3.16 (illustrated with the barrier inverted). A single flexible continuous belt is maintained under tension about rollers or guide elements whose positions can be adjusted to effect controlled planar

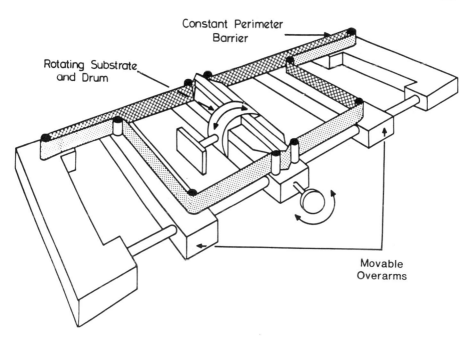

Figure 3.16. Schematic diagram for constant perimeter barrier alternate-layer trough (after Holcroft *et al.*[212]). The barrier system is inverted for clarity.

compression of two separate monomolecular films. The two surface compartments are therefore of the type shown in Figure 3.7b. The compression is achieved independently for the two areas using separate reversible motors linked to pulleys and endless belts. The fixed central barrier is divided to accomodate a cylinder which can oscillate or rotate about a central axis, and thus a suitably clamped substrate can be moved through the floating monolayers. The fact that the substrate enters and leaves the subphase surface at an angle less than 90° does not appear to affect the quality of simple fatty acid layers.[212] Alternate-layer troughs present the possibility of constructing organic "superlattices"; however, the full potential of such equipment has yet to be explored.

ACKNOWLEDGMENTS. We would like to thank O. Albrecht, A. Barraud, B. Blackburn, I. R. Girling, G. Merrington, I. R. Peterson, G. G. Roberts, and A. J. Vickers for useful discussions, and help in the preparation of this chapter.

REFERENCES

1. I. Langmuir, The mechanism of the surface phenomena of flotation, *Trans. Faraday Soc.*, **15,** 62–74 (1920).
2. K. B. Blodgett, Monomolecular films of fatty acids on glass, *J. Am. Chem. Soc.*, **56,** 495 (1934).

3. K. B. Blodgett, Films built by depositing successive monomolecular layers on a solid surface, *J. Am. Chem. Soc.*, **57**, 1007–1022 (1935).

4. J. H. Schulman, R. B. Waterhouse, and J. A. Spink, Adhesion of amphipathic molecules to solid surfaces, *Kolloid Z.*, **146**, 77–95 (1956).

5. I. Langmuir and V. J. Schaefer, Activities of urease and pepsin monolayers, *J. Am. Chem. Soc.*, **60**, 1351–1360 (1938).

6. G. L. Gaines, Jr., *Insoluble Monolayers at Liquid–Gas Interfaces*, Wiley–Interscience, New York (1966).

7. K. Fukuda, H. Nakahara, and T. Kato, Monolayers and multilayers of anthraquinone derivatives containing long alkyl chains, *J. Colloid Interface Sci.*, **54**, 430–438 (1976).

8. N. Nakahara and K. Fukuda, Studies on molecular orientation in multilayers of long-chain anthraquinone derivatives by polarized infrared spectra, *J. Colloid Interface Sci.*, **69**, 24–33 (1979).

9. H. Nakahara and K. Fukuda, Orientation of chromophores in monolayers and multilayers of azobenzene derivatives with long alkyl chains, *J. Colloid Interface Sci.*, **93**, 530–539 (1983).

10. T. Kawaguchi, H. Nakahara, and K. Fukuda, Monomolecular and multimolecular films of cellulose esters with various alkyl chains, *Thin Solid Films*, **133**, 29–38 (1985).

11. D. Day and J. B. Lando, Morphology of crystalline diacetylene monolayers polymerized at the gas–water interface, *Macromolecules*, **13**, 1478–1483 (1980).

12. G. Fariss, J. B. Lando, and S. E. Rickert, Phase controlled surface reaction—Reaction of a monolayer at the gas–water interface, *J. Mater. Sci.*, **18**, 3323–3330 (1983).

13. C. N. Kossi and R. M. Leblanc, Rhodopsin in a new model bilayer membrane, *J. Colloid Interface Sci.*, **80**, 426–436 (1981).

14. G. L. Gaines, Jr., From monolayer to multilayer: Some unanswered questions, *Thin Solid Films*, **68**, 1–5 (1980).

15. A. Barraud, C. Rosilio, and A. Ruaudel-Teixier, Reactivity of organic molecules in monolayers, *Thin Solid Films*, **68**, 7–12 (1980).

16. D. Naegele, J. B. Lando, and H. Ringsdorf, Polymerization of cadmium octadecylfumarate in multilayers, *Macromolecules*, **10**, 1339–1344 (1977).

17. M. Lösche, J. Rabe, A. Fischer, B. U. Rucha, W. Knoll, and H. Möhwald, Microscopically observed preparation of Langmuir–Blodgett films, *Thin Solid Films*, **117**, 269–280 (1984).

18. M. Lösche, C. Helm, H. D. Mattes, and H. Möhwald, Formation of Langmuir–Blodgett films via electrostatic control of the lipid/water interface, *Thin Solid Films*, **133**, 51–64 (1985).

19. I. R. Peterson and G. J. Russell, Deposition mechanisms in Langmuir–Blodgett films, *Br. Polym. J.*, **17**, 364–367 (1985).

20. I. R. Peterson and G. J. Russell, The deposition and structure of LB films of long chain acids, *Thin Solid Films*, **134**, 143–152 (1985).

21. J. H. Clint and T. Walker, Interaction energies between layers of alkyl and partially fluorinated alkyl chains in Langmuir–Blodgett multilayers, *J. Colloid Interface Sci.*, **47**, 172–185 (1974).

22. J. F. Stephens, Mechanisms of formation of multilayers by the Langmuir–Blodgett technique, *J. Colloid Interface Sci.*, **38**, 557–566 (1972).

23. E. P. Honig, Molecular constitution of X- and Y-type Langmuir–Blodgett films, *J. Colloid Interface Sci.*, **43**, 66–72 (1973).

24. H. Hasmonay, M. Vincent, and M. Dupeyrat, Composition and transfer mechanism of Langmuir–Blodgett multilayers of stearates, *Thin Solid Films*, **68**, 21–31 (1980).

25. M. Saint Pierre and M. Dupeyrat, Measurement and meaning of the transfer process energy in the building up of Langmuir Blodgett multilayers, *Thin Solid Films*, **99**, 205–213 (1983).

26. I. Langmuir, V. K. Schaefer, and H. Sobotka, Multilayers of sterols and adsorption of digitonin by deposited monolayers, *J. Am. Chem. Soc.*, **59**, 1751–1759 (1937).

27. E. P. Honig, J. H. Hengst, and D. den Engelsen, Langmuir–Blodgett deposition ratios, *J. Colloid Interface Sci.*, **45**, 92–102 (1973).

28. R. D. Neuman and J. W. Swanson, Multilayer deposition of stearic acid–calcium stearate monomolecular films, *J. Colloid Interface Sci.*, **74**, 244–259 (1980).

29. E. Stenhagen, Built-up films of esters, *Trans. Faraday Soc.*, **34**, 1328–1337 (1938).

30. A. Cemel, T. Fort, Jr., and J. B. Lando, Polymerization of vinyl stearate multilayers, *J. Polym. Sci., A1*, **10**, 2061–2083 (1972).

31. K. Fukuda and T. Shiozawa, Conditions for formation and structural characterization of X-type and Y-type multilayers of long-chain esters, *Thin Solid Films*, **68**, 55–66 (1980).

32. A. E. Alexander, Built-up films of unsaturated and substituted long-chain compounds, *J. Chem. Soc. London*, **1**, 778–781 (1939).

33. Y. K. Levine, A. I. Bailey, and M. H. F. Wilkins, Multilayers of phospholipid bimolecular leaflets, *Nature*, **220**, 577–578 (1968).

34. P. Christie, M. C. Petty, G. G. Roberts, D. H. Richards, D. Service, and M. J. Stewart, The preparation and dielectric properties of polybutadiene Langmuir–Blodgett films, *Thin Solid Films*, **134**, 75–82 (1985).

35. C. Holley and S. Bernstein, Grating space of barium–copper–stearate films, *Phys. Rev.*, **52**, 525 (1937).

36. S. Bernstein, Comparison of X-ray photographs taken with X and Y built-up films. *J. Am. Chem. Soc.*, **60**, 1511 (1938).

37. I. Fankuchen, On the structure of "built-up" films on metals, *Phys. Rev.*, **53**, 909 (1938).

38. R. C. Ehlert, Overturning of monolayers, *J. Colloid Sci.*, **20**, 387–390 (1965).

39. V. Enkelmann and J. B. Lando, Polymerization of ordered tail-to-tail vinyl stearate bilayers, *J. Polym. Sci., Polym. Chem. Ed.*, **15**, 1843–1854 (1977).

40. I. Langmuir, Overturning and anchoring of monolayers, *Science*, **87**, 493–500 (1938).

41. M. W. Charles, Optimization of multilayer soap crystals for ultrasoft X-ray diffraction, *J. Appl. Phys.*, **42**, 3329–3356 (1971).

42. J. P. Green, M. C. Phillips, and G. G. Shipley, Structural investigations of lipid, polypeptide and protein multilayers. *Biochim Biophys. Acta*, **330**, 243–253 (1973).

43. M. Puterman, T. Fort, Jr., and J. B. Lando, The polymerization and structure of mixed monolayers of ethyl and vinyl stearate, *J. Colloid Interface Sci.*, **47**, 705–718 (1974).

44. G. L. Gaines, Jr., Contact angles during monolayer deposition, *J. Colloid Interface Sci.*, **59**, 438–446 (1977).

45. P. S. Vincent, W. A. Barlow, F. T. Boyle, J. A. Finney, and G. G. Roberts, Preparation of Langmuir–Blodgett "built-up" multilayer films of a lightly substituted model aromatic, anthracene, *Thin Solid Films*, **60**, 265–277 (1979).

46. R. Jones, R. H. Tredgold, and P. Hodge, Langmuir–Blodgett films of simple esterified porphyrins, *Thin Solid Films*, **99**, 25–32 (1983).

47. M. F. Daniel, O. C. Lettington, and S. M. Small, Investigations into the Langmuir–Blodgett film formation ability of amphiphiles with cyano head groups, *Thin Solid Films*, **99**, 61–69 (1983).

48. S. Baker, M. C. Petty, G. G. Roberts, and M. V. Twigg, The preparation and properties of stable metal-free phthalocyanine Langmuir–Blodgett films, *Thin Solid Films*, **99**, 53–59 (1983).

49. S. Baker, G. G. Roberts, and M. C. Petty, Phthalocyanine Langmuir Blodgett-film gas detector, *IEE Proc. I*, **130**, 260–263 (1983).

50. G. J. Kovacs, P. S. Vincett, and J. H. Sharp, Stable, tough, adherent Langmuir–Blodgett films: Preparation and structure of ordered, true monolayers of a phthalocyanine, *Can J. Phys.*, **63**, 346–349 (1985).

51. Y. L. Hua, G. G. Roberts, M. M. Ahmad, M. C. Petty, M. Hanack, and M. Rein, Monolayer films of a substituted silicon phthalocyanine, *Phil. Mag. B*, **53**, 105–113 (1986).

52. R. Jones, R. H. Tredgold, A. Hoorfar, and R. A. Allen, Crystal Formation and growth in Langmuir–Blodgett multilayers of azobenzene derivatives: Optical and structural studies, *Thin Solid Films*, **134**, 57–66 (1985).

53. S. J. Mumby, J. F. Rabolt, and J. D. Swalen, Structural characterization of a polymer monolayer on a solid surface, *Thin Solid Films*, **133**, 161–164 (1985).

54. Y. Nishikata, M. Katimoto, A. Morikawa, and Y. Imai, Preparation and characterization of poly(amide–imide) multilayer films, *Thin Solid Films*, **160**, 15–20 (1988).

55. S. B. Hwang, J. I. Korenbrot, and W. Stoeckenius, Structural and spectroscopic characteristics of bacteriorhodopsin in air–water interface films, *J. Membr. Biol.*, **36**, 115–135 (1977).

56. M. T. Flanagan, The deposition of Langmuir–Blodgett films containing purple membrane on lipid- and paraffin-impregnated filters, *Thin Solid Films*, **99**, 133–138 (1983).

57. Y. Kawabata, M. Matsumoto, T. Nakamura, M. Tanaka, E. Manda, H. Takahashi, S. Tamura, W. Tagaki, H. Nakahara, and K. Fukuda, Langmuir–Blodgett Films of Amphiphilic Cyclodextrins *Thin Solid Films*, **159**, 353–358 (1988).

58. L. M. Blinov, N. N. Davydova, V. V. Lazarev, and S. G. Yudin, Spontaneous polarization of Langmuir multimolecular films, *Sov. Phys. Solid State (Engl. Trans.)*, **24**, 1523–1525 (1983).

59. L. M. Blinov, N. V. Dubinin, L. V. Mikhnev, and S. G. Yudin, Polar Langmuir–Blodgett films, *Thin Solid Films*, **120**, 161–170 (1984).

60. J. B. Ketterson, J. B. Peng, B. M. Abraham, and P. Dutta, Contact angle of lead stearate-covered water on mica during the deposition of Langmuir–Blodgett films, *Thin Solid Films*, **134**, 187–193 (1985).

61. R. D. Neuman, Langmuir–Blodgett monolayer deposition on collodion, *J. Colloid Interface Sci.*, **50**, 602–605 (1975).

62. R. D. Neuman, Molecular reorientation in monolayers at the paraffin–water interface, *J. Colloid Interface Sci.*, **63**, 106–112 (1978).

63. A. H. Ellison, Surface pressure–area properties of organic monolayers on mercury, *J. Phys. Chem.*, **66**, 1867–1872 (1962).

64. A. H. Ellison and W. A. Zisman, Force–area properties of films of polyfluoroquaternary ammonium compounds on hydrocarbon liquid substrates, *J. Phys. Chem.*, **59**, 1233 (1955).

65. A. Barraud, J. Leloup, and P. Lesieur, Monolayers on a glycerol subphase, *Thin Solid Films*, **133**, 113–116 (1985).

66. I. Langmuir and V. J. Schaefer, The effect of dissolved salts on insoluble monolayers, *J. Am. Chem. Soc.*, **59**, 2400–2414 (1937).

67. G. Veale and I. R. Peterson, Novel effects of counterions on Langmuir films of 22-tricosenoic acid, *J. Colloid Interface Sci.*, **103**, 178–189 (1985).

68. H. Kuhn, D. Möbius, and H. Bücher, Spectroscopy of monolayer assemblies, in: *Techniques of Chemistry* (A. Weissberger and B. W. Rossiter, eds.), Vol. I, Part IIIB, Wiley, New York (1972).

69. O. Albrecht, The construction of a microprocessor-controlled film balance for precision measurement of isotherms and isobars. *Thin Solid Films*, **99**, 227–234 (1983).

70. H. Bücher, O. V. Elsner, D. Möbius, P. Tillmann, and J. Wegand, *Nachweis der Entmischung Monomolekularer Farbstoff-Arachinsäure-Filme mit der Energiewanderungsmethode*, *Z. Phys. Chem.*, **65**, 152–169 (1969).

71. P. W. Morrison (ed.), *Environmental Control in Electronic Manufacturing*, Van Nostrand Reinhold, New York (1973).

72. J. H. Schenkel and J. A. Kitchener, Contamination of surfaces by conductivity water from ion-exchange resins, *Nature*, **182**, 131 (1958).

73. G. L. Gaines, Jr., Observations on resin-deionized water as a substrate for monolayer studies, *J. Phys. Chem.*, **63**, 1322–1324 (1959).

74. S. Baker, Phthalocyanine Langmuir–Blodgett films and their associated devices, Ph.D. thesis, University of Durham (1985).

75. J. H. Brooks and B. A. Pethica, Properties of ionized monolayers. Part 6.—Film pressures for ionized spread monolayers at the heptane/water interface. *Trans. Faraday Soc.*, **60**, 208–215 (1964).

76. R. R. Highfield, R. K. Thomas, P. G. Cummins, D. P. Gregory, J. Mingins, J. B. Hayter, and O. Schärpf, Critical reflection of neutrons from Langmuir–Blodgett films on glass, *Thin Solid Films*, **99**, 165–172 (1983).

77. L. Holland, *The Properties of Glass Surfaces*, Chapman and Hall, London (1964).

78. H. W. Fox and W. A. Zisman, Some advances in techniques for the study of adsorbed monolayers at the liquid–air interface, *Rev. Sci. Instrum.*, **19**, 274 (1948).

79. A. H. Ellison and W. A. Zisman, Surface activity at the organic liquid air interface, *J. Phys. Chem.*, **60**, 416–421 (1956).

80. T. A. Mann, Jr., and R. S. Hansen, Automatic recording surface balance II, *Rev. Sci. Instrum.*, **34**, 702–703 (1963).

81. C. Sucker, Eine Neuartige Filmwaage zur Automatischen Messung des Filmdruckes F und des Molekularen Flachenbedarfs A Gespreiteter Monomolekularer Filme, *Kolloid Z.*, **190**, 146–153 (1963).

82. R. M. Mendenhall and A. L. Mendenhall, Jr., Surface balance for production of rapid changes of surface area with continuous measurement of surface tension, *Rev. Sci. Instrum.*, **34**, 1350–1352 (1963).

83. J. C. Watkins, The surface properties of pure phospholipids in relation to those of lung extracts, *Biochim. Biophys. Acta*, **152**, 293–306 (1968).

84. R. M. Mendenhall, A surface balance for the study of surfactant movement, *Rev. Sci. Instrum.*, **42**, 878–880 (1971).

85. M. H. Nathoo, Preparation of Langmuir films for electrical studies, *Thin Solid Films*, **16**, 215–216 (1973).

86. P. Fromherz, Instrumentation for handling monomolecular films at an air–water interface, *Rev. Sci. Instrum.*, **46**, 1380–1385 (1975).

87. S. A. Tabak and R. H. Notter, Modified technique for dynamic surface pressure and relaxation measurements at the air–water interface, *Rev. Sci. Instrum.*, **48**, 1196–1201 (1977).

88. O. Albrecht and E. Sackmann, A precision Langmuir film balance measuring system, *J. Phys. E*, **13**, 512–515 (1980).

89. P. Dutta, K. Halperin, J. B. Ketterson, J. B. Peng, and G. Schaps, A wide temperature range hermetically sealed Langmuir–Blodgett apparatus, *Thin Solid Films*, **134**, 5–12 (1985).

90. I. R. Peterson, A fully automated high performance LB trough, *Thin Solid Films*, **134**, 135–141 (1985).

91. T. G. Jones, B. A. Pethica, and D. A. Walker, A new interfacial balance for studying films spread at the oil–water interface, *J. Colloid Sci.*, **18**, 485–488 (1963).

92. A. Barraud and M. Vandevyver, A trough for continuous fabrication of Langmuir–Blodgett films, *Thin Solid Films*, **99**, 221–225 (1983).

93. L. Blight, C. W. N. Cumper, and V. Kyte, Manipulation of insoluble films at an oil/water interface, *J. Colloid Sci.*, **20**, 393–399 (1965).

94. J. H. Brooks and F. MacRitchie, An alternative method for the compression of surface films, *J. Colloid Sci.*, **16**, 442 (1961).

95. D. S. Johnston, E. Coppard, G. V. Parera, and D. Chapman, Langmuir film balance study of interactions between carbohydrates and phospholipid monolayers, *Biochemistry*, **23**, 6912–6919 (1984).

96. J. H. Clint, Partial deposition of mixed monolayers onto solid surfaces, *J. Colloid Interface Sci.*, **43**, 132–143 (1973).

97. D. A. Brandreth, W. M. Riggs, and R. E. Johnson, Detection of metal ions in stearic acid monolayers, *Nat., Phys. Sci.*, **236**, 11 (1972).

98. D. W. Dwight and W. M. Riggs, Fluoropolymer surface studies, *J. Colloid Interface Sci.*, **47**, 650–660 (1974).

99. P. Somasundaran, M. Danitz, and K. J. Mysels, A new apparatus for measurements of dynamic interfacial properties, *J. Colloid Interface Sci.*, **48**, 410–416 (1974).

100. D. den Engelsen, J. H. Hengst, and E. P. Honig, An automated Langmuir trough for building monomolecular layers, *Philips Tech. Rev.*, **36**, 44–46 (1976).

101. I. H. Sher and J. D. Chanley, New technique for compressing surface films, *Rev. Sci. Instrum.*, **26**, 266–268 (1955).

102. U. Schoeler, Reibungsarmer Langmuir-Trog zum Spreiten und Übertragen von Monomolekularen Lipidfilmen bei Geringem Oberflächenschub, *Thin Solid Films*, **23**, S25–S27 (1974).

103. L. M. Walpita, Langmuir films in integrated optics, Ph.D. thesis, University of London (1977).

104. J. Boyle III and A. J. Mautone, A new surface balance for dynamic surface tension studies, *Colloids and Surfaces*, **4**, 77–85 (1982).

105. T. Smith, A controlled atmosphere Langmuir trough with simultaneous automatic recording of ellipsometric, contact potential, and surface tension measurements, *J. Colloid Interface Sci.*, **26**, 509–517 (1968).

106. P. Fromherz, A new technique for investigating lipid protein films, *Biochim. Biophys. Acta*, **225**, 382–387 (1971).

107. P. Fromherz and D. Marcheva, Enzyme kinetics at a lipid protein monolayer, induced substrate inhibition of trypsin, *FEBS Lett.*, **49**, 329–333 (1975).

108. D. B. Zilversmit. A method for compressing monomolecular films at oil–water interfaces, *J. Colloid Sci.*, **18**, 794–798 (1963).

109. J. A. Clements, Surface phenomena in relation to pulmonary function, *Physiologist*, **5**, 11–28 (1962).

110. R. M. Mendenhall, A. L. Mendenhall, Jr., and J. H. Tucker, A study of some biological surfactants, *Ann. N. Y. Acad. Sci.*, **130**, 902–919 (1966).

111. W. A. Barlow and G. Merrington, U. K. Patent 8501352.

112. G. Munger and R. M. Leblanc, New method of studying the low surface pressures of monolayer at air/water interface, *Rev. Sci. Instrum.*, **51**, 710–714 (1980).

113. I. Langmuir, The constitution and fundamental properties of solids and liquids. II. liquids, *J. Am. Chem. Soc.*, **39**, 1848–1906 (1917).

114. H. J. Trurnit and W. E. Lauer, Automatic recording film balance system, *Rev. Sci. Instrum.*, **30**, 975–981 (1959).

115. J. A. Mann, Jr. and R. S. Hansen, Simple technique for automatic recording of monolayer compression characteristics, *Rev. Sci. Instrum.*, **31**, 961–963 (1960).

116. L. Vroman, S. Kanor, and A. L. Adams, Surface film pressure recording system, *Rev. Sci. Instrum.*, **39**, 278–279 (1968).

117. N. L. Gershfeld, R. E. Pagano, W. S. Friauf, and J. Fuhrer, Millidyne sensor for the Langmuir film balance, *Rev. Sci. Instrum.*, **41**, 1356–1358 (1970).

118. B. R. Malcolm and S. R. Davies, A film balance for use with the Langmuir trough, *J. Sci. Instrum.*, **42**, 359–360 (1965).

119. L. Wilhelmy, Ueber die Abhängigkeit der Capillaritäts-Constanten des Alkohols von Substanz und Gestalt des Benetzten Festen Körpers, *Ann. Phys. Chem.*, **119**, 177–217 (1863).

120. D. O. Jordan and J. E. Lane, A thermodynamic discussion of the use of a vertical-plate balance for the measurement of surface tension, *Aust. J. Chem.*, **17**, 7–15 (1964).

121. T. Takenaka, Effect of electrolyte on the molecular orientation in monolayers adsorbed at the liquid–liquid interface: Studies by resonance Raman spectra, *Chem. Phys. Lett.*, **55**, 515–518 (1978).

122. D. Möbius, H. Bücher, H. Kuhn, and J. Sondermann, Reversible Änderung der Fläche und des Grenzflächenpotentials Monomolekularer Filme eines Photochromen Systems, *Ber. Bunsenges. Phys. Chem.*, **73**, 845–850 (1969).

123. A. Kleinschmidt, Lineare Makromoleküle in Protein-Mischfilmen, Proc. Third Int. Cong. Surface Active Substances, Vol. II, 138–143 (1961).

124. G. L. Gaines, Jr., On the use of filter paper Wilhelmy plates with insoluble monolayers, *J. Colloid Interface Sci.*, **62**, 191–192 (1977).

125. D. Möbius, personal communication.

126. B. R. Malcolm, The flow and deformation of synthetic polypeptide monolayers during compression, *J. Colloid Interface Sci.*, **104**, 520–529 (1985).

127. B. R. Malcolm, Studies of the flow of molecular monolayers during compression and the effect of a plateau in the pressure–area curve. *Thin Solid Films*, **134**, 201–208 (1985).

128. This system was developed by workers at ICI plc.

129. V. K. Agarwal, H. Mitsuhashi, and H. Takasaki, A new dipping and raising device for Langmuir film deposition, *J. Phys. E*, **10**, 237–240 (1977).

130. L. S. Miller, D. E. Hookes, P. J. Travers, and A. P. Murphy, A new type of Langmuir–Blodgett trough, *J. Phys. E: Sci. Instrum.*, **21**, 163–167 (1988).

131. H. L. Brockman, C. M. Jones, C. J. Schwebke, J. M. Smaby, and D. E. Jarvis, Application of a microcomputer-controlled film balance system to collection and analysis of data from mixed monolayers, *J. Colloid Interface Sci.*, **78**, 502–512 (1980).

132. A. J. Vickers, R. H. Tredgold, P. Hodge, E. Khoshdel, and I. R. Girling, An investigation of Liquid-crystal side-chain polymer Langmuir–Blodgett films as optical waveguides, *Thin Solid Films*, **134**, 43–48 (1985).

133. M. B. Biddle, S. E. Rickert, and J. B. Lando, Constructing a processing window for a Langmuir–Blodgett film, *Thin Solid Films*, **134**, 121–134 (1985).

134. Such systems have been used extensively for LB film deposition facilities at University College, London and at the University of Durham, UK.

135. J. K. Severn and D. N. Batchelder, Vibration isolation of a Langmuir trough, *J. Phys. E*, **17**, 113–115 (1984).

136. The LB Films incorporated in electronic devices reported by researchers at the University of Durham, UK were deposited in a class 10,000 clean room.

137. S. E. Rickert, C. D. Fung, and J. B. Lando, Process control in the production of integrated Langmuir devices, Proc. Int. Symp. Future Electron Devices, Tokyo, 25–28 (1985).

138. G. Golian, R. S. Hales, J. G. Hawke, and J. M. Gebicki, An automatic surface balance for measurements in a radiation field, *J. Phys. E*, **11**, 787–790 (1978).

139. G. Lieser, R. Tieke, and G. Wegner, Structure, phase transitions and polymerizability of multi-layers of some diacetylene monocarboxylic acids, *Thin Solid Films*, **68**, 77–90 (1980).

140. D. Möbius, Molecular organization and chemical reactivity in monolayers and monolayer systems, *Mol. Cryst. Liq. Cryst.*, **96**, 319–334 (1983).

141. K. B. Blodgett and I. Langmuir, Built-up films of barium stearate and their optical properties, *Phys. Rev.*, **51**, 964–982 (1937).

142. I. Langmuir and V. J. Schaefer, Rates of evaporation of water through compressed monolayers on water, *J. Franklin Inst.*, **235**, 119–162 (1943).

143. F. C. Goodrich, Molecular interaction in mixed monolayers, Proc. 2nd Int. Congress on Surface Activity, Vol. I. 33–39 (1957).

144. J. Mingins, N. F. Owens, and D. H. Iles, Properties of monolayers at the air–water interface I. The effect of spreading solvent on the surface pressure of octadecyltrimethylammonium bromide, *J. Phys. Chem.*, **73**, 2118–2126 (1969).

145. O. Albrecht, D. S. Johnston, C. Villaverde, and D. Chapman, Stable biomembrane surfaces formed by phospholipid polymers, *Biochim. Biophys. Acta*, **687**, 165–169 (1982).

146. D. M. Taylor and M. G. B. Mahboubian-Jones, The electrical properties of synthetic phospholipid Langmuir Blodgett films, *Thin Solid Films*, **87**, 167–169 (1982).

147. L. R. McLean, A. A. Durrani, M. A. Whittam, D. S. Johnston, and D. Chapman, Preparation of stable polar surfaces using polymerizable long-chain diacetylene molecules, *Thin Solid Films*, **99**, 127–131 (1983).

148. G. Colacicco, Applications of monolayer techniques to biological systems: Symptoms of specific lipid–protein interactions, *J. Colloid Interface Sci.*, **29**, 345–364 (1969).

149. D. Möbius, Designed monolayer assemblies, *Ber. Bensenges. Phys. Chem.*, **82**, 848–858 (1978).

150. S. C. Steele, M. N. Wybourne, and D. Möbius, Polarization effects observed in monolayer Langmuir–Blodgett films, *Thin Solid Films*, **99**, 117–118 (1983).

151. R. A. Hann, W. A. Barlow, J. H. Steven, B. L. Eyres, M. V. Twigg, and G. G. Roberts, Langmuir–Blodgett films of substituted aromatic hydrocarbons and phthalocyanines, Proc. 2nd. Int. Workshop Molecular Electron Devices, Washington, April, 1983.

152. G. G. Roberts, M. C. Petty, S. Baker, M. T. Fowler, and N. J. Thomas, Electronic devices incorporating stable phthalocyanine Langmuir–Blodgett films, *Thin Solid Films*, **132**, 113–123 (1985).

153. G. L. Gaines, Jr., On the retention of solvent in monolayers of fatty acids spread on water surfaces, *J. Phys. Chem.*, **63**, 382–383 (1961).

154. G. T. Barnes, A. J. Elliot, and E. C. M. Grigg, The retention of hydrocarbons on octadecanol monolayers, *J. Colloid Interface Sci.*, **26**, 230–232 (1968).

155. J. Sharp, unpublished data: it has been found that for certain dimensions of the constant perimeter barrier trough system, monolayers do not readily spread into the barrier "legs."

156. V. K. LaMer and G. T. Barnes, The effects of spreading technique and purity of sample on the evaporation resistance of monolayers, *Proc. Natl. Acad. Sci. U.S.A.*, **45**, 1274–1280 (1959).

157. H. J. Trurnit, A theory and method for the spreading of protein monolayers, *J. Colloid Sci.*, **15**, 1–13 (1960).

158. T. M. Ginnai, An investigation of electron tunnelling and conduction in Langmuir films, Ph.D. thesis, Leicester Polytechnic (UK) (1982).

159. R. W. Roberts and G. L. Gaines, Jr., Stability of fatty monolayers in vacuum, Trans. 9th. Natl. Vac. Symposium, Los Angeles, 515–518 (1962).

160. J. A. Spink, The transfer ratio of Langmuir–Blodgett monolayers for various solids, *J. Colloid Interface Sci.*, **23**, 9–26 (1967).

161. J. J. Bikerman, On the formation and structure of multilayers, *Proc. R. Soc. London, Ser. A*, **170**, 130–144 (1939).

162. O. Albrecht, A. Laschewsky, and H. Ringsdorf, Polymerizable built up multilayers on polymer supports, *Macromolecules*, **17**, 937–940 (1984).

163. G. L. Gaines, Jr., Some observations on monolayers of carbon-14 labelled stearic acid, *J. Colloid Sci.*, **15**, 321–339 (1960).

164. T. Smith, The hydrophilic nature of a clean gold surface, *J. Colloid Interface Sci.*, **75**, 51–55 (1980).

165. G. L. Gaines, Jr., On the water wettability of gold, *J. Colloid Interface Sci.*, **79**, 295 (1981).

166. See, for example, D. E. Aspnes and A. A. Studna, Chemical etching and cleaning procedures for Si, Ge, and some III–V compound semiconductors, *Appl. Phys. Lett.*, **39**, 316–318 (1981).

167. See, for example, D. T. Clark, T. Fok, G. G. Roberts, and R. W. Sykes. An investigation by electron spectroscopy for chemical analysis of chemical treatments of the (100) surface of n-type InP epitaxial layers for Langmuir film deposition, *Thin Solid Films*, **70**, 261–283 (1980).

168. R. W. Sykes, G. G. Roberts, T. Fok, and D. T. Clark, p-Type InP/Langmuir film M.I.S. diodes, IEE Proc. I, **127**, 137–139 (1980).

169. D. den Engelsen, Ellipsometry of anisotropic films, *J. Opt. Soc. Am.*, **61**, 1460–1466 (1971).

170. G. Fariss, J. Lando, and S. Rickert, Electron beam resists produced from monomer–polymer Langmuir–Blodgett films. *Thin Solid Films*, **99**, 305–315 (1983).

171. I. R. Peterson and G. J. Russell, An electron diffraction study of ω-tricosenoic acid Langmuir–Blodgett films, *Phil. Mag. A*, **49**, 463–473 (1984).

172. J. P. Lloyd, M. C. Petty, G. G. Roberts, P. G. LeComber, and W. E. Spear. Langmuir–Blodgett films in amorphous silicon MIS structures, *Thin Solid Films*, **89**, 395–399 (1982).

173. N. J. Thomas, M. C. Petty, G. G. Roberts, and H. Y. Hall, GaAs/LB film MISS switching device, *Electron. Lett.*, **20**, 838–839 (1984).

174. R. H. Tredgold and C. S. Winter, Langmuir–Blodgett monolayers of preformed polymers on n-type GaP, *Thin Solid Films*, **99**, 81–85 (1983).

175. R. H. Tredgold and G. W. Smith, Surface potential studies on Langmuir–Blodgett multilayers and adsorbed monolayers, *Thin Solid Films*, **99**, 215–220 (1983).

176. J. Batey, G. G. Roberts, and M. C. Petty, Electroluminescence in GaP/Langmuir–Blodgett film metal/insulator/semiconductor diodes, *Thin Solid Films*, **99**, 283–290 (1983).

177. J. Batey, M. C. Petty, G. G. Roberts, and D. R. Wight, GaP/Phthalocyanine Langmuir–Blodgett film electroluminescent diode, *Electron. Lett.*, **20**, 489–491 (1984).

178. K. K. Kan, G. G. Roberts, and M. C. Petty, Langmuir–Blodgett film metal/insulator/semiconductor structures on narrow band gap semiconductors, *Thin Solid Films*, **99**, 291–296 (1983).

179. G. G. Roberts, M. C. Petty, and I. M. Dharmadasa, Photovoltaic properties of cadmium telluride/Langmuir-film solar cells, *IEE Proc. I*, **128**, 197–201 (1981).

180. M. C. Petty, unpublished data.

181. J. Batey, Electroluminescent MIS structures incorporating Langmuir–Blodgett films, Ph.D. thesis, University of Durham (1983).

182. M. Fowler, Optical and photoelectrical applications of Langmuir–Blodgett films, Ph.D. thesis, University of Durham (1985).

183. C. Huh and L. E. Scriven, Hydrodynamic model of steady movement of a solid/liquid/fluid contact line, *J. Colloid Interface Sci.*, **35**, 85–101 (1971).

184. I. R. Peterson, G. J. Russell, and G. G. Roberts, A new model for the deposition of ω-tricosenoic acid Langmuir–Blodgett film layers, *Thin Solid Films*, **109**, 371–378 (1983).

185. I. R. Peterson, Optical observation of monomer Langmuir–Blodgett film structure, *Thin Solid Films*, **116**, 357–366 (1984).

186. G. Veale, I. R. Girling, and I. R. Peterson, A comparison of deposition speed, epitaxy and crystallinity in Langmuir–Blodgett films of fatty acids, *Thin Solid Films*, **127**, 293–303 (1985).

187. M. R. Buhaenko, J. W. Goodwin, R. M. Richardson and M. F. Daniel, The influence of shear viscosity of spread monolayers on the Langmuir–Blodgett process, *Thin Solid Films*, **134**, 217–226 (1985).

188. J. G. Petrov, H. Kuhn, and D. Möbius, Three-phase contact line motion in the deposition of spread monolayers, *Ber. Bunsenges. Phys. Chem.*, **82**, 884 (1978).

189. J. G. Petrov, H. Kuhn, and D. Möbius, Three phase contact line motion in the deposition of spread monolayers, *J. Colloid Interface Sci.*, **73**, 66–75 (1980).

190. M. F. Daniel and J. T. T. Hart, The effect of surface flow on the morphology of Langmuir Blodgett films, *J. Mol. Electron.*, **1**, 97–104 (1985).

191. E. F. Porter and J. Wyman, Jr., Further studies on the electrical properties of stearate films deposited on metal, *J. Am. Chem. Soc.*, **60**, 2855–2869 (1938).

192. F. Grunfeld and C. W. Pitt, Diacetylene Langmuir–Blodgett layers for integrated optics, *Thin Solid Films*, **99**, 249–255 (1983).

193. J. W. Ellis and J. L. Pauley, The infrared determination of the composition of stearic acid multilayers deposited from salt substrata of varying pH, *J. Colloid Sci.*, **19**, 755–764 (1964).

194. C. W. Pitt and L. M. Walpita, Lightguiding in Langmuir–Blodgett films, *Thin Solid Films*, **68**, 101–127 (1980).

195. I. Langmuir and V. J. Schaefer, Improved methods of conditioning surfaces for adsorption, *J. Am. Chem. Soc.*, **59**, 1762–1763 (1937).

196. M. Breton, Formation and possible applications of polymeric Langmuir–Blodgett films. A review, *J. Macromol. Sci., Rev. Macromol. Chem.*, **C21**, 61–87 (1981).

197. A. Banerjie and J. B. Lando, Radiation-induced solid state polymerization of oriented ultrathin films of octadecylacrylamide, *Thin Solid Films*, **68**, 67–75 (1980).

198. K. Fukuda, Y. Shibasaki, and H. Nakahara, Effects of molecular arrangement on polymerization reactions in Langmuir–Blodgett films, *Thin Solid Films*, **99**, 87–94 (1983).

199. B. Tieke, G. Lieser, and K. Weiss, Parameters influencing the polymerization and structure of long-chain diynoic acids in multilayers, *Thin Solid Films*, **99**, 95–102 (1983).

200. C. Bubeck, K. Weiss, and B. Tieke, Sensitized photoreaction of diacetylene multilayers, *Thin Solid Films*, **99**, 103–107 (1983).

201. A. Barraud, Polymerization in Langmuir–Blodgett films and resist applications, *Thin Solid Films*, **99**, 317–321 (1983).

202. A. K. Kapil and V. K. Srivastava, Deposition of different monomolecular films in the same trough, *J. Colloid Interface Sci.*, **72**, 342–343 (1979).

203. M. F. Daniel, J. C. Dolphin, A. J. Grant, K. E. N. Kerr, and G. W. Smith, A trough for the fabrication of non-centrosymmetric Langmuir–Blodgett films, *Thin Solid Films*, **133**, 235–242 (1985).

204. A. Barraud and J. Leloup, Procede et dispositif pour la realisation de couches monomoleculaires alternees, French Patent No. 83.03578 (1983).

205. A. Barraud, J. Leloup, A. Gouzerh and S. Palacin, An automatic trough to make alternate layers, *Thin Solid Films*, **133**, 117–123 (1985).

206. J. J. Bikerman and J. H. Schulman, On the structure of "built-up" films on metals, *Phys. Rev.*, **53**, 909 (1938).

207. J. J. Bikerman, On the potentials of "built-up" multilayers on metals, *Trans. Faraday Soc.*, **34**, 800–803 (1938).
208. H. E. Ries, Jr., and J. Gabor, Monolayer transfer to a rotating cylinder: Surface flow patterns, *Nature*, **212**, 917–918 (1966).
209. H. E. Ries, Jr., Transfer of monolayers to a rotating cylinder: Surface flow patterns, Proc. 5th. Int. Cong. Surface Active Substances, Vol. 2, 443–453 (1969).
210. I. R. Girling and D. R. J. Milverton, A method for the preparation of an alternating multilayer film, *Thin Solid Films*, **115**, 85–88 (1984).
211. B. Blackburn, B. Holcroft, M. C. Petty, and G. G. Roberts, U.K. Patent No. 8428593 (1984).
212. B. Holcroft, M. C. Petty, G. G. Roberts, and G. J. Russell, A Langmuir trough for the production of organic superlattices, *Thin Solid Films*, **134**, 83–88 (1985).
213. W. M. Reichert, C. J. Bruckner, and J. Joseph, Langmuir–Blodgett films and black lipid membranes in biospecific surface-selective sensors, *Thin Solid Films*, **152**, 345–376 (1987).
214. T. Richardson, G. G. Roberts, M. E. C. Polywka, and S. G. Davies, Preparation and characterization of organotransition metal Langmuir–Blodgett films, *Thin Solid Films*, **160**, 231–239 (1988).
215. J. Tsibouklis, J. Cresswell, N. Kalita, C. Pearson, P. J. Maddaford, H. Ancelin, J. Yarwood, M. J. Goodwin, N. Carr, W. J. Feast, and M. C. Petty, Functionalized diarylalkynes: A new class of Langmuir-Blodgett film materials for non-linear optics, *J.Phys.D:Appl.Phys.*, **22** (1989), in press.
216. J. C. Loulergue, M. Dumont, Y. Levy, P. Robin, J. P. Pocholle, and M. Papuchon, Linear electro-optic properties of Langmuir–Blodgett multilayers of an organic azo dye, *Thin Solid Films*, **160**, 399–405 (1988).
217. I. Ledoux, D. Josse, P. Fremaux, J-P Piel, G. Post, J. Zyss, T. McLean, R. A. Hann, P. F. Gordon, and S. Allen, Second-harmonic generation in alternate non-linear Langmuir–Blodgett films, *Thin Solid Films*, **160**, 217–230 (1988).
218. M. R. Buhaenko and R. M. Richardson, Measurement of the forces of emersion and immersion and contact angles during Langmuir–Blodgett deposition, *Thin Solid Films*, **159**, 231–238 (1988).
219. J. Mingins and N. F. Owens, Experimental considerations in insoluble spread monolayers, *Thin Solid Films*, **152**, 9–28 (1987).
220. B. R. Malcolm, The flow of molecular monolayers in relation to the design of Langmuir troughs and the deposition of Langmuir–Blodgett films, *J. Phys.E:Sci.Instrum.*, **21**, 603–607 (1988).
221. W. M. Heckl, A. Miller, and H. Möhwald, Electric-field-induced domain movement in phospholipid monolayers, *Thin Solid Films*, **159**, 125–132 (1988).
222. C. Bubeck, Reactions in monolayers and Langmuir–Blodgett films, *Thin Solid Films*, **160**, 1–14, 1988).
223. T. Kato, Development of a microcomputer-controlled instrument for preparing complex(hetero-) Langmuir–Blodgett films fully automatically, *Jpn.J.Appl.Phys.*, **26**, L1377–L1380 (1987).

Characterization and Properties

M. C. PETTY

4.1. INTRODUCTION

This chapter concentrates on the characterization of Langmuir–Blodgett (LB) monolayer and multilayer films on solid surfaces. Experimental techniques to study the properties of monolayers floating on the surface of a subphase have been discussed by Gaines[1] and are, to some extent, covered in the earlier chapters of this book. A review of the properties of LB films could be arranged in a number of ways: subdivisions according to the type of material or to the nature of the experimental technique are probably the most straightforward approaches. In this chapter the latter plan is adopted. It is hoped that this will not only aid the researcher wishing to investigate monolayer and multilayer systems, but also keep the work relevant as more novel LB materials are synthesized. Examples used have generally been chosen from published data on the simpler types of LB assemblies (e.g., long-chain fatty acids).

Investigations into monolayer and multilayer LB films fall, broadly, into three categories: structural, optical, and electrical. To date a number of review articles have been published which cover one or more of these topics. The work of Kuhn, Möbius, and Bücher[2] concentrates on the spectroscopic properties of monolayer assemblies containing various arrangements of chromophoric groups, while Srivastava[3] has reviewed the optical, electrical, and structural properties of LB films up to 1972.

M. C. PETTY • Molecular Electronics Research Group, School of Engineering and Applied Science, University of Durham, Durham DH1 3LE, England.

More recently, a comprehensive review on optical properties has been presented by Swalen[522] while Khanarian[523] and Allen[524] have both concentrated on nonlinear optical effects in mono- and multilayers. Work on the electrical behavior and transport properties of LB layers up to 1974 has been discussed by Agarwal[4] and, more recently, by Vincett and Roberts.[5] The specialized area of the properties of LB films on semiconducting substrates has been covered by Roberts.[6] Tredgold[525] has discussed the physics of LB films. The structure of multilayers, with particular emphasis on X-ray and electron diffraction techniques, has been reviewed by Feigin et al.[526]. Numerous papers devoted to the assessment and characterization of LB films are to be found in a special issue of Thin Solid Films[7] and also in the Proceedings of four International Conferences on LB Films, held in 1982 in Durham, UK, in 1985 in Schenectady, USA.[8,9] in 1987 in Göttingen, W. Germany, and in 1989 in Tsukuba, Japan.[8,9,527,528] This chapter briefly reviews the early work and attempts to assimilate this with some of the more recent techniques and ideas. Where appropriate, details of the specimen preparation are given.

4.2. EVALUATION OF FILM THICKNESS

One of the main attractions of the LB technique is the ability to deposit organic layers with an ultrafine control of the layer thickness. The evaluation of this parameter is therefore of paramount importance in any LB film study. A list of techniques that are commonly used has been given by Pitt and Walpita[10]; some of these are discussed by Vandevyver.[529] It should be noted that many methods do not give an independent measurement of layer thickness; other physical parameters must also be determined accurately. For instance, some optical techniques require the refractive index of the material to be known; electrical measurements (e.g., reciprocal capacitance versus film thickness) require knowledge of the permittivity.

4.2.1. Interference Techniques

In Blodgett's original work,[11] the film thickness was determined from interference fringes produced by reflection of monochromatic light from the film. For normal incidence, the film thickness, t, giving an interference minimum is simply given by

$$t = m\lambda/4n \qquad (4.1)$$

where λ is the wavelength, n is the refractive index of the film, and m is an integer. Blodgett determined the refractive index by measuring the polarizing (Brewster) angle for the films. However, Blodgett and Langmuir[12] subsequently discovered that the multilayer films were birefringent, invalidating the original refractive index values which were obtained using the extraordinary ray (this does not obey Snell's

Law). Therefore an alternative technique was used to obtain n; this was based on measuring the relative intensities of the interference fringes produced in glasses with nearly the same refractive index as the film.[12,13] A similar approach to that of Blodgett and Langmuir has been used by Drexhage[14]; the main difference was that a quantitative measurement of light intensities was used instead of visual observation. In their method, Jenkins and Norris[15] kept the angle of incidence constant and measured the change in wavelength necessary to preserve an intensity minimum as the thickness of the film was increased by the addition of further monolayers; the measurements were carried out using the ordinary ray. Michelson interferometry has also been used to determine film thickness.[16] Once again, this technique yields the optical thickness; the refractive index of the organic layer must be known in order to calculate the metric thickness. Experiments which do give independent thickness and refractive index values have been described by Hartman $et\ al.$[17,18] and by Mattuck $et\ al.$[19,20] The former method is based on white light interference fringes and may be used for thick multilayers when these can be built up on chromium plated slides in a series of steps. However, problems can arise if this technique is used for films of molecular thickness.[21] The Mattuck technique permits measurement of the thickness and refractive index of very thin (1 nm $< t <$ 25 nm) films deposited on a suitable optical interference gauge. Both the Hartman and Mattuck methods have been used by Bateman and Covington[22] to investigate the optical properties of barium–copper stearate films. It was found that the Hartman multi-layers showed directional properties with respect to the direction of "dipping" during monolayer transfer, and that the apparent refractive indices found by the Mattuck method varied considerably with film pressure during transfer. These findings were interpreted quantitatively in terms of molecular tilt imposed by the process of monolayer transfer, an idea used by other workers from interpretations of X-ray and infrared data (see Section 4.3 below).

Multiple-beam techniques for measuring film thickness are capable of yielding much higher accuracy than two-beam fringe methods because the interference can lead to extreme sharpening of the fringes.[23] White light fringes (fringes of equal chromatic order) are generally preferred to the monochromatic (Fizeau) fringes for such measurements. In 1950, Courtney-Pratt[24,25] reported thickness measurements on monomolecular layers of fatty acids spread by the droplet retraction technique. However, the method of measurement, using the transmitted light from a doubly-silvered mica interference system, still required knowledge of the refractive index of the organic material. The reflection multiple beam method[23,26] permits the true metric thickness of the film to be determined. The application of this to LB film studies has been described by Srivastava and Verma.[27,28] A clean glass slide is first vacuum-coated with a layer of silver. Over the entire surface of the silvered film, a suitable number ($\sim$20) of layers of the film under study are deposited to form a base film. A second multilayer film is then deposited over part of the slide; this forms a sharp step, the height of which is the film thickness to be measured. The base film provides the same material structure on both sides and thus avoids any differential phase changes in reflection from the two sides of the step. An opaque coating of silver is finally deposited by thermal evaporation in a vacuum. This silvered film is now matched, at a small angle, against a reference optical flat which is also silvered. The interference system is illuminated with normally incident, collimated, white light to obtain fringes of equal chromatic order. Using reflection multiple beam interference, Srivastava and Verma[28] have determined the thickness of various fatty acid salt films to an accuracy of approximately ± 0.1 nm. It should be noted that this multiple beam interferometry technique requires the monolayer or multilayer to be coated with a highly reflective metal layer; the organic film must therefore remain stable in vacuum and not be affected by the evaporation process.

4.2.2. Ellipsometry

The technique of ellipsometry relies on the fact that linearly polarized light becomes elliptically polarized on reflection from a metal surface. The presence of a surface film alters the ratio, ψ, of the electric vectors vibrating in the plane of incidence and perpendicular to it, as well as their difference of phase Δ. The theory of ellipsometry, first presented by Drude,[29] correlates the parameters ψ and Δ with the optical thickness of the layer and the optical constants of the metal. In 1934 Feachem and Tronstad[30,31] applied the technique to investigate the properties of monomolecular films of fatty acids on a mercury surface. Some years later, Rothen and Hanson[32] used barium stearate multilayer films to investigate the validity of Drude's formulas (which are first approximations, and are valid only for a film whose thickness is small compared to the wavelength of light). The general formula of Drude expressing the phase shift as a function of the film thickness was found to represent to a fair approximation the experimental results obtained with multilayer films whose thicknesses were up to the order of the wavelength of the light. The application of Drude's equation to less than a monolayer coverage has been discussed by Smith.[33] In ellipsometry measurements there will invariably be an additional (oxide) layer present between the metal substrate and the LB film.[34,35] However, it has been demonstrated that the presence of such a layer can be satisfactorily accounted for in a thickness determination by means of its effect upon the apparent optical constants of the substrate.[35] Thus a metal surface coated with a few monolayers is often taken as the reference in ellipsometry measurements.[34−36] A number of extensions and simplifications to Drude's treatment have been proposed; in many cases, in order to test the validity of the theories, LB multilayer systems have been used as calibration standards.[34,37] In most studies it is assumed that the films are nonabsorbing and effectively isotropic for small thicknesses.[34−44] However, the theory of ellipsometry has been extended to anisotropic films by Engelsen,[45] by Dignam et al.,[46] and by Tomar and Srivastava.[47] Although the theoretical reflectances and transmittances of LB films depend on whether they are assumed to behave in an isotropic or in an anisotropic manner,[48] the allowance for anisotropy does not appear to appreciably affect the ellipsometric thickness values for simple fatty acid layers. For example, using his theory, Engelsen[45] reported that the thickness of a cadmium arachidate monolayer was 2.68 ± 0.02 nm, compared to a value of 2.70 ± 0.03 nm which was obtained if the films were assumed to be optically isotropic. Nevertheless, the full anisotropic theoretical treatments are frequently used for studies on floating monolayers and LB films.[49-52] A theoretical treatment of a transmission ellipsometry technique has also been presented by Engelsen,[53] and the application of this to both nonabsorbing (fatty acid) and absorbing (mixed fatty acid/cyanine dye) LB films has been reported.

4.2.3. X-Ray Diffraction

X-ray diffraction techniques have been used extensively in order to determine the monolayer thickness of LB films. Most of the work has concentrated on long chain (saturated and unsaturated) fatty acids[16,28,54-91,530-532]; however, long chain esters,[92-99] substituted aromatics,[89,100-107] preformed polymers,[102] and biological materials[108-114] have also been investigated. Multilayer assemblies have even been used for gratings in X-ray spectroscopy systems[115-117]; this application is discussed further in Chapter 7. Since the scattering of X-rays from carbon and hydrogen atoms can be assumed to be very small compared to that from the heavier metal ions, the lattice spacing (normal to the film) measured by X-ray diffraction for simple fatty acids corresponds to the distance between adjacent planes containing metal ions. This led originally to the discovery that some X-deposited films possessed essentially the same structure as Y-type layers, and *vice versa* (see Chapter 3, Section 3.1).

Figure 4.1 shows X-ray diffraction data from a film of 43 layers of perdeuterated manganese stearate on a substrate of single-crystal silicon.[80] The experimental values are shown as points; the solid curve is based on calculation and is displaced from the data points to avoid obscuring them. Many orders of Bragg peaks from the basal planes (001) are clearly visible, giving information about the stacking of the planes of LB layers. However, under certain circumstances it is possible to obtain details of the in-plane structure using X-ray techniques.[84,86,94,95,98] X-ray studies are usually carried out on multilayer films, although it is possible to observe diffraction data from a single monolayer.[77] For small (<20) numbers of layers, secondary maxima are often observed between the Bragg reflections[61,64,72,77,78,80]; the intensity and angular position of these depend on the exact number of deposited layers. The phenomenon is in fact analogous to the diffraction of light waves from gratings of only a few slits. A detailed analysis of the subsidiary maxima has been undertaken by Pomerantz and Segmüller.[77] Kapp and Wainfan[64] have reported a further set of X-ray fringes in the nonspecularly scattered radiation from a barium stearate multilayer film; however, no unambiguous explanation of these fringes could be offered.

The *d*-spacings obtained for simple long-chain fatty acid materials are generally found to correspond to those from known crystalline modifications. Where it has been possible to directly compare the X-ray *d*-spacing to values obtained from optical techniques, good agreement, within experimental error, is usually obtained.[28] Table 4.1 lists the monolayer thicknesses obtained for the barium and

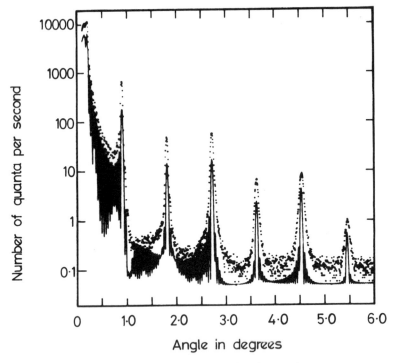

Figure 4.1. X-ray diffraction from a film of 43 layers of perdeuterated manganese stearate on a Si substrate. Experimental values are shown as points. The solid curve is based on calculation and is displaced from the data points. X-ray wavelength, 0.154 nm. (After Nicklow *et al.* [80])

cadmium salts of a number of fatty acids, using the techniques of X-ray diffraction, interferometry, and ellipsometry. The thickness of various fatty acid films as a function of temperature has been reported by Fukui *et al.*[76] Up to about −40 °C, the thickness varies only slightly, with a thermal expansion coefficient of approximately 10^{-5} K^{-1}. However, at temperatures higher than −20 °C, the thickness tends to decrease and the expansion coefficient becomes negative.[70]

The X-ray data listed in Table 4.1 are shown in graphical form in Figure 4.2, the full line being a least-squares fit to the experimental points. The *d*-spacings are close to those calculated for the lengths of the molecules, inferring that the hydrocarbon chains in transferred monolayers are oriented with their hydrocarbon chains almost at right angles to the substrate. However, some workers have noted X-ray *d*-spacings that are significantly less than those expected from the molecular length; such evidence points to a tilt in the hydrocarbon chain. For example, Alexander[60] reports large tilts for dihydroxybehenic acid. The tilting of the molecular chains has also been invoked to explain X-ray data for some long chain esters,[94–97] di-

Table 4.1. Monolayer Thicknesses for Various Fatty Acid LB Films

Material	Interferometric measurement (nm)	X-ray measurement (nm)	Ellipsometry (nm)
Barium palmitate	2.32 ± 0.11[a]	2.33 ± 0.05[a]	2.34[d]
Cadmium palmitate (C_{16})	—	2.265 ± 0.005[b]	
Barium margarate	2.41 ± 0.11[b]	2.40 ± 0.05[a]	2.45[d]
Cadmium margarate (C_{17})	—	2.390 ± 0.005[b]	
Barium stearate	2.62 ± 0.03[a]	2.53 ± 0.04[a]	
		2.555 ± 0.025[f]	
Cadmium stearate (C_{18})	—	2.515 ± 0.005[b]	—
		2.58[c]	
Cadmium nonadecylate (C_{19})	—	2.640 ± 0.005[b]	—
		2.69[c]	
Cadmium arachidate (C_{20})	—	2.755 ± 0.005[b]	2.68 ± 0.02[e]
		2.80[c]	
Barium behenate (C_{22})	3.04 ± 0.13[a]	2.91 ± 0.05[a]	3.08[d]
		3.02[c]	

[a] After Srivastava and Verma.[28]
[b] After Matsuda et al.[73]
[c] After Mann and Kuhn.[70]
[d] After Tomar and Strivastava.[40]
[e] After den Englesen.[45]
[f] After Kapp and Wainfan.[64]

acetylenes,[75,79,81−83] and certain alternate layer structures.[89,526,531] In some cases, models in which the molecules in consecutive layers interpenetrate have been proposed.[100,104] The angle of tilt appears to depend on the LB material and also on the deposition conditions. For instance, simple fatty acid films (as opposed to fatty acid salt films) invariably possess tilted structures. Enkelmann and Lando[96] report that X-type vinyl stearate monomer multilayers exhibit a monoclinic structure with the long axis inclined at 28° to the substrate normal. However, by polymerizing the LB layer under the surface of the subphase, a Y-type bilayer structure could be produced; this was shown to be oriented with the molecules perpendicular to the substrate. Naegele et al.[97] have identified two different phases for cadmium octadecylfumarate LB layers, the dominant phase depending upon the total film thickness. Both phases possess a Y-type structure but differ in the angle of tilt of the paraffin chain. Such effects are examined further in Section 4.3.

The X-ray technique has also been used to investigate fatty acid films of mixed (within the layer) materials[67,73]: in one case an intermediate d-spacing between the individual fatty acids was obtained[67]; but in the other, the d-spacing appeared to be determined by just one of the components.[73] As well as the determination of film thickness, X-ray diffraction data may be used to obtain electron density profiles across (i.e., normal to the substrate plane) multilayer films.[104,526,531,532]

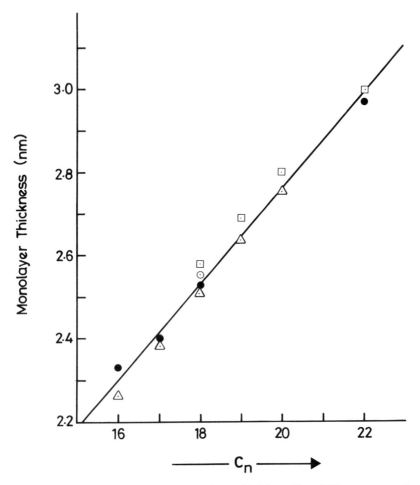

Figure 4.2. X-ray data listed in Table 4.1, plotted in graphical form. The solid line represents a least-squares fit to the experimental points: ●, Srivastava and Verma[28]; □, Mann and Kuhn[70]; △, Matsuda et al.[73]

4.2.4. Neutron Diffraction and Reflection

Neutron diffraction from manganese stearate LB layers has been reported by Nicklow et al.[80] Owing to the large cell spacing of the multilayer structure (~5 nm), a relatively long neutron wavelength of 0.415 nm was used in order to improve the resolution of the Bragg reflections. Neutron diffraction was observed from as few as three monolayers of manganese stearate on a 4 cm² plate. The experimental results were shown to agree well with those calculated from a model structure. However, the theoretical fits to the neutron data were not found to be as good as those to X-ray data. Neutron diffraction data for various fatty acid LB films, prepared under different conditions, have also been presented by Buhaenko *et al.*[530]

Results of the critical reflection of neutrons from LB multilayers of cadmium arachidate-d_{39} on glass have been reported by Highfield *et al.*[118] The deuterated compound was chosen because its neutron refractive index was suitable for the experiment. Interference fringes from the organic layer were observed; the spacing of these gave a measure of the overall thickness of the film and hence the average thickness of a molecular layer. Using a model which treated the film as a uniform slab, the average thickness per layer, 2.46 ± 0.02 nm, was found to be less than the observed neutron Bragg spacing (2.67 ± 0.01 nm). An alternative model based on (1) a zone of slightly reduced scattering density in the glass surface and (2) a decrease in the thickness of the first few layers was shown to give a good fit to all the interference and Bragg diffraction data obtained for multilayer samples prepared on hydrophobic plates. However, multilayers prepared on untreated (hydrophilic) glass seemed to conform to the uniform slab model.

4.2.5. Alternative Methods

Other techniques for measuring LB film thickness include the use of a mechanical probe (e.g., a Talystep) and electrical (e.g., capacitance) measurements. The former method requires a mechanical stylus to be moved across a step presented by the film to be measured: in this way Walpita and Pitt[119] have measured the thickness of approximately 200 layers of fatty acid salts, obtaining a value of 2.44 nm for the thickness of a cadmium stearate monolayer. The method has also been used by other workers.[103,107,120] The electrical technique involves the fabrication of a series of parallel plate capacitors from the LB multilayers. The method essentially yields the dielectric thickness (i.e., metric thickness ÷ relative permittivity) of the monolayer; the permittivity (at the frequency of measurement) must be known in order to evaluate the true metric thickness. This is discussed below in Section 4.5.5. Another method of probing the thickness of LB layers is to excite the film as an optical waveguide[121]; several modes can be guided along a sufficiently thick film (~500 nm) and the wave velocities of the mode can be determined.[10,119,122,533] The surface plasmon resonance technique[522] (see also Section 4.4.1) may also be used to determine the thickness of multilayer films.

4.3. FILM STRUCTURE

Structural information from monolayer and multilayer films can be obtained in a number of ways: X-ray diffraction, electron diffraction, optical birefringence, infrared, and Raman spectroscopy have all been exploited. Such techniques can provide information about the average orientation of chemical groups in the film. The use of X-ray and neutron beam diffraction techniques to determine the thickness of monolayers has been mentioned in the previous section. For highly ordered films, electron diffraction is one of the most useful techniques, combining speed of measurement with ease of interpretation.

4.3.1. Electron Microscopy

Electron diffraction was first used by Havinga and de Wael[123,124] and Germer and Storks[125] for investigating monolayer and multilayer films of fatty acids and their salts. More recent work using this technique has included studies on biological and polymeric materials, and LB films containing aromatic groups.[68,74,79,81–83, 95–97,126–146,526,534–538] Both transmission electron diffraction (TED) and reflection high-energy electron diffraction (RHEED) have been used. Some workers have simply reported on micrographs obtained using scanning or transmission techniques.[147–163,539] Samples for transmission work were originally produced by depositing films onto thin supporting organic foils.[125] However, more recent workers in this field have used the sample preparation technique of Walkenhorst[164] and Zingsheim[165]: this involves deposition of the LB film onto a previously anodically oxidized aluminum substrate; the LB film on its alumina support is then removed by etching the aluminum layer in a mercuric chloride solution. Alternatively, LB films may be floated from glass substrates in highly diluted hydrofluoric acid and picked up on a carbon-coated copper microscope grid,[79] or simply transferred to carbon-coated microscope grids by raising the latter through the condensed floating monolayer.[139] For materials which sublime in the decreased pressure of an electron microscope, special techniques have been devised. Day and Lando[131] describe such a method for use with multilayers of vinyl stearate in which the multilayer structure is essentially sandwiched between layers of cellulose acetate; the diffraction pattern from the cellulose acetate is so diffuse and weak that it does not obstruct the diffraction of the LB material. Specimens for study using electron microscopy can also be prepared using standard replication techniques.[127,150,152,154,156–158] Barraud et al.[162] have taken advantage of the chemical reactivity in LB films to introduce heavy ions (silver) into the polar planes of the layers and thus decorate these for direct-transmission electron microscope observations. RHEED is less demanding on sample preparation: LB layers can be deposited onto a variety of substrates and then rapidly studied.

For fatty acid type LB films, the electron diffraction experiments reveal the packing of the C_2H_4 subunits in the aliphatic chains. For infinite chains, there are three crystal structures with similar subcell packing densities: orthorhombic (R), monoclinic (M), and triclinic (T). For finite molecules there are a number of different variations of each subcell symmetry, and these can be conveniently distinguished by the Miller indices of the interface plane between layers of molecules. For example, a structure presented by M(001) would have monoclinic packing with the (001) plane parallel to the substrate.[136,166,167] The possible packing forms are listed in Table 4.2, together with the angles made by the (001) plane normal and the chain axis with the substrate normal.[136]

Electron diffraction investigations have shown that the packing and state of order

Table 4.2. Angles Made by the (001) Plane
Normal and the Chain Axis with the Substrate
Normal for All Possible Packing Forms, for All
Substances with Close-Packed Aliphatic Chains
(After Peterson and Russell[136])

Packing	Angle to substrate normal (deg)	
	(001) Plane	Chain axis
R(001)	0	0
R(011)	19	19
R(101)	27	27
R(111)	32	32
R(021)	34	34
M(001)	0	0
M(0$\bar{1}$1)	32	32
M($\bar{1}$01)	33	33
T($\bar{1}$02)	18	19
T($\bar{1}$22)	24	39

of LB multilayers differ widely. In the original work of Germer and Storks,[125] the hydrocarbon chains of barium stearate molecules were found to form hexagonal arrays with their axes normal to the supporting surface and separated by 0.485 nm. Apart from the first monolayer, multilayers of stearic acid were discovered to have their hydrocarbon chains approximately in a plane defined by the surface normal and the dipping direction, and pointing upward from the substrate surface. Stephens and Tuck-Lee[129] have reported on the structures of lead stearate multilayers: the unit cell of the lead stearate was shown to be either monoclinic or orthorhombic, the former being considered the more likely. RHEED patterns from the work of Russell et al.[134] showed an orthorhombic packing, R(001), for cadmium stearate films deposited onto single-crystal InP. Transmission electron diffraction data obtained by Garoff et al.[146] revealed that cadmium stearate monolayers possessed a stable bond-orientation but poor translational order. With hexagonal symmetry the monolayer was composed of unit cells containing a single hydrocarbon chain with significant rotational motion about its long axis.

Peterson and coworkers have investigated the structure of 22-tricosenoic acid LB films in some detail, using both TED and RHEED.[135,136,138,142,535,536] Figure 4.3 shows a transmission electron diffraction picture obtained from a 21-layer film; the diffraction spots correspond to an orthorhombic packing of the C_2H_4 subcells. Despite the apparent presence of a single matrix of spots, the pattern can be indexed as arising from a twinned structure (twinning has also been reported in multilayers of vinyl stearate[131] and diacetylenes[79]). As a result of further work it was discovered that LB films of 22-tricosenoic acid were in fact polymorphic, with their structure depending on the precise deposition details. In all, four different packing arrangements were identified. Of these, the R(001) structure was produced

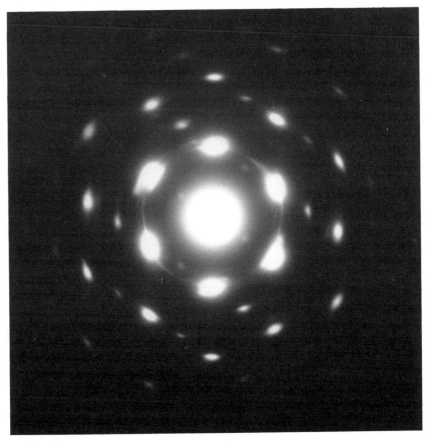

Figure 4.3. 100 kV transmission electron diffraction pattern for a 22-tricosenoic acid LB film. (After Peterson and Russell.[136])

over a wide range of deposition conditions; the other structures were T($\bar{1}$02), R(011), and R(110). The latter packing is unusual in that it does not conform to the conventional model of LB film deposition; it is possibly the result of an uneven transfer of the monolayers from the subphase surface to the substrate. Figure 4.4 shows indexed RHEED patterns for two of the observed structures: R(001)—Figure 6.4a, and T($\bar{1}$02)—Figure 4.4b. The structure of the 22-tricosenoic acid was found to vary with the subphase conditions: an acidified metal-free subphase resulted in a film with a tilted packing of molecules [R(001) or T($\bar{1}$02)], while a neutral cadmium-containing subphase produced a film with an upright packing of molecules.[142] Thus there is some agreement with the original observations of Germer and Storks.[125]

Further work using RHEED has revealed that the molecular tilt of 22-tricosenoic acid is a continuous and monotonic function of the deposition surface

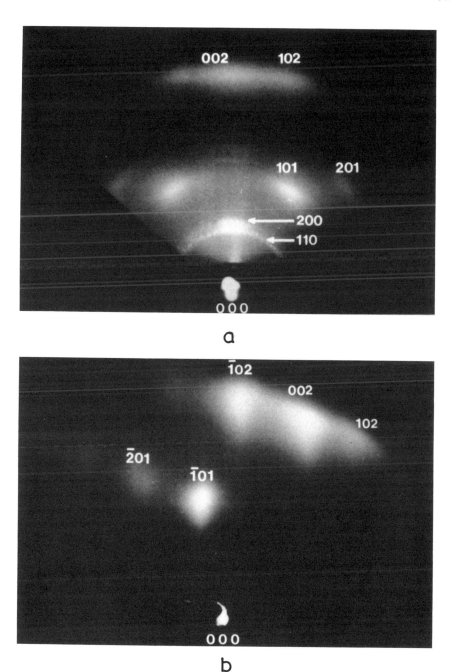

Figure 4.4. 100 kV RHEED pattern for a 22-tricosenoic acid LB film showing (a) R(001) packing and (b) T($\bar{1}$02) packing. (After Peterson and Russell.[136])

pressure.[535,536] Robinson *et al.*[535] have noted that the angle of inclination of the molecules cannot therefore be a characteristic of the subcell; a model consisting of very small grains of a mixture of subcell packings was used to explain this observation.

There is increasing evidence that the structure of the first few monolayers is different from that of subsequent layers. For example, RHEED studies by Jones *et al.*,[534] using 22-tricosenoic acid, have revealed that for a few monolayers of the fatty acid deposited onto hydrophilic silicon, a granular structure is observed in which the molecules are equally tilted by ~20° with respect to the substrate normal, but where the azimuthal angle of the tilt can take any value. For thicker films, the structure of the first few layers is retained, although subsequent layers develop a single tilt in the upward direction with respect to the substrate normal. Such a difference in the packing of the hydrocarbon chains in ultra-thin LB films has been noted by other workers using electron diffraction[141] and is also observed with other techniques, particularly with infrared spectroscopy (see Section 4.3.3).

Electron diffraction has also been used to study polymeric LB layers and to investigate phase changes as a result of the polymerization process. In many cases the electron diffraction data were used in conjunction with X-ray diffraction (and infrared spectroscopy) information in order to gain a more complete picture of the multilayer structure. Materials such as long chain esters and diacetylene monocarboxylic acids have been studied in some detail in these ways. Diffraction patterns of octadecylacrylamide and vinyl stearate monomer and polymer layers exhibited a hexagonal pattern caused by the packing of the hydrocarbon chains.[95,96,99,131,132] X-type multilayers of the vinyl stearate monomer were packed as in the bulk crystalline state (monoclinic) with the hydrocarbon chains inclined to the substrate normal at 28°.[96] This structure was preserved up to a conversion of about 50% to the polymer form. Increasing the polymerization resulted in a transition to a hexagonal packing of side chains which were oriented perpendicular to the substrate; the transition partly destroyed the order of the layers. In contrast, studies of octadecylacrylamide multilayers showed that no phase change occurred as a result of the polymerization.[99] Hexagonal packing has been observed for thin (<10) layers of the cadmium salt of the octadecyl monoester of fumaric acid.[97] Thicker layers showed an additional monoclinic structure, as observed in the bulk crystal. The monoclinic crystals were arranged in a hexagonal symmetry resulting from a transition from the original hexagonal phase. Both phases could be photopolymerized with the preservation of the integrity of the layer planes.

Early work with diacetylene carboxylic acids showed that the crystal structures of both the monomeric and polymeric forms were mainly controlled by the packing of the side chains.[130] Tieke *et al.*[79,83] reported that the monomeric multilayer possessed an orthorhombic subcell with zigzag planes of nearest-neighbor paraffin chains oblique to one another. The subcell of the corresponding polymer was also found to be orthorhombic. The polymerization proceeded without destruction of the LB assembly, the polymer chains growing one dimensionally within the layer plane to form a domain structure. If the original monomer was heated to over 60 °C inside

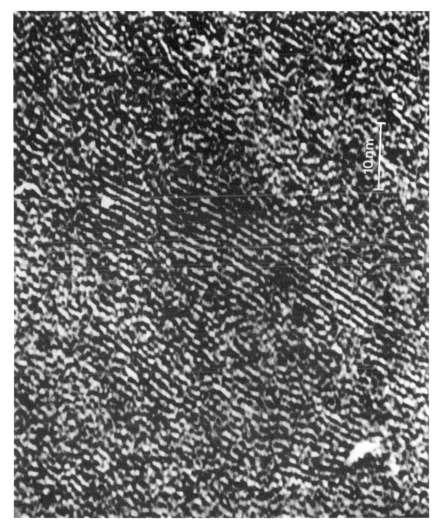

Figure 4.5. Electron microscope image of one monolayer of copper tetra-*t*-butylphthalocyanine. (After Fryer *et al.*[139])

the electron microscope, an irreversible change in the diffraction pattern was observed. This was thought to be the result of a phase transformation to a second monomeric form. The diffraction patterns for this monomer and its corresponding polymer were interpreted in terms of a triclinic subcell.

There has been some TED work on substituted phthalocyanine LB films[139,140,143,144]; these molecules do not possess the long hydrocarbon chains of classical materials. As a result, the order in LB films is found to be significantly lower. Electron diffraction data from a tetra-*t*-butylphthalocyanine have shown that the molecules are lying in columns in the plane of the film with an intermolecular separation of 0.33 nm and an intercolumnar separation of 1.9 nm.[139] A typical transmission electron micrograph (after image enhancement) obtained for just one monolayer of this material is given in Figure 4.5.[139,143] The lines of molecules are well ordered but general bending within the domain takes place, thereby highlighting the additional information available from direct imaging (in electron diffraction, a bend within a domain would be indistinguishable from two domains at an angle to one another). In Figure 4.5 there is only approximate alignment of the domains, implying that this LB film should be regarded as an amorphous matrix containing embedded crystallites.[139] Recent improvements in the quality of deposited films of phthalocyanine are discussed in Chapter 7.

The technique of electron diffraction has also been used to investigate the orientations of both the chromophores and the alkyl chains in LB layers possessing nonlinear optical properties.[145,537,538] The long axes of such molecules can exhibit large angles of tilt with respect to the substrate normal.[145]

Finally, Barraud *et al.*[539] have described scanning electron microscopy performed on in-plane conducting LB films. The application of a transverse electric field allowed the identification of electrical defects and the visualization of electrically active grain boundaries.

4.3.2. Optical Microscopy

A rapid assessment of the structure in an LB film may be made by observing it between crossed polarizers in an optical microscope.[10,83,138,142,168−172] For reasonable sensitivity this technique requires film thicknesses of roughly 500 nm. In such films, essentially just one packing and one orientation of the crystallites can be resolved. For fatty acids, birefringence behavior is ascribed to the presence of one of the tilted structures.[142] The precise orientation is partly determined by the initial monolayer, which fixes the set of packings and orientations available for subsequent epitaxial enhancement; it is also affected by the deposition process.[138,142] Epitaxial deposition has been noted by Bird *et al.*[173] for dye layer systems, and by Peterson and co-workers for a number of long-chain fatty acids.[170,172] LB layers of 22-tricosenoic acid normally exhibit crystallites of the order 100 μm in size. Figure 4.6 shows a photomicrograph of a film of this material deposited from an acidified subphase onto an HF-etched silicon substrate when viewed between crossed polarizers. Most of the area of the substrate shown in the figure was initially coated with a bilayer of stearic acid, onto which 170 layers of 22-tricosenoic acid were subsequently deposited.[138] The region in the top left-hand corner of the photograph was not covered with the stearic acid, providing clear evidence that the 22-tricosenoic acid had deposited epitaxially onto the initially deposited stearic acid bilayer. Such epitaxial development did not occur if the initial layer was produced by a chemisorption process, implying a very different structure for a chemisorbed and a LB-deposited monolayer.[142]

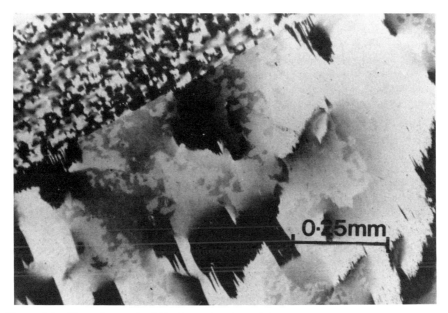

Figure 4.6. Photomicrograph of 22-tricosenoic acid (170 layers) deposited onto HF-etched silicon. The area of the substrate, apart from that shown in the top left-hand corner, was initially covered with a bilayer of stearic acid. (After Peterson and Russell.[138])

The ability of LB films to align liquid-crystal molecules has been studied by Hiltrop and Stegemeyer.[174] The effect has been exploited by Staromylnska *et al.*[175] in order to study defects in multilayers. In this technique the LB film (and substrate) is used as one of the plates in a liquid-crystal cell and then viewed in transmission or reflection using a polarizing microscope. This seems to offer a quick and simple method for studying damage in LB layers, and its dependence on the preparation. Moreover, the technique can be used with just a few monolayers. Gaines and Ward[160] have also noted the use of dark-field illumination in an optical microscope to study defects in thin fatty acid layers.

Another technique which has received some attention is fluorescence microscopy; the equipment for this has been described by Lösche and Möhwald.[176] The incorporation of dye probes into monolayers enables the observation of two-dimensional crystals on the subphase surface and also when transferred to solid supports.[137,163,176–181,540] It has been shown that monolayers manipulated on a water surface can be transferred to a substrate without grossly altering the surface texture.[178,179]

4.3.3. IR and Visible Spectroscopy

Infrared (IR) spectroscopy has proved to be a powerful tool for investigating the orientation of both the aliphatic chain component and other functional groups of

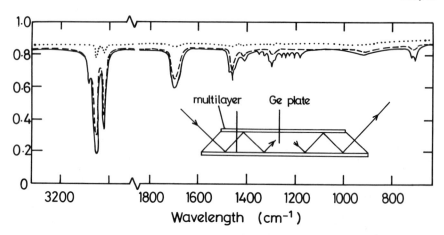

Figure 4.7. Infrared spectra of stearic acid LB films (33 layers). ATR spectra: ———, $R\|$; ------, $R\perp$. · · ·, Transmission spectrum. (After Takenaka et al.[184])

LB assemblies relative to the substrate surface.[98,141,182–210] Since the absorption of the infrared beam in the vicinity of a molecular vibration frequency depends on the relative orientation of the infrared electric field and the dipole transition moment, it is possible to obtain the details regarding the orientation of the molecule by using polarized radiation. A theoretical treatment of the absorption of linearly polarized IR radiation by organic monolayers has been given by Chollet.[193] Although sophisticated signal processing techniques (e.g., fast Fourier transforms) are now available for IR studies, one of the problems encountered initially by experimentalists was the lack of signal from such a small amount ($\leq 10^{16}$ molecules) of material. Francis and Ellison[182] overcame this by the deposition of LB monolayers directly onto two silvered mirrors and then employing a multiple reflection technique. Another approach was proposed by Takenaka et al.[184] in which attenuated total reflection (ATR) was used to obtain polarized IR spectra of 33 layers of stearic acid deposited on a germanium ATR plate. Hjortsberg et al.[211] have described a method which uses the resonant enhancement of surface plasmons at a metal/multilayer interface. Figure 4.7 shows the results of the ATR experiment; the inset depicts the sample configuration. The solid line in Figure 4.7 refers to an orientation of the electric vector parallel to the plane of incidence, and the broken line to the electric vector perpendicular to the plane of incidence. It is apparent that the ATR spectra bear a close resemblance to the simple transmission spectrum (dotted line) except for the peak intensity. The spectra reveal both the localized modes in the CH stretching region (2800–3000 cm^{-1}) and a number of band progressions involving a mixture of —CH$_2$— wagging, twisting, and rocking vibrations in the 1100–1500 cm^{-1} region. Also apparent in Figure 4.7 is a strong absorption at about 1700 cm^{-1} due to the stretching of the C=O bond. This band is not detected for multilayers of fatty acids deposited from subphases containing divalent ions (see, for example, Rabolt et al.[202]), confirming that the monolayers are transferred to the substrate as the fatty acid salt. The observation of this band has also proved useful in obtaining information concerning proton transfer in fatty acid/fatty amine alternate-layer films.[541,542] The infrared spectra of fatty acids may be additionally complicated owing to the existence of two possible configurations for the hydrogen-bonded carboxylic acid dimers.[210,212]

Although perfect polarization of the optical field perpendicular to the germanium plate cannot be achieved in an ATR experiment, from the data in Figure 4.7 Takenaka et al. estimated a tilt angle of 24° to 35° between the stearic acid hydrocarbon chain and the surface normal. In addition, polarized IR spectra of a multilayer sample obtained in transmission suggested that the molecules in the films were randomly oriented within the plane of the monolayer, irrespective of the direction of withdrawal during the film preparation. The uniaxial orientation of the molecules relative to the plane normal was found to be absent in mixed monolayer films of stearic acid and barium stearate[189]; in this case the orientation of the hydrocarbon chains was found to depend on the polishing direction of the Ge substrate during its cleaning procedure. In further work Kimura et al.[210] have reported on Fourier transform ATR spectra for LB films containing less than ten monolayers of stearic acid deposited on Ge plates. Examination of the CH_2 scissoring band (1468 cm^{-1}) suggested that the hydrocarbon chain of the stearic acid in the first monolayer was in a hexagonal or pseudohexagonal subcell packing in which each hydrocarbon chain was freely rotated about its axis oriented approximately perpendicularly to the surface; such a conclusion has also been reached by Bonnerot et al.[141] In contrast, for films thicker than two monolayers, the molecules in layers other than the first crystallized in monoclinic form. A striking feature in this study was the absence of the C=O stretching band in the first monolayer. The phenomenon was interpreted in terms of a short-range image field for the oscillating dipole at the Ge surface. Davies and Yarwood[543] have reported Fourier transform IR studies on small numbers of 22-tricosenoic acid layers deposited onto silicon. In this case the C=O stretching band was observed for a single monolayer. However, the first layer interactions were found to profoundly influence the C=O band shape and height.

Many studies of fatty acid monolayers using polarized infrared techniques suggest that the fatty acid tail is inclined at an angle of between 8° and 25° to the normal of the surface.[141,191,197,205] On the other hand, there is an increasing amount of evidence showing that the molecules in fatty acid *salt* multilayers are almost perpendicular (within experimental error) to the substrate.[141,190,202] These observations are consistent with the electron diffraction studies noted in Section 4.3.1, and also with the packing arrangements given in Table 4.2.

For tilted multilayers, some workers have found a preferential direction of tilt, which is related to the direction of withdrawal of the substrate during monolayer transfer[141,189,205] (see also Section 4.3.1). This preferential direction depends on the nature of the substrate, the precise conditions of the deposition, and on the number of monolayers transferred; it is mostly upward in the direction of substrate withdrawal and disappears for highly compact monolayers.[141,205] Chollet and Messier[205] have suggested that the hydrodynamic force that occurs during the transfer process is responsible for this, although Peterson[170] has reported some conflicting evidence.

The infrared spectra for small numbers of fatty acid salt monolayers, and the effect of evaporating metals onto these, have been reported by Knoll et al.[198] Data for cadmium arachidate, for the C—H stretching mode region between 2800 cm^{-1} and 3000 cm^{-1}, are reproduced in Figure 4.8. On the left-hand side a schematic diagram is given of the monolayer assembly whose IR spectrum is shown on the right. Figure 4.8a shows the infrared absorption for six monolayers deposited onto a flat Ag surface. Five bands are clearly visible, at 2851 cm^{-1}, 2874 cm^{-1}, 2918 cm^{-1}, 2933 cm^{-1}, and 2962 cm^{-1}. Figure 4.8b shows the spectrum for an alumina substrate dipped into a 10^{-3} M solution of arachidate acid and air dried. Radioactive label studies have shown that coverages of approximately one monolayer or less are obtained in this fashion. However, this layer will not be closely packed and consequently the spectrum shown in Figure 4.8b may be regarded as representing a completely disordered cadmium arachidate surface layer. Figures 4.8c and 4.8d show the infrared spectra for one monolayer of cadmium arachidate deposited on silver in the tail-down and head-down configurations. While the spectrum in Figure 4.8d is almost identical to that in Figure 4.8a, the relative peak intensities in Figure 4.8c are approximately intermediate in value to those in Figures 4.8a and 4.8b. This may be attributed to some structural differences in the tail-down and head-down monolayers. Although the head-down monolayer seems to be the more ordered configuration, the tail-down monolayer does not affect subsequent additional layers since Figure 4.8a is more similar in appearance to Figure 4.8d than to Figure 4.8c. The effect of evaporating a 10 nm Ag film on top of the tail-down and head-down monolayer samples is shown in Figures 4.8e and 4.8f, respectively (under these conditions an island Ag film is expected). There is essentially no change in the spectra after this treatment.

Infrared spectroscopy has also been used to study polymerization processes in LB layers.[94–99,130,133,207,213–219] In 1972, Cemel et al.[94] discussed the solid-state polymerization of multilayers of vinyl stearate. The kinetics of the reaction were followed by ATR IR spectroscopy and were found to follow a typical free-radical-type solid-state chain reaction. Subsequently Puterman et al.[95] extended the work to mixed multilayers of ethyl and vinyl stearate. In 1980, Banerjie and Lando[99] reported on the radiation-induced (^{60}Co γ) solid-state polymerization of oriented ultrathin films of octadecylacrylamide. IR spectra have also been used to monitor the polymerization processes of α-amino acids and a long-chain alkyl acrylate.[216,219] It was found that a regular arrangement with some rotational freedom of the monomer molecules in the layered structure markedly accelerated the polymerization reaction, polymerization rates being much faster in the spread monolayers than those in the built-up multilayers.[216] It has been suggested that the optimum packing of monomer molecules for polymerization resembles that of a smectic liquid crystalline state.

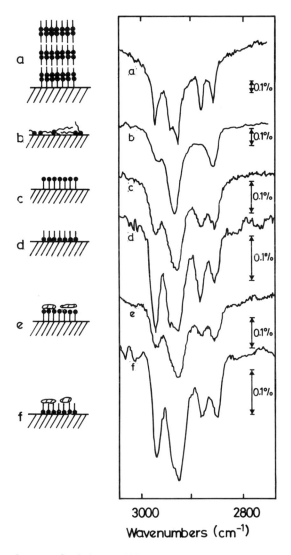

Figure 4.8. Infrared spectra of cadmium arachidate monolayer assemblies deposited onto a variety of surfaces. See text for a detailed description. (After Knoll *et al.*[198])

The electron-induced polymerization of 22-tricosenoic acid multilayers has been investigated by Barraud *et al.*[213] Such ω-bonded fatty acids undergo a particularly easy polymerization due to the degree of freedom left by the weak intermolecular Van der Waals forces.[220] An ionic mechanism was suggested to be responsible for the polymerization. Further work showed that the lattice-controlled propagation was stress-limited; thus it always stopped after the same number of links, giving rise to a nondispersed polymer.[221]

A number of workers have used IR and visible spectroscopy to study other chemical and photochemical reactions in LB multilayers.[222-231,544] A review of such processes has been presented by Bubeck.[230] Films of some spiropyrans have received some attention in this respect.[224,225,227,229] Changes observed in the spectra after chemical or light treatments were associated with structural changes and possible reorientation of the molecules in the organic films. The thermochromic behavior of a merocyanine dye LB film has been monitored using visible spectroscopy by Fukui et al.[232] IR techniques have also been used to study melting processes in cadmium arachidate multilayers.[233] The results indicate that a progressive disordering to the hydrocarbon tails occurred over a range of temperatures, beginning at 40–50 °C and extending up to the melting point. Such disorder was reversible, but not completely in the time frame (6–10 h) of the measurements. Thus, some questions are raised concerning the uses of conventional annealing techniques to remove disorder and defects from LB films.[233]

As already noted, the use of IR techniques to study LB films of biological materials is receiving some attention.[194,195,201,216,219,228,545-548] Such work may reveal information concerning fundamental processes in biological membrane structures. For instance, in the work of Howarth et al.[546,547] it has been possible to study the complexation and decomplexation reactions of the ionophore valinomycin with potassium.

The application of visible and ultraviolet (UV) spectroscopy to complex dye monolayer and multilayer systems is discussed in detail in Chapter 5. This technique has also been used to obtain structural information about fatty acid materials, in particular UV-polymerization processes in long chain diacetylenes.[75,79,81-83,130,217,234-238] The polymerization can be followed directly since the conjugated diacetylene backbone exhibits a strong absorption in the visible. The different phases of the monomer and polymer forms have already been noted in Section 4.3.1. Tieke et al.[82,83] have investigated various parameters influencing the polymerization process. They found the position of the diacetylene unit in the aliphatic chain to be significant in determining the photoreactivity and monolayer stability: acids with the diacetylene unit amid the aliphatic chain were highly photoreactive, but the monomolecular layers were of low stability.[82] The photopolymerization of diacetylenes can be sensitized for visible light by the use of surface-active dyes embedded in the layers.[217,239,240] Polarized visible spectroscopy has also been used to obtain information regarding the orientation of molecules in phthalocyanine LB layers.[241,242]

4.3.4. Raman Scattering

Raman scattering is an inherently weak process, so an enhancement mechanism is usually needed to directly observe LB layers.[198,243-257] However, unenhanced Raman spectra for fatty acid salt films have been reported by Dierker et

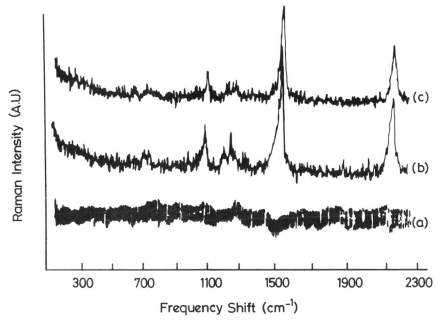

Figure 4.9. Raman spectra of polydiacetylene films: (a) one bilayer on a smooth Ag film; (b) one monolayer on a rough Ag island film; (c) one bilayer on a rough Ag island film. (After Chen et al.[254])

al.[549]; these workers have developed a high sensitivity Raman scattering instrument based on a charge-coupled-device multichannel detector. Signal enhancement can usually be achieved by depositing films onto the surfaces of noble metals such as Au or Ag (surface enhanced Raman). Although the phenomenon is not completely understood, it is generally accepted that a major portion of the signal increase can be attributed to the field enhancement of the incident and scattered radiation.[258] An example is given in Figure 4.9, in which data taken by Chen et al.[254] are reproduced. The spectra are for: (a) one bilayer of a polydiacetylene monomer deposited onto a smooth silver film; (b) one monolayer deposited onto a rough silver island film; (c) one bilayer deposited onto a rough silver film. We note that no Raman scattering was observed for the smooth substrate, but reasonably strong spectra were obtained for the LB films on the rough silver island films; these were attributed to the surface enhanced Raman effect. The two main peaks at 1521 cm^{-1} and 2123 cm^{-1} are due to the C=C and C≡C stretching modes of the diacetylene backbone, respectively. For multilayer polydiacetylene LB samples, shifts were observed in these two stretching frequencies, implying a modification in the film structure. Raman spectra have been measured by Knoll et al.[198] for cadmium arachidate assemblies deposited on flat and sinusoidal (grating) surfaces, on island films, and sandwiched between flat or grating surfaces and an island film. However, the spectra were less detailed than the IR spectra measured by the same workers (Figure 4.8). The intensity of the Raman signal was found to be independent of the number of layers deposited, supporting the view that the signal originates from a very localized (surface) set of molecules.

Enhancement of Raman signals from LB layers can be demonstrated dramatically by the interaction with surface plasmons. This is shown in Figure 4.10 in which a metal grating is used to excite the surface plasmon with high efficiency.[249,250] These spectra were taken under identical conditions, except for the angle

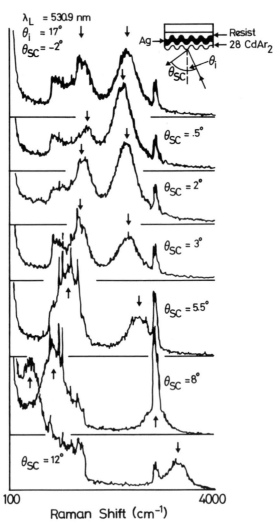

Figure 4.10. Luminescence and Raman scattering from a silver grating coated with 28 monolayers of cadmium arachidate for different scattering angles θ_{SC}. The arrows indicate the positions of the surface plasmon luminescence peaks. (After Knoll *et al.*[249,250])

at which the inelastically scattered light was collected. The sample was fabricated by coating a silver grating with 28 monolayers of cadmium arachidate. The spectra consist of very broad peaks, indicated by the arrows, and a collection of sharp bands corresponding to the Raman vibrational modes of the cadmium arachidate assembly. The two broad emission bands are due to surface plasmon luminescence; these first move together and then apart as the scattering angle is increased. The Raman bands of the LB material clearly become more intense whenever the direction of the Raman scattered light also corresponds to the direction of the luminescence. The effect is seen most clearly in the CH stretching region near 2900 cm^{-1}.

For materials with electronic transitions in the visible, the resonance Raman effect can be utilized to enhance certain bands.[243−245,251,252,255−257] Investigations have been reported for dyes adsorbed at the liquid–liquid interface,[243] for floating dye monolayers,[244] diacetylene multilayers,[245,252,255] and for a number of synthetic dye materials in LB film form.[251,256,257] Surface enhanced resonance Raman spectroscopy of various dye layers has been noted by Uphaus *et al.*[257] Vandevyver *et al.*[251] have suggested that polarized Raman scattering can be used to determine the orientation of the molecules in LB film form.

The application of Fourier transform Raman spectroscopy to LB layers of dye molecules has been described by Zimba *et al.*[550,551] Since no resonance enhancement occurs when excitation in the near IR is used, Fourier transform Raman allows vibrations not associated with the electronic chromophore to be observed.

4.3.5. Surface Analytical Techniques

Surface analytical techniques such as Auger electron spectroscopy (AES),[259−263] X-ray photoelectron spectroscopy (XPS),[43,66,78,190,263−272,552,553] and secondary-ion mass spectrometry (SIMS),[273] have found some applications in LB film studies. The first two methods involve the perturbation of the sample using either photons or electrons, and the characterization of the resultant emission.

In Auger electron spectroscopy the disturbance is produced by bombarding the specimen with low-energy electrons (1–10 keV). Some of the atoms within the sample are ionized and electron rearrangement takes place. The energy of an electron transferring to a lower energy level may be transferred to another electron (via an Auger process) which is then emitted from the solid. The energies of these Auger electrons are low (typically 20–1000 eV) so that, although they may be produced from as far within the sample as the original electron beam penetrates, only those which are generated within the first few atomic layers below the surface can escape with their original energies unchanged. Thus the technique has immense surface sensitivity. Figure 4.11 shows two AES spectra, obtained by Ginnai,[263] for a single layer of calcium stearate deposited on a tin substrate. The most noticeable feature is the complete absence of any structure which could be associated with calcium in the monolayer. It was suggested that cation exchange had occurred during monolayer transfer, resulting in the substrate actually becoming coated with tin stearate. In the case of spectra obtained from calcium stearate films deposited

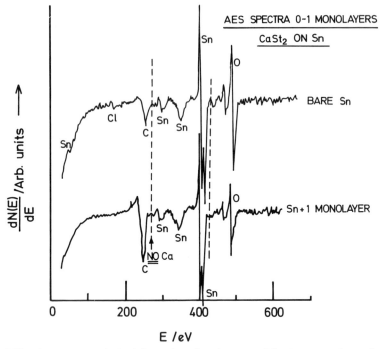

Figure 4.11. Auger spectra obtained from a base Sn substrate, and Sn + one monolayer of calcium stearate, illustrating cation exchange with a reactive metal substrate. (After Ginnai.[263])

onto gold substrates, no ion exchange is possible and evidence for calcium is observed in the spectra in a proportion which is in approximate agreement with that expected. Such effects were originally noted by Barraud *et al.*,[259,261] who presented AES data for calcium stearate films on aluminum substrates.

In the X-ray photoelectron spectroscopy experiment, the sample surface is irradiated by a source of low-energy X-rays; photoionization takes place in the sample producing photoelectrons of a characteristic energy distribution. Because X-rays do not normally cause appreciable surface damage, XPS is usually preferred as an analytical technique for organic materials. It has been used by Brandreth *et al.*[266] to show direct evidence for the incorporation of metal ions (from the sub-phase) into the transferred LB layers. Data from cadmium arachidate LB films deposited on Au, Ag, and In substrates have been reported by Brundle *et al.*[269] Figure 4.12 shows how the intensity of the $4f_{7/2}$ substrate core levels varies along a stepped multilayer, demonstrating quite clearly the reproducible nature of the signal from the underlying substrate for a given number of monolayers.

Attempts have been made to determine electron mean free paths in LB layers using AES and XPS.[43,263,269,270] However, although these values were found to exhibit the expected energy dependence, they were greater than those previously

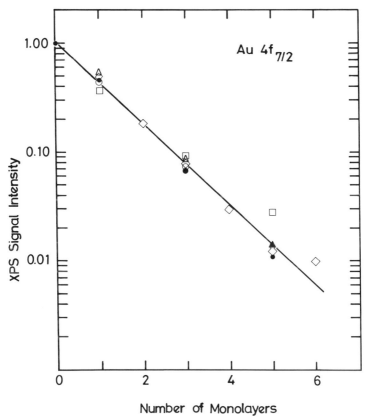

Figure 4.12. XPS signal intensity as a function of the number of monolayers deposited onto an Au substrate. (After Brundle *et al.*[269])

determined for other organic materials. It was suggested that the presence of small patches (0–4%) of bare substrate and/or surface roughness caused an overestimation of the mean free path.[269] Ginnai[263] has noted that film vaporization in vacuum for Au or Ag substrates could also account satisfactorily for the discrepancy.

Synchrotron radiation studies have been described by a number of workers.[85,87,274,275,554–556] Iida *et al.*[85,87] have noted the presence of an X-ray standing wave field during Bragg reflection in multilayer films. This could be exploited for structural analysis of superlattice systems. Angle-resolved photoemission using synchotron radiation has allowed Ueno *et al.*[247] to determine the intramolecular energy band dispersion for the valence bands of the $(CH_2)_{18}$ chain in LB films of cadmium arachidate. Comparison with similar results for long-chain alkane films and theoretical calculations for polyethylene have confirmed the existence of one-dimensional energy bands along the individual molecules. Oyanagi *et al*.[275] have reported data for merocyanine dye LB films using both X-ray absorption near edge structure (XANES) and extended X-ray absorption fine structure (EXAFS) spectroscopies. These workers have also presented EXAFS data for phthalocyanine films.[554] Near edge X-ray absorption fine structure (NEXAFS) has been used to obtain orientation information for monolayers of fatty acids and fatty acid salts.[555,556]

4.3.6. Other Characterization Methods

For systems containing paramagnetic species, electron spin resonance (ESR) may be used.[78,199,276–284,557–563] ESR provides angular distribution data con-

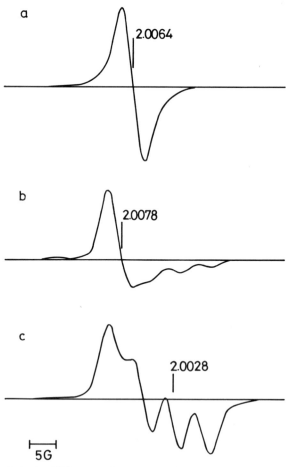

Figure 4.13. First derivative ESR spectra for a merocyanine dye system at room temperature. The external magnetic field is normal to the film plane in (a); it lies in the film plane with its direction parallel and perpendicular to the direction of the displacement of the sample during deposition in (b) and (c), respectively. Numbers show *g* values corresponding to the field positions. (After Kuroda *et al*.[283])

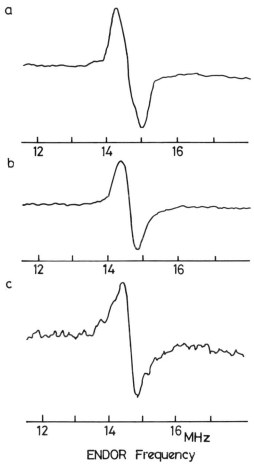

Figure 4.14. The anisotropy of proton ENDOR spectra for a merocyanine dye system at 130 K. The directions of the external magnetic field of (a) to (c) correspond to those of Figures 4.13a to 4.13c, respectively. (After Kuroda *et al.*[283])

cerning the magnetic ions in the multilayer structure. This method has been applied to simple fatty acid systems,[78,276,278] to dyes[199,277,279,281–284,557,558] to biological layers,[280] and to charge-transfer salts.[559–563] As will be discussed in Chapter 7, the technique is particularly valuable for studying phase transitions in two-dimensional arrays of magnetic atoms.[78,278] Electron nuclear double resonance (ENDOR) spectra have been reported by Kuroda *et al.*[283] for merocyanine dyes; Figures 4.13 and 4.14 show both the ESR and ENDOR data obtained by these workers. For Figure 4.13, the microwave power was 2 mW. The external magnetic field was normal to the film plane in Figure 4.13a, while it lay within the film plane in Figures 4.13b and 4.13c. The difference between the latter two figures is that the external field was parallel and perpendicular to the direction of sample deposition,

respectively. The numbers in the figures show corresponding *g* values. The anisotropy both in-plane and out-of-plane is evident, inferring the existence of an oriented radical species. Figure 4.14 shows the orientation dependence of the proton
ENDOR spectra. Figures 4.14a to 4.14c correspond to the external field directions
for Figures 4.13a to 4.13c, respectively; the spectra were recorded at the external
field positions marked by the *g* values in Figure 4.13. A field dependence study
revealed that these ENDOR signals were associated with the single spin species
observed by ESR. The line shape is also anisotropic, showing the radical species to
be preferentially oriented in the film.

Scanning tunnelling microscopy (STM) is potentially a very useful technique
for the structural study of LB films. It can produce images of surfaces with lateral
and vertical resolutions of less than 0.3 nm and less than 0.02 nm, respectively.
Furthermore, the electron energies are usually less than 3 eV, thus avoiding the film
degradation that can be a problem with other methods. A number of reports of STM
images on LB film systems have been published[564–566]; however, because of
substrate effects, care must be exercised in the interpretation of such data.

Other characterization techniques that have been exploited include photoacoustic spectroscopy,[285,286,567] and differential scanning calorimetry.[233,287,288]

4.4. OPTICAL PROPERTIES

4.4.1. Refractive Index

The refractive index of materials is determined by the degree of polarization of
the atoms or molecules of which they are comprised; this, in turn, is dependent not
only on the electric field of the incident electromagnetic wave, but also on that of
the electric dipoles produced in neighboring atoms or molecules. Thus the refractive
index is determined by the direction of the incident electric vector, the crystal
structure, and its density. In the case of multilayer LB films, it is therefore unlikely
that the refractive indices will be equal to those of the bulk material. Furthermore,
the values may vary with deposition conditions and also with the number of layers
in the assembly. One attempt to calculate the refractive index for barium stearate has
been made by Khanna and Srivastava[289]; however, the value of 1.13 obtained for
thick layers is considerably less than figures reported by most workers.

Many of the optically based techniques for determining the thickness of LB
layers also yield refractive index values; methods based on interference and ellipsometric phenomena have been discussed in some detail in Section 4.2, and will not
be repeated here. Measurements of the optical properties of cadmium arachidate
layers have been reviewed by Drexhage,[14] who reported literature values for the
ordinary (in the plane of the film) and the extraordinary index of refraction obtained at
specific wavelengths. The refractive index values are generally found to exhibit a
small frequency variation.[290] Figure 4.15 shows the dispersion curves for the two

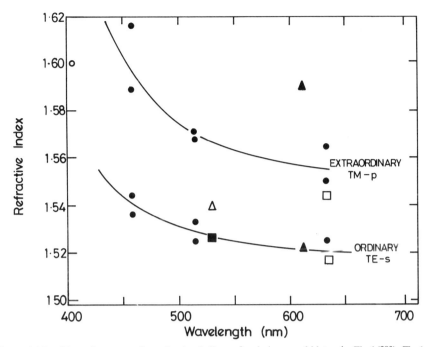

Figure 4.15. Dispersion curves for refractive indices of cadmium arachidate: ▲, Fleck[293]; □, den Engelsen[45]; ■, Drexhage[294]; △, Blodgett[295]; ○, Foster[296]; ●, Swalen *et al.*[291] The solid curve represents a smooth average. (After Swalen *et al.*[291])

refractive indices of cadmium arachidate, reproduced from the work of Swalen *et al.*[291] The graphs show values found using integrated optical techniques,[292] together with figures from other workers. Using similar methods, Pitt and Walpita[10] have reported on the refractive indices of cadmium stearate. At a wavelength of 633 nm, the values obtained were 1.527 ± 0.003 for the ordinary ray and 1.568 ± 0.017 for the extraordinary ray, in excellent agreement with the data for cadmium arachidate in Figure 4.15. However, in common with other workers (see, for example, Blodgett and Langmuir[12]) Pitt and Walpita discovered that the refractive indices depended on the amount of the fatty acid salt in the film. The values quoted above were obtained for films deposited at a subphase pH of 5.6. If the pH was adjusted to 5.0, then the two refractive index values for the transferred films would each be reduced by approximately 0.02. Marked decreases in the refractive indices of fatty acid salt layers can also be achieved by the process of skeletonization,[1,12,13] i.e., heating, or soaking the LB array in a suitable solvent to remove the free fatty acid. Measurements of the refractive indices of skeletonized films have been made by Tomar.[52] For barium margarate, the refractive indices were found to be 1.28 (ordinary ray) and 1.44 (extraordinary ray); these are of the same order as figures obtained for barium stearate by Blodgett and Langmuir.[12] Honig and de Koning[41] have reported on the change in refractive index with time during the skeletonization process.

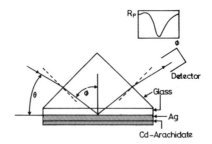

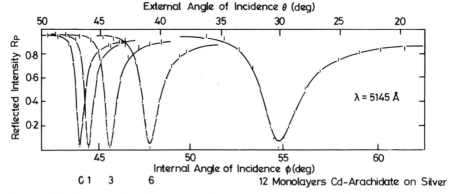

Figure 4.16. Top: experimental arrangement for the observation of surface plasmon resonance. Bottom: surface plasmon resonance data for silver films coated with different numbers of cadmium arachidate monolayers. Symbols are experimental points; solid lines are theoretical curves. (After Pockrand *et al.*[298])

The optical constants of LB layers may also be determined by using a reflection spectroscopy method.[297−302,568,569] In this experiment surface plasmons at a metal/dielectric boundary are excited optically by evanescent waves from the base of a prism (Kretschmann configuration). Figure 4.16 shows the experimental arrangement together with reflection data for cadmium arachidate monolayers deposited onto Ag.[298] The optical constants for the organic material were determined from these curves by using a least-squares fit to the exact Fresnel equations. Calculations were performed for both an isotropic and a uniaxial anisotropic refractive index. However, the anisotropic calculation did not appear to fit the experimental reflectivity curves any better than the isotropic calculation. Using the isotropic fit, the refractive index of cadmium arachidate was found to be 1.53 at a wavelength of 633 nm. In common with workers using ellipsometry (Section 4.2.2), problems were encountered if the uncoated Ag was taken as a reference (in this case, the refractive index of the LB assembly appeared to decrease with increasing numbers of layers). A modification was therefore made, taking into account an interfacial layer between the Ag and the LB film. The surface plasmon technique has also been used to evaluate the optical constants of LB dye monolayers.[30–302] The method might also be used as the basis for obtaining high contrast images.[570] More detailed discussion of these experiments and their potential use in sensor applications are given in Chapters 5 and 7.

4.4.2. Nonlinear Effects

The regularity of LB films and their controllable thickness permit nonlinear optical properties to be studied by transmission or reflection measurements. The macroscopic polarization, P, in a material may be expressed generally in the form[303]

$$P = P_0 + \chi^{(1)}E + \chi^{(?)}E^2 + \chi^{(3)}E^3 + \cdots \tag{4.2}$$

where P_0 is a constant, E is the applied field, and $\chi^{(n)}$ is the nth order susceptibility tensor. The $\chi^{(2)}$ susceptibility is responsible for nonlinear phenomena, such as second harmonic generation (frequency doubling) and the linear electrooptic (Pockels) effect; $\chi^{(3)}$ governs third harmonic generation and the quadratic electrooptic (Kerr) effect. Nonlinear properties are normally measured on macroscopic samples that consist of many individual molecules. On the molecular level, a molecule placed in an electric field experiences a polarizing effect measured through the dipole moment, μ; its relation to E is expressed in terms of the linear polarizability α (which is the origin of refractive index), and of the higher-order hyperpolarizabilities β, γ,. . . . Thus

$$\mu = \mu_0 + \alpha E + \beta E^2 + \gamma E^3 + \cdots \tag{4.3}$$

where μ_0 is a constant; α, β, γ, etc. are, of course, tensor quantities.

Conjugated molecules possess π-electrons which, under certain circumstances, are loosely bound to the molecules and can contribute considerably to the molecular polarizabilities. As a consequence some organic materials exhibit the largest known nonlinear susceptibility coefficients, often considerably larger than those of the more conventional inorganic dielectrics.[303,304] However, an important consequence of crystal symmetry is that, in a material possessing a center of inversion, the lowest-order nonlinear susceptibility $\chi^{(2)}$ is zero. Second-order nonlinear optical effects are therefore not expected for Y-type multilayers. However, there have been reports of second harmonic generation in floating monolayers,[305] and in LB films deposited either as single monolayers[306–308,523,524] or as Y-type alternate-layer assemblies.[309–311,523,524] Second order processes have been noted for Z-type films, although such layers are not always of high quality.[523,524,571–574] Electric-field-induced second-harmonic (EFISH) experiments have also been reported on centrosymmetric multilayers.[312]

Third harmonic generation does not require a centrosymmetric structure and can therefore be observed in conventional Y-type films. The third-order nonlinear susceptibility, $\chi^{(3)}$, is large in the polydiacetylenes and there have been a number of investigations into the nonlinear optical properties of multilayer films of this material.[217,313-317] An LB material that may exhibit significant second and third order effects has been described by Tsibouklis et al.[575] A full review of the nonlinear data is given in Chapter 7.

Changes in the optical properties of organic materials may also be produced by applying large electric fields. Electroabsorption is one technique that has been used to study multilayer films. This is a particular branch of the group of experimental methods known as modulation spectroscopy, and involves monitoring small changes in a sample's optical transmission which result from the application of an external field. For molecular materials, these changes are usually a result of the Stark effect. The change in potential energy, ΔU, which arises when a molecule is placed in an electric field may be written as[318]

$$\Delta U = -\mu E \cos\theta + \frac{\alpha}{2} E^2 \tag{4.4}$$

where the dipole moment μ makes an angle θ to the applied field, and α is the polarizability tensor. The first term on the right-hand side of equation (4.4) represents the linear Stark effect, and the second term the quadratic effect. In general, crystals with polar space-groups exhibit a first-order Stark effect. This is usually recognized in electroabsorption spectra as the second derivative of the zero-field absorption curve. Blinov et al.[319,320,576] have reported a study of the linear Stark effect in multilayers of an amphiphilic azo compound. By varying the deposition conditions, X-, Y-, and Z-type multilayers of this material could be fabricated. Figure 4.17 shows the linear Stark spectra for the X- and Z-type layers. The reversal of the sign of the signal in changing from X- to Z-type layers provides direct evidence for the polar nature of these films. Furthermore, the intensity of the signal for the Y-type layers (not shown in Figure 4.17) is approximately 50 times less than that for the X- or Z-type films.

Crystal structures having nonpolar space groups may still exhibit a second-order Stark effect. This process is recognized in electroabsorption as a first-order derivative of the zero-field absorption curve and has been reported by Roberts et al.[321] for LB films of a substituted anthracene. At the high fields necessary to obtain electroabsorption ($>10^8$ V m^{-1}), electroluminescence was also observed from the LB multilayer. This was attributed to double injection effects at the metallic electrodes (see Section 4.5.3).

Electric-field-induced changes in the optical absorption spectra for some LB film systems have been observed. For instance, Bücher et al.[322] have noted both linear and quadratic electrochromic effects in a merocyanine dye/cadmium arachidate assembly. However, quantitative measurements were hindered by the lack of information regarding the orientation of the chromophore.

The change in the refractive index on application of an electric field to a nonlinear optical LB material (Pockels effect) has been described by a number of workers[572,573,577] and is discussed in Chapter 7.

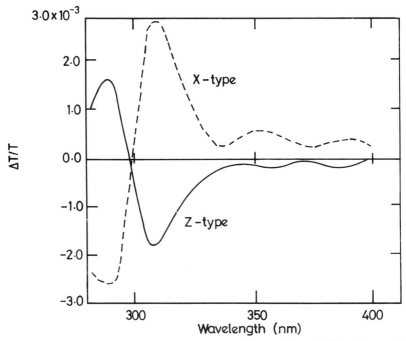

Figure 4.17. Spectra of linear Stark effect $\triangle T/T$ for 60 X-type and Z-type LB films of an amphiphilic azo compound. (After Blinov *et al.*[319])

4.5. ELECTRICAL CHARACTERIZATION

4.5.1. Specimen Preparation

The first attempts to investigate the electrical properties of LB films were made in the late 1930s, a few years after Katharine Blodgett reported the technique to build up multilayer structures on solid plates. Early workers quickly encountered many problems such as poor quality films, the difficulty of making reliable electrical contacts of known area, polarization in the LB layers, and the lack of sensitivity of the measuring equipment.[323–325] Electrical measurements on LB layers (particularly monolayers) are probably the most stringent test of film quality and possibly the most difficult of all the available characterization techniques. Although there have been many reports on the structural and optical properties of LB layers throughout the second part of this century, it is only comparatively recently that reproducible and reliable electrical data have started to be obtained; this is almost certainly due to improved film preparation and handling techniques. However, there still remain many problems in this area: the effects of oxide layers on the electrode surfaces, of pinholes and other defects in the organic layer,[326,327,578–581] and the relationship between the conductivity and film structure[328] have yet to be fully explored.

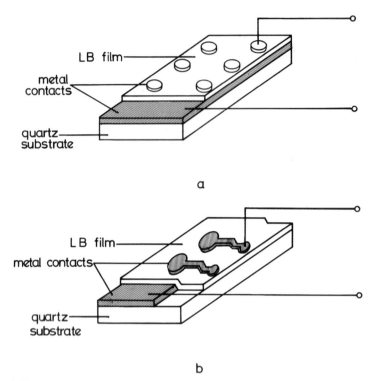

Figure 4.18. Alternative electrode arrangements to enable investigations of the electrical properties of LB layers.

Nearly all experiments to investigate the electrical properties of LB layers require the films to be contacted directly by two electrodes. There are however exceptions to this; for instance, the measurement of surface potential uses a vibrating second electrode at a small distance from the surface of the multilayer. Although some early studies involved the use of electrolytes and mercury to establish electrical contacts, solid metallic electrodes are generally favored by more recent workers. The solid substrate on which the LB film structure is deposited can usually be used as one electrode (e.g., aluminized glass); different approaches have been explored in order to make the second contact. Two typical electrode configurations that have been fabricated are shown in Figure 4.18. The structure in Figure 4.18a is based on a simple sandwich geometry; this is straightforward to produce, but it is not always easy to make a reliable electrical contact to the top electrode without affecting the LB layer. Therefore some workers[70,329–333] favor the arrangement shown in Figure 4.18b; contact to both electrodes can now be readily accomplished by direct soldering. However, in this case not all of the lines of the electric field are perpendicular to the plane of the LB layer; this might well lead to problems in the interpretation of conductivity data for highly anisotropic materials.

The electrical conductivity of LB films has been found by many workers to depend upon the ambient.[334-336] Higher conductivities are invariably found if the measurements are made in air, probably due to the presence of moisture. In fact, it has been proposed to exploit this phenomenon in an LB-film-based hygrometer.[337] Internal voltages, generated by electrochemical reactions between the metal electrodes and the LB film material, may also cause problems in interpreting electrical data.[335,336,338-341] However, such effects can be minimized if the measurements are taken in a dry environment.[335,336] Some workers[342] have reported that, to obtain reproducibility, electrical measurements on LB films needed to be taken in a high vacuum (about 10^{-5} mbar). It should be noted, however, that some LB films may be volatile in such conditions. Using carbon 14-labeled stearic acid, Roberts and Gaines[343] have shown that the stability of monolayer films in vacuum is very dependent upon the monolayer composition, the conditions of deposition, and the nature of the substrate.

The following subsections summarize the results and conclusions of many researchers who have investigated the electrical conduction properties of monolayer assemblies. However, in only a few cases is there sufficient detail for the reader to be entirely convinced of the soundness of the interpretation. This stems from the fact that aluminum, with its attendant oxide layer, has normally been the favored substrate. The effects of macroscopic defects, such as grain boundaries and pinholes, have also been largely ignored. Nevertheless, for completeness all available electrical data have been included. With the development of more stable LB film materials, and greater attention to experimental detail, it is hoped more reliable conductivity data will become available.

4.5.2. Quantum Mechanical Tunneling

The first report of quantum mechanical tunneling through LB layers was due to Miles and McMahon[344] in 1961. These workers used a monolayer of barium stearate sandwiched between tin and lead electrodes. However, problems with device reproducibility and stability were encountered. Subsequently a number of other investigations into the dc conductivity of monolayers were reported, including those of Clark,[345] Beck,[346] and Scala and Handy.[347,348] Although there was evidence to support the view that tunneling was a significant conduction process in LB films only a few monolayers thick, the poor reproducibility of the electrical data precluded the formation of any definite conclusions. In 1968 Horiuchi et al.[349] reported further measurements on the conductivity of barium stearate layers, identifying both quantum mechanical tunneling and Schottky emission as the dominant conduction processes. However, in order to interpret the experimental results, the effect of pinholes in the transferred monolayers needed to be taken into account.

The first quantitative evidence for tunneling through LB layers came, in 1971, from the work of Mann et al.[69] and Mann and Kuhn.[70] The data obtained by these workers were shown to fit the tunneling model proposed by Sommerfeld and Bethe[350]; for small values of applied voltage the tunneling conductivity, σ_t, may be written in the form

$$\sigma_t = \left(\frac{e}{h}\right)^2 (2m\phi)^{1/2} \exp\left[\left(\frac{-2\pi t}{h}\right)(2m\phi)^{1/2}\right] \qquad (4.5)$$

where e is the charge and m the mass of an electron, h is Planck's constant, t is the thickness of the insulating layer, and ϕ is the difference between the work function of the metal and the electron affinity of the LB layer (i.e., the metal/insulator barrier height). Mann and Kuhn investigated a range of cadmium salts of fatty acid LB monolayers. In these experiments the fatty acids were deposited onto aluminum substrates. Mann and Kuhn assumed that their experiments measured a total conductivity, σ, which was comprised of an intrinsic contribution, σ_i, that was independent of film thickness, in addition to the tunneling component, σ_t (i.e., $\sigma = \sigma_i + \sigma_t$). Therefore experimental values of σ needed to be corrected for σ_i (obtained from measurements on multilayer structures) in order to obtain the true tunneling conductivity. Values of electron affinities of 2.25 ± 0.25 eV for the fatty acids were found to fit well with photoelectric data. The height of the Al/cadmium stearate barrier of 1.94 ± 0.25 eV was also found to be in good agreement with the value obtained from the photoelectric experiment (2.3 ± 0.1 eV). Furthermore, the effects of changing the top electrode material were as predicted from the differences between the vacuum work functions of the metals used. In order to account for the excellent correspondence between theory and experiment, Mann and Kuhn concluded that the number of defects in their monolayers was less than one part in 10^8. Kuhn subsequently noted that quantum mechanical tunneling might be a significant electronic conduction process in the lipids of biological membranes.[351]

Similar tunneling data have been obtained by Yamamoto et al.[352] for a variety of fatty acid monolayers. These workers found that the current densities obtained through their films were approximately two orders of magnitude smaller than those measured by Mann and Kuhn, although the shapes of the current–voltage curves were similar. After correcting their conductivity data in a similar fashion to Mann and Kuhn, a value of 2.53 eV was obtained for the height of the Al/barium stearate barrier.

In 1977 Polymeropoulos[333] published an extension to the work of Mann et al. Experimental data on monolayer thicknesses from 2.1 nm (1 : 1 cholesterol : myristic acid) to 2.8 nm (arachidic acid) were reported; these are shown in Figure 4.19. The tunneling conductivity (measured at an applied voltage of 10 mV) dependence upon film thickness is as predicted by equation (4.5). The conductivities measured in this work were, once again, smaller than those of Mann and Kuhn and were assumed to be the true tunneling conductivities as no intrinsic contribution could be measured. Vincett and Roberts[5] have suggested that this might be due to the fact that the

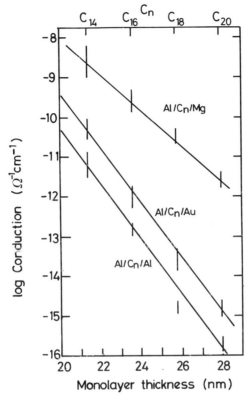

Figure 4.19. Logarithm of the conduction versus monolayer thickness, C_n, for different metal/monolayer/metal structures. (After Polymeropoulos.[333])

experiments of Polymeropoulos were made in a dry nitrogen ambient. As already noted, other workers have also observed that the conductivity of LB layers decreases if devices are stored for a period in a dry ambient.[334–336,353,354] The effect is probably associated with the removal of water from the LB layer and also from the oxide (e.g., Al_2O_3) on the base electrode, although Tredgold and Winter[353] propose that this process results in the annealing out of defects in the monolayer. The value of the electron affinity of the fatty acid monolayers obtained from measurements of all three types of junction shown in Figure 4.19 was found to be 2.3 ± 0.7 eV, in good agreement with the value obtained by Mann and Kuhn. The film thickness range in Figure 4.19 has been extended even further by Polymeropoulos and Sagiv[355] by the use of adsorbed monolayers. These layers were found to be as stable under the influence of a dc electric field as monolayers deposited by the LB technique. However, the films required cooling to 77 K in order to observe the tunneling.

At applied voltages of 50–100 mV the current–voltage behavior of Al/mono-layer/Al sandwich structures usually becomes nonlinear[263,338,356] and deviates from the relationship given by equation (4.5). This behavior is inconsistent with theoretical predictions based on film thickness and on values of barrier height obtained by photoelectric measurements, and has been attributed to the presence of a small potential barrier in the aluminum oxide film caused by impurities.[356] Ginnai[263] has suggested that the nonlinearity might be caused by the LB film itself.

As the applied voltage is further increased, the dc conductivity data for mono-layer films have been shown to follow the theoretically predicted current–voltage relationship for quantum mechanical tunneling. For example, using the Stratton model for a symmetrical rectangular barrier,[357] the tunneling current density, J, for $V \le \phi$, can be expressed by the following equation:

$$J = J_0 \frac{\pi C k T}{\sin(\pi C k T)} \cdot \exp(-BV^2) 2\sinh\left(\frac{CV}{2}\right) \tag{4.6}$$

where J_0, B, and C are constants related to the barrier height ϕ. Conductivity data for an Al/barium stearate monolayer/Al structure, obtained by Ginnai,[263] are shown in Figure 4.20. The characteristics obtained at various fixed temperatures have been normalized to a single curve, using the procedure suggested by Hill,[358] in order to demonstrate that a common characteristic is being exhibited. The cur-rent–voltage curves were found to be symmetrical with respect to the polarity of the applied field. The full line in Figure 4.20 was obtained by fitting equation (4.6) using a least-squares routine. An excellent agreement with the experimental points, over almost four decades of the current, is obtained; however, a low value (0.65 eV) is revealed for the Al/barium stearate barrier. Ginnai has noted that if the tunneling distance used in his analysis were to be increased to include the thickness of the native oxide on the aluminum electrodes, then a barrier height much closer to that obtained by Mann and Kuhn[70] and by Polymeropoulos[333] would be obtained. Fitting of tunneling theories to conductivity data has also been undertaken by Careem and Hill.[359] At 300 K the log (current) versus voltage curves were found to follow a hyperbolic sine curve up to an applied voltage of about 0.5 V; a value of approximately 0.5 eV for the Al/stearic acid barrier height was inferred from this.

A technique that has been used by some authors to directly obtain the tunneling barrier height is to plot the logarithm of the value of $d\ln J/dV$ against voltage; a maximum in this curve is predicted for $eV = \phi$.[360] A further test of electron tunneling between metal electrodes is to make at least one of these electrodes a superconductor. The theory of Bardeen, Cooper, and Schrieffer postulates the exis-tence of a forbidden energy gap centered at the Fermi level of a superconductor[361]; as a consequence a deviation from the low voltage ohmic tunneling characteristic obtained using normal metal electrodes [i.e., equation (4.5)] can be expected. Such evidence has been presented by Leger et al.[338] who studied monolayers of cyanine with two grafted stearic chains. The appearance of the nonlinearity at low tem-

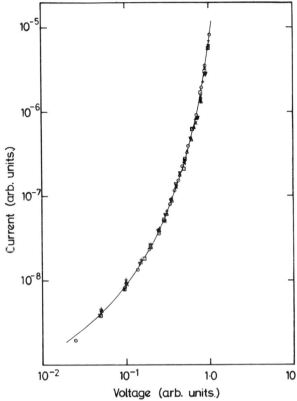

Figure 4.20. Normalized current–voltage characteristics for an Al/barium stearate monolayer/Al structure. Data have been obtained over the temperature range 80 K to 300 K and have been normalized using the procedure of Hill.[358] Voltage range approximately 10^{-2} V to 1 V. (After Ginnai.[263])

peratures due to the superconducting energy gap of the Pb was a clear indication of quantum mechanical tunneling. Similar effects have been observed by Polymeropoulos[362] and by Ginnai[263] for fatty acid monolayers. Josephson tunnel junctions based on polymerizable LB films have also been fabricated.[363–365,582] However, while Larkins et al.[364] suggest that their experiments (using polyvinyl stearate) support a tunneling mechanism through the organic layer. Hao et al.[365] propose that filamentary conductors spanning their LB film (a long chain oxiran) are significant. Iwamoto et al.[582] have found that the metallic pathways in their LB films have a diameter that is smaller than the London penetration depth of the metal contacts. There is ample evidence for defects in LB structures from other investigations.[327,578–580]

Although most of the electrical data published to date on LB monolayers strongly indicate that quantum mechanical tunneling is an important conduction process, some of the interpretations can be criticized on the grounds that not enough importance has been attached to the role of the interfacial oxide layers associated with the metallic electrodes. For instance, Ginnai[263] has demonstrated that it is possible to explain some of his data by assuming a conduction process which is dominated by electron tunneling through the oxide and then via pinholes and other defects in the monolayer. It is interesting to note that if some of the current–voltage data obtained by Ginnai et al.[263,366] are replotted in a log (current) versus voltage$^{0.25}$ form, excellent straight lines are obtainable. This current versus voltage relationship was first reported by Roberts et al.[367] for a monolayer of cadmium arachidate, and also by Lloyd et al.[368] and Tredgold et al.[328] for multilayer LB films. The latter workers have suggested that this might be the result of a defect conductivity mechanism, proposing that the evaporated Au electrode diffuses down defects (e.g., grain boundaries) in the film. A plausible interpretation[369] is that the monolayer film simply improves the ability of the underlying oxide to sustain a large voltage. This then makes it possible to reach the high field strengths where one can observe the effect of image forces on the charge flow through a Schottky barrier.

The disagreement between various workers on the magnitude of the aluminum/fatty acid barrier height could well be a direct result of the presence of the Al_2O_3 layer. Unfortunately it has not been possible to fabricate sandwich structures in which both electrodes are completely free from native oxide layers. Although monolayers can be deposited onto Au, attempts to evaporate metallic top electrodes invariably result in devices which are short-circuited.[348] This has also been found to be the case for adsorbed monolayers,[355] and suggests that the oxide layer plays an important role in determining the electrical properties of monolayer films. However, in view of the large quantity of experimental data, showing that the tunneling conductivity depends strongly upon monolayer thickness, it seems extremely unlikely that the insulating properties of metal/monolayer/metal sandwich structures are derived from the oxide layer(s) alone. A more probable explanation for the short-circuited Au/LB film/Au structure is that the monolayers are less strongly bound to the gold surface and consequently can evaporate in the high vacuum during deposition of the top electrode.[343,348]

Recent work by Geddes et al.[370,583,584] has demonstrated that silver–LB film–magnesium devices can be fabricated which are substantially free from oxide layers on the electrodes. The use of such a structure might well enable some of the above problems to be resolved.

One experimental result which seems to support the view that quantum mechanical tunneling does occur in monomolecular LB films is an inelastic tunneling spectrum (IET), obtained by Ginnai et al.[262,263,367] In addition to the normal elastic tunneling of electrons (i.e., tunneling through a potential barrier to a state at the same energy level), it is possible for a small proportion, typically $<1\%$, of the electrons to tunnel inelastically, losing energy to excite molecular vibrations in the barrier.[371] Thus the observation of this small number of relatively rare inelastic tunneling events would seem to be strong evidence for tunneling through the LB layer. The inelastic tunneling spectra are obtained by electronically measuring d^2V/dI^2 as a function of junction bias. The onset of inelastic events associated with

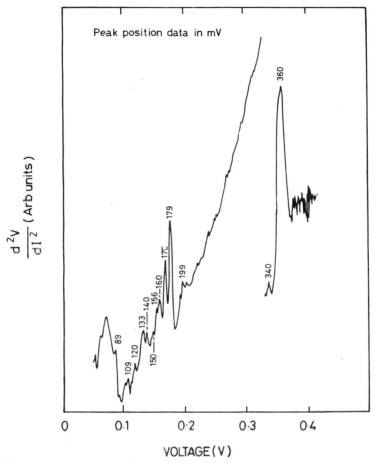

Figure 4.21. Inelastic electron tunneling spectrum obtained from an Sn/barium stearate monolayer/Pb device at 4.2 K. (After Ginnai.[263,366])

the excitation of molecular oscillations is revealed at a particular bias as a local peak in d^2V/dI^2. Figure 4.21 shows an IET obtained for a Sn/barium stearate mono-layer/Pb sandwich at 4.2 K. The spectrum was very different to that obtained for an Al/Al$_2$O$_3$/Pb device; furthermore, the positions of the various peaks were found to correlate well with infrared and Raman spectra for stearic acid.

4.5.3. Direct Current Conduction through Multilayers

For fatty acid LB films comprising more than about five monolayers, quantum mechanical tunneling is not expected to be a dominant conduction process. Some possible physical mechanisms that might therefore account for the conductivity of such multilayer structures have been discussed by Honig[340] and by Careem and

Hill.[359] These are: (1) ohmic (electronic) conductivity; (2) space-charge limited conductivity; (3) Schottky emission; (4) the Poole–Frenkel effect; (5) Poole or hopping conduction; and (6) ionic conductivity.

It is sometimes difficult to distinguish experimentally between the Schottky and Poole–Frenkel processes: both occur at high values of the applied electric field ($>10^7$ V m^{-1}) and exhibit current density–electrical field relationships of the form

$$J \propto \exp (\beta E^{0.5}/kT) \tag{4.7}$$

where E is the electrical field and β is a constant (generally a different value for the two effects) related to the dielectric constant of the material. However, it should be noted that Poole–Frenkel conduction is essentially a bulk effect and should therefore show little dependence upon electrodes or upon the polarity of the field.

Despite some preliminary investigations by Scala and Handy[347,348] and by Horiuchi et al.,[349] systematic investigations into the dc conductivity mechanisms operating in multilayer films did not begin until the 1970s. Mann and Kuhn,[70] in 1971, reported that the current density for 5–21 layers of fatty acid salts was proportional to voltage and inversely proportional to the film thickness (in fact, this was termed the "intrinsic" conductivity, σ_i, and used to extract the tunneling conductivity, σ_t, from their experimental data). Also in 1971, Nathoo and Jonscher[372] reported conductivity data for cadmium stearate films in the thickness range 6–40 monolayers. In order to obtain reproducible results, the samples needed to be kept in a vacuum before and during the measurements. Long time constants (several days) were reported on the application of a step voltage. Therefore the dc characteristics were obtained by subtracting the charging current from the discharging current, both taken three minutes after switching on or off. At fields less than 10^7 V m^{-1} the current density–electric field relationship was found to be approximately $J \propto E^{1.25}$, and for greater fields the characteristics were found to obey Poole's law,[358] i.e.,

$$J \propto \exp (esE/kT) \tag{4.8}$$

where s is the average separation of the ionized centers. Poole's law is based on the classic calculation of ionic conductivity. However, Hill[358] has demonstrated that the same J–V characteristic is obtained when the density of ionized centers in a film is high and, as a result, the coulombic potentials, which give rise to the Poole–Frenkel effect, overlap. Consequently the peak in the potential barriers then becomes pinned midway between the ionized centers. Using temperature measurements, Nathoo and Jonscher found that the distance s fell between one or two molecular lengths. Honig[340] reported on the dc conductivity of a variety of sandwich structures of lead stearate (1–15 layers). At low voltages (<0.1 V) the current–voltage relationship was linear and the conductivity was found to be inversely proportional to N^3, where N is the number of monolayers. For larger applied voltages the current–voltage characteristics were strongly nonlinear and were dependent upon the polarity of the applied bias, even when similar electrode materials were used, suggesting an electrode-dependent conductivity mechanism. Although some of the data revealed straight lines when plotted in the form of log (current) versus bias$^{0.5}$, neither the Schottky emission nor the Poole–Frenkel theories fitted the experimental points precisely; the Schottky equation in fact showed the slightly better fit. In 1978 Careem and Hill[359] reported on measurements of cadmium stearate films up to 21 monolayers in thickness. As with the work of Nathoo and Jonscher,[372] a long time constant was observed on the application of a step bias. However, a delay period of only 100 s was needed in order to ensure that even the most resistive specimens were in a state of quasi-equilibrium. For LB films with the number of layers in the range $3 < N < 9$, the log (current) was inversely proportional to the square root of film thickness, while for thicker films a simple reciprocal relationship appeared to be applicable. This suggested in the mid-thickness range that either a Poole–Frenkel or a Schottky effect was being observed, while for thicker layers Poole's law was dominant. In the former case a monolayer thickness close to the accepted value (2.5 nm) was evaluated using the theoretical value for the Schottky constant, implying that the high field conduction was electrode- rather than bulk-limited. Examination of the Poole conductivity data as a function of temperature revealed that the separation between the ionized centers was essentially equal to the molecular length. Roberts et al.[367] have reported log (current) versus bias$^{0.5}$ behavior for cadmium stearate multilayers deposited on both Al and single-crystal InP base electrodes. The results were essentially independent of the polarity of the voltage, even when dissimilar electrodes were used. Hence the data were interpreted in terms of a barrier-limited conduction process through the bulk of the LB film (i.e., the Poole–Frenkel mechanism). Simpson and Reucroft[373] have noted a log (current) versus voltage$^{0.5}$ relationship for multilayers of chlorophyll.

There have been many other investigations in which monolayers and multilayers have been deposited onto semiconductor surfaces[369,374−396] in order to produce metal–insulator–semiconductor (MIS) structures. This has enabled the surface conductivity of the semiconductor to be varied with an applied voltage, and in some cases has resulted in improved photovoltaic,[379,381−383,394] electroluminescent,[386,388,391,394,396] or switching characteristics.[392,393] For such devices, the applied (or generated) voltage is distributed between the LB layer, the semiconductor surface (oxide) layer, and the space-charge region of the semiconductor. The conduction processes in these devices are quite complex and are not yet fully understood.[394] In many instances both electron and hole currents must pass through the organic layer in order for the MIS device to operate. More details of this subject are given in Chapter 7.

Most of the dc conductivity experiments on LB films have been undertaken using long chain fatty acids, and data have been presented for moderate to high electric field strengths (i.e., $>10^7$ V m^{-1}). Table 4.3 summarizes the results obtained by different research groups. Unfortunately, there is no clear picture that emerges from the numerous investigations that have taken place. Most of the possible dc conduction mechanisms expected to be observed in thin insulating layers

Table 4.3. Interpretation of Direct Current Conductivity for Fatty Acid LB Films

Workers	Film material	Measurement conditions	J/V and J/N dependencies	Interpretation
Mann and Kuhn[70] (1971)	5–21 Layers cadmium salts of fatty acids	—	$J \propto V$ $J \propto N^{-1}$	Impurity conduction
Honig[340] (1976)	1–15 Layers lead stearate	Dry ($< 1\%$ humidity) N_2 atmosphere	< 0.1 V: $J \propto V$ $J \propto N^{-4}$ > 0.1 V: $\ln J \propto V^{1/2}$	Schottky emission
Careem and Hill[359] (1978)	1–21 Layers stearic acid	Vacuum $< 10^{-5}$ mbar	$3 < N < 9$: $\ln J \propto N^{-1/2}$ $N > 9$: $\ln J \propto N^{-1}$ $J \propto \sinh(AV)$ ($A = $ constant)	Schottky emission Poole's law
Roberts et al.[367] (1978)	7 Layers cadmium stearate 3 Layers cadmium arachidate	Vacuum or under low pressure of nitrogen	$\ln J \propto V^{1/2}$	Poole–Frenkel
Tredgold et al.[328] (1984)	15 Layers cadmium stearate	Continuous flow of dry nitrogen	$\ln J \propto V^{1/4}$	Defect conduction (e.g., due to Au electrode material diffusing along grain boundaries in film)

have been reported by different workers. The conclusion that must inevitably be reached is that high-field dc conduction in multilayer structures is determined by defects in the LB layer which, in turn, are dependent on the details of the film and electrode preparation. This situation is found for many other thin-film systems and is exemplified by the work of Tredgold et al.,[328] who have correlated the conductivity of cadmium stearate layers with their method of preparation and subsequent heat treatment. Clearly a much better understanding and control of the structural properties of multilayer systems are required before substantial progress can be made in this area.

There have been a number of reports of multilayer structures which possess semiconducting properties.[140,143,144,242,397–414] Such systems are based on dyes with or without long hydrocarbon chains. These materials are expected to behave differently (electrically) from the simple fatty acid multilayers.

For example, space-charge limited conductivity has been observed by Roberts et al.[398] in LB layers of anthracene derivatives. This is characterized by a power law relationship between current and applied electric field. Data for charge carrier flow normal to the plane of a specimen of a substituted anthracene 51 layers thick are shown in Figure 4.22. In this direction a linear dependence was found between current and field up to electric fields of about 10^8 V m^{-1}, followed by a region where the current varied as the square of the voltage. In some anthracene LB layers, Roberts et al. reported a much steeper dependence of current upon applied bias; this was accompanied by blue light emission from the LB layers. The cause of this electroluminescence was assumed to be double carrier injection into the anthracene layer from the two electrodes (holes injected via a gold contact and electrons via an Al contact). A marked anisotropy in the conductivity of this anthracene derivative was also noted, the in-plane value exceeding that in a direction perpendicular to the layers by a factor of 10^8. Anisotropies in the photoconductivity of dye-sensitized LB layers have also been reported by Sugi and Iizima.[407]

The conductivity of various phthalocyanines in LB film form has been studied.[140,143,144,242,399–402] The dc conductivity for these materials has been shown to be ohmic, although the absolute conductivity values can vary over a wide range, depending on the exact material; figures from 10^{-9} $(\Omega$ m$)^{-1}$ to 10^{-4} $(\Omega$ m$)^{-1}$ have been quoted.[143,242,399–402] The conductivity was also found to be very sensitive to the presence of certain ambient gases and vapor, which is possibly the basis for a sensing device.[400–402]

Extensive studies into the electrical properties of other dye layers have also been undertaken.[403–414] The possibility of using such materials to fabricate superlattice structures in which appropriate chromophores are spaced at precise distances from one another is an exciting prospect.[279] Both p-type and n-type LB films of less than ten monolayers thickness have been identified: merocyanine and pyrene derivative films were characterized as p-type, while crystal violet and paraquat derivatives were n-type.[405,411–413] Molecular p–n junctions have been made with such materials.[412–414] Although rectifying effects were observed, the current–voltage and capacitance–voltage data for these devices did not seem to follow the diode theory developed for conventional inorganic semiconductors.

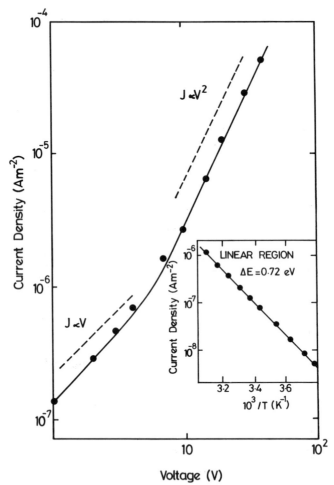

Figure 4.22. Current density normal to film plane versus voltage for a substituted anthracene (51 monolayers) film deposited onto an Al/Al$_2$ O$_3$ substrate. The inset shows the activation energy in the low-field linear region. (After Roberts *et al.*[398])

A considerable effort has been devoted to the synthesis of charge-transfer LB materials, in order to produce highly conducting thin films (highly conducting in the film plane). This work has been reviewed by Vandevyver.[585] Substituted derivatives of tetracyanoquinodimethane (TCNQ) complexes,[101,415,416,560,563,586−593] tetrathiafulvalene (TTF) and its derivatives[594−596] and TCNQ/TTF complexes[559,561,597−601] have all been studied. Alternate-layers of TCNQ and TTF materials have also been fabricated.[602] For some of these films, doping (e.g., with iodine) is required in order to achieve a high conductivity. Generally ohmic behavior is observed for these systems, with room temperature conductivities, measured by d.c. techniques, in the range $10^{-4}–10^2$ $(\Omega$ m$)^{-1}$. The conductivities are invariably thermally activated (the conductivity increasing with increasing temperature) with thermal activation energies of 0.1–0.3 eV. Thus the materials exhibit behavior seen for (doped) inorganic semiconductors.

It should be noted, however, that higher conductivities than those above are often inferred from high frequency (optical) measurements,[563,588] implying that "grain boundaries" in the films may well be limiting the d.c. conductivity values. The highest d.c. room temperature conductivity for layers of this kind (although, in this instance the lifting method was used to transfer the film) appears to be that from a tridecylmethylammonium-Au (dmit)$_2$ complex.[603] 1 : 1 mixed films with icosanoic acid could be converted into conductors by oxidation. The films then exhibited a room temperature conductivity of approximately $2.5 \times 10^3 (\Omega$m$)^{-1}$; moreover, a metallic temperature dependence (conductivity decreasing with increasing temperature) was observed around room temperature.

A number of polymeric LB film systems have also been explored as possible high conductivity materials[604−608]; however, these layers are not yet as well characterized as those formed from the charge-transfer salts.

4.5.4. Conductance

Direct current conduction and dielectric phenomena are essentially separate and, for the most part, independent processes. However, at low frequencies, both can contribute to a measured conductance. The frequency-dependent conductivity, $\sigma(\omega)$, of a sample may be expressed in the form

$$\sigma(\omega) = \sigma_0 + \sigma'(\omega) \tag{4.9}$$

where σ_0 is the dc conductivity and $\sigma'(\omega)$ is the "true ac" component of the conductivity. The latter quantity is simply related to the imaginary part of the dielectric permittivity $\epsilon(\omega) = \epsilon'(\omega) - i\epsilon''(\omega)$ by

$$\sigma'(\omega) = \omega\epsilon''(\omega) \tag{4.10}$$

As with tunneling measurements, interpretation of ac data must be subject to caution because of the effect of the native oxide layer(s) on the electrodes.[329] The

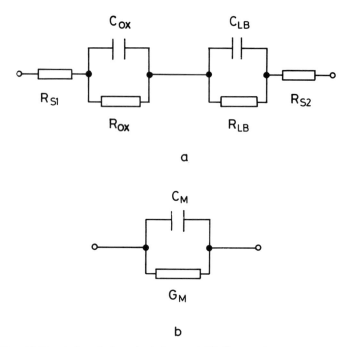

Figure 4.23. (a) Electrical equivalent circuit for metal/LB film/metal structure. (b) Circuit actually measured.

conductance is usually measured together with the capacitance of the sample, as an admittance. The electrical equivalent circuit of the metal/LB multilayer/metal sandwich structure is shown in Figure 4.23a; the circuit that is actually measured is shown in Figure 4.23b. Here C_{ox} and R_{ox} represent the capacitance and resistance of the oxide layer(s), C_{LB} and R_{LB} the capacitance and resistance of the LB layer, and R_{S1} and R_{S2} the resistances of the two metallic electrodes. The measured quantities, C_M and G_M, therefore do not simply reflect the conductance and capacitance of the LB layers, and changes in the measured conductance and capacitance values as a function of frequency are not necessarily related directly to polarization processes in the LB layer, but may simply result from Maxwell–Wagner interfacial polarization effects. Figure 4.24 shows a good example of this: these capacitance and conductance data were taken for seven monolayers of lead stearate between Al and Hg electrodes.[417] A careful examination of the conductance curve reveals that the slope of the straight line changes from a value slightly less than unity to approximately 2 at a frequency close to 20 Hz. This can be attributed to the presence of the interfacial oxide layer. If the circuit shown in Figure 4.23a is simplified by neglecting the oxide capacitance and the metallic contact resistances, then above a critical frequency the measured conductance, G_M, is given by[418]

$$G_M = \omega^2 C_{LB}^2 R_{ox} \tag{4.11}$$

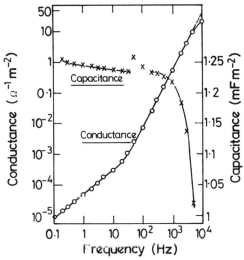

Figure 4.24. ac conductance (left scale) and capacitance (right scale) as a function of frequency for 7 layers of lead stearate between an Al and an Hg electrode. (After Honig and de Koning.[417])

In this frequency range, if C_{LB} is independent of frequency then $G_M \propto \omega^2$ and its magnitude is proportional to the series resistance. However, if the capacitance of the sample under study displays dispersion effects such that $C_{LB} \propto \omega^{n-1}$, then it is important to note that G_M will be approximately proportional to ω^{2n}. A series resistance of about 9×10^{-3} Ω m^2 can account for the conductance curve transition shown in Figure 4.24 and also for the rapid decrease in capacitance with frequency above 100 Hz. Series resistance problems have also been noted by Taylor and Mahboubian-Jones,[335] Roberts et al.,[398] and Marc and Messier.[329]

Once electrode effects have been minimized, a power law relationship between the measured LB film conductance and frequency is often observed[368,372,419–423]; this takes the form

$$\sigma'(\omega) \propto \omega^n \qquad (4.12)$$

and is in agreement with Jonscher's "Universal" response of dielectrics.[424] Figure 4.25a shows some conductance data for seven-layer cadmium palmitate (A), cadmium stearate (B), and cadmium arachidate (C) assemblies at 77 K.[421] For lower frequencies, the conductance becomes frequency independent and in each case is in agreement with equation (4.9). Above 10 Hz, Figure 4.25b (7 layers of cadmium stearate measured at room temperature) reveals that the conductance–frequency relationship takes the form of equation (4.12) with an exponent n of approximately unity. For fatty acid materials, this exponent has been reported by Sugi et al.[420] to be $n = 0.87 \pm 0.2$ over the frequency range 10^2–10^4 Hz. Careem and Jonscher[422] have studied the frequency and temperature dependence of n in some detail. At room temperature and above a few Hz it was found that $n = 0.98 \pm 0.2$ but at lower

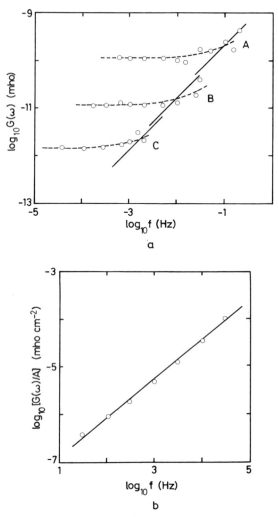

Figure 4.25. (a) Conductance, $G(\omega)$, at 77 K for 7-layer assemblies of C_{16} (A), C_{18} (B), and C_{20} (C) fatty acid assemblies. (b) An example of the almost linear variation of $G(\omega)$ at room temperature for a 7-layer C_{18} film. (After Sugi et al.[421])

frequencies $n = 0.6$–0.8, according to the LB film thickness (the larger values corresponding to the thinner films). The high-frequency response was attributed to the effects of the stearic acid lattice (arising mostly from the COOH) groups, while the low-frequency data were thought to be the result of hopping carriers, most probably injected from the electrodes. Further experimental evidence to support these ideas was provided by photogenerated dielectric response measurements.[425–

428) Measurements by Taylor and Mahboubian-Jones[335] on phospholipid LB films have also revealed a power law dependence for the conductance over the frequency range 10^2 to about 10^4 Hz; however, this was found to depend strongly on ambient conditions and was attributed to the presence of water in the films. Marc and Messier[429] have also noted the presence of a dipolar relaxation process below 100 Hz in poorly dried calcium behenate films.

A detailed investigation into possible low-frequency hopping conduction in LB layers has been undertaken by Sugi's group[330,404,430−435]; a review of this work has also been published.[421] In 1975, Sugi et al.[330] derived the following expression for the dc and low-frequency conductance of multilayer films:

$$\sigma = e^2 N(E_F)(2\alpha)^{3/2} \, l^{5/2} \tau_0^{-1} \exp\left\{ -2\alpha l - \left[\frac{4\alpha}{\pi N(E_F) l k T} \right]^{0.5} \right\} \quad (4.13)$$

where $N(E_F)$ is the two-dimensional density of states around the Fermi level E_F, α is the damping constant of the wave function, l is the hopping distance, and τ_0 is a constant. Equation (4.13) was developed assuming the dominance of one-layer hops and starting from a Miller–Abrahams-type expression for the hopping relaxation time.[436,437] Thus an electron near the Fermi level E_F hops into one of the localized states distributed with a density $N(E_F)$ on the neighboring interface. The dominant thickness dependence in equation (4.13) is due to the term $\exp(-2\alpha l)$, while the second term in the exponent is a minor correction.

Experimental data obtained by Sugi et al.[330] exhibited good agreement with equation (4.13). For a homogeneous assembly the dc conductance also scaled according to Ohm's law.[421] However, the dc conductivity values were shown to exhibit large deviations (compared to the capacitance data), and statistical means were essential to avoid misinterpretation. Figure 4.26 shows how the measured dc

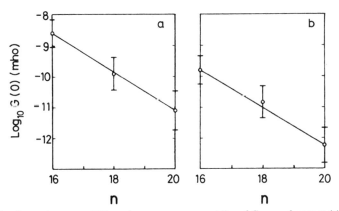

Figure 4.26. dc conductance of 7-layer homogeneous assemblies of C_n monolayers at (a) room temperature and (b) 77 K. (After Sugi et al.[421])

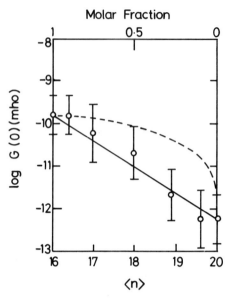

Figure 4.27. dc conductance of 7-layer mixed system $(C_{16})_x–(C_{20})_{1-x}$. The solid and broken lines correspond to the homogeneous mixing and phase-separated cases, respectively. (After Iizima and Sugi.[421,431])

conductance for the cadmium salts of fatty acids depends upon the chain length of the materials.[421] The conductance was obtained for electric fields less than 5×10^7 V m^{-1} as the frequency-independent component of the ac admittance by analyzing Lissajou's figures taken at frequencies ranging from 10^{-1} Hz down to 10^{-5} Hz. The exponential dependence of the conductance upon monolayer thickness, predicted by equation (4.13), is seen at both room temperature and at 77 K from Figure 4.26. The slopes of the lines can be identified with the wave-function damping constant $2\alpha = 0.1149 \pm 0.013$ nm obtained by Mann and Kuhn.[70] The $T^{-0.5}$ law predicted by equation (4.13) has also been verified by Sugi et al.[432] Using a least-squares fitting routine to the data, values of $2\alpha = 0.163$ nm and $N(E_F) = 2.25 \times 10^{19}$ eV^{-1} m^{-2} were obtained.

Iizima and Sugi[431] have extended these conductivity studies to investigate mixed LB multilayer films of cadmium arachidate/cadmium palmitate. The conductivity of the mixed layers was found to decrease exponentially with the weighted average of chain lengths of the fatty acids, indicating that these two molecular species can be mixed very homogeneously. Figure 4.27 shows these experimental results in the form of a similar plot to that of Figure 4.26; more than 50 samples were prepared for each composition and geometrical averages were taken, allowing for the exponential dependence of the conductivity upon the hopping distance. The lines in Figure 4.27 were based on calculations for homogeneous mixing of the two molecular species (solid line) and for the phase-separated case (dashed line). The two situations will give rise to different hopping relaxation times. Hence a conclusion derived from these experiments is that two species of fatty acid salts can be mixed homogeneously.

In 1977, Sugi and Iizima[433] developed a theoretical model for hopping conductivity in heterogeneous multilayer assemblies. A frequency dispersion was predicted as characteristic of each superstructure in the layer sequence, while the homogeneous assemblies with identical layers were shown to exhibit a frequency-independent conductivity. Experimental data showing these effects were subsequently published[434]; a system based on one cadmium arachidate monolayer in a cadmium palmitate layer multilayer array was shown to exhibit frequency dispersion of the conductance below 10^{-1} Hz. Maxwell–Wagner-type interfacial polarization effects were eliminated as a possible dispersion mechanism. The good agreement between the theory and experiment suggested that the conductance of one single monolayer could be evaluated as a bulk property without the influence of electrode materials, if the monolayer were to be placed in an appropriate multilayer system. An example showing the calculation of the single-layer conductance of a cadmium behenate monolayer incorporated into a cadmium stearate multilayer was reported by Sugi and Iizima.[435] The value of this conductance was found to fall on the same exponential line as shown in Figure 4.26, indicating that C_{22} monolayers are associated with the same hopping mechanism as the other fatty acid films. Severn et al.[609] have recently reported on the electron hopping time in a porphyrin LB film; the interlayer hop time was found to vary from 50 ns to less than 5 ns, depending on the film preparation.

4.5.5. Permittivity

The real part of the permittivity (ϵ') of LB layers is usually determined by measuring the capacitance of a metal/LB film/metal sandwich structure. Problems of electrode and oxide resistance can therefore complicate the interpretation of measured capacitance data, as discussed in the previous section. However, assuming that the electrode resistances can be neglected, the equivalent circuit of Figure 4.23a may be reduced to that of C_{ox} in series with C_{LB}. The total measured capacitance, C_M, of the structure is then related to the permittivity of the LB film material by the equation

$$\frac{1}{C_M} = \frac{1}{A}\left(\frac{Nt_{LB}}{\epsilon'_{LB}} + \frac{t_{OX}}{\epsilon'_{OX}}\right) \qquad (4.14)$$

where the permittivity and thickness of the oxide layer are given by ϵ'_{ox} and t_{ox}, respectively; ϵ'_{LB} and t_{LB} represent the corresponding values for one LB monolayer; N is the total number of monolayers; and A is the electrode area. Equation (4.14) shows clearly that the slope of a straight line plot of C_M^{-1} versus N gives the dielectric thickness ($t_{LB}\epsilon_0/\epsilon'_{LB}$) for each monolayer while the intercept yields similar information about the oxide layer(s) on the metallic electrode(s). If the silver–LB

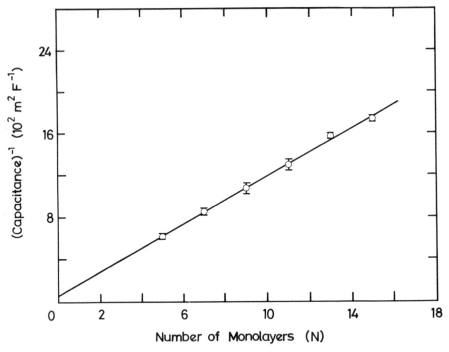

Figure 4.28. Reciprocal capacitance versus number of monolayers for a series of Au/cadmium stearate/Al structures. (After Batey.[438])

film–magnesium structure of Geddes *et al.*[584] is used (i.e., negligible "oxide" layers on the electrodes), then the C^{-1} versus N curve passes through the origin. Some caution is required when one of the electrodes in this experiment is a semiconductor. The space-charge capacitance associated with the semiconductor (which may well vary with N) will add an additional term to equation (4.14).[438] A plot in the form of equation (4.14) for cadmium stearate layers sandwiched between metallic electrodes, obtained by Batey,[438] is shown in Figure 4.28. The linear dependence of C_M^{-1} versus N demonstrates clearly the reproducibility of a monolayer capacitance and hence for one monolayer to the next. Assuming a molecular chain length of 2.5 nm, a relative permittivity of 2.55 is obtained from the slope of this graph; the intercept on the ordinate yields a value for t_{ox} of approximately 4.5 nm (assuming $\epsilon'_{ox} = 8.5\epsilon_o$). Figures obtained by other workers are shown for comparison in Table 4.4. An interesting point is that the values obtained by Race and Reynolds[325] in 1939 are still consistent with data taken by more recent workers. The higher permittivities obtained for the stearates in Table 4.4 can be attributed to the stronger influence of the highly polarizable carboxyl groups in the shorter chain length material. These values are also substantially independent of frequency. Using the above capacitance method, dielectric constants for other materials have been

Table 4.4. Relative Permittivity Values for LB Films of Long Chain Fatty Acids

Workers	Measurement frequency	Film material	Relative permittivity
Race and Reynolds[325] (1939)	40–10^6 Hz	Cadmium stearate	2.59 ± 0.08
		Cadmium arachidate	2.49 ± 0.06
Mann and Kuhn[70] (1971)	10^3 Hz	Cadmium stearate	2.71 ± 0.17
		Cadmium arachidate	2.52 ± 0.07
Sugi et al.[420] (1973)	10^3 Hz	Cadmium arachidate	2.6
Agarwal and Ichijo[439] (1977)	3.5 MHz	Stearic acid	2.71
Polymeropoulos[333] (1977)	10^3 Hz	Cadmium stearate	2.7 ± 0.3
		Cadmium arachidate	2.5 ± 0.2
Roberts et al.[367] (1978)	10^2–10^5 Hz	Cadmium stearate	2.52
		Cadmium arachidate	2.45
Honig and de Koning[417] (1978)	1–30 Hz	Lead stearate	2.54
Sugi et al.[421] (1979)	10^{-3} Hz	Cadmium stearate	2.77 ± 0.07
		Cadmium arachidate	2.47 ± 0.07

evaluated, for example, poly(vinyl benzoate),[347] anthracene derivatives,[398] orthophenathioline,[339440] phospholipids,[441] polypeptides,[442] and phthalocyanines.[140,143,144,242,400,443]

Attempts to calculate the static dielectric constant of barium stearate films have been made by Khanna and co-workers.[444−446] These were the basis for the refractive index calculations noted in Section 4.4.1. The approach was along simple classical lines, and yielded data which showed some agreement with experimental values (e.g., a value of 3.0 for the dielectric constant of barium stearate multilayers). The calculations also suggested that monolayers should exhibit lower values of dielectric constant, because of the reduced intermolecular interaction effects in thinner films. Measurements by Kapur and Srivastava[447] have in fact shown that the dielectric constant of barium stearate films can be a function of the film thickness, but this effect has been attributed to the porosity of the films in the lower thickness range. Other workers have suggested that the first monolayer in a multilayer assembly of fatty acid molecules may well have different electrical properties from subsequent layers[5,368]; there may also be some connection with structural differences noted in the first layer.[141,210,534]

The capacitance of monolayer and multilayer assemblies has been shown to depend upon the applied voltage,[338,448] the fractional change in capacitance being proportional to the square of the electric field in the organic layer. Explanations based on compression associated with the electrostatic pressure on the capacitance plates have been proposed in order to account for this.

4.5.6. Permanent Polarization

There have been many reports of significant voltages being measured when high input impedance voltmeters are connected across metal/LB film/metal structures.[338,339,354,449−455] Leger et al.[338] have even fabricated a 50 V voltage source by connecting 226 metal–LB film–metal junctions in series. The phenomenon has been investigated by Srivastava et al.,[449−455] who have reported that the voltages developed across their metal–LB film–metal structures were dependent on the work functions of the two electrodes and also on the number of organic layers; in general the voltage was found to decrease with increasing film thickness. For structures with even numbers of monolayers, the voltages were originally assumed to result from the differences in the work functions of the metallic electrodes. However, it was subsequently shown that voltages even existed across symmetrical Al/barium stearate/Al systems[454]; this was thought to arise from the oxide layer on the base electrode. For symmetrical structures containing an odd number of layers the voltages were thought to be associated with the dipole moment of the individual molecules.[451] Vincett and Roberts[5] have noted that such arguments are not entirely satisfactory. Electrochemical processes must also be occurring, possibly under the influence of an internal electric field.

The polarization associated with monolayer and multilayer films can be measured directly using the Kelvin probe technique.[456,457] This method has also been widely used to study floating monolayers.[1] The experiment requires an LB layer to be deposited onto a conducting surface; a second vibrating electrode is then positioned at a small distance above the organic film. The displacement of the top electrode generates a variation in the capacitance of the system, and thus produces a variation of charge; this can be detected as a voltage across a suitable measuring resistor. An external dc voltage source is included in series with the circuit and adjusted until no potential is dropped across the measuring resistor. The value of this voltage then provides a direct measure of the contact potential difference between the top surface of the LB layer and the base electrode.

In the original studies of Porter and Wyman,[458,459] an increasing surface potential with the number of X-type layers of calcium stearate was noted (up to 8.6 V for 170 layers). In contrast, the potentials associated with Y-type films were independent of the film thickness. These data seem reasonable when one considers the expected dipole alignment for X- and Y-type layers (see Chapter 3, Figure 3.2). However, X-type layers of fatty acids are known to revert to a Y-type structure during, or shortly after, deposition (see Chapter 3, Section 3.1). The data of Porter and Wyman are therefore somewhat surprising. Langmuir[460] considered this problem and pointed out that the contact potential could be due to one or more of the following factors: (1) internal dipoles, (2) surface charges, or (3) volume charges. On the basis of Porter and Wyman's experiments with alternating X- and Y-type films,[459] Langmuir argued that the surface potential values were probably due to a surface charge on the uppermost monolayer, rather than to the presence of dipoles or volume charges within each layer. This hypothesis was also proposed by Harkins and Mattoon.[461] However, some further data obtained by Porter and Wyman[462] were difficult to fully interpret using this idea. An alternative explanation comes from the work of Goranson and Zisman[463,464] who reported measurements for various fatty acid films deposited on ebonite rods and suspended in a Faraday cage connected to an electrometer. These experiments suggested that the X-type multi-layers were positively charged and the Y-type films were uncharged. It was concluded that the electrical charges originated from the adsorption of ions from the subphase onto the carboxyl groups of the floating monolayer. The surface potentials of multilayers may be influenced by exposure to X rays.[459,465]

Figure 4.29 shows Kelvin probe data for stepped structures of alternate layers of a fatty acid and a long chain amine.[466,467] The only difference between the two sets of data is the order in which the two materials were deposited. The values in Figure 4.29a were obtained for a structure in which a layer of fatty amine was first deposited on the substrate (aluminized glass); for the data in Figure 4.29b, the acid was transferred first. Thus the surface potential appears to reflect (qualitatively) the

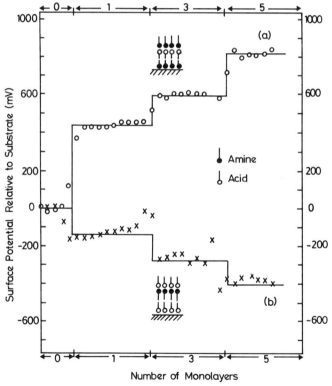

Figure 4.29.   Surface potential of 22-tricosenoic acid/docosylamine alternate layers as a function of the number of layers, for two complementary orientations of the polar axis. (After Christie.[466,467])

true dipole orientations in the two multilayer assemblies. The effect of ion incorporation in the film was observed if the acid/amine system was produced from a subphase containing SO_4^{2-} ions. In this case the potentials for both types of assembly increased in the same direction.[466] Thus surface potential measurements on LB layers are likely to depend on the precise details of film deposition.

The surface potential technique has been used by other workers to gain information regarding new monolayer materials.[468-471] Data concerning fixed charge and/or dipoles in LB films may also be obtained from capacitance–voltage measurements of MIS structures.[472,473,610] The slope of the flat-band voltage versus insulator thickness graph gives the sign and magnitude of any charges incorporated in the insulating film.[369,474,475,610] The use of the thermally stimulated current (TSC) technique has also been reported to investigate the relaxation processes of permanent dipoles in LB films.[331,476,611]

The permanent electrical polarization in X-type and Z-type LB layers and in alternate layers of different molecules may be expected to give rise to pyroelectric and/or piezoelectric effects. This indeed has been found to be the case.[89,319,320,466,467,477,541,611-614] Drabble et al.[478] have described ultrasonic transducer action in Y-type fatty acid layers when a dc bias was applied. However, it is not clear whether this was associated with any piezoelectric activity in the LB film. Such effects are discussed further in Chapter 7.

4.5.7. Dielectric Breakdown

The dielectric breakdown strength of LB films is an important practical consideration and therefore has been the subject of some study.[332,479-492] Breakdown fields for fatty acid LB layers are generally of the order 10^8–10^9 V m^{-1}.[479,483] Agarwal and Srivastava have identified two types of breakdown occurrence: an "onset breakdown" at which a sudden increase in the current was observed, and destructive breakdown at much higher fields.[485-492] Both these breakdown strengths were found to be approximately proportional to the film thickness for all the fatty acid systems investigated. Agarwal and Srivastava have used known theories of breakdown to fit their experimental data. However, the presence of oxide layers on the Al electrodes appears to have been ignored by these workers.[5]

From their studies Barraud and Rosilio[332] have observed three steps in the breakdown process: bubble formation, a sharp current increase, and the breakdown proper. The first phenomenon appears to be restricted to monolayer samples or to samples which were measured in air, and has been attributed to water expulsion by electrostatic pressure. The second process is thought to be associated with electrostatic striction, which results in the LB film thinning down at spots where the monolayers are not compact. Only the third process is believed to be an intrinsic mechanism.

The high field ($>10^8$ V m^{-1}) conduction through one monolayer of calcium behenate has been investigated by Furtlehner and Messier.[493] The effect of the monolayer was found to reduce the electric field present in the native oxide layer because of electron trapping at the surface or in the organic layer. Under vacuum the monomolecular layer was found to support an electric field of 5×10^8 V m^{-1}. The use of LB layers to improve the dielectric breakdown strength of underlying insulating layers has also been noted by Roberts et al.[494] and by Batey.[438] In the former case, fatty acid layers deposited onto Si/SiO$_2$ structures enabled voltage-controlled changes to be observed in the population of electron traps at the Si/SiO$_2$ interface. However, the role of LB layers in such composite dielectrics may not be to improve the intrinsic (proper) breakdown strength of the underlying material; they may simply decrease significantly the dc leakage current.[332] This may also explain some of the data obtained for LB film/semiconductor MIS structures (see Section 4.5.3), where the presence of the organic layer appears to effectively decouple the Fermi level in the semiconductor from the Fermi level of the metallic top electrode; hence relatively large voltages can be dropped across the LB film/semiconductor native "oxide" layer combination.

Some workers have postulated the formation of conducting filaments in LB multilayers as a result of a large dc field. Gundlach and Kadlec[495,496] have used the existence of these to explain differential negative resistance effects in MIM and MIMIM sandwich structures, and Agarwal[492] has suggested that they might even explain the dielectric breakdown data. Couch et al.[580] have reported on a switching effect which, on the application of a voltage pulse, allows the conduction through the filaments to be turned on or off.

4.6. OTHER PHENOMENA

4.6.1. Mechanical Properties

There are comparatively few reports of the mechanical properties of monolayer and multilayer films. However, the frictional behavior of monolayers of fatty acids has been investigated by a number of workers.[206,497–499,615] Studies can be conveniently made by depositing films onto molecularly smooth mica sheets and bringing the layers into contact in crossed-cylinder formation. The possible transfer of monolayers from one surface to another in such situations has been discussed.[500,503] This technique was first used by Bailey and Courtney-Pratt[497] to measure the shear strength of calcium stearate monolayers. More recently such work has been extended to much higher values of contact pressure (up to 500 MPa) by Briscoe and Evans.[499] A variety of monolayer chain lengths from myristic to behenic was reported. In the range of contact conditions studied, the specific frictional force, or shear strength, was found to vary linearly with pressure, temperature, and the logarithm of sliding velocity. Figure 4.30 shows data obtained by these workers; a linear relation between the contact pressure and the shear strength is clearly observed for stearic acid monolayers. Throughout the experiments the sliding motion was usually smooth and the surfaces very robust, even up to the highest pressures. The shear strength increased markedly with pressure but de-

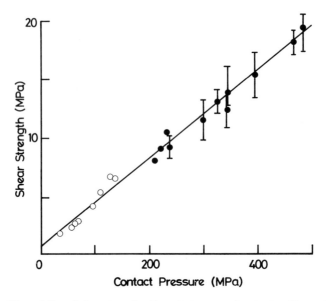

Figure 4.30. The variation of shear strength with contact pressure for stearic acid monolayers. (After Briscoe and Evans.[499])

creased with temperature. The results were relatively insensitive to the chain length of the fatty acid but significant changes were induced by saponification and partial fluorination: the pressure coefficient of shear strength was greatly reduced by saponification but considerably increased by the incorporation of fluorine. LB layers have subsequently been used to lower the coefficient of friction between magnetic recording tapes and a steel rod.[206] Blank et al.[504–507] have studied the thickness dependence of the surface and bulk properties of floating protein films. Five surface properties (yield, shear resistance, potential, viscosity, and elastic modulus) showed extrema or saturation effects at 5 to 8 nm thickness. Two bulk properties (viscosity and elastic modulus) became independent of thickness after about 8 nm (4 monolayers), while the bulk residual yield decreased with thickness after one monolayer.

Elastic constants of LB layers may be measured using Brillouin scattering. Zanoni et al.[508,509] have described such an investigation with cadmium arachidate films. Compressional constants were found to be comparable to those obtained for oriented polymer films. In contrast, the resistance to shear strains was small, as evidenced by the small shear elastic constants. Such properties were similar to those expected from a Smectic-B liquid crystal. The propagation of ultrasound through arachidic acid layers using surface acoustic wave techniques has also been noted by Jain and Jericho.[510]

One parameter that is extremely important, particularly for LB film applications, is the adhesion of the first monolayer to the underlying substrate and of subsequent monolayers to each other. Although some surface analytical methods have been used to study the chemical bonding involved (see Section 4.3.5), little attention seems to have been given to the macroscopic effects. A technique described by Rothen[511] and Baker,[140] although qualitative, may be used to provide some indication of the film adhesion. A strip of "cellotape" is simply placed over the multilayer sample, and subsequently torn off. A spectroscopic measurement before and after this process gives a measure of the amount of film removed by the procedure. Baker simply monitored the visible transmission spectrum for various phthalocyanine derivatives. However, the method could in principle be used in conjunction with other characterization techniques. Kovacs et al.[512] have also reported on the use of an abrasion wheel in a similar test. Related freeze-fracture experiments on stearate bilayers have been described by Deamer and Branton[513]: these revealed cleavage to occur almost entirely within the hydrocarbon plane of the model membrane system.

4.6.2. Permeability Studies

There are a number of reports discussing material transport through floating and adsorbed monolayers.[1,514−516] The permeability of LB layers to various atoms, ions, and molecules is a different matter. Using radiolabel methods, Adam and Zull[517] have reported on the transfer of calcium through a bilayer deposited onto glass. It was found that the calcium moved readily through the bilayer, but that the ion movement was not related to an exchange of the lipid molecules in one layer with those in the other. Attempts to use cross-linked monolayers as hyperfiltration membranes have been made by Heckmann et al.[518] Gas transport through supported multilayers has been investigated by Albrecht et al.,[161] by Gaines and Ward,[160] and by Rose and Quinn.[519,520] In the latter case mass transfer coefficients for LB multilayers deposited onto permeable support films were measured for three gases: carbon dioxide, nitrogen, and helium. For stearate multilayers varying in thickness from 4 to 48 layers (10 nm to 120 nm) the permeabilities were found to decrease with increasing film thickness. For all the deposited monolayers the CO_2 transfer rate was consistently greater than that for either the He or N_2, although the molecular diameter of CO_2 is greater than that of the other two gases. This would seem to indicate that the mass transfer is not primarily a molecular sieve process, but that it must depend on both the solubility and diffusivity of the gases in the deposited monolayers. Gaines and Ward[160] have suggested that substantially all the gas flow in their fatty acid films was through defects. The permeation of water through fatty acid LB layers has been investigated by Osiander et al.[567] using a photoacoustic technique. This work demonstrated that the permeation resistance of the water molecules was independent of temperature below 50°C, but increased at high temperatures. In addition, the resistance was found not to depend on layer thickness. The exploitation of LB layers for molecular separation has been noted by Kowel et al.[521] and is discussed further in Chapter 7.

REFERENCES

1. G. L. Gaines, Jr., *Insoluble Monolayers at Liquid–Gas Interfaces,* Wiley–Interscience, New York (1966).
2. H. Kuhn, D. Möbius, and H. Bücher, Spectroscopy of monolayer assemblies, in *Techniques of Chemistry* (A. Weissberger and B. W. Rossiter, eds.), Vol. 1, Part IIIB, pp. 577–702, Wiley, New York (1972).
3. V. K. Srivastava, Built-up molecular films and their applications, *Phys. Thin Films,* **7,** 311–397 (1973).
4. V. K. Agarwal, Electrical behaviour of Langmuir films: A review, Parts I and II, *Electrocomponent Sci. Technol.,* **2,** 1–31, 75–107 (1975).
5. P. S. Vincett and G. G. Roberts, Electrical and photoelectrical transport properties of Langmuir–Blodgett films and a discussion of possible device applications, *Thin Solid Films,* **68,** 135–171 (1980).
6. G. G. Roberts, Langmuir–Blodgett films on semiconductors, Proc. 2nd. Int. Conf., INFOS 81, pp. 56–67, Erlangen, Springer-Verlag (1981).
7. W. A. Barlow (ed.), Langmuir Blodgett Films, *Thin Solid Films,* **68** (1980).
8. G. G. Roberts and C. W. Pitt (eds.), Proceedings of the First International Conference on Langmuir–Blodgett Films, *Thin Solid Films,* **99** (1983).
9. G. L. Gaines, Jr. (ed.), Proceedings of the Second International Conference on Langmuir–Blodgett Films, *Thin Solid Films,* **132, 133,** and **134** (1985).
10. C. W. Pitt and L. M. Walpita, Lightguiding in Langmuir–Blodgett films, *Thin Solid Films,* **68,** 101–127 (1980).
11. K. B. Blodgett, Films built by depositing successive monomolecular layers on a solid surface, *J. Am. Chem. Soc.,* **57,** 1007–1021 (1935).
12. K. B. Blodgett and I. Langmuir, Built-up films of barium stearate and their optical properties, *Phys. Rev.,* **51,** 964–982 (1937).
13. K. B. Blodgett, Properties of built-up films of barium stearate, *J. Phys. Chem.,* **41,** 975–984 (1937).
14. K. H. Drexhage, Interaction of light with monomolecular dye layers, *Prog. Opt.,* **12,** 163–232 (1974).
15. G. I. Jenkins and A. Norris, Thickness of built-up films, *Nature,* **144,** 441 (1939).
16. C. Holley, X-ray and optical measurements of multi-molecular films, *Phys. Rev.,* **51,** 1000 (1937).
17. R. E. Hartman, Fringes of equal reflection coefficient ratio and their application to the determination of the thickness and refractive index of monomolecular films. I. Theory, *J. Opt. Soc. Am.,* **44,** 192–196 (1954).
18. R. E. Hartman, R. S. Hartman, K. Larson, and J. B. Bateman, Fringes of equal reflection coefficient ratio and their application to the determination of the thickness and refractive index of monomolecular films. II. Determination of the thickness and refractive index of barium stearate double layers, *J. Opt. Soc. Am.,* **44,** 197–198 (1954).
19. R. D. Mattuck, Stepped interference reflector for determining the optical constants of nonabsorbing unimolecular films. I. Theory, *J. Opt. Soc. Am.,* **46,** 621–628 (1956).
20. R. D. Mattuck, R. D. Petti, and J. B. Bateman, Stepped interference reflector for determining the optical constants of nonabsorbing unimolecular films. II. Experimental, *J. Opt. Soc. Am.,* **46,** 782–789 (1956).
21. R. D. Mattuck, Recent attempts to determine the thickness and refractive index of unimolecular films, *J. Opt. Soc. Am.,* **46,** 615–620 (1956).
22. J. B. Bateman and E. J. Covington, Molecular tilt in fatty acid multilayers, *J. Colloid Sci.,* **16,** 531–548 (1961).
23. S. Tolansky, *Multiple-Beam Interferometry of Surfaces and Films,* Clarendon Press, London (1948).

24. J. S. Courtney-Pratt, Direct optical measurement of the length of organic molecules, *Nature*, **165**, 346–348 (1950).

25. J. S. Courtney-Pratt, An optical method of measuring the thickness of adsorbed monolayers, *Proc. R. Soc. London, Ser. A*, **212**, 505–508 (1952).

26. G. D. Scott, T. A. McLauchlan, and R. S. Sennet, The thickness measurement of thin films by multiple beam interferometry, *J. Appl. Phys.*, **21**, 843–846 (1950).

27. V. K. Srivastava and A. R. Verma, Multiple beam interferometric study of "built-up" films of barium stearate, *Proc. Phys. Soc.*, **80**, 222–225 (1962).

28. V. K. Srivastava and A. R. Verma, Interferometric and X-ray diffraction study of "built-up" molecular films of some long chain compounds, *Solid State Commun.*, **4**, 367–371 (1966).

29. P. Drude, *Theory of Optics*, Longmans, New York (1902).

30. C. G. P. Feachem and L. Tronstad, An optical examination of thin films. II.—The behaviour of thin films of fatty acids on mercury, *Proc. R. Soc. London, Ser. A*, **145**, 127–135 (1934).

31. L. Tronstad, The validity of Drude's optical method of investigating transparent films on metals, *Trans. Faraday Soc.*, **31**, 1151–1158 (1935).

32. A. Rothen and M. Hanson, Optical properties of surface films. II, *Rev. Sci. Instum.*, **20**, 66–72 (1949).

33. T. Smith, Ellipsometry for measurements at and below monolayer coverage, *J. Opt. Soc. Am.*, **58**, 1069–1079 (1968).

34. F. Partovi, Theoretical treatment of ellipsometry, *J. Opt. Soc. Am.*, **52**, 918–925 (1962).

35. F. P. Mertens, P. Theroux, and R. C. Plumb, Some observations on the use of elliptically polarized light to study metal surfaces, *J. Opt. Soc. Am.*, **53**, 788–796 (1963).

36. R. Steiger, Studies of oriented monolayers on solid surfaces by ellipsometry, *Helv. Chim. Acta*, **54**, 2645–2658 (1971).

37. J. A. Faucher, G. M. Mc.Manus, and H. J. Trurnit, Simplified treatment of ellipsometry, *J. Opt. Soc. Am.*, **48**, 51–54 (1958).

38. J. P. Lloyd, C. Pearson, and M. C. Petty, SPR studies of gas effects in phthalocyanine Langmuir-Blodgett films, *Thin Solid Films*, **160**, 431–443 (1988).

39. E. Hofmeister, Bestimmung des Brechungsindex Monomolekularer Fettsäureschichten, *Z. Phys.*, **136**, 137–151 (1953).

40. M. S. Tomar and V. K. Srivastava, Ellipsometric studies of "built-up" molecular films, *Thin Solid Films*, **12**, S29–S30 (1972).

41. E. P. Honig and B. R. de Koning, Ellipsometric investigation of the skeletonization process of Langmuir–Blodgett films, *Surf. Sci.*, **56**, 454–461 (1976).

42. A. C. Hall, A study of Langmuir–Blodgett layers on metal surfaces by the method of reflected polarized light, *J. Phys. Chem.*, **69**, 1654–1659 (1965).

43. S. M. Hall, J. D. Andrade, S. M. Ma, and R. N. King, Photoelectron mean free paths in barium stearate layers, *J. Electron Spectrosc. Relat. Phenom.*, **17**, 181–189 (1979).

44. D. Ducharme, C. Salesse, and R. M. Leblanc, Ellipsometric studies of rod outer segment phospholipids at the nitrogen–water interface, *Thin Solid Films*, **132**, 83–90 (1985).

45. D. den Engelsen, Ellipsometry of anisotropic films, *J. Opt. Soc. Amer.*, **61**, 1460–1466 (1971).

46. M. J. Dignam, M. Moskovits, and R. W. Stobie, Specular reflectance and ellipsometric spectroscopy of oriented molecular layers, *Trans. Faraday Soc.*, **67**, 3306–3317 (1971).

47. M. S. Tomar and V. K. Srivastava, Anisotropic effects in the ellipsometry of "built-up" films and determination of their optical constants, *Thin Solid Films*, **15**, 207–215 (1973).

48. M. S. Tomar and V. K. Srivastava, Validity of Schopper's formulas for anisotropic films, *J. Appl. Phys.*, **45**, 1849–1851 (1974).

49. E. P. Honig, J. H. T. Hengst, and D. den Engelsen, Langmuir–Blodgett deposition ratios, *J. Colloid Interface Sci.*, **45**, 92–102 (1973).

50. D. den Engelsen and B. de Koning, Ellipsometric study of organic molecules. Part I.—Condensed monolayers, *J. Chem Soc., Faraday Trans. 1*, **70**, 1603–1614 (1974).

51. D. den Engelsen and B. de Koning, Ellipsometry of spread monolayers. Part 2.—Coloured

systems: Chlorophyll *a,* carotenoic acid, rhodamine 6G, and a cyanine dye, *J. Chem. Soc., Faraday Trans. 1,* **70,** 2100–2112 (1974).

52. M. S. Tomar, Skeletonized films and measurement of their optical constants, *J. Phys. Chem.,* **78,** 947–950 (1974).

53. D. den Engelsen, Transmission ellipsometry and polarization spectrometry of thin layers, *J. Phys. Chem.,* **76,** 3390–3397 (1972).

54. G. L. Clark, R. R. Sterrett, and P. W. Leppla, X-ray diffraction studies of built-up films of long-chain compounds, *J. Am. Chem. Soc.,* **52,** 330–331 (1935).

55. G. L. Clark and P. W. Leppla, X-ray diffraction studies of built up films, *J. Am. Chem. Soc.,* **58,** 2199–2201 (1936).

56. C. Holley and S. Bernstein, X-ray diffraction by a film of counted molecular layers, *Phys. Rev.,* **49,** 403 (1936).

57. C. Holley and S. Bernstein, Grating space of barium–copper–stearate films, *Phys. Rev.,* **52,** 525 (1937).

58. S. Bernstein, Comparison of X-ray photographs taken with X and Y built-up films, *J. Am. Chem. Soc.,* **60,** 1511 (1938).

59. I. Fankuchen, On the structure of "built-up" films on metals, *Phys. Rev.,* **53,** 909 (1938).

60. A. E. Alexander, Built-up films of unsaturated and substituted long-chain compounds, *J. Chem. Soc. London,* **1,** 777–781 (1939).

61. D. C. Bisset and J. Iball, X-ray diffraction from built-up multilayers consisting of only a few monolayers, *Proc. Phys. Soc. London A,* **67,** 315–322 (1954).

62. B. L. Henke, X-ray fluorescence analysis for sodium, fluorine, oxygen, nitrogen, carbon and boron, *Adv. X-Ray Anal.,* **7,** 460–488 (1964).

63. R. C. Ehlert, Overturning of monolayers, *J. Colloid Sci.,* **20,** 387–390 (1965).

64. D. S. Kapp and N. Wainfan, X-ray interference structure in the scattered radiation from barium stearate multilayer films, *Phys. Rev.,* **138,** 1490–1495 (1965).

65. M. W. Charles and B. A. Cooke, Optimization of lead stearate crystals for the diffraction of ultra soft X-rays, *J. Sci. Instrum.,* **44,** 976–982 (1967).

66. K. Larsson, M. Lundquist, S. Ställberg-Stenhagen, and E. Stenhagen, Some recent studies on the structural arrangements of lipids in surface layers and interphases, *J. Colloid Interface Sci.,* **29,** 268–278 (1969).

67. M. W. Charles, Optimization of multilayer soap crystals for ultrasoft X-ray diffraction, *J. Appl. Phys.,* **42,** 3329–3356 (1971).

68. M. W. Charles, Lead stearate formation in monomolecular films; A new insight using X-ray spectrometry, *J. Colloid Interface Sci.,* **35,** 167–170 (1971).

69. B. Mann, H. Kuhn, and L. v. Szentpály, Tunnelling through fatty acid monolayers and its relevance to photographic sensitization, *Chem. Phys. Lett.,* **8,** 82–84 (1971).

70. B. Mann and H. Kuhn, Tunnelling through fatty acid salt monolayers, *J. Appl. Phys.,* **42,** 4398–4405 (1971).

71. W. Lesslauer, X-ray diffraction from fatty-acid multilayers. Angular width of reflexions from systems with a few unit cells, *Acta Crystallogr., Sect. B,* **30,** 1932–1937 (1974).

72. M. Pomerantz, F. H. Dacol, and A. Segmüller, X-ray diffraction from ordered films of a few molecular thicknesses, *Bull. Am. Phys. Soc.,* **20,** 477 (1975).

73. A. Matsuda, M. Sugi, T. Fukui, S. Iizima, M. Miyahara, and Y. Otsubo, Structure study of multilayer assembly films, *J. Appl. Phys.,* **48,** 771–774 (1977).

74. T. Fukui, A. Matsuda, M. Sugi, and S. Iizima, Structure of Langmuir films, *Bull. Electrotech. Lab. Jpn.,* **41,** 423–436 (1977).

75. B. Tieke, G. Lieser, and G. Wegner, Polymerization of diacetylenes in multilayers, *J. Polym. Sci., Polym. Chem. Ed.,* **17,** 1631–1644 (1979).

76. T. Fukui, M. Sugi, and S. Iizima, Temperature dependence of the thickness of Langmuir multilayer assembly films, *Phys. Rev. B,* **22,** 4898–4899 (1980).

77. M. Pomerantz and A. Segmüller, High resolution X-ray diffraction from small numbers of Langmuir–Blodgett layers of manganese stearate, *Thin Solid Films,* **68,** 33–45 (1980).

78. M. Pomerantz, Studies of literally two-dimensional magnets of manganese stearate, in: *Phase Transitions in Surface Films* (J. G. Dash and J. Ruvalds, eds.), pp. 317–346, Plenum, New York (1980).

79. G. Lieser, B. Tieke, and G. Wegner, Structure, phase transitions and polymerizability of multi-layers of some diacetylene monocarboxylic acids, *Thin Solid Films*, **68**, 77–90 (1980).

80. R. M. Nicklow, M. Pomerantz, and A. Segmüller, Neutron diffraction from small numbers of Langmuir–Blodgett monolayers of manganese stearate, *Phys. Rev. B*, **23**, 1081–1087 (1981).

81. B. Tieke and G. Lieser, Polymerization of diacetylenes in mixed multilayers, *J. Colloid Interface Sci.*, **83**, 230–239 (1981).

82. B. Tieke and G. Lieser, Influences of the structure of long-chain diynoic acids on their polymerization properties in Langmuir–Blodgett multilayers, *J. Colloid Interface Sci.*, **88**, 471–486 (1982).

83. B. Tieke, G. Lieser, and K. Weiss, Parameters influencing the polymerization and structure of long-chain diynoic acids in multilayers, *Thin Solid Films*, **99**, 95–102 (1983).

84. M. Prakash, P. Dutta, J. B. Ketterson, and B. M. Abraham, X-ray diffraction study of the in-plane structure of an organic multilayer ("Langmuir–Blodgett") film, *Chem. Phys. Lett.*, **111**, 395–398 (1984).

85. A. Iida, T. Matsushita, and T. Ishikawa, Observation of X-ray standing wave field during Bragg reflection in multilayer of lead stearate, *Jpn. J. Appl. Phys.*, **24**, L675–L678 (1985).

86. M. Prakash, J. B. Ketterson, and P. Dutta, Study of in-plane structure in lead–fatty acid Langmuir–Blodgett films using X-ray diffraction, *Thin Solid Films*, **134**, 1–4 (1985).

87. T. Nakagiri, K. Sakai, A. Iida, T. Ishikawa, and T. Matsushita, X-ray standing wave method applied to structural study of Langmuir–Blodgett films, *Thin Solid Films*, **133**, 219–225 (1985).

88. P. Fromherz, U. Oelschlägel, and W. Wilke, Medium X-ray scattering of Langmuir–Blodgett films of cadmium salts of fatty acids, *Thin Solid Films*, **159**, 421–427 (1988).

89. G. W. Smith, M. F. Daniel, J. W. Barton, and N. Ratcliffe, Pyroelectric activity in non-centrosymmetric Langmuir–Blodgett multilayer films, *Thin Solid Films*, **132**, 125–134 (1985).

90. S. Hirota, U. Itoh, and M. Sugi, Polymerization and optical properties of mixed Langmuir–Blodgett films of β-parinaric acid and stearic acid, *Thin Solid Films*, **134**, 67–74 (1985).

91. Y. M. Lvov and L. A. Feigin, Small-angle X-ray investigation of the structure of LB molecular films, *Stud. Biophys.*, **112**, 221–224 (1986).

92. E. Stenhagen, Built-up films of esters, *Trans. Faraday Soc.*, **34**, 1328–1337 (1938).

93. G. Knott, J. H. Schulman, and A. F. Wells, On the structure of multilayers. Part I, *Proc. R. Soc. London, Ser. A*, **176**, 534–542 (1940).

94. A. Cemel, T. Fort, Jr., and J. B. Lando, Polymerization of vinyl stearate multilayers, *J. Polym. Sci., A1*, **10**, 2061–2083 (1972).

95. M. Puterman, T. Fort, Jr., and J. B. Lando, The polymerization and structure of mixed multilayers of ethyl and vinyl stearate, *J. Colloid Interface Sci.*, **47**, 705–718 (1974).

96. V. Enkelmann and J. B. Lando, Polymerization of ordered tail-to-tail vinyl stearate bilayers, *J. Polym. Sci.*, **15**, 1843–1854 (1977).

97. D. Naegele, J. B. Lando, and H. Ringsdorf, Polymerization of cadmium octadecylfumarate in multilayers, *Macromolecules*, **10**, 1339–1344 (1977).

98. K. Fukuda and T. Shiozawa, Conditions for formation and structural characterization of X-type and Y-type multilayers of long-chain esters, *Thin Solid Films*, **68**, 55–66 (1980).

99. A. Banerjie and J. B. Lando, Radiation-induced solid state polymerization of oriented ultrathin films of octadecylacrylamide, *Thin Solid Films*, **68**, 67–75 (1980).

100. P. S. Vincett and W. A. Barlow, Highly organized molecular systems using Langmuir–Blodgett films: Structure, optical properties and probable epitaxy of anthracene-derivative multilayers, *Thin Solid Films*, **71**, 305–326 (1980).

101. A. Barraud, A. Ruaudel-Teixier, M. Vandevyver, and P. Lesieur, Sel d'ion-radical a transfert de charge conducteur en films de Langmuir–Blodgett, *Nouv. J. Chim.*, **9**, 365–367 (1985).

102. R. H. Tredgold, A. J. Vickers, A. Hoorfar, P. Hodge, and E. Khoshdel, X-ray analysis of some porphyrin and polymer Langmuir–Blodgett films, *J. Phys. D*, **18**, 1139–1145 (1985).

103. R. Jones, R. H. Tredgold, A. Hoorfar, and R. A. Allen, Crystal formation and growth in

Langmuir–Blodgett multilayers of azobenzene derivatives: Optical and structural studies, *Thin Solid Films*, **134**, 57–66 (1985).

104. B. Belbeoch, M. Roulliay, and M. Tournarie, Evidence of chain interdigitation in Langmuir–Blodgett films, *Thin Solid Films*, **134**, 89–99 (1985).

105. M. J. Cook, A. J. Dunn, M. F. Daniel, R. C. O. Hart, R. M. Richardson, and S. J. Roser, Fabrication of ordered Langmuir–Blodgett multilayers of octa-*n*-alkoxy phthalocyanines, *Thin Solid Films*, **159**, 395–404 (1988).

106. Y. Yoshioka, H. Nakahara, and K. Fukuda, Photopolymerization in Langmuir–Blodgett films of monoacids containing phenyl and diacetylene groups simultaneously, *Thin Solid Films*, **133**, 11–19 (1985).

107. Y. Kawabata, T. Sekiguchi, M. Tanaka, T. Nakamura, H. Komizu, M. Matsumoto, E. Manda, M. Saito, M. Sugi, and S. Iizima, Formation and deposition of super-monomolecular layers by means of surface pressure control, *Thin Solid Films*, **133**, 175–180 (1985).

108. W. T. Astbury, F. O. Bell, E. Gorter, and J. Van Ormondt, Optical and X-ray examination and direct measurement of built-up protein films, *Nature*, **142**, 33–34 (1938).

109. Y. K. Levine, A. I. Bailey, and M. H. F. Wilkins, Multilayers of phospholipid bimolecular leaflets, *Nature*, **220**, 577–578 (1968).

110. Y. K. Levine and M. H. F. Wilkins, Structure of oriented lipid bilayers, *Nat., New Biol.*, **230**, 69–72 (1971).

111. J. P. Green, M. C. Phillips, and G. G. Shipley, Structural investigations of lipid, polypeptide and protein multilayers, *Biochim. Biophys. Acta*, **330**, 243–253 (1973).

112. N. P. Franks and K. A. Snook, The structure and permeability of reconstituted membranes, *Thin Solid Films*, **99**, 139 (1983).

113. T. Furuno, H. Sasabe, R. Nagata, and T. Akaike, Studies of Langmuir–Blodgett films of poly(1-benzyl-L-histine)–stearic acid mixtures, *Thin Solid Films*, **133**, 141–152 (1985).

114. V. V. Erokhin, R. L. Kayushina, Y. M. Lvov, and L. A. Feigin, Protein Langmuir–Blodgett films as sensing elements, *Stud. Biophys.*, **131**, 120–127 (1989).

115. C. L. Andrews, Concave spherical crystals of barium–copper–stearate for use in long wave-length X-ray spectrometers, *Rev. Sci. Instrum.*, **11**, 111–114 (1940).

116. J. B. Nicholson and D. B. Wittry, A comparison of the performance of gratings and crystals in the 20–115 Å region, *Adv. X-Ray Anal.*, **7**, 497–511 (1964).

117. B. L. Henke, Application of multilayer analyzers to 15–150 Å fluorescence spectroscopy for chemical and valence band analysis, *Adv. X-Ray Anal.*, **9**, 430–440 (1966).

118. R. R. Highfield, R. K. Thomas, P. G. Cummins, D. P. Gregory, J. Mingins, J. B. Hayter, and O. Schärpf, Critical reflection of neutrons from Langmuir–Blodgett films on glass, *Thin Solid Films*, **99**, 165–172 (1983).

119. L. M. Walpita and C. W. Pitt, Measurement of Langmuir-film properties by optical waveguide probe, *Electron. Lett.*, **13**, 210–212 (1977).

120. Y. Kawabata, T. Sekiguchi, M. Tanaka, T. Nakamura, H. Komizu, K. Honda, and E. Manda, Supermolecular structure in the Langmuir–Blodgett films of a surface-active dye-fatty acid mixed system, *J. Am. Chem. Soc.*, **107**, 5270–5271 (1985).

121. P. K. Tien, R. Ulrich, and R. J. Martin, Modes of propagating light waves in thin deposited semiconductor films, *Appl. Phys. Lett.*, **14**, 291–294 (1969).

122. A. J. Vickers, R. H. Tredgold, P. Hodge, E. Khoshdel, and I. Girling, An investigation of liquid crystal side chain polymeric Langmuir–Blodgett films as optical waveguides, *Thin Solid Films*, **134**, 43–48 (1985).

123. E. Havinga and J. de Wael, Untersuchung Monomolekularer Filme mit Hilfe von Elektronenstrahlen, *Recl. Trav. Chim. Pays-Bas*, **56**, 375–381 (1937).

124. J. de Wael and E. Havinga, Untersuchung Monomolekularer Filme mit Hilfe von Elektronenstrahlen. II., *Recl. Trav. Chim. Pays-Bas*, **59**, 770–776 (1940).

125. L. H. Germer and K. H. Storks, Arrangement of molecules in a single layer and in multiple layers, *J. Chem. Phys.*, **6**, 280–293 (1938).

126. G. F. Mattei and G. Parravano, Ricerche sui multistrati.—Nota I., *Gazz. Chim. Ital.*, **73**, 291–300 (1943).

127. H. T. Epstein, On the structure of monolayers and multilayers of polar hydrocarbon molecules on solid substrates, *J. Phys. Colloid Chem.*, **54**, 1053–1069 (1950).

128. B. R. Malcolm, Molecular structure and deuterium exchange in monolayers of synthetic polypeptides, *Proc. R. Soc. London, Ser. A*, **305**, 363–385 (1968).

129. J. F. Stephens and C. Tuck-Lee, The structure of a multilayer of lead stearate, *J. Appl. Crystallogr.*, **2**, 1–10 (1969).

130. B. Tieke, H. J. Graf, G. Wegner, B. Naegele, H. Ringsdorf, A. Banerjie, D. Day, and J. B. Lando, Polymerization of mono- and multilayer forming diacetylenes, *Colloid Polym. Sci.*, **255**, 521–531 (1977).

131. D. Day and J. B. Lando, Electron diffraction studies of multilayers of vinyl stearate, *J. Polym. Sci., Polym. Chem. Ed.*, **16**, 1431–1434 (1978).

132. D. Day and J. B. Lando, Structure determination of a poly(diacetylene) monolayer, *Macromolecules*, **13**, 1483–1487 (1980).

133. M. Sarkar and J. B. Lando, Polymerization of two amphiphilic diacetylenes in multilayer films, *Thin Solid Films*, **99**, 119–126 (1983).

134. G. J. Russell, M. C. Petty, I. R. Peterson, G. G. Roberts, J. P. Lloyd, and K. K. Kan, A RHEED study of cadmium stearate Langmuir–Blodgett films, *J. Mater. Sci.*, **3**, 25–28 (1984).

135. I. R. Peterson, G. J. Russell, and G. G. Roberts, A new model for the deposition of ω-tricosenoic acid Langmuir–Blodgett film layers, *Thin Solid Films*, **109**, 371–378 (1983).

136. I. R. Peterson and G. J. Russell, An electron diffraction study of ω-tricosenoic acid Langmuir–Blodgett films, *Philos. Mag. A*, **49**, 463–473 (1984).

137. A. Fischer, M. Lösche, H. Möhwald, and E. Sackmann, On the nature of the lipid monolayer phase transition, *J. Phys. (Paris), Lett.*, **45**, L785–L791 (1984).

138. I. R. Peterson and G. J. Russell, Deposition mechanisms in Langmuir–Blodgett films, *Br. Polym. J.*, **17**, 364–367 (1985).

139. J. R. Fryer, R. A. Hann, and B. L. Eyres, Single organic monolayer imaging by electron microscopy, *Nature*, **313**, 382–384 (1985).

140. S. Baker, Phthalocyanine Langmuir–Blodgett films and their associated devices, Ph.D. thesis, University of Durham (1985).

141. A. Bonnerot, P. A. Chollet, H. Frisby, and M. Hoclet, Infrared and electron diffraction studies of transient stages in very thin Langmuir–Blodgett films, *Chem. Phys.*, **97**, 365–377 (1985).

142. I. R. Peterson and G. J. Russell, The deposition and structure of LB films of long-chain acids, *Thin Solid Films*, **134**, 143–152 (1985).

143. R. A. Hann, S. K. Gupta, J. R. Fryer, and B. L. Eyres, Electrical and structural studies on copper tetra-tert butyl phthalocyanine Langmuir–Blodgett films, *Thin Solid Films*, **134**, 35–42 (1985).

144. G. G. Roberts, M. C. Petty, S. Baker, M. T. Fowler, and N. J. Thomas, Electronic devices incorporating stable phthalocyanine LB films, *Thin Solid Films*, **132**, 113–123 (1985).

145. J. D. Earls, I. R. Peterson, G. J. Russell, I. R. Girling, and N. A. Cade, An electron diffraction study of optically nonlinear hemicyanine LB films, *J. Mol. Elec.*, **2**, 85–94 (1986).

146. S. Garoff, H. W. Deckman, J. H. Dunsmuir, M. S. Alvarez, and J. M. Bloch, Bond-orientational order in Langmuir–Blodgett surfactant monolayers, *J. Phys. (Paris)*, **47**, 701–709 (1986).

147. H. E. Ries, Jr., and W. A. Kimball, Monolayer structure as revealed by electron microscopy, *J. Phys. Chem.*, **59**, 94–95 (1955).

148. H. E. Ries, Jr., and W. A. Kimball, Structure of fatty acid monolayers and a mechanism for collapse, *Proc. 2nd Int. Congr. Surface Activity*, **1**, 75–84 (1957).

149. H. E. Ries, Jr. and W. A. Kimball, Electron micrographs of monolayers of stearic acid, *Nature*, **181**, 901 (1958).

150. H. E. Ries, Jr., and D. C. Walker, Films of mixed horizontally and vertically oriented compounds, *J. Colloid Sci.*, **16**, 361–374 (1961).

151. H. J. Trurnit and G. Schidlovsky, Thin cross-sections of artificial stacks of monomolecular films, *Proc. Eur. Regional Conf. Electron Microscopy*, **2**, 721–725 (1961).

152. E. Sheppard, R. P. Bronson, and N. Tcheurekdjian, Monolayer studies I. Electron microscopy of stearic acid and barium stearate films, *J. Colloid Sci.*, **19**, 833–837 (1964).

153. E. Sheppard, R. P. Bronson, and N. Tcheurekdjian, Monolayer studies II. Electron microscopy of normal paraffinic polar compounds, *J. Colloid Sci.*, **20**, 755–765 (1965).

154. J. A. Spink, The transfer ratio of Langmuir–Blodgett monolayers for various solids, *J. Colloid Interface Sci.*, **23**, 9–26 (1967).

155. W. Baumeister and M. Hahn, Atomic resolution electron microscopy of model-membranes, *Cytobiologie*, **7**, 244–267 (1973).

156. F. Kopp, U. P. Fringeli, K. Mühlethaler and H. H. Günthard, Instability of Langmuir–Blodgett layers of barium stearate, cadmium arachidate and tripalmitin, studied by means of electron microscopy and infrared spectroscopy, *Biophys. Struct. Mech.*, **1**, 75–96 (1975).

157. R. D. Neuman, Calcium binding in stearic acid monomolecular films, *J. Colloid Interface Sci.*, **53**, 161–171 (1975).

158. R. D. Neuman, Stearic acid and calcium stearate monolayer collapse, *J. Colloid Interface Sci.*, **56**, 505–510 (1976).

159. S. B. Hwang, J. I. Korenbrot, and W. Stoeckenius, Structural and spectroscopic characteristics of bacteriorhodopsin in air–water interface films, *J. Membr. Biol.* **36**, 115–135 (1977).

160. G. L. Gaines, Jr. and W. J. Ward III, Visualization of defects in built-up soap multilayers by optical microscopy, *J. Colloid Interface Sci.*, **60**, 210–213 (1977).

161. O. Albrecht, A. Laschewsky, and H. Ringsdorf, Polymerizable built up multilayers on polymer supports, *Macromolecules*, **17**, 937–940 (1984).

162. A. Barraud, J. Leloup, P. Maire, and A. Ruaudel-Teixier, Microdefect decoration and visualization in Langmuir–Blodgett films, *Thin Solid Films*, **133**, 133–139 (1985).

163. W. M. Heckl, M. Lösche, and H. Möhwald, Langmuir–Blodgett films containing proteins of the photosynthetic apparatus, *Thin Solid Films*, **133**, 73–81 (1985).

164. W. Walkenhorst, Ein Einfaches Verfahren zur Herstellung Strukturloser Trägerschichten aus Aluminiumoxyd, *Naturwissenschaften*, **34**, 373 (1947).

165. H. P. Zingsheim, STEM as a tool in the construction of two-dimensional molecular assemblies, *Scanning Electron Microsc.*, **1**, 357–364 (1977).

166. A. I. Kitaigorodski, *Organic Chemical Crystallography*, Consultants Bureau, New York (1961).

167. E. von Sydow, The normal fatty acids in solid state, *Ark. Kem.*, **9**, 231–254 (1956).

168. S. C. Steele, M. N. Wybourne, and D. Möbius, Polarization effects observed in monolayer Langmuir–Blodgett films, *Thin Solid Films*, **99**, 117–118 (1983).

169. F. Grunfeld and C. W. Pitt, Diacetylene Langmuir–Blodgett layers for integrated optics, *Thin Solid Films*, **99**, 249–255 (1983).

170. I. R. Peterson, Optical observation of monomer Langmuir–Blodgett film structure, *Thin Solid Films*, **116**, 357–366 (1984).

171. G. Veale and I. R. Peterson, Novel effects of counterions on Langmuir films of 22-tricosenoic acid, *J. Colloid Interface Sci.*, **103**, 178–189 (1985).

172. G. Veale, I. R. Girling, and I. R. Peterson, A comparison of deposition speed, epitaxy and crystallinity in Langmuir–Blodgett films of fatty acids, *Thin Solid Films*, **127**, 293–303 (1985).

173. G. R. Bird, G. Debuch, and D. Möbius, Preparation of a totally ordered monolayer of a chromophore by rapid epitaxial attachment, *J. Phys. Chem.*, **81**, 2657–2663 (1977).

174. K. Hiltrop and H. Stegemeyer, Alignment of liquid crystals by amphiphilic monolayers, *Ber. Bunsenges. Phys. Chem.*, **82**, 884–889 (1978).

175. J. Staromylnska, F. C. Saunders, and G. W. Smith, A technique for the characterization of Langmuir–Blodgett films, *Mol. Cryst. Liq. Cryst.*, **109**, 233–243 (1984).

176. M. Lösche and H. Möhwald, Fluorescence microscope to observe dynamical processes in monomolecular layers at the air/water interface, *Rev. Sci. Instrum.*, **55**, 1968–1972 (1984).

177. M. Lösche, E. Sackmann, and H. Möhwald, A fluorescence microscopic study concerning the phase diagram of phospholipids, *Ber. Bunsenges. Phys. Chem.*, **87**, 848–852 (1983).

178. M. Lösche, J. Rabe, A. Fischer, B. U. Rucha, W. Knoll, and H. Möhwald, Microscopically observed preparation of Langmuir–Blodgett films, *Thin Solid Films*, **117**, 269–280 (1984).

179. M. Lösche, C. Helm, H. D. Mattes, and H. Möhwald, Formation of Langmuir–Blodgett films via electrostatic control of the lipid/water interface, *Thin Solid Films*, **133**, 51–64 (1985).

180. A. Miller, W. Knoll, H. Möhwald, and A. Ruaudel-Teixier, Langmuir–Blodgett films containing porphyrins in a well-defined environment, *Thin Solid Films*, **133**, 83–91 (1985).

181. H. Möhwald, Direct characterization of monolayers at the air–water interface, *Thin Solid Films*, **159**, 1–15 (1988).

182. S. A. Francis and A. H. Ellison, Infrared spectra of monolayers on metal mirrors, *J. Opt. Soc. Am.*, **49**, 131–138 (1959).

183. L. H. Sharpe, Observation of molecular interactions in oriented monolayers by infrared spectroscopy involving total internal reflection, *Proc. Chem. Soc.*, 461–463 (1961).

184. T. Takenaka, K. Nogami, H. Gotoh, and R. Gotoh, Studies on built-up films by means of the polarized infrared ATR spectrum. I. Built-up films of stearic acid, *J. Colloid Interface Sci.*, **35**, 395–402 (1971).

185. T. Takenaka, K. Nogami, and H. Gotoh, Studies of built-up films by means of the polarized infrared ATR spectrum. II. Mixed films of stearic acid and barium stearate, *J. Colloid Interface Sci.*, **40**, 409–416 (1972).

186. U. P. Fringeli, H. G. Müldner, H. H. Günthard, W. Gäschc, and W. Leuzinger, The structure of lipids and proteins studied by attenuated total-reflection (ATR) infrared spectroscopy, *Z. Naturforsch.*, **27B**, 780–796 (1972).

187. J. Breton, M. Michel-Villaz, G. Paillotin, and M. Vandevyver, Application of linear dichroism to the study of the distribution of pigments in monomolecular layers, *Thin Solid Films*, **13**, 351–357 (1972).

188. P. A. Chollet, J. Messier, and C. Rosilio, Infrared determination of the orientation of molecules in stearamide monolayers, *J. Chem. Phys.*, **64**, 1042–1050 (1976).

189. Y. Koyama, M. Yanagishita, S. Toda, and T. Matsuo, The relation between the crystal axis and the dipping direction of built-up films of octadecanoic acid as revealed by polarized infrared transmission spectroscopy, *J. Colloid Interface Sci.*, **61**, 438–445 (1977).

190. T. Ohnishi, A. Ishitani, H. Ishida, N. Yamamoto, and H. Tsubomura, X-ray photoelectron spectra and Fourier transform infrared spectra of mono- and multilayer films of cadmium arachidate, *J. Phys. Chem.*, **82**, 1989–1991 (1978).

191. P. A. Chollet, Determination by infrared absorption of the orientation of molecules in monomolecular layers, *Thin Solid Films*, **52**, 343–360 (1978).

192. H. Nakahara and K. Fukuda, Studies on molecular orientation in multilayers of long-chain anthraquinone derivatives by polarized infrared spectra, *J. Colloid Interface Sci.*, **69**, 24–33 (1979).

193. P. A. Chollet, IR determination of the orientation of molecules in polycrystalline monolayers, *Thin Solid Films*, **68**, 13–19 (1980).

194. T. Takenaka, K. Harada, and M. Matsumoto, Structural studies of poly-γ-benzyl-L-glutamate monolayers by infrared and transmission spectra, *J. Colloid Interface Sci.*, **73**, 569–577 (1980).

195. F. Takeda, M. Matsumoto, T. Takenaka, and Y. Fujiyoshi, Studies of poly-γ-methyl-L-glutamate monolayers by infrared ATR and transmission spectroscopy and electron microscopy, *J. Colloid Interface Sci.*, **84**, 220–227 (1981).

196. D. L. Allara and J. D. Swalen, An infrared reflection spectroscopy study of oriented cadmium arachidate monolayer films on evaporated silver, *J. Phys. Chem.*, **86**, 2700–2704 (1982).

197. P. A. Chollet and J. Messier, Studies of oriented Langmuir–Blodgett multilayers by infrared linear dichroism, *Chem. Phys.*, **73**, 235–242 (1982).

198. W. Knoll, M. R. Philpott, and W. G. Golden, Surface infrared and surface enhanced Raman vibrational spectra of monolayer assemblies in contact with rough metal surfaces, *J. Chem. Phys.*, **77**, 219–225 (1982).

199. M. Vandevyver, A. Barraud, A. Ruaudel-Teixier, P. Maillard, and C. Gianotti, Structure of porphyrin multilayers obtained by the Langmuir–Blodgett technique, *J. Colloid Interface Sci.*, **85**, 571–585 (1982).

200. W. G. Golden, C. D. Snyder, and B. Smith, Infrared reflection–absorption spectra of ordered and disordered arachidate monolayers on aluminium, *J. Phys. Chem.*, **86**, 4675–4678 (1982).

201. F. Takeda, M. Matsumoto, T. Takenaka, Y. Fujiyoshi, and N. Uyeda, Surface pressure dependence of monolayer structure of poly-ε-benzyloxycarbonyl-L-lysine, *J. Colloid Interface Sci.*, **91**, 267–271 (1983).

202. J. F. Rabolt, F. C. Burns, N. E. Schlotter, and J. D. Swalen, Anisotropic orientation in molecular monolayers by infrared spectroscopy, *J. Chem. Phys.*, **78**, 946–952 (1983).

203. H. Nakahara and K. Fukuda, Orientation control of chromophores in monolayer assemblies of long-chain dyes and the effects on some physical properties, *Thin Solid Films*, **99**, 45–52 (1983).

204. H. Nakahara and K. Fukuda, Orientation of chromophores in monolayers and multilayers of azobenzene derivatives with long alkyl chains, *J. Colloid Interface Sci.*, **93**, 530–539 (1983).

205. P. A. Chollet and J. Messier, IR dichroism of anisotropic Langmuir–Blodgett multilayers, *Thin Solid Films*, **99**, 197–204 (1983).

206. J. Seto, T. Nagai, C. Ishimoto, and H. Watanabe, Frictional properties of magnetic media coated with Langmuir–Blodgett films, *Thin Solid Films*, **134**, 101–108 (1985).

207. J. P. Rabe, J. F. Rabolt, C. A. Brown, and J. D. Swalen, Polymerization of two unsaturated fatty acid esters in Langmuir–Blodgett films as studied by IR spectroscopy, *Thin Solid Films*, **133**, 153–159 (1985).

208. S. J. Mumby, J. F. Rabolt, and J. D. Swalen, Structural characterization of a polymer monolayer on a solid surface, *Thin Solid Films*, **133**, 161–164 (1985).

209. T. Kawaguchi, H. Nakahara, and K. Fukuda, Monomolecular and multimolecular films of cellulose esters with various alkyl chains, *Thin Solid Films*, **133**, 29–38 (1985).

210. F. Kimura, J. Umemura, and T. Takenaka, FTIR-ATR studies on Langmuir–Blodgett films of stearic acid with one to nine monolayers, *Langmuir*, **2**, 96–101 (1986).

211. A. Hjortsberg, W. P. Chen, E. Burstein, and M. Pomerantz, Infrared surface EM-wave prism spectroscopy of Langmuir–Blodgett Mn-stearate layers on Ag, *Opt. Commun.*, **25**, 65–68 (1978).

212. S. Hayashi and J. Umemura, Infrared spectroscopic evidence for the coexistence of two molecular configurations in crystalline fatty acids, *J. Chem. Phys.*, **63**, 1732–1790 (1975).

213. A. Barraud, C. Rosilio, and A. Ruaudel-Teixier, Solid-state electron-induced polymerization of ω-tricosenoic acid multilayers, *J. Colloid Interface Sci.*, **62**, 509–523 (1977).

214. A. Barraud, C. Rosilio, and A. Ruaudel-Teixier, Selective photo and radiation induced polymerization of butadiene derivatives in monolayers, *Polym. Prepr., Am. Chem. Soc., Div. Polym. Chem.*, **19**, 179–182 (1978).

215. M. Breton, Formation and possible applications of polymeric Langmuir–Blodgett films. A review, *J. Macromol. Sci., Rev. Macromol. Chem.*, **C21**, 61–87 (1981).

216. K. Fukuda, Y. Shibasaki, and H. Nakahara, Effects of molecular arrangement on polymerization reactions in Langmuir–Blodgett films, *Thin Solid Films*, **99**, 87–94 (1983).

217. F. Kajzar and J. Messier, Solid state polymerization and optical properties of diacetylene Langmuir–Blodgett films, *Thin Solid Films*, **99**, 109–116 (1983).

218. G. Fariss, J. B. Lando, and S. E. Rickert, Phase controlled surface reaction—Reaction of a monolayer at the gas–water interface, *J. Mater. Sci.*, **18**, 3323–3330 (1983).

219. K. Fukuda, Y. Shibasaki, and H. Nakahara, Molecular arrangement and polymerizability of amino acid derivatives and dienoic acid in Langmuir–Blodgett films, *Thin Solid Films*, **133**, 39–49 (1985).

220. A. Barraud, Polymerization in Langmuir–Blodgett films and resist applications, *Thin Solid Films*, **99**, 317–321 (1983).

221. A. Barraud, C. Rosilio, and A. Ruaudel-Teixier, Polymerized monomolecular layers: A new class of ultrathin resins for microlithography, *Thin Solid Films*, **68**, 91–98 (1980).

222. J. W. Ellis and J. L. Pauley, The infrared determination of the composition of stearic acid multilayers deposited from salt substrata of varying pH, *J. Colloid Interface Sci.*, **19**, 755–764 (1964).

223. J. Bagg, M. B. Abramson, M. Fichman, M. D. Haber, and H. P. Gregor, Composition of stearic acid monolayers from calcium-containing substrates, *J. Am. Chem. Soc.*, **86**, 2759–2763 (1964).

224. E. E. Polymeropoulos and D. Möbius, Photochromism in monolayers, *Ber. Bunsenges. Phys. Chem.*, **83**, 1215–1222 (1979).

225. M. Morin, R. M. Leblanc, and I. Gruda, Spectral and photochromic properties of two long-chain spiropyranindoline monolayers at the air–solid interface, *Can. J. Chem.*, **58**, 2038–2043 (1980).

226. G. L. Gaines, Jr., Langmuir–Blodgett films of long-chain amines, *Nature*, **298**, 544–545 (1982).

227. C. B. McArdle, H. Blair, A. Barraud, and A. Ruaudel-Teixier, Positive and negative photochromism in thin organic Langmuir–Blodgett films, *Thin Solid Films*, **99**, 181–188 (1983).

228. C. Chapados and R. M. Leblanc, Aggregation of chlorophylls in monolayers V. The effect of water on chlorophyll *b* in monolayer and multilayer arrays, *Biophys. Chem.*, **17**, 211–244 (1983).

229. E. Ando, J. Miyazaki, K. Morimoto, H. Nakahara, and K. Fukuda, J-aggregation of photochromic spiropyran in Langmuir–Blodgett films, *Thin Solid Films*, **133**, 21–28 (1985).

230. C. Bubeck, Reactions in monolayers and Langmuir–Blodgett films, *Thin Solid Films*, **160**, 1–14 (1985).

231. A. Yabe, Y. Kawabata, H. Niino, M. Matsumoto, A. Ouchi, H. Takahashi, S. Tamura, and W. Tagaki, Photoisomerization of the Azobenzenes induced in Langmuir–Blodgett films of cyclodextrins, *Thin Solid Films*, **160**, 33–41 (1988).

232. T. Fukui, M. Saito, M. Sugi, and S. Iizima, Thermochromic behaviour of merocyanine Langmuir–Blodgett films, *Thin Solid Films*, **109**, 247–254 (1983).

233. C. Naselli, J. F. Rabolt, and J. D. Swalen, Order–disorder transitions in Langmuir–Blodgett monolayers. 1. Studies of two dimensional melting by infrared spectroscopy, *J. Chem. Phys.*, **82**, 2136–2140 (1985).

234. D. R. Day, H. Ringsdorf, and J. B. Lando, Polymerization of surface active diacetylene monolayers at the gas–water interface, *Polym. Prepr., Am. Chem. Soc., Div. Polym. Chem.*, **19**, 176–178 (1978).

235. D. R. Day and H. Ringsdorf, The monolayer polymerization of 10,12-nonacosadiynoic acid studied by a sepctroscopic technique, *Makromol. Chem.*, **180**, 1059–1063 (1979).

236. J. P. Fouassier, B. Tieke, and G. Wegner, The photochemistry of the polymerization of diacetylenes in multilayers, *Isr. J. Chem.*, **18**, 227–232 (1979).

237. D. Day, H. H. Hub, and H. Ringsdorf, Polymerization of mono- and bi-functional diacetylene derivatives in monolayers at the gas–water interface, *Isr. J. Chem.*, **18**, 325–329 (1979).

238. O. Albrecht, D. S. Johnston, C. Villaverde, and D. Chapman, Stable biomembrane surfaces formed by phospholipid polymers, *Biochim. Biophys. Acta*, **687**, 165–169 (1982).

239. C. Bubeck, B. Tieke, and G. Wegner, Cyanine dyes as sensitizers of the photopolymerization of diacetylenes in multilayers, *Ber. Bunsenges. Phys. Chem.*, **86**, 499–504 (1982).

240. C. Bubeck, K. Weiss, and B. Tieke, Sensitized photoreaction of diacetylene multilayers, *Thin Solid Films*, **99**, 103–107 (1983).

241. D. W. Kalina and S. W. Crane, Langmuir–Blodgett films of soluble copper octa (dodecoxymethyl)-phthalocyanine, *Thin Solid Films*, **134**, 109–119 (1985).

242. Y. L. Hua, G. G. Roberts, M. M. Ahmad, M. C. Petty, M. Hanack, and M. Rein, Monolayer films of a substituted silicon phthalocyanine, *Philos. Mag. B*, **53**, 105–113 (1986).

243. T. Takenaka, Effect of electrolyte on the molecular orientation in monolayers adsorbed at the liquid–liquid interface: Studies by resonance Raman spectra, *Chem. Phys. Lett.*, **55**, 515–518 (1978).

244. T. Takenaka and H. Fukuzaki, Resonance Raman spectra of insoluble monolayers spread on a water surface, *J. Raman Spectrosc.*, **8**, 151–154 (1979).

245. B. Tieke and D. Bloor, Raman spectroscopic studies of the solid-state polymerization of diacetylenes, 3. UV-polymerization of diacetylene Langmuir–Blodgett multilayers, *Makromol. Chem.*, **180**, 2275–2278 (1979).

246. A. Girlando, J. G. Gordon II, D. Heitmann, M. R. Philpott, H. Seki, and J. D. Swalen, Raman spectra of molecules on metal surfaces, *Surf. Sci.*, **101**, 417–424 (1980).

247. J. F. Rabolt, R. Santo, and J. D. Swalen, Raman measurements on thin polymer films and organic monolayers, *Appl. Spectrosc.*, **34**, 517–521 (1980).

248. J. F. Rabolt, R. Santo, N. E. Schlotter, and J. D. Swalen, Integrated optics and Raman scattering: Molecular orientation in thin polymer films and Langmuir–Blodgett monolayers, *IBM J. Res. Dev.*, **26**, 209–216 (1982).

249. W. Knoll, M. R. Philpott, J. D. Swalen, and A. Girlando, Surface plasmon enhanced Raman spectra of monolayer assemblies, *J. Chem. Phys.*, **77**, 2254–2260 (1982).

250. M. R. Philpott, A. Girlando, W. G. Golden, W. Knoll, and J. D. Swalen, Surface plasmon enhanced vibrational spectra of monolayer assemblies on gratings and rough metal surfaces, *Mol. Cryst. Liq. Cryst.*, **96**, 335–351 (1983).

251. M. Vandevyver, A. Ruaudel-Teixier, L. Brehamet, and M. Lutz, Polarized resonance Raman spectroscopy of Langmuir–Blodgett films, *Thin Solid Films*, **99**, 41–44 (1983).

252. D. N. Batchelder, D. Bloor, and I. R. J. Lyall, A study of monolayer and multilayer films containing polydiacetylenes using resonance Raman spectroscopy, *Thin Solid Films*, **99**, 118 (1983).

253. W. Knoll, J. Rabe, M. R. Philpott, and J. D. Swalen, Ellipsometry and reflection, luminescence and Raman spectroscopies of monolayer assemblies on solid substrates, *Thin Solid Films*, **99**, 173–179 (1983).

254. Y. J. Chen, G. M. Carter, and S. K. Tripathy, Study of Langmuir–Blodgett polydiacetylene polymer films by surface enhanced Raman scattering, *Solid State Commun.*, **54**, 19–22 (1985).

255. I. R. J. Lyall and D. N. Batchelder, Resonance Raman spectroscopy of polydiacetylene Langmuir–Blodgett films, *Br. Polym. J.*, **17**, 372–376 (1985).

256. D. P. DiLella, W. R. Barger, A. W. Snow, and R. R. Smardzewski, Resonance Raman spectra of Langmuir–Blodgett monolayers, *Thin Solid Films*, **133**, 207–217 (1985).

257. R. A. Uphaus, T. M. Cotton, and D. Möbius, Surface-enhanced resonance Raman spectroscopy of synthetic dyes and photosynthetic pigments in monolayer and multilayer assemblies, *Thin Solid Films*, **132**, 173–184 (1985).

258. C. A. Murray, *Recent Adv. Laser Spectrosc.* (B. A. Garetz and J. Lombardi, eds.) **4**, John Wiley, New York (1983).

259. A. Barraud, A. Ruaudel-Teixier, and C. Rosilio, Réactions chimiques a l'état solide dans les couches monomoléculaires organiques, *Ann. Chim.*, **10**, 195–200 (1975).

260. C. Mori, H. Noguchi, M. Mizuno, and T. Watanabe, A range measurement of low energy electrons using radioisotope auger electrons, *Jpn. J. Appl. Phys.*, **19**, 725–732 (1980).

261. A. Barraud, C. Rosilio, and A. Ruaudel-Teixier, Reactivity of organic molecules in monolayers, *Thin Solid Films*, **68**, 7–12 (1980).

262. T. M. Ginnai, R. E. Thurstans, and D. P. Oxley, The structure and electron transport properties of Langmuir films, *Vide Couches Minces*, **201**(suppl.), Proc. Int. Vac. Congs., 435–438 (1980).

263. T. M. Ginnai, An investigation of electron tunneling and conduction in Langmuir films, Ph.D. thesis, Leicester Polytechnic, UK (1982).

264. R. G. Steinhardt, Jr. and E. J. Serfass, Surface analysis with the X-ray photoelectron spectrometer, *Anal. Chem.*, **25**, 697–700 (1953).

265. K. Larsson, C. Nordling, K. Siegbahn, and E. Stenhagen, Photoelectron spectroscopy of fatty acid multilayers, *Acta Chem. Scand.*, **20**, 2880–2881 (1966).

266. D. A. Brandreth, W. M. Riggs, and R. E. Johnson, Detection of metal ions in stearic acid monolayers, *Nat., Phys. Sci.*, **236**, 11–12 (1972).

267. M. Pomerantz and R. A. Pollak, Spin state of manganese in monolayer films of Mn arachidate, *Chem. Phys. Lett.*, **31**, 602–604 (1975).

268. H. R. Anderson, Jr. and J. D. Swalen, X-ray photoelectron spectroscopic studies of the bonding of organic monolayers on various oxidized surfaces, *J. Adhes.*, **9**, 197–211 (1978).

269. C. R. Brundle, H. Hopster, and J. D. Swalen, Electron mean-free path lengths through monolayers of cadmium arachidate, *J. Chem. Phys.*, **70**, 5190–5196 (1979).

270. D. T. Clark, Y. C. T. Fok, and G. G. Roberts, Electron mean free paths in Langmuir–Blodgett multilayers, *J. Electron Spectrosc. Relat. Phenom.*, **22**, 173–185 (1981).

271. F. C. Burns and J. D. Swalen, X-ray photoelectron spectroscopy of cadmium arachidate monolayers on various metal surfaces, IBM Research Report (1983).

272. C. A. Brown, F. C. Burns, W. Knoll, J. D. Swalen, and A. Fischer, Unusual monolayer behaviour of a geminally distributed fatty acid. Characterization via surface plasmons and X-ray photoelectron spectroscopy study, *J. Phys. Chem.*, **87**, 3616–3619 (1983).

273. L. Laxhuber, H. Möhwald, and M. Hashmi, Secondary-ion mass spectrometry of organized organic model systems, *Int. J. Mass Spectrom. Ion Phys.*, **51**, 93–110 (1983).

274. N. Ueno, W. Gädeke, E. E. Koch, R. Englehardt, R. Dudde, L. Laxhuber, and H. Möhwald, One-dimensional energy band dispersion in Langmuir–Blodgett films determined by angle-resolved photoemission with synchrotron radiation, *J. Molec. Electron.*, **1**, 19–23 (1985).

275. H. Oyanagi, M. Sugi, S. Kuroda, S. Iizima, T. Ishiguro, and T. Matsushita, Polarized X-ray absorption spectra of LB films: Local structure studies of merocyanine dyes, *Thin Solid Films*, **133**, 181–188 (1985).

276. J. Messier and G. Marc, Étude par résonance paramagnétique électronique de la structure de couches monomoléculaires de stérate de cuivre, *J. Phys. (Paris)*, **32**, 799–804 (1971).

277. P. A. Chollet, EPR studies on the structure of copper dioctadecyl-dithiocarbamate monolayers, *J. Phys. C*, **7**, 4127–4134 (1974).

278. M. Pomerantz, F. H. Dacol, and A. Segmüller, Preparation of literally two-dimensional magnets, *Phys. Rev. Lett.*, **40**, 246–249 (1978).

279. J. Cunningham, E. E. Polymeropoulos, and D. Möbius, Photoinitiated electron transfer between donors and acceptors in monomolecular assemblies—EPR and optical studies, in *Magnetic Resonance in Colloid and Interface Science* (J. P. Fraissard and H. A. Resing, eds.), pp. 603–608, Reidel, Dordrecht (1980).

280. P. Jost, L. J. Libertini, V. C. Herbert, and O. H. Griffith, Lipid spin labels in lecithin multilayers. A study of motion along fatty acid chains, *J. Mol. Biol.*, **59**, 77–98 (1971).

281. S. Kuroda, M. Sugi, and S. Iizima, Electron spin resonance in Langmuir films of merocyanine dyes, *Thin Solid Films*, **99**, 21–24 (1983).

282. T. Iwasaki, H. Wakabayashi, T. Ishii, and K. Iriyama, Light-induced electron spin resonance in a Langmuir–Blodgett film of pure merocyanine dye, *Appl. Phys. Lett.*, **45**, 1089–1090 (1984).

283. S. Kuroda, M. Sugi, and S. Iizima, Origin of stable spin species in Langmuir–Blodgett films of merocyanine dyes studied by ESR and ENDOR, *Thin Solid Films*, **133**, 189–196 (1985).

284. K. Iriyama, M. Yoshiura, Y. Ozaki, T. Ishii, and S. Yasui, Preparation of merocyanine dye Langmuir–Blodgett films and some of their physiochemical properties, *Thin Solid Films*, **132**, 229–242 (1985).

285. H. Hediger and R. Steiger, Studies of organized monolayer assemblies by photoacoustic spectroscopy, *J. Colloid Interface Sci.*, **103**, 343–353 (1985).

286. A. Désormeaux and R. M. Leblanc, Electronic and photoacoustic spectroscopies of chlorophyll *a* in mono and multilayer arrays, *Thin Solid Films*, **132**, 91–99 (1985).

287. C. Naselli, J. P. Rabe, J. F. Rabolt, and J. D. Swalen, Thermally induced order–disorder transitions in Langmuir–Blodgett films, *Thin Solid Films*, **134**, 173–178 (1985).

288. W. Barger, J. Dote, M. Klusty, R. Mowery, R. Price, and A. Snow, Morphology and properties of Langmuir films containing tetrakis (cumylphenoxy) phthalocyanines, *Thin Solid Films*, **159**, 369–378 (1988).

289. U. Khanna and V. K. Srivastava, Refractive index of "built-up" barium stearate films, *Thin Films*, **2**, 153 157 (1972).

290. H. Hasmonay, M. Dupeyrat, and R. Dupeyrat, Stearate thin films of adjustable refractive index and some practical applications, *Opt. Acta*, **23**, 665–677 (1976).

291. J. D. Swalen, K. E. Rieckhoff, and M. Tacke, Optical properties of arachidate monolayers by integrated optical techniques, *Opt. Commun.*, **24**, 146–148 (1978).

292. J. D. Swalen, M. Tacke, R. Santo, and J. Fischer, Determination of optical constants of polymeric thin films by integrated optical techniques, *Opt. Commun.*, **18**, 387–390 (1976).

293. M. Fleck, Dissertation, University of Marburg, Germany (1966).

294. K. H. Drexhage, Habilitahon-Schrift, University of Marburg, Germany (1966).

295. K. B. Blodgett, Use of interference to extinguish reflection of light from glass, *Phys. Rev.*, **55**, 391–404 (1939).

296. H. Foster, Diplomarbeit, University of Marburg, Germany (1966).
297. J. G. Gordon II and J. D. Swalen, The effect of thin organic films on the surface plasma resonance on gold, *Opt. Commun.*, **22**, 374–376 (1977).
298. I. Pockrand, J. D. Swalen, J. D. Gordon II, and M. R. Philpott, Surface plasmon spectroscopy of organic monolayer assemblies, *Surf. Sci.*, **74**, 237–244 (1977).
299. G. Wähling, Arachidate layers on Ag and Au substrates detected by the ATR method, *Z. Natur-forsch.*, **33a**, 536–539 (1978).
300. I. Pockrand and J. D. Swalen, Anomalous dispersion of surface plasma oscillations, *J. Opt. Soc. Am.*, **68**, 1147–1151 (1978).
301. I. Pockrand, J. D. Swalen, R. Santo, A. Brillante, and M. R. Philpott, Optical properties of organic dye monolayers by surface plasmon spectroscopy, *J. Chem. Phys.*, **69**, 4001–4011 (1978).
302. G. Wähling, H. Raether, and D. Möbius, Studies of organic monolayers on thin silver films using the attenuated total reflection method, *Thin Solid Films*, **58**, 391–395 (1979).
303. J. Zyss, New organic molecular materials for non linear optics, *J. Non-Cryst. Solids*, **47**, 211–226 (1982).
304. D. J. Williams, Organic polymeric and non-polymeric materials with large optical non linearities, *Angew. Chem., Int. Ed. Engl.*, **23**, 690–703 (1984).
305. T. Rasing, Y. R. Shen, M. W. Kim, P. Valint, Jr. and J. Bock, Orientation of surfactant molecules at a liquid–air interface measured by optical second-harmonic generation, *Phys. Rev. A*, 537–539 (1985).
306. O. A. Aktsipetrov, N. N. Akhmediev, E. D. Mishina, and V. R. Novak, Second-harmonic generation on reflection from a monomolecular Langmuir layer, *JETP Lett.*, **37**, 207–209 (1983).
307. I. R. Girling, N. A. Cade, P. V. Kolinsky, and C. M. Montgomery, Observation of second-harmonic generation for a Langmuir–Blodgett monolayer of a merocyanine dye, *Electron. Lett.*, **21**, 169–170 (1985).
308. Z. Chen, W. Chen, J. Zheng, W. Wang, and Z. Zhang, Surface-enhanced SHG study on Lang-muir–Blodgett mono-molecular layers, *Opt. Commun.*, **54**, 305–310 (1985).
309. I. R. Girling, P. V. Kolinsky, N. A. Cade, J. D. Earls, and I. R. Peterson, Second harmonic generation from alternating Langmuir–Blodgett films, *Opt. Commun.*, **55**, 289–292 (1985).
310. I. R. Girling, N. A. Cade, P. V. Kolinsky, J. D. Earls, G. H. Cross, and I. R. Peterson, Observation of second harmonic generation from Langmuir–Blodgett multilayers of a hemicyanine dye, *Thin Solid Films*, **132**, 101–112 (1985).
311. D. B. Neal, M. C. Petty, G. G. Roberts, M. M. Ahmad, W. J. Feast, I. R. Girling, N. A. Cade, P. V. Kolinsky, and I. R. Peterson, Second harmonic generation from LB superlattices containing two active components, *Electron. Lett.*, **22**, 460–462 (1986).
312. P. A. Chollet, F. Kajzar, and J. Messier, Electric field induced optical second harmonic generation and polarization effects in polydiacetylene Langmuir–Blodgett multilayers, *Thin Solid Films*, **132**, 1–10 (1985).
313. G. M. Carter, Y. J. Chen, and S. K. Tripathy, Third-order nonlinear susceptibility in multilayers of polydiacetylene, in: *Nonlinear Optical Properties of Organic and Polymeric Materials* (D. J. Williams, ed.), pp. 213–228, Am. Chem. Soc. (1983).
314. G. M. Carter, Y. J. Chen, and S. K. Tripathy, Intensity-dependent index of refraction in multilayers of polydiacetylene, *Appl. Phys. Lett.*, **43**, 891–893 (1983).
315. F. Kajzar, J. Messier, J. Zyss, and I. Ledoux, Non linear interferometry in Langmuir–Blodgett multilayers of polydiacetylene, *Opt. Commun.*, **45**, 133–137 (1983).
316. F. Kajzar and J. Messier, Resonance enhancement in cubic susceptibility of Langmuir–Blodgett multilayers of polydiacetylene, *Thin Solid Films*, **132**, 11–19 (1985).
317. J. Zyss, Nonlinear organic materials for integrated optics: A review, *J. Mol. Electron.*, **1**, 25–45 (1985).
318. See, for example, P. Debye, *Polar Molecules,* Lancaster Press, Lancaster (1929).
319. L. M. Blinov, N. N. Davydova, V. V. Lazarev, and S. G. Yudin, Spontaneous polarization of Langmuir multimolecular films, *Sov. Phys. Solid State (Engl. Transl.)*, **24**, 1532–1525 (1983).

320. L. M. Blinov, N. V. Dubinin, L. V. Mikhnev, and S. G. Yudin, Polar Langmuir–Blodgett films, *Thin Solid Films*, **120**, 161–170 (1984).

321. G. G. Roberts, T. M. McGinnity, W. A. Barlow, and P. S. Vincett, Electroluminescence, photoluminescence, and electroabsorption of a lightly substituted anthracene Langmuir film, *Solid State Commun.*, **32**, 683–686 (1979).

322. H. Bücher, J. Wiegand, B. B. Snavely, K. H. Beck, and H. Kuhn, Electric field induced changes in the optical absorption of a merocyanine dye, *Chem. Phys. Lett.*, **3**, 508–511 (1969).

323. C. E. Buchwald, D. M. Gallagher, C. P. Haskins, E. M. Thatcher, and P. A. Zahl, Measurements of resistance and capacity of monofilms of barium stearate, *Proc. Natl. Acad. Sci. U.S.A.*, **24**, 204–208 (1938).

324. P. A. Zahl, C. P. Haskins, D. M. Gallagher, and C. E. Buchwald, Some electrical properties of deposited layers of calcium stearate, *Trans. Faraday Soc.*, **35**, 308–312 (1939).

325. H. H. Race and S. I. Reynolds, Electrical properties of multimolecular films, *J. Am. Chem. Soc.*, **61**, 1425–1432 (1939).

326. I. R. Peterson, Defect density in a metal–monolayer–metal cell, *Aust. J. Chem.*, **33**, 1713–1716 (1980).

327. I. R. Peterson, The reproducibly perfect Langmuir–Blodgett film, *J. Chim. Physique*, **85**, 997–1001 (1988).

328. R. H. Tredgold, A. J. Vickers, and R. A. Allen, Structural effects on the electrical conductivity of Langmuir–Blodgett multilayers of cadmium stearate, *J. Phys. D*, **17**, L5–L8 (1984).

329. G. Marc and J. Messier, Dielectric losses in organic monomolecular layers, *J. Appl. Phys.*, **45**, 2832–2835 (1974).

330. M. Sugi, T. Fukui, and S. Iizima, Hopping conduction in Langmuir films, *Appl. Phys. Lett.*, **27**, 559–561 (1975); **28**, 240 (1976).

331. J. Tanguy, Effect of phase transition on the dielectric properties of thin organic layers, *J. Appl. Phys.*, **47**, 2792–2799 (1976).

332. A. Barraud and A. Rosilio, Dielectric breakdown in monomolecular layers, *Thin Solid Films*, **31**, 243–251 (1976).

333. E. E. Polymeropoulos, Electron tunnelling through fatty-acid monolayers, *J. Appl. Phys.*, **48**, 2404–2407 (1977).

334. M. H. Nathoo, Preparation of Langmuir films for electrical studies, *Thin Solid Films*, **16**, 215–226 (1973).

335. D. M. Taylor and M. G. B. Mahboubian-Jones, The electrical properties of synthetic phospholipid Langmuir–Blodgett films, *Thin Solid Films*, **87**, 167–179 (1982).

336. L. Y. Wei and B. Y. Woo, Electronic conduction in lipid films with metal contacts, *Biophys. J.*, **13**, 877–889 (1973).

337. A. Barraud, A. A. Rosilio, and J. Messier, Monomolekularschicht-Hygrometer, German Patent No. 2702487 (1977).

338. A. Leger, J. Klein, M. Belin, and D. Defourneau, Properties of metal-insulating Langmuir film–metal junction, *Thin Solid Films*, **8**, R51–R54 (1971).

339. J. Tanguy, Study of metal–organic monolayer–semiconductor structures, *Thin Solid Films*, **13**, 33–39 (1972).

340. E. P. Honig, D.C. conduction in Langmuir–Blodgett films with various electrode materials, *Thin Solid Films*, **33**, 231–236 (1976).

341. J. Holoyda, R. V. McDaniel, C. R. Kannewurf, and J. W. Kauffman, Optical effects in organic layered structures, *Bull. Am. Phys. Soc.*, **24**, 356 (1979).

342. M. A. Careem, Electrical properties of thin stearic acid films, Ph.D. thesis, University of London (1976).

343. R. W. Roberts and G. L. Gaines, Jr., Stability of fatty monolayers in vacuum, Trans. 9th. Nat. Vac. Symposium, Los Angeles, pp. 515–518 (1962).

344. J. L. Miles and H. O. McMahon, Use of monomolecular layers in evaporated-film tunnelling devices, *J. Appl. Phys.*, **32**, 1176–1177 (1961).

345. T. D. Clark, Tunnelling experiments using monolayer tunnelling barriers, Mullard Research Labs. Report No. 512 (1964).

346. K. H. Beck, Photoeffekte an Monomolekularen Farbstoffschichten, Ph.D. thesis, University of Marburg (1966).

347. L. C. Scala and R. M. Handy, Structure and electrical properties of poly(vinyl benzoate) monolayers, *J. Appl. Polym. Sci.,* **9**, 3111–3122 (1965).

348. R. M. Handy and L. C. Scala, Electrical and structural properties of Langmuir films, *J. Electrochem. Soc.,* **113**, 109–116 (1966).

349. S. Horiuchi, J. Yamaguchi, and K. Naito, Electrical conduction through thin insulating Langmuir film, *J. Electrochem. Soc.,* **115**, 634–637 (1968).

350. A. Sommerfeld and H. Bethe, in: *Handbuch der Physik* (H. Geiger and K. Scheel, eds.), Vol. 24/2, p. 450, Springer, Berlin (1933).

351. H. Kuhn, Electron tunnelling effects in monolayer assemblies, *Chem. Phys Lipids,* **8**, 401–404 (1972).

352. N. Yamamoto, T. Ohnishi, M. Hatakeyama, and H. Tsubomura, Electrical properties of fatty acid salts and carotenoic acid, *Bull. Chem. Soc. Jpn.,* **51**, 3462–3465 (1978).

353. R. H. Tredgold and C. S. Winter, Tunnelling currents in Langmuir–Blodgett monolayers of stearic acid, *J. Phys. D,* **14**, L185–L188 (1981).

354. W. L. Procarione and J. W. Kauffman, The electrical properties of phospholipid bilayer Langmuir films, *Chem. Phys. Lipids,* **12**, 251–260 (1974).

355. E. E. Polymeropoulos and J. Sagiv, Electrical conduction through adsorbed monolayers, *J. Chem. Phys.,* **69**, 1836–1847 (1978).

356. K. H. Gundlach and J. Kadlec, The influence of the oxide film on the current in Al–Al oxide–fatty acid monolayer–metal junctions, *Chem. Phys. Lett.,* **25**, 293–295 (1974).

357. R. Stratton, Volt–current characteristics for tunnelling through insulating films, *J. Phys. Chem. Solids,* **23**, 1177–1190 (1962).

358. R. M. Hill, Poole–Frenkel conduction in amorphous solids, *Philos. Mag.,* **23**, 59–86 (1971).

359. M. A. Careem and R. M. Hill, Transport processes in stearic acid layers, *Thin Solid Films,* **51**, 363–371 (1978).

360. K. H. Gundlach, Theory of metal–insulator–metal tunnelling for a simple two-band model, *J. Appl. Phys.,* **44**, 5005–5010 (1973).

361. J. Bardeen, L. N. Cooper, and J. R. Schrieffer, Theory of superconductivity, *Phys. Rev.,* **108**, 1175–1204 (1957).

362. E. E. Polymeropoulos, Electron tunnelling through superconducting Al/monolayer/Pb junctions, *Solid State Commun.,* **28**, 883–885 (1978).

363. G. L. Larkins, Jr., The use of Langmuir–Blodgett films as a barrier layer in Josephson tunnel junctions, M.Sc. Thesis, CASE Western Reserve University, USA (1982).

364. G. L. Larkins, Jr., E. D. Thompson, E. Ortiz, C. W. Burkhart, and J. B. Lando, Langmuir–Blodgett films as barrier layers in Josephson tunnel junctions, *Thin Solid Films,* **99**, 277–282 (1983).

365. S. Hao, B. H. Blott, and D. Melville, Metal/Langmuir–Blodgett/metal junctions using Pb–In superconducting electrodes, *Thin Solid Films,* **132**, 63–68 (1985).

366. T. M. Ginnai, D. P. Oxley, and R. G. Pritchard, Elastic and inelastic tunnelling in single-layer Langmuir films, *Thin Solid Films,* **68**, 241–256 (1980).

367. G. G. Roberts, P. S. Vincett, and W. A. Barlow, AC and DC conduction in fatty acid Langmuir films, *J. Phys. C,* **11**, 2077–2084 (1978).

368. J. P. Lloyd, M. C. Petty, G. G. Roberts, P. G. LeComber, and W. E. Spear, Amorphous silicon/Langmuir–Blodgett film field effect transistor, *Thin Solid Films,* **99**, 297–304 (1983).

369. G. G. Roberts, An applied science perspective of Langmuir–Blodgett films, *Adv. Phys.,* **34**, 475–512 (1985).

370. N. J. Geddes, J. R. Sambles, D. J. Jarvis, and N. R. Couch, Filamentary conduction through ω-tricosenoic acid multilayers, in "Molecular Electronic Devices," F. L. Carter, R. E. Siatkowski, and H. Wohltjen (eds.), Elsevier, 495–506 (1988).

371. P. K. Hansma, Inelastic electron tunnelling, *Phys. Rep.*, **30**, 145–206 (1977).

372. M. H. Nathoo and A. K. Jonscher, High field and ac properties of stearic acid films, *J. Phys. C*, **4**, L301–L304 (1971).

373. W. H. Simpson and P. J. Reucroft, Quantum-mechanical tunnelling in thin films of chlorophyll *a*, *Thin Solid Films*, **6**, 167–174 (1970).

374. G. G. Roberts, K. P. Pande, and W. A. Barlow, InP–Langmuir–Film M.I.S. structures, *Electron. Lett.*, **13**, 581–583 (1977).

375. G. G. Roberts, K. P. Pande, and W. A. Barlow, InP/Langmuir-film M.I.S.F.E.T., *Solid State Electron. Dev.*, **2**, 169–175 (1978).

376. M. C. Petty and G. G. Roberts, CdTe/Langmuir-film MIS structure, *Electron. Lett.*, **15**, 335–336 (1979).

377. M. C. Petty and G. G. Roberts, Analysis of p-type CdTe-Langmuir film interface, *Inst. Phys. Conf. Ser.*, **50**, 186–192 (1980).

378. R. W. Sykes, G. G. Roberts, T. Fok, and D. T. Clark, p-Type InP/Langmuir-film M.I.S. diodes, *IEE Proc. I*, **127**, 137–139 (1980).

379. I. M. Dharmadasa, G. G. Roberts, and M. C. Petty, Cadmium telluride/Langmuir film photovoltaic structures, *Electron. Lett.*, **16**, 201–202 (1980).

380. K. K. Kan, M. C. Petty, and G. G. Roberts, Polymerized Langmuir film MIS structures, in: *Proc. Int. Conf. Phys. MOS Insulators*, pp. 344–348, Pergamon, New York (1980).

381. G. G. Roberts, M. C. Petty, and I. M. Dharmadasa, Photovoltaic properties of cadmium–telluride/Langmuir-film solar cells, *IEE Proc. I*, **128**, 197–201 (1981).

382. R. H. Tredgold and R. Jones, Schottky-barrier diodes incorporating Langmuir-film interfacial monolayers, *IEE Proc. I*, **128**, 202–206 (1981).

383. R. H. Tredgold and G. W. Smith, Schottky photodiodes incorporating monolayers formed by adsorption and by the Langmuir–Blodgett technique, *IEE Proc. I*, **129**, 137–140 (1982).

384. J. P. Lloyd, M. C. Petty, G. G. Roberts, P. G. LeComber, and W. E. Spear, Langmuir–Blodgett films in amorphous silicon MIS structures, *Thin Solid Films*, **89**, 395–399 (1982).

385. R. H. Tredgold and C. S. Winter, Langmuir–Blodgett monolayers of preformed polymers on n-type GaP, *Thin Solid Films*, **99**, 81–85 (1983).

386. J. Batey, G. G. Roberts, and M. C. Petty, Electroluminescence in GaP/Langmuir–Blodgett film metal/insulator/semiconductor diodes, *Thin Solid Films*, **99**, 283–290 (1983).

387. K. K. Kan, G. G. Roberts, and M. C. Petty, Langmuir–Blodgett film metal/insulator/semiconductor structures on narrow band gap semiconductors, *Thin Solid Films*, **99**, 291–296 (1983).

388. J. Batey, M. C. Petty, and G. G. Roberts, Electroluminescent MIS structures incorporating Langmuir–Blodgett films, in: *Insulating Films on Semiconductors* (J. F. Verweij and D. R. Wolters, eds.), pp. 141–144, Elsevier (North-Holland), Amsterdam (1983).

389. C. S. Winter and R. H. Tredgold, Control of Schottky diode barrier height by Langmuir–Blodgett monolayers, *IEE Proc. I*, **130**, 256–259 (1983).

390. C. S. Winter and R. H. Tredgold, Influence of Langmuir–Blodgett monolayers on the Schottky barrier height of gallium phosphide diodes, *J. Phys. D*, **17**, L123–L126 (1984).

391. J. Batey, M. C. Petty, G. G. Roberts, and D. R. Wright, GaP/phthalocyanine Langmuir–Blodgett film electroluminescent diode, *Electron. Lett.*, **20**, 489–491 (1984).

392. N. J. Thomas, M. C. Petty, G. G. Roberts, and H. Y. Hall, GaAs/LB film MISS switching device, *Electron. Lett.*, **20**, 838–839 (1984).

393. N. J. Thomas, G. G. Roberts, and M. C. Petty, Switching characteristics for GaAs/LB film MISS devices, *Proc. Int. Conf. Insulating Films on Semiconductors, INFOS 85*, J. J. Simonne and J. Buxo (eds.), Elsevier, 71–74 (1986).

394. M. C. Petty, J. Batey, and G. G. Roberts, A comparison of the photovoltaic and electroluminescent effects in GaP/Langmuir–Blodgett film diodes, *IEE Proc. I*, **132**, 133–139 (1985).

395. R. H. Tredgold and Z. I. El-Badawy, Increase of Schottky barrier height at GaAs surfaces by carboxylic acid monolayers and multilayers, *J. Phys. D*, **18**, 103–109 (1985).

396. M. T. Fowler, M. C. Petty, G. G. Roberts, P. J. Wright, and B. Cockayne, Forward bias elec-

troluminescence from phthalocyanine Langmuir–Blodgett film/ZnSeS MIS diodes, *J. Mol. Electron.*, **1**, 93–95 (1985).

397. P. S. Vincett, W. A. Barlow, F. T. Boyle, J. A. Finney, and G. G. Roberts, Preparation of Langmuir–Blodgett "built-up" multilayer films of a lightly substituted model aromatic, anthracene, *Thin Solid Films*, **60**, 265–277 (1979).

398. G. G. Roberts, T. M. McGinnity, W. A. Barlow, and P. S. Vincett, A.C. and D.C. conduction in lightly substituted anthracene Langmuir films, *Thin Solid Films*, **68**, 223–232 (1980).

399. S. Baker, M. C. Petty, G. G. Roberts, and M. V. Twigg, The preparation and properties of stable metal-free phthalocyanine Langmuir–Blodgett films, *Thin Solid Films*, **99**, 53–59 (1983).

400. S. Baker, G. G. Roberts, and M. C. Petty, Phthalocyanine Langmuir–Blodgett-film gas detector, *IEE Proc. I*, **130**, 260–263 (1983).

401. H. Wohltjen, W. R. Barger, A. W. Snow, and N. Lynn Jarvis, A vapour-sensitive chemiresistor fabricated with planar microelectrodes and a Langmuir–Blodgett organic semiconductor film, *IEEE Trans. Electron Devices*, **ED-32**, 1170–1174 (1985).

402. R. H. Tredgold, M. C. J. Young, P. Hodge, and A. Hoorfar, Gas sensors made from Langmuir–Blodgett films of porphyrins, *IEE Proc. I*, **132**, 151–156 (1985).

403. U. Schoeler, K. H. Tews, and H. Kuhn, Potential model of dye molecule from measurements of the photocurrent in monolayer assemblies, *J. Chem. Phys.*, **61**, 5009–5016 (1974).

404. M. Sugi, K. Nembach, and D. Möbius, Photoconduction in Langmuir films with periodically arranged dye-sensitizers, *Thin Solid Films*, **27**, 205–216 (1975).

405. E. E. Polymeropoulos, D, Möbius, and H. Kuhn, Monolayer assemblies with functional units of sensitizing and conducting molecular components: Photovoltage, dark conduction, and photoconduction in systems with aluminium and barium electrodes, *Thin Solid Films*, **68**, 173–190 (1980).

406. N. Yamamoto, T. Ohnishi, M. Hatakeyama, and T. Tsubomura, Photoelectric properties of molecular layers of a fatty acid mixed with cyanine dyes, *Thin Solid Films*, **68**, 191–198 (1980).

407. M. Sugi and S. Iizima, Anisotropic photoconduction in dye-sensitized Langmuir films, *Thin Solid Films*, **68**, 199–204 (1980).

408. J. Holoyda, C. R. Kannewurf, and J. W. Kauffman, Photoeffects in chromophore phospholipid Langmuir films, *Thin Solid Films*, **68**, 205–222 (1980).

409. M. Sugi, M. Saito, T. Fukui, and S. Iizima, High field photoconduction in Langmuir films of merocyanine dyes, *Thin Solid Films*, **88**, L15–L17 (1982).

410. M. Sugi, M. Saito, T. Fukui, and S. Iizima, Effect of dye concentration in Langmuir multilayer photoconductors, *Thin Solid Films*, **99**, 17–20 (1983).

411. M. Saito, M. Sugi, T. Fukui, and S. Iizima, Photoelectric effects in dye-sensitized Langmuir–Blodgett film diodes, *Thin Solid Films*, **100**, 117–120 (1983).

412. M. Saito, M. Sugi, and S. Iizima, Evidence for Ambipolar conduction in dye-sensitized p–n junctions of Langmuir–Blodgett films, *Jpn. J. Appl. Phys.*, **24**, 379–380 (1985).

413. K. Sakai, M. Saito, M. Sugi, and S. Iizima, Molecular p–n junction photodiodes of Langmuir multilayer semiconductors, *Jpn. J. Appl. Phys.*, **24**, 864–868 (1985).

414. M. Sugi, K. Sakai, M. Saito, Y. Kawabata, and S. Iizima, Photoelectric effects in heterojunction Langmuir–Blodgett film diodes, *Thin Solid Films*, **132**, 69–76 (1985).

415. A. Barraud, P. Lesieur, A. Ruaudel-Teixier, and M. Vandevyver, Characterization and properties of conducting Langmuir–Blodgett films, *Thin Solid Films*, **134**, 195–199 (1985).

416. A. Barraud, P. Lesieur, A. Ruaudel-Teixier, and M. Vandevyver, Structure and properties of a n-docosylpyridinium-tetracyanoquinodimethane salt in Langmuir–Blodgett multilayers, *Thin Solid Films*, **133**, 125–131 (1985).

417. E. P. Honig and B. R. de Koning, Transient and alternating electric currents in thin organic films, *J. Phys. C*, **11**, 3259–3271 (1978).

418. R. A. Street, G. Davies, and A. D. Yoffe, The square law dependence with frequency of the electrical conductivity of As_2 Se_3, *J. Non-Cryst. Solids*, **5**, 276–278 (1971).

419. A. K. Jonscher and M. H. Nathoo, High-field transient behaviour of stearic acid films, *Thin Solid Films*, **12**, S15–S18 (1972).

420. M. Sugi, K. Nembach, D. Möbius, and H. Kuhn, Evidence for contact effects in frequency-dependent conduction in Langmuir films, *Solid State Commun.*, **13**, 603–606 (1973).

421. M. Sugi, T. Fukui, and S. Iizima, Structure-dependent feature of electron transport in Langmuir multilayer assemblies, *Mol. Cryst. Liq. Cryst.*, **50**, 183–200 (1979).

422. M. Careem and A. K. Jonscher, Lattice and carrier contributions to the dielectric polarization of stearic acid multilayer films, *Philos. Mag.*, **35**, 1489–1502 (1977).

423. H. M. Millany and A. K. Jonscher, Dielectric properties of stearic acid multilayers, *Thin Solid Films*, **68**, 257–273 (1980).

424. A. K. Jonscher, *Dielectric Relaxation in Solids*, Chelsea Dielectrics Press, London (1983).

425. A. K. Jonscher and F. Taiedy, Dark and photo-stimulated dielectric relaxation in Langmuir films, *J. Phys. C*, **8**, L107–L111 (1975).

426. M. A. Careem and A. K. Jonscher, Dark and photo-generated dielectric response in stearic acid films, *Philos. Mag.*, **35**, 1503–1508 (1977).

427. A. K. Jonscher and S. Buddhabadana, The similarity of the dark and photo-stimulated dielectric relaxation in Langmuir films, *Solid State Electron.*, **21**, 991–994 (1978).

428. A. K. Jonscher, F. Meca, and H. M. Millany, Charge-carrier contributions to dielectric loss, *J. Phys. C*, **12**, L293–L296 (1979).

429. G. Marc and J. Messier, Non-linear dielectric properties of aluminium/monomolecular layers of calcium behenate/aluminium structures, *Thin Solid Films*, **68**, 275–288 (1980).

430. M. Sugi, K. Nembach, D. Möbius, and H. Kuhn, Quantum mechanical hopping in one dimensional superstructure, *Solid State Commun.*, **15**, 1867–1870 (1974).

431. S. Iizima and M. Sugi, Electrical conduction in mixed Langmuir films, *Appl. Phys. Lett.*, **28**, 548–549 (1976).

432. M. Sugi, T. Fukui, and S. Iizima, $T^{-1/2}$-Law of dc conductivity in Langmuir Films, *Chem. Phys. Lett.*, **45**, 163–165 (1977).

433. M. Sugi and S. Iizima, Frequency-dependent conductivity in multilayer assemblies, *Phys. Rev. B*, **15**, 574–579 (1977).

434. M. Sugi, T. Fukui, and S. Iizima, Direct evaluation of the hopping rate in Langmuir multilayer assemblies, *Phys. Rev. B*, **18**, 725–732 (1978).

435. M. Sugi and S. Iizima, Single layer conductance of cadmium behenate in the Langmuir multilayer assembly system, *Appl. Phys. Lett.*, **34**, 290–292 (1979).

436. A. Miller and E. Abrahams, Impurity conduction at low concentrations, *Phys. Rev.*, **120**, 745–755 (1960).

437. N. F. Mott and E. A. Davies, *Electronic Processes in Non-Crystalline Materials*, Clarendon Press, Oxford (1979).

438. J. Batey, Electroluminescent MIS structures incorporating Langmuir–Blodgett films, Ph.D. thesis, University of Durham, UK (1983).

439. V. K. Agarwal and B. Ichijo, Determination of dielectric constant of stearic acid films using varying gap immersion method, *Electrocomponent Sci. Technol.*, **4**, 23–28 (1977).

440. A. Barraud, J. Messier, A. Rosilio, and J. Tanguy, Electronic properties of organic monomolecular layers, Colloque AVISM, Versailles, 1971, Soc. Fr. Ing. Tech. Vide Paris, 341–349 (1971).

441. W. L. Procarione and J. W. Kauffman, Capacitance studies of synthetic phospholipid Langmuir films, *Chem. Phys. Lipids*, **18**, 49–61 (1977).

442. C. S. Winter and R. H. Tredgold, Langmuir–Blodgett multilayers of polypeptides, *Thin Solid Films*, **123**, L1–L3 (1985).

443. R. A. Hann, W. A. Barlow, J. H. Steven, B. L. Eyres, M. V. Twigg, and G. G. Roberts, Langmuir–Blodgett films of substituted aromatic hydrocarbons and phthalocyanines, Proc. 2nd. Int. Workshop Molecular Electron Devices, Washington (1983).

444. U. Khanna, V. K. Srivastava, and V. K. Agarwal, Static dielectric constant of "built-up" molecular films of barium stearate, *Thin Films*, **2**, 83–88 (1971).

445. U. Khanna and V. K. Srivastava, Corrected calculation of static dielectric constant of barium stearate films, *Thin Films*, **2**, 167–168 (1972).

446. U. Khanna and V. K. Srivastava, Studies of dielectric constant of some "built-up" molecular films, *Thin Solid Films*, **12**, S25–S28 (1972).

447. U. Kapur and V. K. Srivastava, Thickness dependence of the static dielectric constant of some built-up Langmuir films, *Phys. Status Solidi A*, **38**, K77–K80 (1976).

448. A. Barraud and A. Rosilio, Physical, optical, and electronic properties of organic monomolecular layers, *Colloque Int. Microélectronique Avancée, Paris,* 845–853 (1970).

449. S. K. Gupta, C. M. Singal, and V. K. Srivastava, Diode characteristics of a metal–insulator–metal structure, *Electrocomponent Sci. Technol.,* **3,** 119–120 (1976).

450. A. K. Kapil and V. K. Srivastava, Decay and recovery of internal voltage in MIM sandwich structures using organic molecular films as dielectric, *Proc. Nucl. Phys., Solid State Phys. Symp.,* **20c,** 105 (1977).

451. S. K. Gupta, C. M. Singal, and V. K. Srivastava, Thickness dependence of internal voltage in metal–insulator–metal structure with dissimilar electrodes, *J. Appl. Phys.,* **48,** 2583–2586 (1977).

452. C. M. Singal, S. K. Gupta, A. K. Kapil, and V. K. Srivastava, Intrinsic voltage in insulating films in aluminium–barium–stearate–aluminium structures, *J. Appl. Phys.,* **49,** 3402–3405 (1978).

453. S. K. Gupta, A. K. Kapil, C. M. Singal, and V. K. Srivastava, Measurement of the work function of some metals using internal voltage in MIM structures, *J. Appl. Phys.,* **50,** 2852–2855 (1979).

454. A. K. Kapil, C. M. Singal, and V. K. Srivastava, Internal voltage in symmetric MIM junctions with even number of organic monolayers, *J. Appl. Phys.,* **50,** 2856–2858 (1979).

455. A. K. Kapil, S. K. Gupta, C. M. Singal, and V. K. Srivastava, Electric dipole moment measurements by internal-voltage technique, *J. Appl. Phys.,* **50,** 2896–2898 (1979).

456. N. A. Surplice and R. J. D'Arcy, A critique of the Kelvin method of measuring work functions, *J. Phys. E,* **3,** 477–482 (1970).

457. A. Noblet, H. Ridelaire, and G. Sylin, Measurement of surface potentials, *J. Phys. E,* **17,** 234–239 (1984).

458. E. F. Porter and J. Wyman, Jr., Contact potentials of multilayer films on metal plates, *J. Am. Chem. Soc.,* **59,** 2746–2747 (1937).

459. E. F. Porter and J. Wyman, Jr., Contact potentials of stearate films on metal surfaces, *J. Am. Chem. Soc.,* **60,** 1083–1094 (1938).

460. I. Langmuir, Surface electrification due to the recession of aqueous solutions from hydrophobic surfaces, *J. Am. Chem. Soc.,* **60,** 1190–1194 (1938).

461. W. D. Harkins and R. W. Mattoon, The contact potential of solid films formed by evaporation and by solidification of built-up multilayers on metals, *Phys. Rev.,* **53,** 911–912 (1938).

462. E. F. Porter and J. Wyman, Jr., Further studies on the electrical properties of stearate films deposited on metal, *J. Am. Chem. Soc.,* **60,** 2855–2869 (1938).

463. R. W. Goranson and W. A. Zisman, On the electrical properties of multilayers, *Phys. Rev.,* **54,** 544 (1938).

464. R. W. Goranson and W. A. Zisman, Electrical properties of multilayers, *J. Chem. Phys.,* **7,** 492–505 (1939).

465. F. J. Norton and I. Langmuir, Effect of X-rays on surface potentials of multilayers, *J. Am. Chem. Soc.,* **60,** 1513 (1938).

466. P. Christie, Acentric Langmuir–Blodgett film assemblies, Ph.D. thesis, University of Durham, UK (1985).

467. P. Christie, G. G. Roberts, and M. C. Petty, Spontaneous polarization in organic superlattices, *Appl. Phys. Lett.,* **48,** 1101–1103 (1986).

468. R. H. Tredgold and G. W. Smith, Surface potential studies on Langmuir–Blodgett films, *J. Phys. D,* **14,** L193–L195 (1981).

469. R. H. Tredgold and G. W. Smith, Surface potential studies on Langmuir–Blodgett multilayers and adsorbed monolayers, *Thin Solid Films,* **99,** 215–220 (1983).

470. R. Jones, R. H. Tredgold, and A. Hoorfar, Effects of thickness on surface potential and surface conductivity in non-insulating Langmuir–Blodgett multilayers of porphyrins, *Thin Solid Films,* **123,** 307–314 (1985).

471. P. Christie, M. C. Petty, G. G. Roberts, D. H. Richards, D. Service, and M. J. Stewart, The preparation and dielectric properties of polybutadiene Langmuir–Blodgett films, *Thin Solid Films,* **134,** 75–82 (1985).

472. I. Lundström and D. McQueen, Capacitance measurements on lipid multilayers, *Chem. Phys. Lipids,* **10,** 181–190 (1973).

473. I. Lundström and M. Stenberg, Charge injection and charge storage in lipid multilayers, *Chem. Phys. Lipids*, **12**, 287–302 (1974).

474. T. W. Hickmott, Dipole layers at the gold–SiO_2 interface, *Proc. Int. Conf. Physics MOS Insulators*, pp. 227–231, Pergamon, New York (1980).

475. P. J. Martin, Gas effects on the interface state spectrum of MIS devices, Ph.D. thesis, University of Durham, UK (1980).

476. J. Tanguy and P. Hesto, Thermally stimulated currents in organic monomolecular layers, *Thin Solid Films*, **21**, 129–143 (1974).

477. L. M. Blinov, L. V. Mikhnev, E. B. Sokolova, and S. G. Yudin, Pyroelectric effect in one and several monomolecular layers, *Sov. Tech. Phys. Lett.*, **9**, 640–641 (1983).

478. J. R. Drabble and S. M. Al-Khowaildi, Ultrasonic transducer action of Langmuir–Blodgett films, *Thin Solid Films*, **99**, 271–275 (1983).

479. L. Holt, Langmuir–Blodgett multi-monolayers as thin film dielectrics, *Nature*, **214**, 1105 (1967).

480. V. K. Agarwal and V. K. Srivastava, Thickness dependence of breakdown field in thin films, *Thin Solid Films*, **8**, 377–381 (1971).

481. V. K. Agarwal and V. K. Srivastava, Thickness dependence of breakdown field, *Thin Solid Films*, **13**, S23–S24 (1972).

482. D. K. Agarwal and V. K. Srivastava, Temperature dependence of the breakdown field in barium stearate multilayer films, *Thin Solid Films*, **14**, 367–371 (1972).

483. D. K. Agarwal and V. K. Srivastava, Temperature dependence of d.c. destructive breakdown field in "built-up" barium stearate films, *Solid State Commun.*, **11**, 1461–1466 (1972).

484. V. K. Srivastava, Evidence for Schottky-emission-dominated dielectric breakdown, *Phys. Rev. Lett.*, **30**, 1046–1047 (1973).

485. V. K. Agarwal and V. K. Srivastava, Thickness dependent studies of dielectric breakdown in Langmuir thin molecular films, *Solid State Commun.*, **12**, 829–834 (1973).

486. V. K. Agarwal and V. K. Srivastava, AC breakdown studies of built-up barium stearate films, *J. Appl. Phys.*, **44**, 2900–2901 (1973).

487. V. K. Agarwal and V. K. Srivastava, Destructive D.C. breakdown in "built-up" barium stearate films, *Electrocomponent Sci. Technol.*, **1**, 87–90 (1974).

488. V. K. Agarwal, Interpretation of destructive breakdown in Langmuir molecular films, *Thin Solid Films*, **23**, S9–S12 (1974).

489. V. K. Agarwal, Breakdown conduction in Langmuir films of low thicknesses, *Thin Solid Films*, **23**, S3–S7 (1974).

490. D. K. Agarwal and V. K. Srivastava, Studies of the temperature dependence of the dielectric breakdown field in molecular Langmuir films, *Thin Solid Films*, **27**, 49–62 (1975).

491. V. K. Srivastava, On the nature of dielectric breakdown in ultra-thin Langmuir films, *Electrocomponent Sci. Technol.*, **3**, 117 (1976).

492. V. K. Agarwal, "Electroformed" breakdown in thin insulating films, *Thin Solid Films*, **33**, L27–L30 (1976).

493. J. P. Furtlehner and J. Messier, Conduction in MIM structures with an organic monomolecular layer at high electric fields, *Thin Solid Films*, **68**, 233–239 (1980).

494. G. G. Roberts, M. C. Petty, P. J. Caplan, and E. H. Poindexter, Paramagnetic defects in oxidized silicon wafers coated with Langmuir–Blodgett films, in *Insulating Films on Semiconductors* (J. F. Verweij and D. R. Wolters, eds.), pp. 20–23, Elsevier (North-Holland), Amsterdam (1983).

495. K. H. Gundlach and J. Kadlec, Negative resistance in organic monomolecular layers sandwiched between metal electrodes, *Phys. Status Solidi A*, **10**, 371–379 (1972).

496. K. H. Gundlac and J. Kadlec, Negative resistance in organic monomolecular layers: Experiments on triodes, *Thin Solid Films*, **13**, 225–230 (1972).

497. A. I. Bailey and J. S. Courtney-Pratt, The area of real contact and the shear strength of monomolecular layers of a boundary lubricant, *Proc. R. Soc. London, Ser. A*, **227**, 500–515 (1955).

498. J. H. Schulman, R. B. Waterhouse, and J. A. Spink, Adhesion of amphipathic molecules to solid surfaces, *Kolloid Z.*, **146**, 77–95 (1955).

499. B. J. Briscoe and D. C. B. Evans, The shear properties of Langmuir–Blodgett layers, *Proc. R. Soc. London, Ser. A*, **380**, 389–407 (1982).

500. E. Rideal and J. Tadayon, On overturning and anchoring of monolayers I. Overturning and transfer, *Proc. R. Soc. London, Ser. A*, **225**, 346–356 (1954).

501. E. Rideal and J. Tadayon, On overturning and anchoring of monolayers II. Surface diffusion, *Proc. R. Soc. London, Ser. A*, **225**, 357–361 (1954).

502. G. L. Gaines, Jr., Material transfer in monomolecular layers of a boundary lubricant, *Nature*, **183**, 1110 (1959).

503. G. L. Gaines, Jr., Overturning of stearic acid molecules in monolayers, *Nature*, **186**, 384–385 (1960).

504. M. Blank and L. Soo, The effect of cholesterol on the viscosity of protein–lipid monolayers, *Chem. Phys. Lipids*, **17**, 416–422 (1976).

505. M. Blank, R. G. King, L. Soo, R. E. Abbott, and S. Chien, The viscoelastic properties of monolayers of red cell membrane proteins, *J. Colloid Interface Sci.*, **69**, 67–73 (1979).

506. M. Blank, L. Soo, R. E. Abbott, and U. Cogan, Surface potentials of films of membrane proteins, *J. Colloid Interface Sci.*, **73**, 279–281 (1980).

507. M. Blank, The thickness dependence of properties of membrane protein multilayers, *J. Colloid Interface Sci.*, **75**, 435–440 (1980).

508. R. Zanoni, C. Naselli, J. Bell, G. Stegeman, R. Sprague, C. Seaton, and S. Lindsay, Brillouin spectroscopy of Langmuir–Blodgett films, *Thin Solid Films*, **134**, 179–186 (1985).

509. R. Zanoni, J. Bell, C. Naselli, C. Seaton, and G. Stegeman, Elastic properties of Langmuir–Blodgett films, *Phys. Rev. Lett.*, **57**, 2838–2840 (1986).

510. M. C. Jain and M. H. Jericho, Propagation of ultrasonic surface waves through molecular films, *Appl. Phys. Lett.*, **26**, 491–493 (1975).

511. A. Rothen, Enzymatic reaction across a thin membrane, *J. Phys. Chem.*, **63**, 1929–1934 (1959).

512. G. J. Kovacs, P. S. Vincett, and J. H. Sharp, Stable, tough, adherent, Langmuir–Blodgett films: Preparation and structure of ordered, true monolayers of a phthalocyanine, *Can. J. Phys.*, **63**, 346–349 (1985).

513. D. W. Deamer and D. Branton, Fracture planes in an ice-bilayer model membrane system, *Science*, **158**, 655–657 (1967).

514. M. Blank, L. Soo, and R. E. Abbott, The ionic permeability of adsorbed membrane protein monolayers, *J. Electrochem. Soc.*, **126**, 1471–1474 (1979).

515. M. Blank, L. Soo, and R. E. Abbott, The permeability of adsorbed- and spread-membrane protein (spectrin-actin) films to ions, in: *Bioelectrochemistry: Ions, Surfaces, Membranes* (M. Blank, ed.), Advances in Chemistry, Vol. 188, Chap. 19, Am. Chem. Soc., Washington, D.C. (1980).

516. R. J. VanderVeen and G. T. Barnes, Water permeation through Langmuir–Blodgett monolayers, *Thin Solid Films*, **134**, 227–236 (1985).

517. H. K. Adam and J. E. Zull, Transfer of calcium through a stearic acid bilayer, *J. Colloid Interface Sci.*, **34**, 272–277 (1970).

518. K. Heckmann, C. Strobl, and S. Bauer, Hyperfiltration through cross-linked monolayers, *Thin Solid Films*, **99**, 265–269 (1983).

519. G. D. Rose and J. A. Quinn, Composite membranes: The permeation of gases through deposited monolayers, *Science*, **159**, 636–637 (1968).

520. G. D. Rose and J. A. Quinn, Gas transport through supported Langmuir–Blodgett multilayers, *J. Colloid Interface Sci.*, **27**, 193–207 (1968).

521. S. T. Kowel, R. Selfridge, C. Eldering, N. Matloff, P. Stroeve, B. G. Higgins, M. P. Srinivasan, and L. B. Colman, Future applications of ordered polymeric thin films, *Thin Solid Films*, **152**, 377–403 (1987).

522. J. D. Swalen, Optical properties of Langmuir–Blodgett films, *J. Mol. Elec.*, **2**, 155–181 (1986).

523. G. Khanarian, Langmuir–Blodgett films and nonlinear optics, *Thin Solid Films*, **152**, 265–274 (1987).

524. S. Allen, Langmuir–Blodgett films for nonlinear optical applications, *Inst. Phys. Conf. Ser. No. 103*, (M. H. Lyons, ed.) 163–174 (1989).

525. R. H. Tredgold, The physics of Langmuir–Blodgett films, *Rep. Prog. Phys.*, **50**, 1609–1656 (1987).

526. L. A. Feigin, Y. M. Lvov, and V. I. Troitsky, X-Ray and electron-diffraction of Langmuir–Blodgett films, *Sov. Sci. Rev. A. Phys.*, **11** (in press).

527. Proceedings of Third International Conference on Langmuir–Blodgett Films (D. Möbius, ed.), *Thin Solid Films*, **159** and **160** (1988).

528. Proceedings of Fourth International Conference on Langmuir–Blodgett Films (M. Sugi, ed.) (in press).

529. M. Vandevyver, Characterization methods for Langmuir–Blodgett films, *Thin Solid Films*, **159**, 243–251 (1988).

530. M. R. Buhaenko, M. J. Grundy, R. M. Richardson, and S. J. Roser, Structure and temperature dependence of fatty acid Langmuir–Blodgett films studied by neutron and X-ray scattering, *Thin Solid Films*, **159**, 253–265 (1988).

531. Y. M. Lvov, D. Svergun, L. A. Feigin, C. Pearson, and M. C. Petty, Small angle X-ray analysis of alternate-layer Langmuir–Blodgett films, *Phil. Mag. Lett.*, **59**, 317–323 (1989).

532. M. Von Frieling, H. Bradaczek, and W. S. Durfee, X-ray diffraction of Langmuir–Blodgett films: Establishing a new method for the calculation of electron density distributions from a single set of intensity data, *Thin Solid Films*, **159**, 451–459 (1988).

533. W. L. Barnes and J. R. Sambles, Guided optical waves in Langmuir–Blodgett films of 22-tricosenoic acid, *Surf. Sci.*, **177**, 399–416 (1986).

534. C. A. Jones, G. J. Russell, M. C. Petty, and G. G. Roberts, A reflection high-energy diffraction study of ultra-thin Langmuir–Blodgett films of ω-tricosenoic acid, *Phil. Mag. B*, **54**, L89–L94 (1986).

535. I. Robinson, J. R. Sambles, and I. R. Peterson, A reflection high energy electron diffraction analysis of the orientation of the monoclinic subcell of 22-tricosenoic acid Langmuir–Blodgett bilayers as a function of the deposition pressure, *Thin Solid Films*, **189**, 149–158 (1989).

536. I. R. Peterson, G. J. Russell, J. D. Earls, and I. R. Girling, Surface pressure dependence of molecular tilt in Langmuir–Blodgett films of 22-tricosenoic acid, *Thin Solid Films*, **161**, 325–331 (1988).

537. D. B. Neal, G. J. Russell, M. C. Petty, G. G. Roberts, M. M. Ahmad, and W. J. Feast, A highly ordered Langmuir–Blodgett monolayer of an amido nitrostilbene, *J. Molec. Elec.*, **2**, 135–138 (1986).

538. D. Heard, G. G. Roberts, B. Holcroft, and M. J. Goringe, Electron diffraction studies of Langmuir–Blodgett film structures with potential for optical applications, *Thin Solid Films*, **160**, 491–499 (1988).

539. A. Barraud, J. Richard, A. Ruaudel-Teixier, and M. Vandevyver, Electric field mapping in conducting Langmuir–Blodgett films, *Thin Solid Films*, **159**, 413–419 (1988).

540. M. Flörsheimer and H. Möhwald, Energy Transfer and Aggregation in Monolayers Containing Porphyrins and Phthalocyanines, *Thin Solid Films*, **159**, 115–123 (1988).

541. C. A. Jones, M. C. Petty, G. G. Roberts, G. H. Davies, J. Yarwood, N. M. Ratcliffe, and J. W. Barton, IR studies of pyroelectric Langmuir–Blodgett films, *Thin Solid Films*, **155**, 187–195 (1987).

542. G. H. Davies, J. Yarwood, M. C. Petty, and C. A. Jones, Fourier transform IR studies of alternate layer acid–amine Langmuir–Blodgett films with pyroelectric properties, *Thin Solid Films*, **159**, 461–467 (1988).

543. G. H. Davies and J. Yarwood, Infrared intensity and band shape studies on Langmuir–Blodgett films of ω-tricosenoic acid on silicon, *Spectrochimica Acta*, **43A**, 1619–1623 (1987).

544. A. Ouchi, M. Tanaka, T. Nakamura, M. Matsumoto, Y. Kawabata, S. Tomimasu, and A. Yabe, Photodimerization of anthracene derivative in Langmuir–Blodgett films, *Chem. Lett.*, 1833–1836 (1986).

545. V. A. Howarth, M. C. Petty, G. H. Davies, and J. Yarwood, The deposition and characterization of multilayers of the ionophore valinomycin, *Thin Solid Films*, **160**, 483–489 (1988).

546. V. A. Howarth, M. C. Petty, G. H. Davies, and J. Yarwood, Infrared studies of valinomycin-containing Langmuir–Blodgett films, *Langmuir*, **5**, 330–332 (1989).

547. V. A. Howarth, D. F. Cui, M. C. Petty, H. Ancelin, and J. Yarwood, Phospholipid-based potassium selective Langmuir–Blodgett films, *Thin Solid Films* (in press).

548. D. G. Zhu, M. C. Petty, H. Ancelin, and J. Yarwood, On the formation of Langmuir–Blodgett films containing enzymes, *Thin Solid Films*, **176**, 151–156 (1989).

549. S. B. Dierker, C. A. Murray, J. D. Legrange, and N. E. Schlotter, Characterization of order in Langmuir–Blodgett monolayers by unenhanced Raman spectroscopy, *Chem. Phys. Lett.*, **137**, 453–457 (1987).

550. C. G. Zimba, V. M. Hallmark, J. D. Swalen, and J. F. Rabolt, Fourier transform Raman spectroscopy of long-chain molecules containing strongly absorbing chromophores, *Appl. Spectrosc.*, **41**, 721–726 (1987).

551. C. G. Zimba, V. M. Hallmark, J. F. Rabolt, and J. D. Swalen, Fourier transform Raman spectroscopy of thin films, *Thin Solid Films*, **160**, 311–316 (1988).

552. N. G. Cave, R. A. Cayless, L. B. Hazell, and A. J. Kinloch, Characterisation of the structure of Langmuir–Blodgett films of ω-tricosenoic acid using X-ray photoelectron spectroscopy, *Langmuir* (in press).

553. K. Kobayashi, K. Takanaka, and S. Ochiai, Application of X-ray photoelectron spectroscopy and Fourier transform IR–reflection absorption spectroscopy to studies of the composition of Langmuir–Blodgett films, *Thin Solid Films*, **159**, 267–273, (1988).

554. H. Oyanagi, M. Yoneyama, K. Ikegami, M. Sugi, S-I Kuroda, T. Ishiguro, and T. Matsushita, Langmuir–Blodgett monolayers studied by surface-sensitive X-ray absorption fine structure, *Thin Solid Films*, **159**, 435–442 (1988).

555. J. P. Rabe, J. D. Swalen, D. A. Outka, and J. Stöhr, Near-edge fine structure studies of oriented molecular chains in polyethylene and Langmuir–Blodgett monolayers on Si(111), *Thin Solid Films*, **159**, 275–283 (1988).

556. D. A. Outka, J. Stöhr, J. P. Rabe, and J. D. Swalen, The orientation of Langmuir–Blodgett monolayers using NEXAFS, *J. Chem. Phys.*, **88**, 4076–4087 (1988).

557. S-I. Kuroda, K. Ikegami, K. Saito, M. Saito, and M. Sugi, Characterization of in-plane and out-of-plane molecular orientation in Langmuir–Blodgett films of merocyanine dyes using electron spin resonance, *Thin Solid Films*, **159**, 285–291 (1988).

558. S. Kuroda, K. Ikegami, M. Sugi, and S. Iizima, Direct observation of in-plane molecular alignment in LB films of a merocyanine dye using ESR spectroscopy, *Sol. State Commun.*, **58**, 493–497 (1986).

559. K. Ikegami, S. Kuroda, M. Sugi, M. Saito, S. Iizima, T. Nakamura, M. Matsumoto, Y. Kawabata, and G. Saito, ESR study on LB films of TMTTF-octadecylTCNQ, *Synth. Met.*, **19**, 669–674 (1987).

560. K. Ikegami, S-I. Kuroda, M. Saito, K. Saito, M. Sugi, T. Nakamura, M. Matsumoto, and Y. Kawabata, Electron spin resonance study of Langmuir–Blodgett films of N-docosylpyridinium-di(tetracyanoquinodimethane), *Thin Solid Films*, **160**, 139–143 (1988).

561. K. Ikegami, S-I. Kuroda, K. Saito, M. Saito, M. Sugi, T. Nakamura, M. Matsumoto, Y. Kawabata, and G. Saito, Anomalous low-temperature behaviour of $(TMTTF)_3$-(tetradecylTCNQ)$_2$, *Synth. Met.*, **27**, B587–B592 (1988).

562. K. Ikegami, S-I. Kuroda, M. Saito, K. Saito, M. Sugi, T. Nakamura, M. Matsumoto, and Y. Kawabata, Random-exchange Heisenberg antiferromagnetic chains in Langmuir–Blodgett films of N-docosylpyridinium-bistetracyanoquinodimethane, *Phys. Rev. B*, **35**, 3667–3670 (1987).

563. J. Richard, M. Vandevyver, P. Lesieur, A. Ruaudel-Teixier, A. Barraud, R. Bozio, and C. Pecile, Structural properties of Langmuir–Blodgett films of charge transfer salts: Pristine and iodine doped conducting films of (N-docosyl-pyridinium)TCNQ, *J. Chem. Phys.*, **86**, 2428–2438 (1987).

564. J. H. Coombs, J. B. Pethica, and M. E. Welland, Scanning tunnelling microscopy of thin organic films, *Thin Solid Films*, **159**, 293–299 (1988).

565. H. G. Braun, H. Fuchs, and W. Schrepp, Surface structure investigation of Langmuir–Blodgett films, *Thin Solid Films*, **159**, 301–314 (1988).

566. C. A. Lang, J. K. H. Hörber, T. W. Hänsch, W. M. Heckl, and H. Möhwald, Scanning tunnelling microscopy of Langmuir–Blodgett films on graphite, *J. Vac. Sci. Technol.*, **A6**, 368–370 (1988).

567. R. Osiander, P. Korpiun, C. Duschl, and W. Knoll, Photoacoustic detection of water permeation through ultrathin films, *Thin Solid Films,* **160,** 501–505 (1988).

568. W. L. Barnes and J. R. Sambles, Thin Langmuir–Blodgett films studied using surface plasmon-polaritons, *Surf. Sci.,* **183,** 189–200 (1987).

569. P. N. Prasad, Non-linear optical effects in thin organic polymeric films, *Thin Solid Films,* **152,** 275–294 (1987).

570. B. Rothenhäusler, C. Duschl, and W. Knoll, Plasmon surface polariton fields for the characterization of thin films, *Thin Solid Films,* **159,** 323–330 (1988).

571. L. M. Hayden, B. L. Anderson, J. Y. S. Lam, B. G. Higgins, P. Stroeve, and S. T. Kowel, Second harmonic generation in Langmuir–Blodgett films of hemicyanine-poly(octadecyl methacrylate) and hemicyanine-behenic acid, *Thin Solid Films,* **160,** 379–388 (1988).

572. J. C. Loulergue, M. Dumont, Y. Levy, P. Robin, J. P. Pocholle, and M. Papuchon, Linear electro-optic properties of Langmuir–Blodgett multilayers of an organic azo dye, *Thin Solid Films,* **160,** 399–405 (1988).

573. J. Tsibouklis, J. Cresswell, N. Kalita, C. Pearson, P. J. Maddaford, H. Ancelin, J. Yarwood, M. J. Goodwin, N. Carr, W. J. Feast, and M. C. Petty, Functionalised diarylalkynes: A new class of Langmuir–Blodgett materials for non-linear optics, *J. Phys. D: Appl. Phys.* **22** (1989), in press.

574. T. Richardson, G. G. Roberts, M. E. C. Polywka, and S. G. Davies, Preparation and characterization of organotransition metal Langmuir–Blodgett films, *Thin Solid Films,* **160,** 231–239 (1988).

575. J. Tsibouklis, J. Cresswell, C. Pearson, M. C. Petty, and W. J. Feast, A new non-linear optical Langmuir–Blodgett film material based on a liquid crystalline butadiyne, *Inst. Phys. Conf. Ser. No. 103* (M. H. Lyons, ed.) 187–192 (1989).

576. L. M. Blinov, A. V. Ivanschchenko, and S. G. Yudin, Stark spectroscopy of Langmuir–Blodgett monolayers and multilayers, *Thin Solid Films,* **160,** 271–278 (1988).

577. G. H. Cross, I. R. Girling, I. R. Peterson, N. A. Cade, and J. D. Earls, Optically nonlinear Langmuir–Blodgett films: Linear electro-optic properties of monolayers, *J. Opt. Soc. Am. B,* **4,** 962–967 (1987).

578. I. R. Peterson, J. D. Earls, I. R. Girling, and G. J. Russell, Disclinations and annealing in fatty-acid monolayers, *Mol. Cryst. Liq. Cryst.,* **147,** 141–147 (1987).

579. P. Lesieur, A. Barraud, and M. Vandevyver, Defect characterization and detection in Langmuir–Blodgett films, *Thin Solid Films,* **152,** 155–164 (1987).

580. N. R. Couch, B. Movaghar, and I. R. Girling, Electromigration failure in filaments through Langmuir–Blodgett films, *Solid State Commun.,* **59,** 7–9 (1986).

581. N. A. Cade, G. H. Cross, R. A. Lee, S. Bajic, and R. V. Latham, Field-induced electron emission through Langmuir–Blodgett multilayers, *J. Phys. D: Appl. Phys.,* **21,** 148–153 (1988).

582. M. Iwamoto, S-I. Shidoh, T. Kubota, and M. Sekine, Electrical properties of Langmuir–Blodgett films sandwiched between Pb–Bi superconducting electrodes, *Jap. J. Appl. Phys.,* **27,** 1825–1830 (1988).

583. N. J. Geddes, J. R. Sambles, and D. J. Jarvis, The sticking of magnesium on 22-tricosenoic acid multilayers deposited by the Langmuir–Blodgett technique, *Thin Solid Films,* **167,** 261–268 (1988).

584. N. J. Geddes, W. G. Parker, J. R. Sambles, D. J. Jarvis, and N. R. Couch, A metal (organic bilayer) metal capacitor, *Thin Solid Films,* **168,** 151–156 (1989).

585. M. Vandevyver, Results and prospects in conducting L.B. films, *J. Chim. Physique,* **85,** 1033–1037 (1988).

586. A. Ruaudel-Teixier, M. Vandevyver, and A. Barraud, Novel conducting LB films, *Mol. Cryst. Liq. Cryst.,* **120,** 319–322 (1985).

587. A. Barraud, M. Lequan, R. M. Lequan, P. Lesieur, J. Richard, A. Ruaudel-Teixier, and M. Vandevyver, A novel highly conducting tetracyanoquinodimethane (TCNQ) Langmuir–Blodgett film, *J. Chem. Soc., Chem. Commun.,* 797–798 (1987).

588. A. Barraud, P. Lesieur, J. Richard, A. Ruaudel-Teixier, M. Vandevyver, M. Lequan, and R. M. Lequan, A novel highly conducting phosphonium tetracyanoquinodimethane Langmuir–Blodgett film, *Thin Solid Films,* **160,** 81–85 (1988).

589. T. Nakamura, M. Tanaka, T. Sekiguchi, and Y. Kawabata, Structure-dependent electrical conductivity of alkylpyridinium tetracyanoquinodimethane Langmuir–Blodgett films, *J. Am. Chem. Soc.*, **108,** 1302–1303 (1986).

590. T. Nakamura, M. Matsumoto, F. Takei, M. Tanaka, T. Sekiguchi, E. Manda, and Y. Kawabata, Conducting Langmuir–Blodgett films of 1:2 charge transfer complex, N-docosylpyridinium-$(TCNQ)_2$, *Chem. Lett.*, 709–712 (1986).

591. M. Matsumoto, T. Nakamura, F. Takei, M. Tanaka, T. Sekiguchi, M. Mizuno, E. Manda, and Y. Kawabata, Conductivities of N-docosylpyridinium-$(TCNQ)_n$ (n = 1, 2) in the forms of monolayers at air-glycerin interface and Langmuir–Blodgett films, *Synth. Met.*, **19,** 675–680 (1987).

592. A. S. Dhindsa, M. R. Bryce, J. P. Lloyd, and M. C. Petty, Electroactive Langmuir–Blodgett films of N-octadecylpyridinium-TCNQ charge transfer salt, *Synth. Met.*, **22,** 185–189 (1987).

593. A. S. Dhindsa, G. H. Davies, M. R. Bryce, J. P. Lloyd, Y. M. Lvov, M. C. Petty, and J. Yarwood, Structural investigation into multilayer films of N-octadecylpyridinium-TCNQ, *J. Molec. Elec.*, **5,** 135–142 (1989).

594. A. S. Dhindsa, M. R. Bryce, J. P. Lloyd, and M. C. Petty, A highly conducting tetrathiafulvalene Langmuir–Blodgett film, *Thin Solid Films*, **165,** L97–L100 (1988).

595. A. S. Dhindsa, M. R. Bryce, J. P. Lloyd, and M. C. Petty, Electroactive Langmuir–Blodgett films of hexadecanoyl-TTF, *Synth. Met.*, **27,** B563–B568 (1988).

596. F. Bertho, D. Talham, A. Robert, P. Batail, S. Megtert, and P. Robin, New amphiphilic tetrathiafulvalene (TTF) derivatives: Synthesis and Langmuir–Blodgett film formation, *Mol. Cryst. Liq. Cryst. Inc. Nonlin. Opt.*, **156,** 339–345 (1988).

597. J. Richard, M. Vandevyver, A. Barraud, J. P. Morand, R. Lapouyade, R. Delhaes, J. F. Jacquinot, and M. Roulliay, Preparation of new conducting Langmuir–Blodgett films based on ethylenedithiodecylthiotetrathiafulvalene charge transfer complex, *J. Chem. Soc., Chem. Commun.*, 754–756 (1988).

598. T. Nakamura, F. Takei, M. Tanaka, M. Matsumoto, T. Sekiguchi, E. Manda, Y. Kawabata, and G. Saito, Conducting monolayer of simple charge transfer complex on a glycerin subphase, *Chem. Lett.*, 323–324 (1986).

599. Y. Kawabata, T. Nakamura, M. Matsumoto, M. Tanaka, T. Sekiguchi, H. Komizu, E. Manda, and G. Saito, Intrinsically conducting Langmuir–Blodgett film of TMTTF-octadecylTCNQ, *Synth. Met.*, **19,** 663–668 (1987).

600. A. S. Dhindsa, C. Pearson, M. R. Bryce, and M. C. Petty, Conducting Langmuir–Blodgett films of 1-tetrathiafulvalenyl-octadecan-1-ol-TCNQ complex, *J. Phys. D: Appl. Phys.*, **22** (1989), in press.

601. M. Matsumoto, T. Nakamura, E. Manda, Y. Kawabata, K. Ikegami, S-I. Kuroda, M. Sugi, and G. Saito, Highly conducting Langmuir–Blodgett films of the TMTTF-C_nTCNQ system, *Thin Solid Films,* **160,** 61–66 (1988).

602. C. Pearson, A. S. Dhindsa, M. R. Bryce, and M. C. Petty, Alternate-layer Langmuir–Blodgett films of long-chain TCNQ and TTF derivatives, *Synth. Met.*, **31,** 275–279 (1989).

603. T. Nakamura, K. Kojima, M. Matsumoto, H. Tachibana, M. Tanaka, E. Manda, and Y. Kawabata, Metallic temperature dependence in the conductivity of Langmuir–Blodgett films of tridecylmethylammonium-Au(dmit)$_2$, *Chem. Lett.*, 367–368 (1989).

604. I. Iyoda, M. Ando, T. Kaneko, A. Ohtani, T. Shimidzu, and K. Honda, Electrochemical polymerization in Langmuir–Blodgett film of new amphiphilic pyrrole derivatives, *Tetrahed. Lett.*, **27,** 5633–5636 (1986).

605. K. Hong, and M. F. Rubner, Synthesis of electrically conductive polypyrroles at the air–water interface of a Langmuir–Blodgett film balance, *Thin Solid Films,* **160,** 187–195 (1988).

606. H. Nakahara, J. Nakayama, M. Hoshino, and K. Fukuda, Langmuir–Blodgett films of oligo- and polythiophenes with well-defined structure, *Thin Solid Films,* **160,** 87–97 (1988).

607. T. Shimidzu, T. Iyoda, M. Ando, A. Ohtani, T. Kaneko, and K. Honda, A novel anisotropic conducting thin film having a conducting and insulating layered structure, *Thin Solid Films,* **160,** 67–79 (1988).

608. Y. Nishikata, M. Kakimoto, and Y. Imai, Preparation of a conducting ultrathin multilayer film of poly(p-phenylene vinylene) using a Langmuir–Blodgett technique, *J. Chem. Soc., Chem. Commun.*, 1040–1042 (1988).

609. J. K. Severn, R. V. Sudiwala, and E. G. Wilson, Determination of interlayer electron hop time in a Langmuir–Blodgett multilayer, *Thin Solid Films*, **160**, 171–175 (1988).

610. N. J. Evans, M. C. Petty, and G. G. Roberts, Charge incorporation in 22-tricosenoic acid Langmuir–Blodgett multilayers, *Thin Solid Films*, **160**, 177–185 (1988).

611. C. A. Jones, M. C. Petty, G. Davies, and J. Yarwood, Thermally stimulated discharge of alternate-layer Langmuir–Blodgett film structures, *J. Phys. D: Appl. Phys.*, **21**, 95–100 (1988).

612. C. A. Jones, M. C. Petty, and G. G. Roberts, Langmuir–Blodgett films: A new class of pyroelectric materials, *IEEE Trans. Ultrason. Ferroel. Freq. Contr.*, **35**, 736–740 (1988).

613. C. A. Jones, M. C. Petty, and G. G. Roberts, Polarisation processes in pyroelectric Langmuir–Blodgett films, *Thin Solid Films*, **160**, 117–123 (1988).

614. C. A. Jones, M. C. Petty, G. J. Russell, and G. G. Roberts, Some structural considerations for pyroelectric multilayer films, *J. Chim. Physique*, **85**, 1099–1102 (1988).

615. M. Suzuki, Y. Saotome, and M. Yanagisawa, Characterization of monolayer and bilayer (polymer/monolayer) structures for use as a lubricant, *Thin Solid Films*, **160**, 453–462 (1988).

Spectroscopy of Complex Monolayers

D. MÖBIUS

5.1. INTRODUCTION

The Langmuir trough technique presents a method of constructing simple artificial systems of cooperating molecules on a substrate. Much of the pioneering work in this field has been carried out by Kuhn and his colleagues at Marburg and Göttingen. They first appreciated the possibility of using well-known monolayers, such as those of long-chain fatty acids, as matrices for appropriate dyes. The elegance of their work on various molecular arrangements containing chromophoric groups is evident in their early papers on the subject.[1-3] These complex monolayers consist of various components, and their structure and the properties of particular components must be evaluated before the monolayers can be used as parts of even more complex monolayer organizates. On this basis, a large variety of systems has been organized to investigate intermolecular interactions, and photophysical and photochemical processes.[4,5] New techniques of monolayer organization and manipulation have also been developed.[4,6-8]

This chapter focuses on the many spectroscopic methods that have been employed in studying monolayers and monolayer systems; see the relevant reviews.[9-11] Special instruments for the measurement of transmission and reflection properties of monolayers and multilayer systems have been designed and constructed to meet the particular requirements of monolayer studies. Optical methods based on transmission and reflection techniques are discussed in the first section.

D. MÖBIUS • Max Planck Institute for Biophysical Chemistry (Karl-Friedrich-Bonhoeffer Institute), D-3400 Göttingen-Nikolausberg, Federal Republic of Germany.

Emission spectroscopy is reviewed in the second section, while the third section is concerned with spectroscopic methods which give indirect information about the spectroscopic properties of molecules in monolayers. Very often, the techniques applied to investigation of biological membranes or bilayer lipid membranes can be modified for application to monolayer studies. The renewed interest in artificial complex monolayers[12,13] will lead to further developments in spectroscopic methods in the future, especially for floating monolayers.

5.2. TRANSMISSION, ABSORPTION, AND REFLECTION OF LIGHT

The absorption of light by a dye containing complex monolayer is of the order of a few percent of the incident light, depending on the organization of the monolayer and the density of the dye. Monolayers are microheterogeneous systems in which the molecules may or may not be randomly oriented in space as in homogeneous systems. Therefore all optical measurements must take into account the particular organization of the monolayer. In many cases, the transition moments of the dye molecules are oriented parallel to the monolayer plane, giving rise to an absorption of light at normal incidence. However, when the orientation of the chromophoric part of the dye molecule changes during compression of the monolayer or in the course of a photochemical process, the result of an optical measurement is strongly altered even when no change in dye density occurs.

The experimental method which allows one to determine all optical properties of a monolayer system by reflection of light in a single experiment is ellipsometry[14]; this technique has already been discussed in Section 4.2.2. It requires plane polarized light under oblique incidence and consists of the measurement of the ellipsometric angles ψ, which is related to the variation of the amplitude, and Δ, which is related to the phase of the electric field of the light wave due to reflection. From the change of these two angles when a thin layer is deposited on the reflecting substrate with respect to the clean substrate, the thickness of the layer and its optical properties can be evaluated with the aid of computers. Only a few spectroscopic measurements have been carried out with this method.[15-19] However, the measurement of light intensities is a simple and more direct method to determine the absorption of a monolayer and the average orientation of the transition moments. A detailed theory based on a point-dipole model, of the transmission and reflection of monolayers on subphases, has been given recently[20,21] and is summarized here. A dye monolayer can be treated as an assembly of polarizable dipoles. Each dipole polarized under illumination by an external light source emits radiation and induces a polarization of its neighbors. Therefore, the local electric-field action on a molecule is a superposition of the external field of the incident light and of the fields of all other dipoles. In the usual case of mixed monolayers, the density of chromophores is relatively small and the absorption bands are broad (i.e., the bandwidth is larger than the sum of all dipole–dipole interactions). For systems homogeneous on the length scale of the wavelength of the incident light, the long-range contribu-

tions to the local field can be accounted for by a mean polarizability approximation.[22] In practical situations, the dipole distribution in the monolayer is symmetric around the surface normal. The reflection and transmission depend on the
angle β of incidence of the light, the transition moment of the chromophore, the
resonance frequency and half-band width of the transition, and on the average
orientation of the dipoles.

This can be characterized by a single parameter P, provided the distribution is
symmetric around the surface normal. With the angle of the dipole with the surface
normal θ, and the probability density of the orientation $n(\theta)$, the order parameter P
is given by,

$$P = \langle \cos^2\theta \rangle = \int_0^{\pi/2} \cos^2\theta\, n(\theta)\, d\theta \qquad (5.1)$$

If the dipoles are statistically oriented in the monolayer plane, $P = 0$; for an
isotropic orientation $P = \frac{1}{3}$, and with all dipoles oriented normal to the surface, $P = 1$.

So far, the optical response of the isolated dye monolayer has been examined.
The subphase (for example, water in the case of spread monolayers or transparent
solid supports for monolayer organizates) modifies this response. The long-range
interactions can be rationalized with the classical frame of continuous dielectrics,
where the dye monolayer lies at a distance $d \ll \lambda$ from the interface air/dielectric
half-space. A multireflection scheme should be used to determine the reflection and
transmission amplitudes r and t, respectively, of the combined system of monolayer
and subphase.[20-24]

In some monolayers of densely packed chromophores, the transition moments
may be oriented parallel to each other in domains much larger than the wavelength
of light, e.g., in the case of J-aggregates. The strong interaction gives rise to a
narrow and very intense absorption band. This situation has also been analyzed
theoretically,[21] and the results will be mentioned later in this chapter.

It should be mentioned that the optical response of a dye monolayer on a
subphase has been derived here by using a point dipole model. A model treating the
system as a thin anisotropic dielectric film with anisotropic Fresnel equations is also
appropriate. This approach has been used in the interpretation of IR measurements,[25] where it may be more satisfactory than our model, since the thickness of
the multilayer system investigated becomes comparable to the wavelength of light.

5.2.1. Transmission

Early measurements of the absorption spectra of monolayers involved the
measurement of the transmitted light intensity after a series of passes through the
monolayer.[9] Generally, what is measured in this type of experiment is not the
absorption A, but $(1 - T)$, where T is the transmission. Since light is not only

transmitted or absorbed but also reflected by a dye-containing monolayer, the absorption is

$$A = (1 - T) - R \qquad (5.2.)$$

where R is the reflection of the system comprising dye monolayer plus substrate. The evaluation of the absorption from a measurement involving multiple passes of the light through a monolayer or multilayer system is rather complicated. To overcome this problem, an instrument involving a single pass has been described.[26] Practically, one determines the difference ΔT between the transmissions of the reference system (reference monolayer + substrate), T_R and of the dye system (dye monolayer + substrate), T_M. Normalization to the transmission of the reference system T_R, yields ΔT_R:

$$\Delta T_R = \Delta T / T_R = (\Delta A + \Delta R) / T_R \qquad (5.3)$$

Here ΔA, and ΔR are the differences in absorption, and reflection, respectively, between dye monolayer system and reference system due to the presence of the dye. The reference system has no absorption and the transmission of the reference system is close to 1 (for quantitative evaluation of the orientation of the transition moments one may have to take T_R into account).

In the usual case of a dye monolayer deposited on a glass plate, covered by a few monolayers of arachidate, the enhanced reflection is approximately $\Delta R = 0.15 \times A$ for measurement under normal incidence.[27] A detailed discussion of enhanced reflection is given in Section 5.2.2. The phenomena observed in monolayer systems on glass plates due to the change of reflection at the interface by deposition of multilayers have been described in detail elsewhere.[28]

For the determination of the orientation parameter from transmission measurements with plane polarized light under oblique incidence, it is reasonable to calculate theoretical values of the transmission instead of evaluating the absorption from the transmission measurements for the two polarizations of light under various angles of incidence and orientation parameters.[21]

The theoretical ratios $\Delta T_s / \Delta T_p$ thus obtained may be compared with the experimental results in order to determine the average orientation of the transition moments.

A high-quality monochromator with suitable sources and detectors plus associated signal processing techniques are used to measure the transmitted light intensity. A typical substrate is a glass or quartz plate; in the experimental arrangement used by the author, one half is covered with a film containing the dye monolayer and the other half with the reference film without a dye. Only the side of the plate facing the incoming light beam is coated with the layer system. The substrate is mounted on an oscillating support that moves the plate perpendicular to the light beam in the case of normal incidence. Hence the reference and sample films are alternately moved in and out of the light beam. Care must be taken to eliminate effects due to the sometimes irregular borderline between sample and reference areas. The difference in the signals can be measured using phase-sensitive detection techniques. After suitably calibrating the system, the experiment yields ΔT_R, the transmission difference due to the presence of the dye molecules.

Examples of Transmission Data

a. Low Dye Densities; Normal Incidence. Figure 5.1 shows two spectra ΔT_R of a mixed monolayer containing two different cyanine dyes, I and II (structures given in Table 5.1), each with a surface density 0.05 nm^{-2} in a matrix of methylarachidate and cadmium arachidate. Spectrum 1 (shifted up for clarity) was scanned at a speed of 25 nm min^{-1} and spectrum 2 at 250 nm min^{-1}. The maximum value in the spectrum at 427 nm is $\Delta T_R = 0.0019$, corresponding to an optical density of roughly 8×10^{-4}.

This type of experiment permits the investigation of monolayers with small dye densities and of intermolecular interactions like the formation of heterodimers in monolayer systems.[29] These dimers formed by association of two different chromophores are characterized by a new absorption band at shorter wavelengths with respect to the monomer absorption band of the short-wavelength chromophore and another low intensity absorption band at longer wavelengths with respect to the monomer absorption band of the long wavelength chromophore.

b. Orientation of Chromophores; Polarized Light. The absorption of a dye monolayer depends strongly on the orientation of the transition moments with respect to the direction of the electric vector of the incident light wave. Under normal incidence of light, molecules with transition moments perpendicular to the

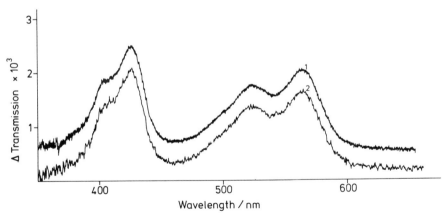

Figure 5.1. Absorption spectrum ΔT of a monolayer system on a glass plate with a mixed monolayer of the cyanine dyes I and II in a matrix of methylarachidate (MA) and cadmium arachidate (CdA), with molar ratios $I:II:MA:CdA = 1:1:50:50$. The scan speed was 25 nmmin^{-1} (curve 1) and 250 nmmin^{-1} (curve 2).

Table 5.1. Dye Structures

No.	Structure

I ClO_4^-

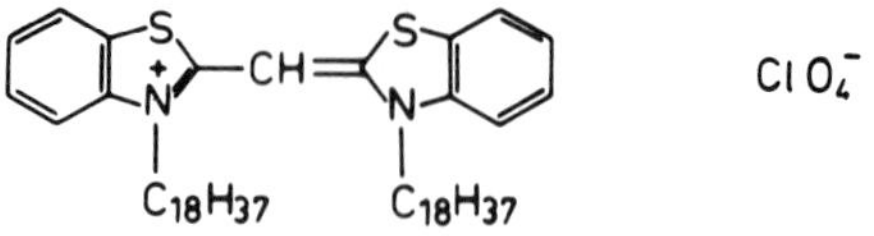

II ClO_4^-

III $4 \ Cl$—⟨ ⟩—SO_3^-

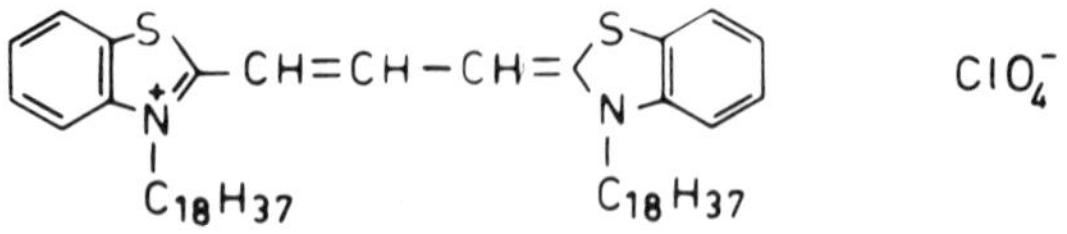

IV ClO_4^-

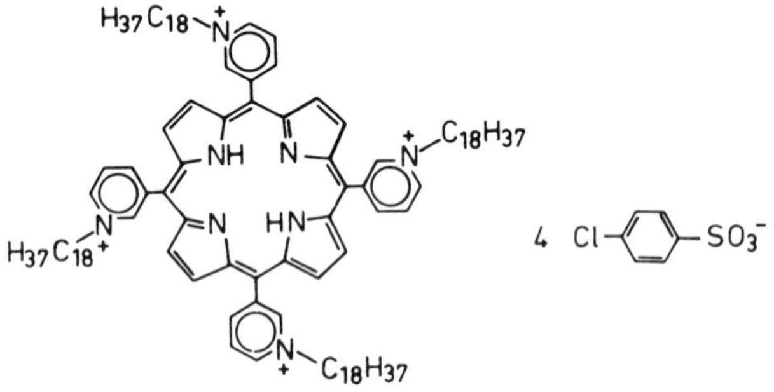

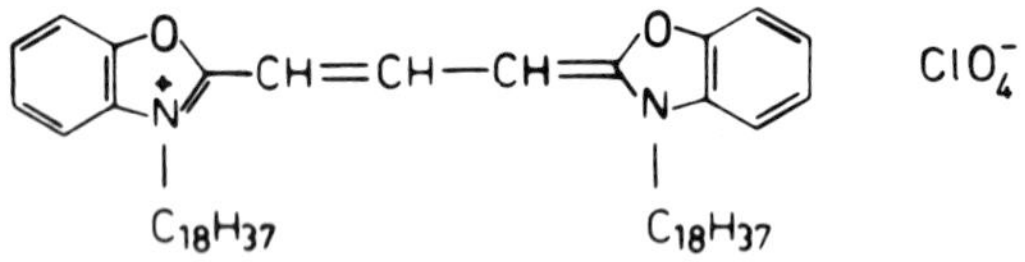

V

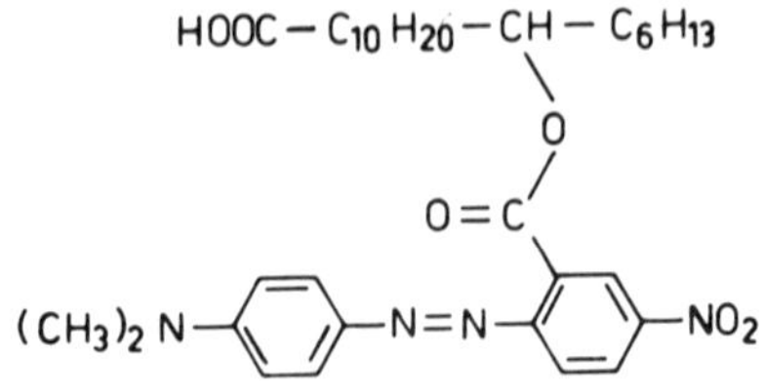

Table 5.1. *(Continued)*

No.	Structure
VI	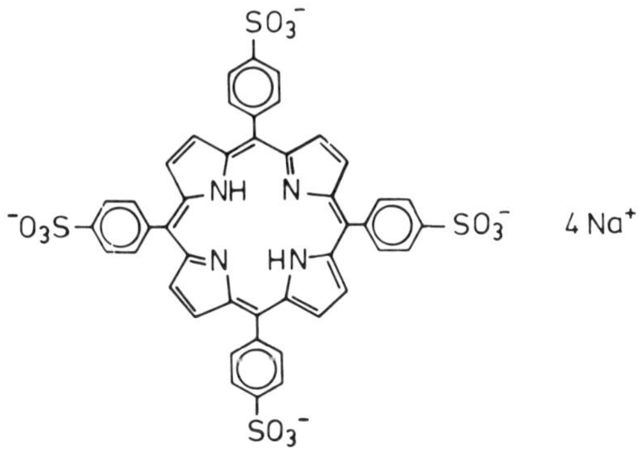 4 Na$^+$
VII	I$^-$
VIII	ClO$_4^-$
IX	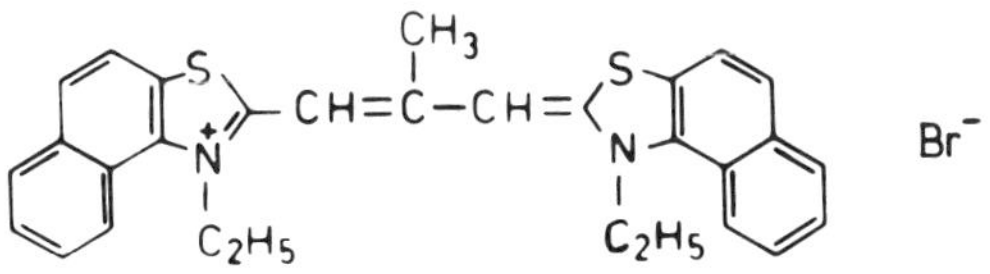 Br$^-$
X	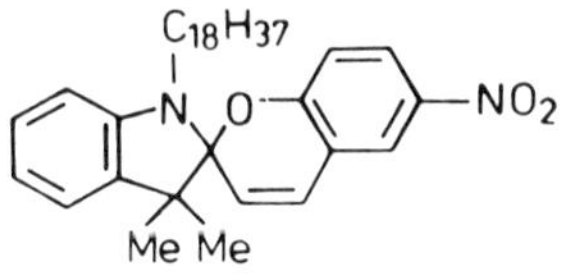
XI	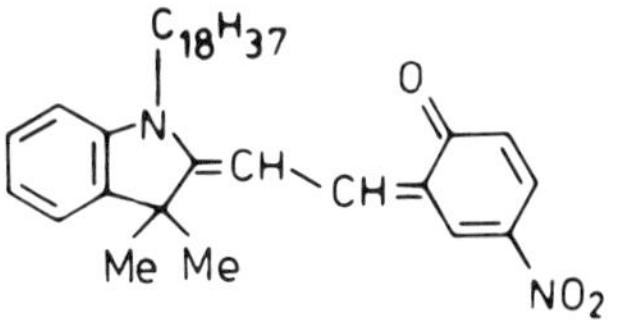

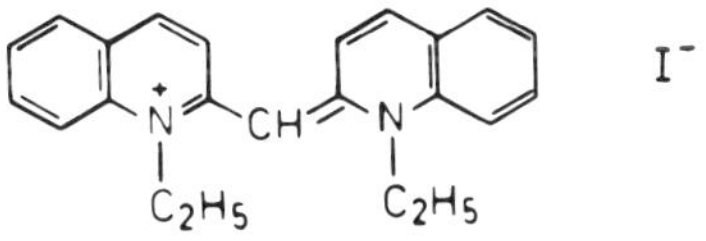

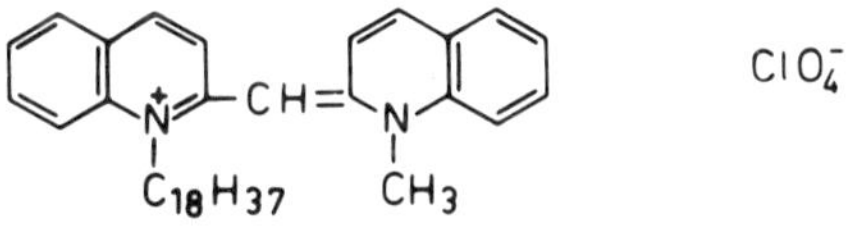

(continued)

Table 5.1.   (*Continued*)

No.	Structure

XII

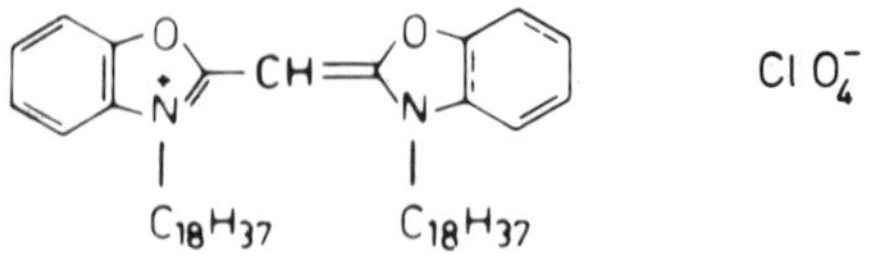

$Cl\,O_4^-$

XIII

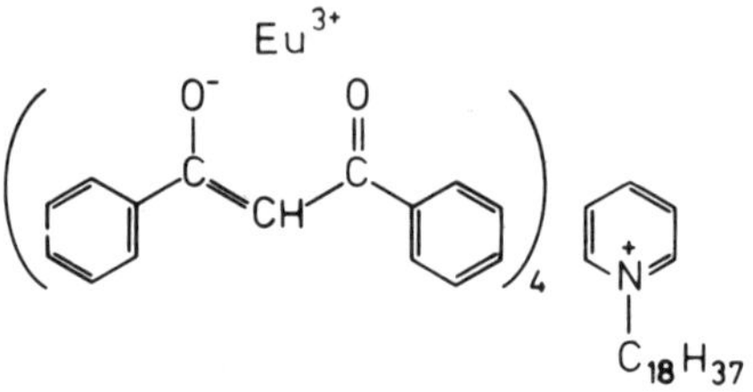

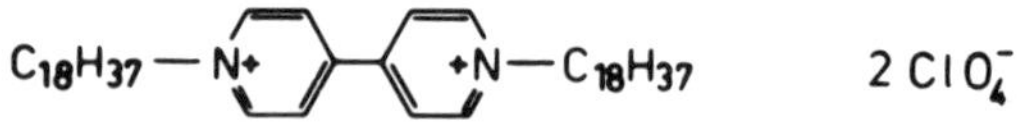

XIV

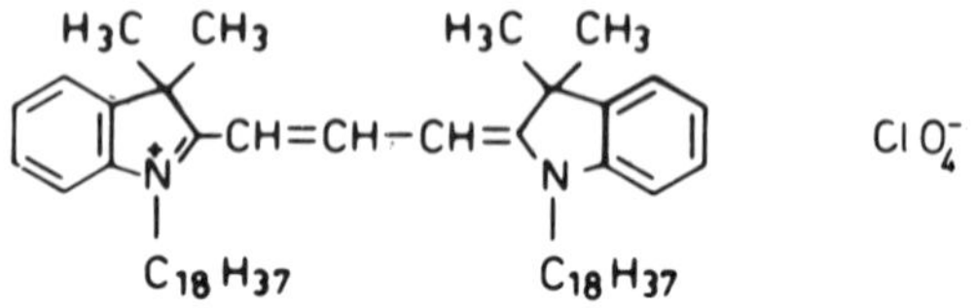

$2\,Cl\,O_4^-$

XV

$Cl\,O_4^-$

XVI

ClO_4^-

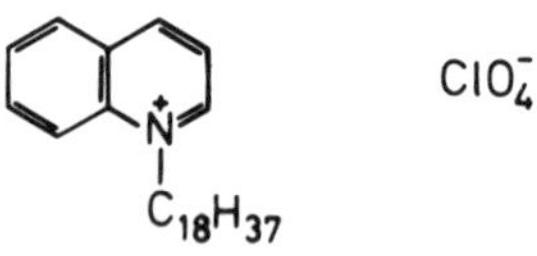

monolayer plane are not detectable, provided the surface of the supporting solid is really flat. It is not a trivial experiment to test this, since most solid surfaces may have considerable disorder in molecular dimensions. With monolayers containing an azo dye it has been possible to demonstrate that, under appropriate conditions, the chromophores are oriented perpendicular to the layer plane.[30]

The average orientation of the transition moments in a complex dye monolayer has been determined by using plane polarized light under oblique incidence.[31–33] The oscillating sample support can be rotated around a vertical axis centered in the light beam in order to have various angles of incidence with respect to the surface normal of the glass plate. As an example, Figure 5.2 shows the absorption spectra of

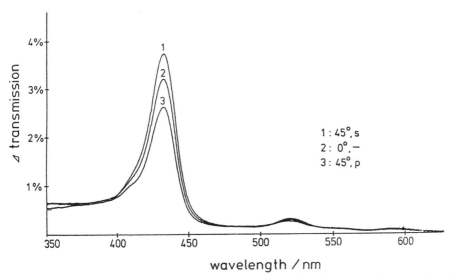

Figure 5.2. Absorption spectra ΔT of a monolayer system on glass plates measured at an angle of incidence $\beta = 45°$ with linearly polarized lights and at normal incidence ($\beta = 0°$, curve 2) for the evaluation of the orientation of the transition moments. Mixed monolayer of the porphyrin III, methylarachidate (MA), arachidic acid (AA), and hexadecane (HD), with molar ratios III : MA : AA : HD = 1 : 10 : 10 : 20; absorption measured with s-polarized (curve 1) and p-polarized (curve 3) light.

a mixed monolayer on glass of the amphiphilic porphyrin III (see Table 5.1), methylarachidate (MA), cadmium arachidate (CdA) and hexadecane (HD), with molar ratios III : MA : CdA : HD = 1 : 10 : 10 : 20. The spectra refer to ΔT_s (45°), curve 1; ΔT_n at normal incidence, curve 2; and ΔT_p (45°), respectively. Absorption is strongest in the case of s-polarization. Comparison of the experimental ratios $\Delta T_s/\Delta T_p$ and $\Delta T_s/\Delta T_n$ for different angles at the maximum of the absorption band, with the theoretical values calculated at orientation parameter $P = 0$ for a statistical orientation of the transition moments in the layer plane (see Table 5.2), shows excellent agreement between theory and experiment.[21]

Nearly all monolayers studied so far with this optical method, display this type of orientation of the transition moments. It should be stressed that the analysis given above is valid only if the dipole distribution is symmetrical around the surface normal. This may not be true in, for example, monolayers of J-aggregates. The validity of this assumption can be verified by measuring the spectrum with plane polarized light under normal incidence at different settings of the polarizer with respect to the dipping direction. The result of such a procedure is shown in Figure 5.3 for a glass plate coated with a mixed monolayer of dye IV, methylarachidate (MA), and cadmium arachidate (CdA), with molar ratios IV : MA : CdA = 1 : 1 : 9.[34] The absorption properties change throughout the entire spectral range studied when the polarizer is rotated. This indicates a preferential orientation of the dye molecules in the layer plane with more molecules oriented parallel to the direction of dipping. This

Table 5.2. Comparison of Measured and Calculated Values of ΔR and ΔT Ratios for Mixed Monolayers of Porphyrin III[a]

$\beta =$	15°	30°	45°
$\Delta R_s/\Delta R_p$	—	1.74 ± 0.04	4.92 ± 0.10
		(1.73)	(4.70)
$\Delta R_s/\Delta R_n$	—	1.26 ± 0.20	1.74 ± 0.30
		(1.28)	(1.75)
$\Delta T_s/\Delta T_p$	1.03 ± 0.10	1.18 ± 0.10	1.51 ± 0.10
	(1.04)	(1.19)	(1.50)
$\Delta T_s/\Delta R_n$	1.00 ± 0.05	1.10 ± 0.05	1.25 ± 0.05
	(1.02)	(1.10)	(1.24)

[a]Flat orientation of the chromophores and a distribution of the transition moments symmetrical around the surface normal are assumed in the calculation.

may be due to rapid monolayer transfer and/or the high viscosity of the monolayer. The naive assumption of a statistical orientation of the chromophores in the mixed monolayer in the layer plane can be shown to be incorrect.

An azo dye chromophore linked chemically to stearic acid forms monolayers with an orientation of the transition moments normal to the surface, if the dye

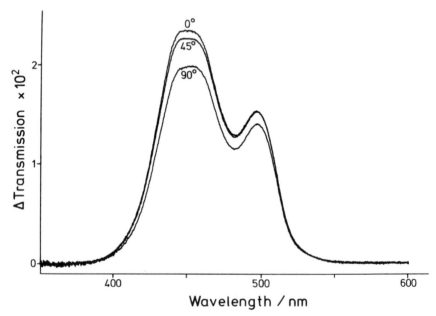

Figure 5.3. Anisotropic distribution of chromophores in the monolayer plane; ΔT at normal incidence of the cyanine dye IV in a mixed monolayer with methylarachidate (MA) and cadmium arachidate (CdA), with molar ratios IV : MA : CdA = 1 : 1 : 9; electric vector of the light parallel to the direction of dipping during monolayer transfer (0°), perpendicular to this direction (90°), and for an intermediate direction (45°).

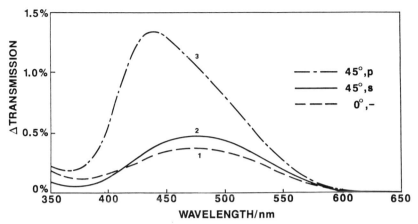

Figure 5.4. Absorption spectra ΔT of a monolayer of azo dye V and hexadecane, with molar ratio 1 : 1, transferred under a surface pressure of 20 mNm^{-1} onto glass plates measured at normal incidence (curve 1, dashed line, ΔT_n) and at angle of incidence $\beta = 45°$ with s-polarized (curve 2, full line, ΔT_s) and p-polarized (curve 3, dot–dash line, ΔT_p) light.

molecules associate to aggregates.[35] The spectra ΔT_s and ΔT_p taken at an angle β = 45° and ΔT_n of a pure monolayer of dye V (see Table 5.1) are shown in Figure 5.4.[21] Under normal incidence (curve 1) and with s-polarized light at $\beta = 45°$ (curve 2), a broad band with a maximum of 470 nm is observed. This band is attributed to dye monomers oriented mainly flat. A different spectrum with a maximum at 410 nm is found with p-polarized light, $\beta = 45°$ (curve 3). Since this band, which is attributed to an aggregate of the azo dye, is missing in the other two spectra, the transition has to be associated with the direction normal to the layer plane. This is confirmed by comparison of experimental and calculated ΔT ratios (see Table 5.3),[21] although the evaluation is somewhat arbitrary due to the partial

Table 5.3. Comparison of Measured and Calculated Values of ΔR and ΔT Ratios for Monolayers of Azo Dye V[a]

| $\beta =$ | V (470 nm)[b] | | | β | V (410 nm) | |
	15°	30°	45°		15°	29.3° (30°)
$\Delta R_s/\Delta R_p$	—	2.04 ± 0.30	5.0 ± 1.5	$\Delta R_p/\Delta R_{p,44}$	—	0.78 ± 0.10
		(1.77)	(4.92)			(0.87)
$\Delta R_s/\Delta R_n$	—	1.17 ± 0.20	2.02 ± 0.40	$\Delta T_p/\Delta T_{p,45}$	—	0.45 ± 0.05
		(1.29)	(1.78)			(0.43)
$\Delta T_s/\Delta T_p$	?	?	?			
	(1.05)	(1.22)	(1.58)			
$\Delta T_s/\Delta T_n$	1.07 ± 0.10	1.17 ± 0.10	1.52 ± 0.10			
	(1.02)	(1.10)	(1.26)			

[a]Flat orientation of the chromophores and a distribution of the transition moments symmetrical around the surface normal are assumed in the calculation for the band at 470 nm and vertical orientation for the band at 410 nm.
[b]In reflection, incidence angles are 29.3° and 44.0°.

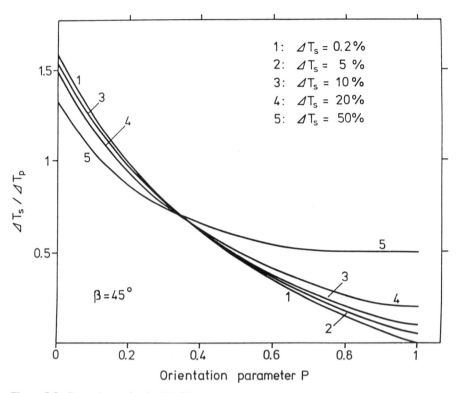

Figure 5.5 Dependence of ratio $\Delta T_s/\Delta T_p$ on orientation parameter P (for definition see text), calculated for angle of incidence $\beta = 45°$ and various values of ΔT_s.

overlap of the two bands. These results confirmed the earlier, more qualitative measurements.[35]

The ratios $\Delta T_s/\Delta T_p$ for a given angle of incidence β and orientation parameter P depend on the magnitude of ΔT, as shown in a plot of $\Delta T_s/\Delta T_p$ vs. P for $\beta = 45°$ (see Figure 5.5).[21] This is particularly important in studying monolayers of J-aggregates with large values of ΔT (such as 20%) at the maximum of the absorption band.

5.2.2. Reflection

The light reflection of a dye monolayer at an interface between two media of different refractive index can be calculated using the same treatment as for transmission of light.[20–24] Using a simple theoretical approach[36] it has been demonstrated that the enhancement of reflection at normal incidence from the interface due to the presence of a dye monolayer can be written as

$$\Delta R = A\sqrt{R_i} + \rho^2 \tag{5.4}$$

where R_i is the reflectivity of the interface in the absence of the monolayer, ρ^2 that of the dye monolayer, and A the absorption of the monolayer. This second term may be negligible as compared to $A\sqrt{R_i}$, especially in the case of very small dye densities; the reflection spectrum ΔR should then correspond to the absorption spectrum.

The reflection change at the air–water interface due to the presence of a dye monolayer has been studied with a multiple pass instrument, and the spectrum of a monolayer of chlorophyll a has been measured in the entire visible range.[37,38] However, no further investigations have been reported.

Spectra of the enhanced reflection from the air–water interface at normal incidence have been measured with an instrument equipped with fiber optic light guides.[36] The arrangement is shown schematically in Figure 5.6. Monochromatic light is directed through fiber optic bundles alternately down to a reference water surface and the surface of the monolayer trough. Half of these bundles are pick-up fibers that guide the light reflected from the reference and monolayer surfaces, respectively, to a second monochromator provided with a photomultiplier tube for measuring the light intensities. The use of a clean water reference surface allows the absolute determination of the enhanced light reflection ΔR, since the reflectivity of the water surface can easily be calculated from the well-known refractive index of water using Fresnel's equations. The spectra of enhanced reflection are therefore evaluated with the aid of a microcomputer, which also controls the stepping motor drive of the two synchronized monochromators. The logarithmic amplifier has the advantage of a larger dynamic range compared with a linear amplifier.

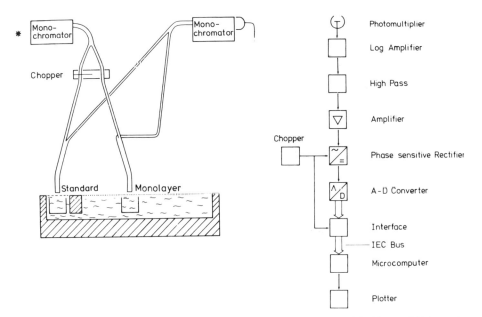

Figure 5.6. Spectrometer for measurement of reflection spectra at normal incidence of light, and schematic of light path and electronic signal processing.

The reflection spectrum is obtained in three steps. First, the asymmetry of the optical arrangement is determined as a function of the wavelength of light and stored in the computer memory. Then, the ratio of light intensities reflected from a reference monolayer without the dye (matrix monolayer) and from the water surface is measured as a function of the wavelength and also stored. These two sets of data are used to evaluate the value of ΔR of complex dye monolayers. In practice, the reflectivity of the reference monolayer is identical to the reflectivity of the clean water surface within experimental error, and this step may be omitted.

Examples of Reflection Data

a. *Enhanced Reflection; A Linear Function of Chromophore Density.* The linear dependence of the enhanced reflection on the surface density of the chromophores has been checked using mixed monolayers of porphyrin III, methylarachidate (MA), and arachidic acid (AA) by varying the density of the porphyrin.[36] The reflection spectrum of this porphyrin free base in monolayers at the air–water interface is characterized by the strong Soret band with a maximum at 430 nm (e.g., Figure 5.2). In Figure 5.7, the enhanced reflection ΔR at the maximum of the Soret

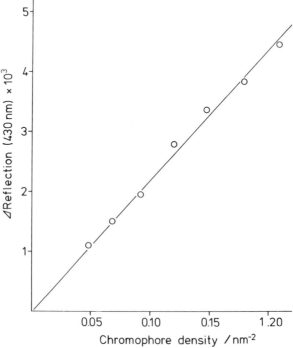

Figure 5.7. Dependence of enhanced light reflection at 430 nm on the density of chromophores in mixed monolayers of porphyrin III in matrices of methylarachidate (MA) and arachidic acid (AA) on water, with molar ratio MA : AA = 1 : 1. The second term on the right side of equation (5.4) can be neglected.

band is plotted versus the density of the porphyrin. It is obvious from this plot that the enhanced reflection due to the porphyrin chromophores is a linear function of the density under these conditions.

b. Formation of Dimers. A set of reflection spectra measured with complex monolayers of the cyanine dye IV in different matrices composed of methylarachidate (MA) and arachidic acid (AA) is shown in Figure 5.8.[39] The molar fraction of the dye in the monolayers is constant (⅙), while the ratio MA : AA varies systematically from 1 : 4 (curve 1), 2 : 3 (curve 2), 3 : 2 (curve 3), 4 : 1 (curve 4), to 5 : 0 (curve 5). The spectra are characterized by two bands, one with a maximum of 445 nm and the other at 500 nm. The spectra cross at 468 nm, a position known as the isosbestic point in absorption spectroscopy. This is observed typically in systems of two forms in equilibrium with each other when the equilibrium is shifted. The cyanine dye IV can form different associates in monolayer systems. In particular, the formation of dimers has been investigated as a function of the molar fraction of this dye in transferred monolayers on glass plates.[3] The band at 500 nm must be attributed to the monomeric dye, and the band at 445 nm to the dimeric form. The set of reflection spectra shows clearly that the equilibrium between the monomeric and dimeric forms of the cyanine dye can be shifted systematically by changing the composition of the matrix monolayer. In the presence of the methylester head group, the dye chromophores are dispersed molecularly. An increasing fraction of carboxyl head groups at the air–water interface leads to increasing association of the chromophores to dimers.

This example demonstrates the potential of the reflection method for investigating the molecular organization of a complex monolayer at the air–water interface. For studying monomer–dimer equilibria, this type of measurement is more

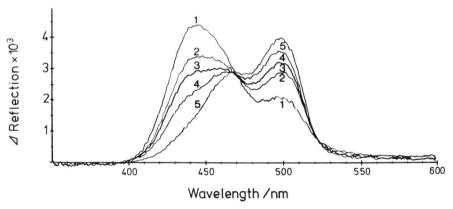

Figure 5.8. Reflection spectra under normal incidence of light of mixed monolayers at the air–water interface containing cyanine dye IV in matrices of methylarachidate (MA) and arachidic acid (AA). Monolayer composition: IV : MA : AA = 1 : 1 : 4 (curve 1), 1 : 2 : 3 (curve 2), 1 : 3 : 2 (curve 3), 1 : 4 : 1 (curve 4), 1 : 5 : 0 (curve 5). The equilibrium between monomeric dye and dimers of the dye in the monolayer is strongly influenced by the composition of the matrix.

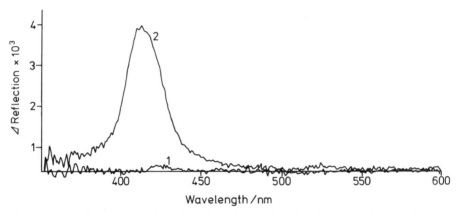

Figure 5.9. Reflection spectra from the surface of a solution of porphyrin VI (10^{-6}) in phosphate buffer (2×10^{-3} M), pH $= 5.0$. No monolayer (curve 1), and monolayer of eicosylamin (EA) and octadecanol (OD), with molar ratio EA : OD $= 1 : 4$, surface pressure 20 mNm^{-1} (curve 2) at the surface. Enhanced light reflection in the presence of the complex porphyrin-matrix monolayer alone formed by controlled adsorption of the porphyrin.

direct and reliable than the measurement of transmission spectra of transferred monolayers. Monolayer states have been studied at the air–water interface by transmission measurements involving two passages of the light beam across the monolayer and reflection at a mirror immersed in the aqueous subphase.[40] The reflection method does not include the bulk subphase. This is of particular value when investigating the formation of complex monolayers by adsorption of molecules from the subphase to a matrix monolayer.[36,41,42]

c. Adsorption of Water-Soluble Dyes to Matrix Monolayers. The adsorption of the water-soluble tetraphenylsulfonatoporphyrin VI to matrix monolayers of eicosylamine (EA) and octadecanol (OD) in various molar fractions has been studied in some detail.[36,42] Figure 5.9 shows the reflection spectrum of a complex monolayer of this porphyrin and a matrix of EA : OD $= 1 : 4$ at a surface pressure of 20 mN m^{-1} on a 10^{-5} M solution of the porphyrin in 2×10^{-3} molar phosphate buffer (curve 2). Without the matrix monolayer at the surface of the solution, no enhanced reflection is observed (curve 1 in Figure 5.9). This result demonstrates clearly that the enhanced reflection measured from the surface of a dye solution is caused by the monolayer formed at the interface. Only those dye molecules that are arranged appropriately combine their scattered light waves with the light wave reflected at the interface. The well-defined phase relation yields the first term in equation (5.4) which is the dominant contribution to the enhanced reflection under these conditions. This is the essential feature of the reflection method which makes it superior for this purpose than other techniques that include the absorption or luminescence of the subphase in the experimental result.

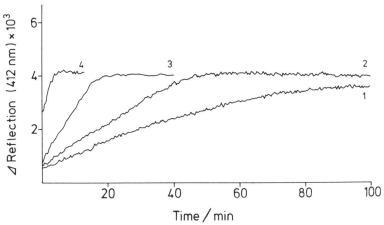

Figure 5.10. Time dependence of enhanced light reflection at the maximum of the Soret band (412 nm) after spreading and compression of a matrix monolayer at the surface of solutions of porphyrin VI in 2×10^{-3} M phosphate buffer, pH = 5.0. Concentration of VI: 10^{-7} M (curve 1), 2×10^{-7} M (curve 2), 5×10^{-7} M (curve 3), 10^{-6} M (curve 4). Matrix monolayer of eicosylamin (EA) and octadecanol (OD), with molar ratio EA:OD = 1:4, surface pressure 20 mNm^{-1}.

d. Transient Effects; Phase Transition. The reflection spectrometer can also be operated in the time mode which allows one to record time-dependent changes in enhanced light reflection with the monochromator set at a fixed wavelength. In this way, it has been possible to follow the adsorption of porphyrin VI to the mixed matrix monolayer of eicosylamine and octadecanol (EA:OD = 1:4) at the surface of the porphyrin solution. A set of adsorption curves is shown in Figure 5.10 for varying concentrations of the porphyrin in the subphase. The experimental results have been rationalized with a model of a Nernst diffusion layer.

Another useful application of the time mode is the study of changes in molecular organization on compression of a complex dye-containing monolayer. Parallel to the measurement of the surface pressure–area isotherm, the enhanced light reflection is recorded. This allows the detection of phase transitions that are correlated with changes in dye association like the formation of J-aggregates, or changes in the orientation of dye molecules. Reorganization of matrix molecules on compression of a mixed monolayer may cause changes in the molecular environment of dye chromophores and lead to the formation of dimers. This effect has been observed with mixed monolayers of the cyanine dye IV and octadecanedicarboxylic acid monomethylester, with molar ratio dye:monoester = 1:5.[39] At low surface pressures, the molecules of the monoester are bound to the water surface by both the carboxyl group and the ester group. At a surface pressure of about 5 mN m^{-1} the ester group leaves the water surface, and the molecules become densely packed in a stretched upright orientation. The dye chromophores, which were surrounded initially by both types of hydrophilic head groups, face the carboxyl groups only when the monolayer has undergone the phase transition at 5 mN m^{-1}. This environment leads to the formation of dimers (see Figure 5.8), and the reflection spectrum changes on compression of the mixed monolayer. The cyanine dye serves as a probe for the density of carboxyl groups at the interface and the reflection technique provides a sensitive detection method.

e. Monitoring of Chemical Reactions. Chemical reactions at the air-water interface can be studied by measuring the enhanced light reflection, if changes occur in the absorption spectra or chromophore orientation. An interesting example is the formation of porphyrin complexes by metallation of the free base and has been investigated in monolayer assemblies on glass plates.[43] Here, the amphiphilic porphyrin III is incorporated in mixed matrix monolayers of methylarachidate (MA) and arachidic acid (AA) at the surface of the solution containing the metal ion. Due to the substitution of the pyridinyl rings, the porphyrin free base has four positive charges per molecule; under appropriate conditions these should prevent the approach of divalent heavy metal ions and subsequent metallation of the free base. These charges are not compensated in the mixed monolayer if the matrix has no negatively charged head groups, e.g., in the case of methylarachidate (MA). Indeed the reflection spectrum of a mixed monolayer of the porphyrin III and methylarachidate, with molar ratio III : MA = 1 : 20, at the surface of an aqueous solution of 10^{-6} M CdCl$_2$ and 5×10^{-5} M NaHCO$_3$ is characterized by the Soret band of the free base with a maximum at 430 nm; see Figure 5.11, curve 1.[44] The local field of the four positive charges at the porphyrin can be compensated by substituting part of

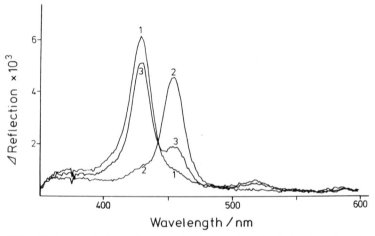

Figure 5.11. Matrix-controlled complex formation by reaction of cadmium ions from the aqueous subphase with free base porphyrin III in a mixed monolayer of methylarachidate (MA) and arachidic acid (AA) at the air–solution interface. Molar ratios III : MA : AA are 1 : 20 : 0 (curve 1), 1 : 10 : 10 (curve 2), 1 : 17 : 3 (curve 3). When the positive charges on porphyrin III are compensated by the carboxyl groups, the band of the free base (455 nm) is observed. Subphase: 10^{-6} M CdCl$_2$ and 5×10^{-5} M NaHCO$_3$; surface pressure 20 mNm^{-1}.

the ester groups by the negatively charged carboxyl groups. The reflection spectrum of a mixed monolayer of the porphyrin III, methylarachidate (MA) and arachidic acid (AA), with molar ratios III : MA : AA $= 1 : 10 : 10$, shows the shifted Soret band of the cadmium complex with a maximum at 455 nm (Figure 5.11, curve 2). Partial compensation and consequently incomplete complex formation under these conditions, is observed with a mixed monolayer of III : MA : AA $= 1 : 17 : 3$, with a reflection spectrum showing both bands at 430 nm and 455 nm (Figure 5.11, curve 3). This result is a demonstrates the environmental control of an interfacial chemical reaction.

    f. Molecular Orientation in Complex Monolayers; Use of Oblique Incidence and Polarized Light. The simple reflection spectrophotometer shown in Figure 5.6 was constructed for measuring the reflection under normal incidence of light. Therefore, a signal is obtained only if the transition moments of the dye molecules have a component parallel to the water surface. Measurement of the reflection of linearly polarized light under oblique incidence allows determination of the average orientation of the transition moments analogous to the transmission measurement described for layer systems on glass plates. The reflectivities for the two polarizations of the incident light can be calculated for various angles of incidence and orientation parameters.

    The orientation of molecules in monolayers is mainly deduced from measurements of surface pressure–area isotherms. This may be acceptable in the case of simple molecules like the usual long-chain fatty acids, amines, and alcohols. Dye molecules, however, may form dimers where the chromophores are lying flat on top of each other. The average area per dye molecule is consequently smaller than that expected for a single flat-lying molecule. In many cases, this reduced area has been erroneously interpreted by assuming a tilted orientation of the dye molecules. The fiber optic reflection technique using polarized light under oblique incidence provides the necessary information.[21] The optical alignment of the instrument is very sensitive to changes in water level and to small deviations in the angle of incidence from the original value due to changes in the leveling of the supporting table. Therefore, the angle of incidence is determined before studying the reflection of monolayers by first measuring the ratio of reflectivities R_s and R_p of the clean water surface.

    The system has been tested with mixed monolayers of porphyrin III, methylarachidate (MA), arachidic acid (AA), and hexadecane (HD), with molar ratios III : MA : AA : HD $= 1 : 10 : 10 : 20$, on water.[20,21] The reflection spectra measured at angles of 0°, 30°, and 45° are shown in Figure 5.12. The enhanced reflection ratios $\Delta R_s / \Delta R_p$ and $\Delta R_s / \Delta R_n$ (where subscripts s, p, and n refer to s-polarization, p-polarization, and normal incidence respectively) obtained from the measured values at the maximum of the Soret band, 430 nm, are compared in Table 5.2 with the theoretical ratios calculated for a flat-lying orientation of the chromophores ($P = 0$). The result indicates that the porphyrin rings are lying flat on the water surface.

    As in the transmission measurements, monolayers of azo dye V were used in

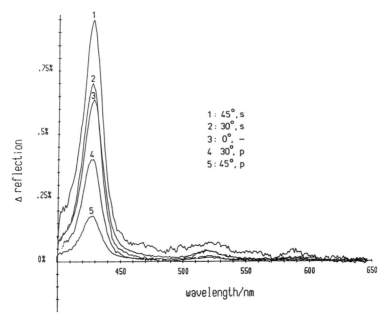

Figure 5.12. Reflection spectra ΔR of a mixed monolayer of porphyrin III, methylarachidate (MA), arachidic acid (AA), and hexadecane (HD), with molar ratios III : MA : AA : HD = 1 : 10 : 10 : 20, with plane polarized light, and angles of incidence indicated in the figure; subphase: water, surface pressure 20 mNm^{-1}. The dependence of ΔR on the polarization indicates flat-lying chromophores.

establishing the reflection method to assess chromophores oriented normal to the water surface. Theoretically, a negative band should be observed for this particular orientation ($P = 1$), i.e., the light reflection from the water surface should be decreased in the presence of the monolayer.[20,21,45] This effect can be observed under oblique incidence with p-polarized light only. The results are shown in Figure 5.13. The broad band with a maximum at 470 nm is found with s-polarized light, $\beta = 45°$, and under normal incidence. The reflection spectrum obtained with p-polarized light, $\beta = 45°$, shows this band also, but an additional negative band around 410 nm is clearly seen. This corresponds to the band attributed to aggregates of the azo dye. The reflection data are summarized in Table 5.3.

The ratio of the reflections, $\Delta R_s / \Delta R_p$ for a given angle and orientation parameter P depends on the magnitude of ΔR_s. This is similar to the situation in the transmission measurements. The ratio $\Delta R_p / \Delta R_s$ calculated for $\beta = 45°$ is plotted vs P in Figure 5.14 for various values of ΔR_s.[21]

g. Structure of Complex Monolayers Formed Using Adsorption Techniques. The structure of the monolayers obtained by matrix-controlled adsorption of porphyrin VI has been investigated in this way. An expansion of the matrix monolayer was observed, depending on the molar fraction of the eicosylamine (EA) in the mixed monolayer of EA and octadecanol.[42] Such an expansion might be in-

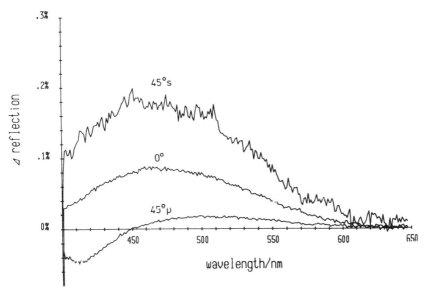

Figure 5.13. Reflection spectra ΔR of a monolayer of azo dye V with plane polarized light at angle of incidence $\beta = 45°$ (polarization indicated on the curves) and normal incidence; subphase: water, surface pressure 20 mNm^{-1}. The negative band around 410 nm indicates orientation of optical transition normal to the water surface.

terpreted as the effect of a penetration of the adsorbed species, porphyrin VI, into the matrix monolayer. Then, the porphyrin ring should be oriented with its plane perpendicular to the water surface. Alternatively, the expansion could be interpreted as a looser packing of the matrix molecules under the influence of the porphyrin in order to match the densities of charges on the porphyrin. In this case, the porphyrin ring would be oriented horizontally. The results of the surface pressure–area measurements have been interpreted assuming porphyrin dimers underneath the matrix monolayer. The results of the reflection measurements indicate an essentially flat orientation of the porphyrin rings with a slight deviation from the horizontal. This example demonstrates again the value of the reflection method for the evaluating the structure of complex monolayers.

The method of organizing complex monolayers by matrix-controlled adsorption has also been used to form highly ordered extended two-dimensional aggregates, in particular the J-aggregates of cyanine dyes.[21,46] In the case of the 2,2'-cyanine chromophore, the aggregates obtained by the adsorption method can be compared with those assembled from a water-insoluble derivative by spreading and compression. The reflection spectra of monolayers of these two types are compared in Figure 5.15. The spectrum with a maximum at 583 nm (full line) was measured for a monolayer of cyanine dye VII which was assembled from an aqueous 10^{-4} M solution of the dye by adsorption to a matrix monolayer of arachidic acid at a surface pressure of 20 mN m^{-1}. The spectrum clearly shows the spectral features of the J-

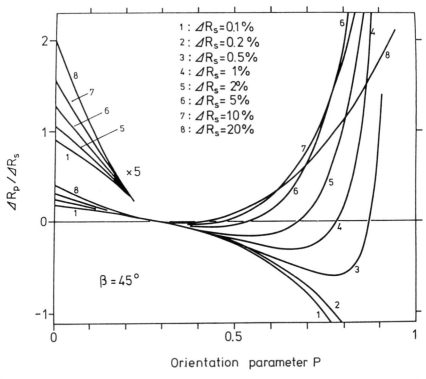

Figure 5.14. Dependence of ratio $\Delta R_p / \Delta R_s$ on orientation parameter P (for definition see equation 5.1) for various values of reflection ΔR_s.

aggregate, i.e., the strong and narrow band shifted to longer wavelengths with respect to the monomer band. The full width at half maximum (FWHM) is 280 cm^{-1}.

An amphiphilic analogue of this cyanine is dye VIII. Mixed monolayers of this dye and hexadecane (HD), with molar ratio VIII : HD = 1 : 1, form J-aggregates at the water surface.[47−52,36,21] The reflection spectrum of such a monolayer at a surface pressure of 20 mN m^{-1} is shown by the dashed line in Figure 5.15. In this case, the intensity of the reflection at the maximum (585 nm) is smaller than in the former case of the dye monolayer formed by adsorption, and the FWHM is somewhat larger, 380 cm^{-1}. The microscopic structure of these differently organized monolayers after transfer to glass plates is different. However, the transferred layers cannot be directly compared since transfer of the "adsorbed" dye layer requires the deposition of a dye layer both on dipping and withdrawal. This Y deposition has not been achieved with monolayers of the amphiphilic cyanine J-aggregates. Therefore, the measured ΔT of the system after transfer of the adsorbed monolayers of cyanine dye VII is 0.42 at the maximum (580 nm) while one mixed monolayer of dye VIII : HD = 1 : 1, coated with a monolayer of arachidic acid, has $\Delta T = 0.21$ at 582 nm.

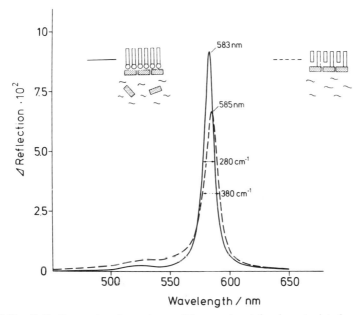

Figure 5.15. Reflection spectra of monolayers of J-aggregates at the air–water interface formed by adsorption from a 10^{-4} M solution of dye VII to a monolayer of arachidic acid (full line) and by spreading and compression of amphiphilic analogue dye VIII and hexadecane (HD), with molar ratio VIII : HD = 1 : 1 (dashed line). Surface pressure 20 mNm^{-1}.

A very interesting phenomenon has been observed in systems of J-aggregates formed by adsorption to a matrix monolayer at the air–water interface. The aggregate of the cyanine dye VII can be replaced entirely by the J-aggregate of another cyanine dye present in the aqueous subphase at much smaller concentration than dye VII.[46] This is demonstrated in Figure 5.16, where the series of reflection spectra was taken with increasing time after formation of the arachidic acid monolayer on top of the dye solution. Due to the high concentration of dye VII, the aggregate of this dye is formed immediately. Then, the reflection at 583 nm (aggregate of the cyanine dye VII) decreases and the band of the new J-aggregate attains a maximum at 650 nm. The spectra seem to indicate that the cyanine dye IX replaces the initial J-aggregate in a cooperative way, since there is no evidence for preadsorption of monomers of dye IX before formation of the new aggregates. Again these investigations illustrate the power of the reflection method in studies of molecular organization at interfaces.

h. Correlation with Microscopic Results. The interpretation of transmission and reflection measurements is particularly difficult in the case of dye monolayers organized in J-aggregates. This is a consequence of the heterogeneous structure of these monolayers composed of extended two-dimensional crystals. The size of the individual aggregates depends strongly on monolayer composition and details of

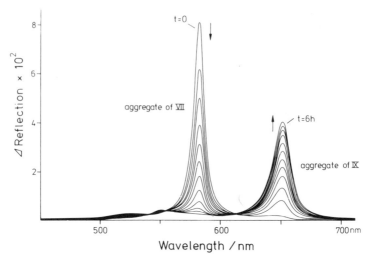

Figure 5.16. Replacement of a monolayer of J-aggregates of dye VII formed by rapid adsorption of dye VII (10^{-4} M) to a monolayer of arachidic acid by the J-aggregate of dye IX (5×10^{-7} M) due to slow approach to the monolayer; surface pressure 20 mNm^{-1}.

Figure 5.17. Photograph of a glass plate with a monolayer of J-aggregates of the cyanine dye, with molar ratio VIII : hexadecane = 1 : 1; monolayer transfer at 20 mNm^{-1} by horizontal contacting, top coating monolayer of arachidic acid. Magnification indicated by the scale.

monolayer formation. Since all dye molecules in an aggregate are oriented parallel to each other, absorption and reflection are anisotropic. This can be seen very easily when a transferred monolayer on a glass plate is inspected in a microscope between crossed polarizers[53]; see Figure 5.17. This is a photograph of the light transmitted by a monolayer of dye VIII and hexadecane (HD), with molar ratio VIII : HD = 1 : 1. The regions of different brightness represent individual crystals with different orientation. Within a crystal, the brightness is quite uniform as expected for a totally ordered array. Between the larger aggregates are smaller entities and some material without preferential orientation, as evidenced by turning one polarizer. The essential information obtained from this micrograph is the average size of the two-dimensional crystals. The length of the aggregates of this particular monolayer is of the order of 50 μm.

This is an important fact with respect to the interpretation of the reflection measurements. The incoming light wave does not see individual oscillators statistically oriented in the layer plane as in the usual case of mixed dye-matrix monolayers but an array of well-ordered oscillators. The polarized absorption and reflection must both be evaluated. The theoretical implications have been studied and tested experimentally with monolayers of J-aggregates of dye VIII.[21] In the model, the local field corrections have been taken into account in a more detailed way than by the mean polarizability approximation used for deriving transmission and reflection of monolayers. With J-aggregates on glass plates oriented preferentially in the layer plane[49,21] having $\Delta T = 45$, the experimental results are in excellent agreement with the theoretical model. It should be mentioned that the second term on the right side of eq. (5.4) can no longer be neglected here.

i. Propagation of Surface Pressure Jumps. The reflection method has been used to study photochemical reactions at the air–water interface like the photopolymerization of monolayers of an amphiphilic diacetylene compound[54] and the isomerization of an amphiphilic spiropyran in a matrix monolayer of octadecanol.[44] A new way of studying the propagation of surface pressure jumps induced in a monolayer by a photochemical reaction has been demonstrated using the reflection method.[55] The photochromic system of the spiropyran (SP) X isomerizing to merocyanine (MC) XI provides the possibility of generating a surface pressure jump in a monolayer, since the merocyanine requires a larger area per molecule than the spiropyran in a mixed monolayer with octadecanol, SP : OD = 1 : 5. The spiropyran is transformed into the merocyanine by irradiation with UV radiation, and the merocyanine can be isomerized to the spiropyran by irradiation with green light. The UV reaction is fast, and a flash exposure of the mixed monolayer of the spiropyran causes a sudden rise of the surface pressure at constant monolayer area. In order to study the propagation of the resulting pressure wave, a method is required by which the arrival and the profile of the surface pressure rise can be detected in a place remote from the site of the photoreaction.

Mixed monolayers of cyanine dye VIII and co-aggregates like stearic acid (SA), with molar ratio of dye : SA = 1 : 1, undergo a transition from monomeric dye

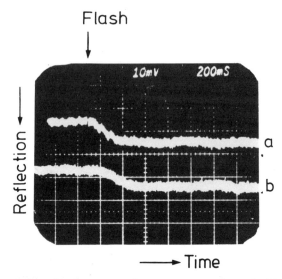

Figure 5.18. Transduction of surface pressure jump across a monolayer. Oscilloscope traces of the photomultiplier output proportional to the light reflection of a detection monolayer. The reflection increases due to molecular reorganization when the surface pressure jump arrives. The delay between the flash and the onset of the change in reflection is about 40 ms (upper trace) for a transmission distance along a monolayer of archidic acid of about 5 cm, and 150 ms (lower trace) for a distance of 20 cm. The shape of the upper trace may be determined by the kinetics of the reorganization; that of the lower trace allows analysis of the spectrum of the interfacial surface pressure jump.

to J-aggregate at a surface pressure of 7 mN m^{-1}. This transition is characterized by the appearance of a new absorption band, and consequently a new band in the reflection spectrum shifted to longer wavelengths with respect to the maximum of the monomer absorption band. When a surface pressure jump travels across a monolayer of this type maintained at a surface pressure slightly below the phase transition, the formation of the J-aggregate will be induced. This should be detectable by measuring the change in the enhanced light reflection ΔR at the wavelength of the maximum of the aggregate band. Indeed, such a change has been observed in a system of three different monolayers in series: the surface pressure jump generating monolayer (SP : OD = 1 : 5), the transducing monolayer (arachidic acid), and the detection monolayer (dye VIII : SA = 1 : 1). This is shown in Figure 5.18, where the output of the detector is plotted versus time for two different experiments. The photocurrent increases with increasing light reflection, and consequently the signal on the storage oscilloscope becomes more negative. The traces clearly show an increase in the output delayed with respect to the signal input (flash) in the photosensitive monolayer. The separation between the input site and the detection site was about 5 cm in the case of trace a and about 20 cm for trace b. The time delays between the flash and the onset of the increase in reflection are about 40 ms (a) and 150 ms (b), respectively. Therefore, the transmission of the mechanical signal across a monolayer of arachidic acid takes place at a speed of the order of 1 m s^{-1}.

The traces in Figure 5.18 contain additional information in the shape of the curves. If the generation of the surface pressure wave is fast on the time scale shown, the shape of trace a may be determined by the kinetics of the transformation from monomeric to aggregated dye chromophores in the detection monolayer. This experiment thus provides the unique possibility of analyzing the kinetics of a two-dimensional phase transition. By comparison with trace a, the shape of trace b is different. Obviously, the high-frequency components of the pressure jump are more strongly damped in case b than in case a. In other words, the kinetics of the phase transition used to detect the pressure jump is fast in case b compared with the rise time of the surface pressure jump. Now the method can be used to analyze the dispersion of the transmission of the two-dimensional pressure jump at the mono-layer-covered interface. Detailed investigation of these phenomena should therefore lead to better understanding of longitudinal waves which are generated strictly at the interface without initial mechanical movement of the underlying bulk phase.

Measurements of the transmission of such waves at the monolayer-covered water surface have been performed with a mechanical sensor.[56] As might be expected, the velocity of the surface pressure jump propagation depends on the applied surface pressure and, for example, is 1.8 m s^{-1} for a monolayer of di-myristoyl-phosphatidylcholin at 20 mN m^{-1}. The dynamic compressional modulus is about four times the static compressional modulus, which can be determined from the surface pressure–area isotherms.

5.2.3. Infrared Spectroscopy

Light absorption in the UV and visible range is caused by the interaction of electrons with the electric field vector of the incident light. Infrared absorption arises from the interaction of a dipole moment with the electric vector. The intensities of infrared absorption bands are therefore much smaller than those in the UV and visible range. Consequently, the application of infrared spectroscopy to the investigation of monolayer systems requires arrangements with multiple pass or multiple internal reflection of light even more than spectroscopy in the UV and visible.[57] However, infrared spectra can be obtained from single monolayers on metal substrates using conventional Fourier Transform Infrared (FTIR) techniques to analyze the IR radiation reflected from the sample.[58–60] The FTIR method is often combined with grazing incidence reflection.[61–63] As was discussed in Section 4.3.3, infrared studies of monolayer systems have been performed to investigate the structure of LB film systems[25,58–60,62–75] or to follow chemical reactions.[63,76–78] A recent review on the application of FTIR spectroscopy to characterize thin films is given elsewhere.[71]

Earlier in this chapter, results were described showing how visible and UV light can be used to measure the orientation of molecules. This can also be deduced from the measurement of IR spectra using linearly polarized radiation. An example is given in Figure 5.19, which shows infrared transmission spectra of 2 × 49 layers of behenic acid on a CaF$_2$ substrate.[25] In the figure, i is the angle of incidence in

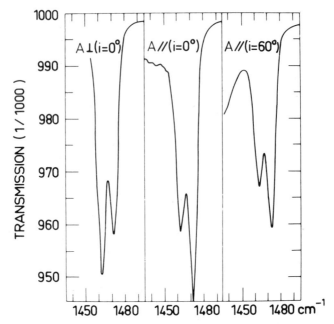

Figure 5.19. Infrared transmission spectra of 2 × 49 layers of behenic acid on a CaF$_2$ substrate. Absorption coefficients for s-polarized (A⊥) and p-polarized (A//) light, angle of incidence i.

the air, A$_{//}$ is the absorption coefficient for s-polarized and A⊥ for p-polarized radiation, respectively. The IR bands at 1462 cm^{-1} and 1471 cm^{-1} are attributed to CH$_2$ deformation. The splitting of the band into two sharp bands has been interpreted as being indicative of a crystalline structure for the acid layer. The oblong crystallites are formed on the aqueous subphase. Using data for different polarizations, it was concluded that the hydrocarbon chains of the behenic acid molecules are tilted with respect to the normal by an angle of 23°30'. The analysis of IR spectra of mixed porphyrin–behenic acid monolayers provides a more detailed picture of the structure of these complex monolayers.[25] The molecular interactions between the matrix molecules and the embedded chromophores can also be studied by Raman spectroscopy.

Surface chemical reactions can be used to form monolayers on a solid substrate like glass[79] or metal oxides.[80] Chemically modified electrodes[81] have gained much interest as parts of electrochemical devices for solar energy conversion. In most cases, however, disordered layers are formed on the solid surfaces and cannot be compared with transferred monolayers. Layers obtained by reacting octadecyltrichlorosilane seem to be densely packed, according to an investigation of electron transport across such layers.[82] As described in Section 2.6.1, a new method of constructing multilayers based on repetitive surface chemical reaction has been developed, and the course of the process can be followed by infrared spectroscopy.[83]

5.3. EMISSION

Transmission or absorption measurements of complex monolayers and mono-layer systems require the determination of a small difference between two large signals. The situation is different in the measurement of light emission by dye molecules incorporated in monolayers. The total light intensity detected is carrying the information (if the background of scattered exciting light can be neglected). This is the reason for the higher sensitivity of the emission techniques as compared to transmission or reflection methods.

The emission is due to the radiative deactivation of an excited singlet state (fluorescence) or the triplet state (phosphorescence) to the ground state. The mo-lecular emitter can be described as an electric dipole, which is the most prominent case, or as a magnetic dipole or an electric quadrupole. The emission of excited molecules is characterized by the spectral distribution, the quantum yield, the lifetime, and the polarization.

These experimentally accessible properties of excited molecules depend strongly on the environment. A molecule in a vacuum may emit light even from a higher vibronic level of the excited state. In monolayers or monolayer systems, which must be regarded as condensed media, thermal equilibration rapidly removes the excess vibronic energy, and the molecule may emit from the lowest vibronic level of the excited state. The pronounced influence of the microenvironment, for example, on the association of dye molecules has been demonstrated in Section 5.2.2. As a result of this sensitivity of excited molecules to the nature and structure of their environment, appropriate molecules have been used as fluorescent probes in complex monolayers[10,11] in lipid bilayer membranes, BLMs,[84] or in membranes of biological origin.[85,86] These probes must only be a small fraction of the entity in order to avoid changes of the host structure. The problem of probe-induced changes of the structure is not addressed here. It should be mentioned, however, that emission techniques usually require a smaller account of probe material than do transmission or absorption methods.

Emission measurements are not only a means of evaluating the structure of a complex system but are also useful in following the course of a chemical reaction or in studying molecular interactions involving the fluorescent species as an active component. This function is far beyond that of a probe. An example of this is the investigation of the energy transfer from an excited cyanine dye aggregate to an acceptor[87] as a model for the energy-harvesting photosynthetic unit.

The measurement of fluorescence intensities at a fixed wavelength in order to follow structural or chemical changes of a system very often does not provide sufficient information for a quantitative analysis. It must be kept in mind that the fluorescence intensity depends on the excitation which varies with changes of the absorption spectrum, on the fluorescence spectrum, the radiative rate, the fluores-cence quantum yield, and molecular orientation. It has to be verified that the observed changes in fluorescence intensity are really due to the anticipated struc-tural or chemical changes. Even then, the quantitative analysis may still be quite complex.

5.3.1. Steady-State Fluorescence

The steady-state fluorescence emitted upon excitation of components in a spread monolayer at the air–water interface, or incorporated in a complex monolayer assembly on a solid substrate, is widely used to obtain information on the structure of the investigated system and/or on the dynamics of chemical or structural changes.

5.3.1.1. Fluorescence Experiments at the Air–Water Interface

Special instruments have been constructed for the measurement of monolayer fluorescence, but commercially available equipment can be modified according to particular needs. The fluorescence of dyes in complex monolayers can be excited by light incident from the air, e.g., by using a collimated beam or quartz fiber optics, or from the aqueous subphase, preferentially under total reflection of the exciting light. An instrument of this type has been used to study the fluorescence spectra and concentration quenching of chlorophyll *a* monolayers.[88]

Phase transitions in monolayers[89] have been studied by measuring the fluorescence of the film forming material or by using fluorescent probes. An interesting example is the investigation of monolayers of chlorophyll *a* (Chla) at the air–water interface by measuring the fluorescence spectra and the relative fluorescence yield.[90] The particular feature in this investigation was the admixture of hexadecane to the Chla, a procedure which has been used for the formation of homogeneous monolayers in cases where segregation into two phases has been observed.[91,92] The fluorescence spectrum of the Chla–hexadecane monolayers depends strongly on the average area per Chla molecule, and compression of the monolayers leads to association and reorientation of the porphyrin rings. In order to support these conclusions of the fluorescence data, more direct measurement of the enhanced reflection of linearly polarized light under oblique incidence (see Section 5.2.2) would be helpful. The organization and fluorescent behavior of mixed monolayers of Chla and dimyristoylphosphatidylcholin (DMPC) have been studied, including transmission measurements with a multiple reflection instrument.[93] The fluorescence data have been interpreted by an increase of Chla dimers in the DMPC matrix when the phase transition occurs due to reorientation of the Chla molecules. This should bring the planes of the porphyrin rings into closer contact by aligning the planes perpendicular to the surface. A detailed discussion of the structure and properties of monolayers of chlorophyll *a*[93–98] is not given here, but the subject is discussed further in Chapter 6.

Appropriate cyanine dyes associate in monolayers at the air–water interface on compression to large two-dimensional aggregates (J-aggregates). The appearance of the typical emission of the aggregate which is only slightly Stokes shifted with respect to the narrow absorption band of the aggregate has been monitored during compression of a mixed monolayer of cyanine dye XII (see Table 5.1) and stearic acid as co-aggregate by excitation in the range of the monomer absorption band.[99]

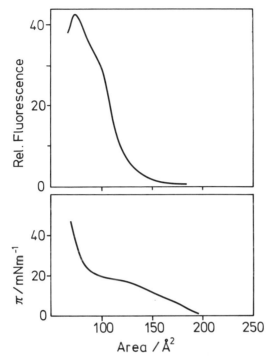

Figure 5.20. Relative fluorescence intensity at 408 nm (excitation at 365 nm) and surface pressure vs area per dye molecule of a mixed monolayer of dye XII and stearic acid (SA), with molar ratio XII : SA = 1 : 1. Strong rise of aggregate emission at the phase transition in the monolayer. Subphase: bidistilled water pH = 5.6.

Figure 5.20 shows a plot of the surface pressure and of the aggregate fluorescence intensity versus the area per dye molecule. It is obvious that the onset for aggregate formation coincides with the onset of the phase transition, as indicated by the nearly horizontal part of the isotherm. The results have been confirmed by measuring the fluorescence spectra of the monolayer at the relevant areas per dye molecule. However, these measurements are not sufficient for a quantitative evaluation of the fraction of the dye organized in the aggregate, since the aggregate fluorescence intensity is not proportional to the amount of aggregate in the monolayer. Reflection measurements are required and it has to be taken into account that the reflecting entities are large compared to the wavelength of the incident light.

Structural changes in the head group region of phospholipid monolayers have been deduced from fluorescence measurements. The investigations of phase transitions in phospholipid monolayers at the air-water interface are an illustrative example of the difficulties arising in the interpretation of the experimental results. Various probes added to the phospholipid in molar fractions of the order of 0.01 show changes in fluorescence intensity on compression of the monolayer in the range of the area/molecule, where the π/A isotherm indicates a phase transition.[100–105]

Such changes have been attributed to a modification of the probe environment and/or to a geometrical reorientation of the probe.[100,103] Environmental changes lead frequently to changes in the fluorescence spectrum and/or fluorescence quantum yield. A convincing analysis must therefore be based on the measurement of the fluorescence spectra of the probe in various interesting situations. The determination of the relative fluorescence quantum yield requires knowledge of the orientation of the probe molecules during the absorption and emission of light.

The measurement of the fluorescence intensity during excitation with linearly polarized light under oblique incidence provides information on these orientations. The use of polarized light spectroscopy in the evaluation of the orientation and mobility of molecules in membranes has been analyzed in detail.[106] However, an unambiguous evaluation in the case of spread monolayers is quite difficult due to the following reasons: (i) The orientation of the transition moment during absorption may differ from that during emission due to depolarization by convective motion, lateral or rotational diffusion, or due to energy transfer during the lifetime of the excited state. (2) In the reported measurements,[103] the aperture of the detection system was not taken into account, i.e., an average over a large spatial angle for the emission must be considered, not only emission under a particular angle. Again, the average orientation of the absorption dipoles can be directly obtained from reflection measurements (see Section 5.2.2). This removes a considerable uncertainty in the evaluation of the fluorescence data. Conclusions regarding structural changes of a particular monolayer region based on the measurement of fluorescence intensities should therefore be supported by additional experimental evidence.

Another difficulty in the interpretation of fluorescence data may arise from a possible segregation of the probe and formation of separate monolayer phases, either when the material is spread or on compression of the monolayer. This has been demonstrated very clearly in microfluorimetric studies of phospholipid monolayers containing various fluorescent probes[105,107−113] or proteins of the bacterial photosynthetic apparatus.[114] The growth of crystalline phospholipid domains in monolayers has been analyzed with a model involving fractal dimensions.[115] The domains can be moved laterally by asymmetric electrical fields.[116] A phase separation of dye and matrix molecules may lead to the formation of dimers or other associates, and this dye association should be revealed by measurement of the fluorescence spectra or reflection spectra. The homogeneous distribution of the probe in the host matrix is a common assumption in the evaluation of fluorescence data and requires experimental verification.

The diffusion of fluorescent components can be investigated systematically in monolayers at the air–water interface and may be related to lateral or translational diffusion of components in biological systems. Diffusion constants of different probes in different host monolayers have been determined,[100−103,110] either by measuring the fluorescence recovery after photobleaching[117] or by studying the monomer–excimer dynamics,[118] a technique formerly applied to bilayer lipid membranes.[119] The reported values of the diffusion coefficient of probes range between smaller than 2×10^{-9} cm^2 s^{-1} and 2×10^{-6} cm^2 s^{-1}. The small values refer to monolayers of dipalmitoylphosphatidylcholin in the crystalline state. A serious problem in the study of diffusion in fluid monolayers, i.e., normally at small surface pressures, is the flow of monolayer material due to convection in the bulk phases induced by temperature gradients.[120] This flow obviously leads to large overestimates of the diffusion coefficients, but can be minimized in diffusion measurements using fluorescence microphotolysis[110] by measuring the fluorescence of the monolayer in a small compartment in contact with the bulk monolayer.

5.3.1.2. Fluorescence Experiments on Solid Substrates

Systems of complex monolayers built up on solid substrates provide the possibility of arranging dye molecules in planes of well defined separation.[1,10] The distance dependence of intermolecular interactions can be systematically investigated. Steady-state fluorescence methods are widely used in such studies. More and more complex systems with ingenious combinations of the different interactions between various components via photon, electron, and proton transfer have been devised and studied.[5,8,49,121] Assemblies of this type can be very valuable as models for investigating complex processes by systematic variation of the relevant parameters. One example is the evaluation of the relative contributions of the electron injection and the energy transfer mechanisms in the spectral sensitization of silver halides for particular materials.[122−124]

The instrumentation for steady-state fluorescence measurements of monolayers transferred onto glass or quartz plates poses no particular difficulties. Commercially available instruments have been used. If normal incidence is used, it must be kept in mind that this situation is ineffective if the transition moments of the dye molecules in the organizate are oriented perpendicular to the layer plane. In such a case, the sample must be turned in order to provide oblique incidence. Another particular characteristic in monolayer systems in addition to this effect of molecular order is the high density or small separation of molecules which can strongly affect the observed phenomena.

The nature of the emitting state of an excited molecule can be evaluated from the interaction of the field emitted by the molecule in the direction of the observer with the field of the same source reflected from a metal mirror (or any reflecting surface). An example of this effect is shown in Figure 5.21.[28] The two emission bands of europium complex XIII are altered in different ways by the presence of the gold mirror, showing that the band at 592 nm is not electric dipolar in nature. The interference causes a characteristic pattern of the angular dependence of the fluorescence (or phosphorescence) intensity of a dye monolayer kept at a certain distance from the mirror.[2,28,125−129] It is amazing that changes in the environment at distances of several wavelengths of light can influence the deactivation process of an excited molecule.[130−132]

A particularly short-range intermolecular interaction that has been extensively studied in monolayer organizates with steady-state fluorescence is the photoinduced electron transfer. It is observed by fluorescence quenching of the excited species

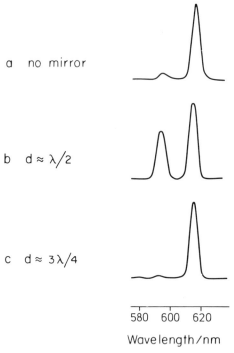

a   no mirror

b   d ≈ λ/2

c   d ≈ 3λ/4

580   600   620

Wavelength/nm

Figure 5.21. Emission spectrum of europium complex XIII in monolayers in the absence of a metal mirror (a) and in the presence of a mirror at distance λ/2 (b) and 3 λ/4 (c), respectively. The different effect of the mirror on the intensities of the two bands indicates a different nature of the corresponding transitions.

which may act as an electron donor by transfer of the excited electron from the LUMO (lowest unoccupied molecular orbital) to an appropriate acceptor molecule. Since the excitation of a molecule leaves a vacancy in the HOMO (highest occupied molecular orbital), the excited molecule can act as an electron acceptor by filling this vacancy. The fundamental difference between these two electron transfer processes is sometimes overlooked. In the first process the electron is transferred at a high level of energy but in the second, at low energy. This causes quite drastic differences in the range of these electron transfer processes if thin energy barriers are present between the donor and acceptor molecules. The energetic position of the valence band rather than the conduction band has to be considered if hole transfer processes are involved.

Fluorescence quenching due to electron transfer from excited cyanine dye XII to bipyridinium ion XIV, both arranged in mixed monolayers at the same hydrophilic interface, can be described by a hard disk model[133] by analogy to the hard sphere model in solution. According to this model, the relative fluorescence intensity I_A/I of the donor (I_A in the presence of the acceptor, and I in the absence) is given by

$$I_A/I = \exp(-\pi R_{DA}\sigma^2) \tag{5.5}$$

where σ is the acceptor density and R_{DA} is the critical radius for electron transfer in this two-dimensional system. In Figure 5.22, the ratio I_A/I is plotted versus the acceptor density. The experimental results (bars) demonstrate that I_A/I decreases exponentially with increasing acceptor density according to this model for a situation without donor–donor energy transfer (the mole fraction of the cyanine dye in the mixed monolayer is only $1/1000$). From the slope of the straight line in Figure 5.22 the critical radius for electron transfer under these particular conditions is calculated to be $R_{DA} = 1.32$ nm.

More interesting than the case of the statistical distribution of donor and acceptor molecules at the same interface is the situation where the planes of the cyanine chromophores and the acceptor π-electron systems are separated by one monolayer of well defined thickness. Such arrangements with cyanine dyes like XII as donors and the electron acceptor XIV require contacts between the hydrophilic

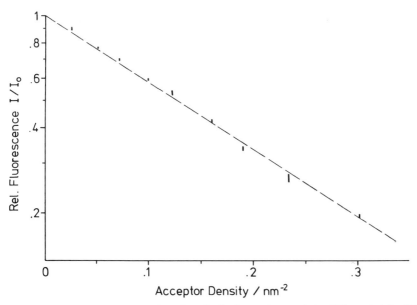

Figure 5.22. Relative fluorescence intensity of mixed monolayers of dye XII, methylylarchidate (MA), and arachidic acid (AA), with molar ratios XII : MA : AA = 1 : 100 : 900, in contact with mixed monolayers of the electron acceptor XIV, MA, and AA in various molar ratios vs the density of the electron acceptor. Bars correspond to experimental results, and the dotted line to a fit according to a hard disk model of electron transfer.

groups of one monolayer and the hydrophobic surface of the barrier monolayer. With similar assemblies, although prepared differently, rearrangement of the structure has been observed.[134-137] Indeed, complex monolayers containing appropriate dyes can be used to study the conditions under which undesired rearrangements can be avoided.[8,138]

Fluorescence quenching of the excited donor monolayer has been studied in such systems, in particular with cyanine dye I as donor.[49,139] The thickness of the barrier monolayer can be varied by using fatty acids with different chain lengths for the barrier monolayers. The rate constant k_{DA} of the electron transfer as determined from steady-state measurements decreases exponentially with increasing barrier thickness. This type of dependence is typical for a tunneling mechanism of photoinduced electron transfer in these organizates. This subject is also discussed in Chapter 7 when "tunneling" applications of LB films are considered.

In order to rule out the influence of possible defects in the layer structure, an additional energy acceptor E competing with the electron acceptor A has been incorporated in the layer system. The energy transfer from the donor to this acceptor E can be investigated in separate systems by measuring the relative donar fluorescence intensity I_E/I as a function of the density of E at constant distance between D and E. In the presence of this additional energy acceptor, the donor fluorescence intensity $I_{E,A}$ is related to the intensities I_E and I by the equation

$$(I_E/I_{E,A}) - 1 = k_{DA}(I_E/I) \tag{5.6}$$

However, this is true only when the rate of electron transfer is of the same order of magnitude as the rate of energy transfer to the competing energy acceptor. If the fluorescence quenching is due to contacts of some acceptor molecules with the donor layer, the much higher rate of contact electron transfer would not be affected by the additional energy transfer deactivation process. In this case, the ratio $I_{E,A}/I_E$ must be independent of I_E/I. In systems containing electron donor dye I, electron acceptor XIV and the competing energy acceptor dye IV, the ratio $I_{E,A}/I_E$ was found to depend on I_E/I according to equation (5.6), indicating that the electron transfer rate was observed in an intact layer system.[140]

In a similar way it has been shown that the spectral sensitization of the photographic process in silver bromide layers by dye molecules that are separated by two or more fatty acid spacer layers from the AgBr surface must be due to energy transfer.[124] This result seemed somewhat exotic to most scientists working in the field of spectral sensitization of semiconductors. Recently, conditions have been selected under which the energy transfer mechanism seems to be the only operating way of spectral sensitization.[141] Although unimportant in practical systems, where electron injection predominates, it has been a scientific challenge for nearly a century to evaluate the relative contributions of the two mechanisms. Due to the advanced monolayer technique, including steady-state fluorescence measurements of organized assemblies to provide the necessary information, this question can now be regarded as answered. These examples of investigations of complex processes have demonstrated the potential of energy transfer experiments. Via energy transfer, appropriate acceptor molecules can be used as probes for photophysical[142−149,51] or photochemical processes. Steady-state fluorescence measurements permit a study of such processes where transmission measurements could not be useful. An example is the detection of the formation and disappearance of a persistent form of the violen radical after photoinduced electron transfer from cyanine dye XII excited with UV radiation to bipyridinium sale XIV. The probe dye XV monolayer emits a fluorescence in the spectral range of the radical absorption band at 600 nm. This red fluorescence decreases when the radical is formed due to energy transfer, and recovers when the UV radiation that causes radical formation is turned off.[91] This fluorescence recovery indicates a decrease in radical density caused presumably by reaction with O_2. At room temperature, the absorption of the persistent radical can also be directly measured in a modified spectrometer of the type described in Section 5.1.[50] Other photoinduced electron transfer processes, like the sensitized polymerization of diacetylene derivatives in monolayers[150] or processes at interfaces involving organic crystals[151−153] or semiconductors and electron donors or acceptors, have been studied in monolayer systems by absorption and fluorescence measurements.[121,154−158] The reversible photochemical system of spiropyran X and merocyanine XI which emits a red fluorescence has been studied in a similar manner.[8] The signal obtained from the merocyanine was amplified using a monolayer of cyanine dye I as energy donor for the merocyanine formed by the photoreaction from the spiropyran. The stronger excitation of the system due to the larger absorption of the cyanine gave rise to an enhanced merocyanine fluorescence after energy transfer.

The skilled use of energy transfer experiments provides information on the path of deactivation of excited molecules. Evidence has been found for population of the triplet state of a cyanine dye, at least partly, via a higher vibronic level of the first excited singlet state, by using selective energy acceptors for the fluorescence and phosphorescence, respectively.[145]

The use of a nonfluorescent excited azo dye as a donor for energy transfer to a fluorescent acceptor has provided information on the probability of emission from the potentially luminescent state. Indeed, a sensitized fluorescence could be measured in this system.[143] Measurement of the excitation spectrum of the cyanine dye fluorescence proved the sensitization effect of the azo dye. The fluorescence excitation spectrum is particularly important here, since a small amount of a fluorescent impurity can lead to sensitized fluorescence of the acceptor dye. The coincidence of the fluorescence excitation spectrum with the azo dye absorption spectrum (in the spectral window of the cyanine dye) confirmed the assumption of energy transfer from the azo to the cyanine dye.

5.3.2. Fluorescence Decay

Steady-state fluorescence measurements of monolayer systems are relatively simple to perform and provide a large amount of structural and dynamic information. It must, however, be kept in mind that the monolayer organizates are micro-heterogeneous systems and therefore quite different from homogeneous systems such as solutions. A measurement of the fluorescence decay is required in many studies in order to obtain a complete description of the system.

Two principal techniques are used for the determination of fluorescence life-times.[159–164] The phase fluorimeter is based on the phase shift between the periodic sinusoidal excitation and the corresponding fluorescence signal. Evaluation of the lifetime with modern techniques is no longer based on the assumption of an exponential decay of the fluorescence. The other method uses a flash of short duration and follows the decay of the fluorescence intensity with time. This flash method can be used with a low excitation intensity and photon counting techniques. The time resolution has been extended into the picosecond range with mode-locked cavity dumped lasers of low intensity.[165,166] Alternatively, a high-intensity flash, such as the output of a mode-locked laser, is used in combination with a streak camera. The high intensity can give rise to undesired effects (nonlinear effects, local sample heating, etc.). Only a few measurements of the fluorescence decay function of monolayers at the gas–water interface[98,167–169] and monolayer systems[170] have been performed. The well-known cyanine dyes have lifetimes in the nanose-cond range and can be determined using a very stable flash lamp with deconvolution of the time distribution of the photons counted after a large number of flashes.[164]

An example of the use of fluorescence lifetime measurement is the investiga-tion of exciplex formation in monolayer organizates.[170] These complexes between a donor and an acceptor molecule, one of which is in the excited state, are well known in solutions.[171–174] The characteristic feature of exciplexes is their in-

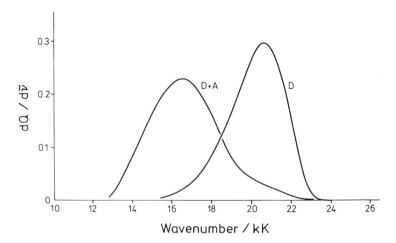

Figure 5.23. Evidence for exciplex formation in monolayer organizates. Fluorescence spectra of a mixed monolayer of dye I, methylarachidate (MA), and arachidic acid (AA), with molar ratios I : MA : AA = 1 : 2 : 18, obtained in the presence of electron acceptor XVI in the adjacent monolayer (D + A) and in the absence of XVI (D). The spectra are normalized to give $Q = 1$. Excitation of dye I at 405 nm; composition of the layer of A, XVI : AA = 1 : 5.

stability in the ground state, i.e., the complexes dissociate after deactivation. In many cases, an emission is observed which allows measurement of the decay function of the exciplex. The spectrum of the exciplex emission differs strongly from that of the excited donor. This is shown in Figure 5.23 for the emission of an exciplex formed in a monolayer assembly by cyanine dye I and acceptor XVI (curve D + A) as compared to the emission of the cyanine in the absence of acceptor XVI (curve D). The emission of the exciplex is shifted to longer wavelengths with respect to the emission of the cyanine dye. The position of the maximum of the exciplex emission depends on the donor, as demonstrated with a series of structurally similar cyanine dyes.[170] In addition, the position of the exciplex emission band can be influenced by variation of the microenvironment; the latter is performed in monolayer assemblies, e.g., by varying the matrix composition in which the acceptor and the donor are embedded.[170] This shows again the potential of using organized monolayers in systematic investigations of photophysical processes.

The fluorescence decay function measured with a nanosecond flash lamp using photon counting techniques at the maximum of the emission of a mixed monolayer of cyanine dye I is represented in Figure 5.24 by the dots around curve 3. The rise

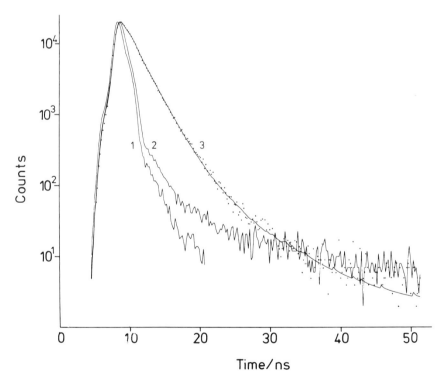

Figure 5.24. Fluorescence decay at 485 nm after excitation with a nanosecond flash. Curve 1: flash; curve 2: monolayer system of dye I and acceptor XVI at the same interface; curve 3: monolayer system without XVI, dots represent the experiment, and the line here is a reconvoluted decay. Monolayer composition: I : MA : AA = 1 : 2 : 18; XVI : AA = 1 : 5.

and decay of the flash is given by curve 1 in this diagram. A reconvoluted fluorescence function based on a decay with two exponentials is represented by the smooth line (curve 3) through the experimental points. It is plausible that the measured decay cannot be described by a single exponential function, since dimers of the cyanine dye should be present in addition to monomeric dye. The major component has a lifetime of $\tau_1 = 0.27$ ns and for the second component, $\tau_2 = 2.30$ ns. In the presence of acceptor XVI, the fluorescence of the cyanine dye is strongly quenched: this is immediately seen from curve 2 in Figure 5.24. Essentially, one lifetime of the excited cyanine dye of $\tau_A = 0.31$ ns is obtained, which is attributed to the quenched long-lived species of the donor fluorescence.

When the decay of the emission is followed at 565 nm, i.e., near the maximum of the exciplex emission, curve 2 in Figure 5.25 is obtained in the presence of acceptor XVI. A new long-lived emission is observed with a lifetime of $\tau_{EX} = 11.46$ ns. The fast component of the decay must be attributed to the emission of the cyanine dye at this wavelength. That the cyanine emission can be observed at this wavelength is shown by curve 3 in Figure 5.25, obtained in the absence of acceptor XVI. The lifetimes $\tau_1 = 0.22$ ns and $\tau_2 = 2.50$ ns observed for this emission are in

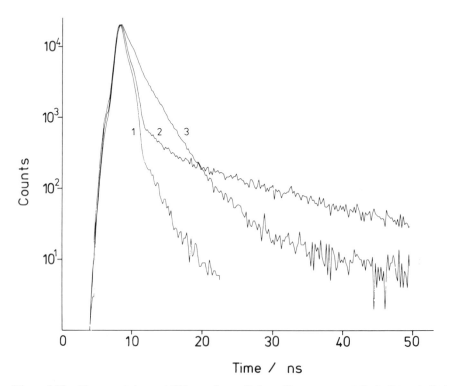

Figure 5.25. Fluorescent decay at 565 nm after excitation with a nanosecond flash. Curve 1: flash; curve 2: monolayer system of dye I and acceptor XVI; curve 3: monolayer system without XVI. The long-lived exciplex emission shows up in curve 2. Monolayers as in Figure 5.24.

agreement with those obtained from curve 3 in Figure 5.24 (decay at 485 nm), within experimental error. Curve 1 in Figure 5.25 represents the rise and decay of the nanosecond flash at 565 nm. The cyanine dye was excited at 405 nm in the measurements. This example of an investigation on formation and decay of an exciplex in monolayer organizates demonstrates the potential of time-resolved fluorescence experiments. It would also be interesting to extend these studies to record emission spectra, at defined delay after excitation.

5.3.3. Phosphorescence

Phosphorescence is emitted by deactivation of a triplet state to the ground state. Since this requires a change of the electron spin, this transition is forbidden. The lifetime of the phosphorescence can be many orders of magnitude larger than the fluorescence lifetimes. The phosphorescence spectrum is displaced to larger wavelengths with respect to the fluorescence as a consequence of the lower energy of the triplet state as compared with the excited singlet state.

The measurement of phosphorescence is normally done by blocking the fast fluorescence either mechanically (by a chopper) or electronically (by setting a window). The complexity of the instrumentation is similar to that required for fluorescence. The measurement of the decay function is similar to that of the fluorescence and involves excitation with a short pulse and then following the time dependence of the phosphorescence intensity with a photomultiplier or by photon counting techniques.

The investigation of the pathway of population of the triplet state mentioned in Section 5.3.2 requires the measurement of phosphorescence lifetime, which is shortened due to energy transfer from the triplet state to an appropriate energy acceptor.[145] This energy transfer leads to a single–singlet transition of the acceptor and follows the Förster relationship. It is therefore different from the well-known triplet–triplet energy transfer in solution based on an exchange mechanism with a different distance dependence.

The decay of the phosphorescence is influenced by a dielectric mirror in the remote environment of the molecule in the triplet state.[125] This seems surprising at first glance if energy transfer can be ruled out, since only the immediate environment normally affects the deactivation. The dependency of the lifetime on the distance of the monolayer of phosphorescent europium complex XIII from the interface is an oscillating function (see Figure 5.26), and this behavior has been rationalized using a classical picture.[130] These investigations are an illustrative example of the use of complex monolayers to study the nature of excited states and fundamental photophysical processes.

5.4. INDIRECT OPTICAL METHODS

Several optical techniques to characterize LB films are discussed in Chapter 4. Valuable results concerning monolayer organizates can be achieved using, for ex-

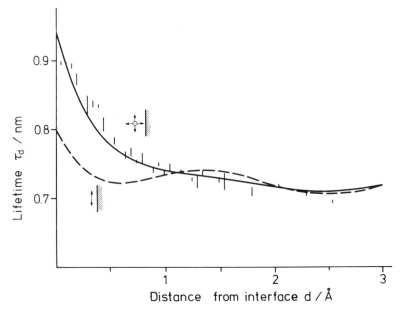

Figure 5.26. Variation in the lifetime of europium complex XIII with distance from the interface monolayer system (air). Bars denote experimental results; full curve calculated with isotropic distribution of emission dipoles, dashed curve calculated for statistical distribution of dipoles in monolayer plane.

ample, secondary ion mass spectrometry,[175] X-ray photoelectron spectroscopy,[176,177] and electron paramagnetic resonance.[178,179] ESR probes can be used as in bilayer membranes[180] to evaluate the structure and dynamics of thin film systems. To end this chapter we shall mention three techniques recently applied, or about to be applied, to complex dye films.

5.4.1. Holographic Techniques

The products of a photochemical reaction in monolayers may be detected by measuring the change in transmission of the sample (absorption change, either persistent or transient) as described in Section 5.2. The disadvantage of the transmission method is the measurement of a small difference between two large signals. The holographic method of following a chemical reaction[181–185] is essentially a zero background method. Although there is not yet any report on a holographic study of complex monolayers in the literature, the method seems to be adequate and powerful. The main features are shown in Figure 5.27.

The radiation from a laser which causes the chemical change is split into two beams, a and b, which form an interference pattern on the sample. The product is formed according to this spatial distribution. The result is a periodic change in optical density or refractive index of the sample which can be used as a diffraction

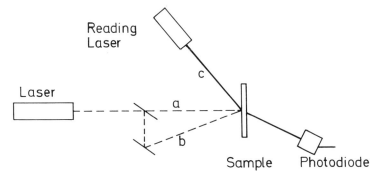

Figure 5.27. Arrangement for measuring photochemical changes in a thin film by a holographic technique. A grating is built photochemically by interference of coherent beams a and b, and the light from a reading laser c is deflected accordingly.

grating for a detection laser beam c. The part of this beam diffracted by the sample is measured with a photodiode or a photomultiplier tube. The undiffracted part of the detection laser beam is not seen by the photodiode. Therefore, the holographic technique is a zero background method.

The growth of the photochemical product with time has been studied in solid solutions in this way and the time and intensity dependencies have been interpreted.[181] A transient technique can also be applied by using a pulsed light source. The time dependence of the holographic grating efficiency yields information on the formation and decay of photoproducts.

5.4.2. Photoacoustic Spectroscopy

The measurement of photoacoustic spectra (PAS) is based on the transformation of light energy into heat due to radiationless deactivation processes like thermal equilibration and internal conversion. The sample is enclosed in a small cell, and the ambient gas is periodically heated when the sample is periodically illuminated. This causes a periodic change in gas pressure, which is detected by a sensitive microphone. The effect was discovered more than 100 years ago,[186] but has been developed extensively in the past 10 years.[187−189]

The PAS technique provides a means for bookkeeping of the energy in the cycle of excitation and deactivation. This has been nicely demonstrated in a study of the decay of optically excited surface plasmons at the silver–air interfaces (not involving monolayers).[190] It will be shown that this is not at all trivial, although the determination of absolute fluorescence quantum yields should seem sufficient to account for the absorbed energy. The situation, however, is different if chemical processes are involved.

The photoinduced electron transfer from cyanine dyes to the bipyridinium salt XIV in monolayer organizates has been studied to a large degree by steady-state fluorescence quenching measurements. In particular systems the formation of the

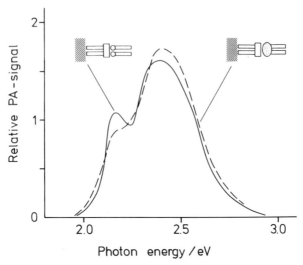

Figure 5.28. Relative photoacoustic signal of monolayer systems with dye IV in the absence (full line) and presence (dashed line) of electron acceptor XIV in the adjacent monolayer. Although the fluorescence of dye IV is strongly quenched by XIV, the photoacoustic signal is not increased in the presence of XIV due to chemical storage of energy absorbed by dye IV.

reduced acceptor, the violen radical, has been observed and followed by transmission measurements.[50] Systems containing cyanine dye IV as a donor and acceptor XIV have also been investigated by measuring the photoacoustic spectra.[191] Figure 5.28, taken from this study, shows the photoacoustic spectra of a system with the electron acceptor (dashed line) and of a similar system without the acceptor (full line). The surprising result is that the spectra differ only slightly, although the cyanine fluorescence is practically completely quenched in the system with the electron acceptor. In this type of electron transfer reactions, it is generally assumed that the transferred electron returns rapidly into the vacancy in the ground state level of the photooxidized donor. This cannot be true in the system described in Figure 5.28 for, in such a process, the light energy absorbed initially would have been entirely transformed into heat. The observed effect has been explained in terms of energy storage in products formed in secondary reactions. The account of heat-producing deactivation processes using photoacoustic spectroscopy seems to become particularly important in the investigation of photochemical processes in organized systems.

5.4.3. Coupling to Surface Plasmons

It was mentioned in Section 4.4.1 that surface plasmons can exist at the surface of a metal or semiconductor with electrons behaving like an electron gas. The surface plasmons are produced by excitation of oscillations of surface charges by an external electric field, and their electromagnetic field is concentrated around the boundary.[192–194] Therefore, surface plasmons are a sensitive probe for changes in the immediate environment of the metal surface. The reflection of light from the metal surface in an appropriate arrangement is strongly diminished due to the excitation of surface plasmons. The angular position of these resonances is changed when a monolayer or monolayer assemblies are deposited on the metal surface; see Figure 4.16. Calculated shifts of the reflection minima based on the thickness of the multilayer systems and the refractive index of these layers are in excellent agreement with the observed shifts.

The interactions of dye-containing monolayers with surface plasmons are very interesting. The complex monolayers on top of the metal cause shifts in the surface plasmons resonances.[147,195,196] The optical properties of the dye-containing monolayer can be derived from a theoretical fit to the experimental resonances and halfwidths of the attenuated reflection curves.

A different type of interaction is the transfer of excitation energy from the dye monolayer to the metal.[130,197] This effect has been investigated with monolayers of a J-aggregate on silver and gold films.[51,148,149,198] The surface plasmons can be excited via energy transfer from excited dye molecules in competition with energy transfer that causes the formation of rapidly decaying electron–hole pairs in the metal. Therefore, the dye-sensitized surface plasmon emission has maximum efficiency when the dye monolayer is not in contact with the metal.[149]

Delocalized excited states at the surface of organic crystals can cause phenomena similar to those observed with surface plasmons. Exciton surface polaritons have been investigated by the technique of attenuated total reflection of a organic dye crystal[199] and by measuring high-resolution, low-temperature reflectivity, fluorescence, and fluorescence excitation spectra of anthracene crystals.[200,201] New interesting ways of using the interactions between dye molecules at surfaces and surface plasmons or exciton surface polaritons can be expected in the near future.

REFERENCES

1. H. Kuhn, *Pure Appl. Chem.*, **11**, 345 (1965).
2. K. H. Drexhage and H. Kuhn, in: *Basic Problems in Thin Film Physics* (R. Niedermayer and H. Mayer, eds.), p. 339, Vandenhoeck & Ruprecht, Göttingen (1966).
3. H. Bücher, K. H. Drexhage, M. Fleck, H. Kuhn, D. Möbius, F. P. Schäfer, J. Sondermann, W. Sperling, P. Tillmann, and J. Wiegand, *Mol. Cryst.*, **2**, 199 (1967).
4. H. Kuhn, *Pure Appl. Chem.*, **53**, 2105 (1981).
5. H. Kuhn, *Thin Solid Films*, **99**, 1 (1983).
6. O. Inacker, H. Kuhn, D. Möbius, and G. Debuch, *Z. Phys. Chem.*, *N.F.*, **101**, 337 (1976).
7. D. Möbius and G. Debuch, *Ber. Bunsenges. Phys. Chem.*, **80**, 1180 (1976).
8. D. Möbius, in: *Colloids and Surfaces in Reprographic Technology* (M. Hair and M. D. Croucher, eds.), ACS Symp. Ser., No. 200, p. 93 (1982).
9. G. L. Gaines, Jr., *Insoluble Monolayers at Liquid-Gas Interfaces*, Interscience, New York, 1965.
10. H. Kuhn, D. Möbius, and H. Bücher, in: *Physical Methods of Chemistry* (A. Weissberger and B. Rossiter, eds.), Vol. I, Pt. 3B, p. 577, Wiley, New York (1972).
11. J. Heesemann and P. Zingsheim, in: *Membrane Spectroscopy* (E. Grell, ed.), p. 172, Springer-Verlag, Berlin (1981).

12. W. E. Barlow (ed.), *Thin Solid Films,* **68** (1980), Special issue on Langmuir–Blodgett films.
13. G. G. Roberts and C. W. Pitt (eds.), *Thin Solid Films,* **99** (1983), Papers presented at the First International Conference on Langmuir–Blodgett Films.
14. R. M. A. Azzam and N. M. Bashara, *Ellipsometry and Polarized Light,* North-Holland, Amsterdam (1977).
15. R. Steiger, *Helv. Chim. Acta,* **54,** 2645 (1971).
16. D. den Engelsen, *J. Opt. Soc. Am.,* **61,** 1460 (1971).
17. E. P. Honig, J. H. Th. Hengst, and D. den Engelsen, *J. Colloid Interface Sci.,* **45,** 92 (1973).
18. D. den Engelsen and B. de Koning, *J. Chem. Soc., Faraday Trans. 1,* **70,** 2100 (1974).
19. D. den Engelsen, *J. Phys. Chem.,* **76,** 3390 (1972).
20. D. Möbius, M. Orrit, H. Grüniger, and H. Meyer, *Thin Solid Films,* **132,** 41 (1985).
21. M. Orrit, D. Möbius, U. Lehmann, and H. Meyer, *J. Chem. Phys.,* **85,** 5966 (1986).
22. V. M. Agranovich, *Sov. Phys. Usp.,* **17,** 103 (1974).
23. O. S. Heavens, *Optical Properties of Thin Films,* Dover, New York (1965).
24. M. Born and E. Wolf, *Principles of Optics,* 4th ed., pp. 51–70, Pergamon Press, Oxford (1970).
25. M. Vandevyver, A. Barraud, A. Ruaudel-Teixier, P. Maillard, and Ch. Gianotti, *J. Colloid Interface Sci.,* **85,** 571 (1982).
26. Ref. 10, p. 676.
27. Ref. 10, p. 678.
28. K. H. Drexhage, in: *Progress in Optics* (E. Wolf, ed.), Vol. XII, p. 165, North-Holland, Amsterdam (1974).
29. V. Czikkely, G. Dreizler, H. D. Försterling, H. Kuhn, J. Sondermann, P. Tillmann, and J. Wiegand, *Z. Naturforsch.,* **24a,** 1921 (1969).
30. J. Heesemann, Ph.D. thesis, University of Göttingen (1976).
31. Ref. 10, p. 625, Table 7.3 contains an error: the last line referring to an orientation of the chromophores of ½ statistical in layer plane, ½ normal to layer plane has a factor of ⅓ in all columns. The correct factor is ¾. The influence of the reflection was not taken into account correctly in this review.
32. J. Breton, M. Michel-Villaz, G. Paillotin, and M. Vandevyver, *Thin Solid Films,* **13,** 351 (1972).
33. K. Matsuki, Y. Nagahira, and H. Fukutome, *Bull. Chem. Soc. Jpn.,* **53,** 1817 (1980).
34. D. Möbius (unpublished).
35. J. Heesemann, *J. Am. Chem. Soc.,* **102,** 2167 (1980).
36. H. Grüniger, D. Möbius, and H. Meyer, *J. Chem. Phys.,* **79,** 3701 (1983).
37. A. G. Tweet, *Rev. Sci. Instrum.,* **34,** 1412 (1963).
38. A. G. Tweet, General Electric Res. Labs. Report No. 63-RL-3434G (1963).
39. V. Vogel and D. Möbius, *Thin Solid Films,* **132,** 205 (1985).
40. P. Fromherz, *Z. Naturforsch.,* **28c,** 144 (1972).
41. D. den Engelsen, *J. Colloid Interface Sci.,* **45,** 1 (1973).
42. D. Möbius and H. Grüniger, *Bioelectrochem. Bioenerg.,* **12,** 375 (1984).
43. R. H. Schmehl, G. L. Shaw, and D. G. Whitten, *Chem. Phys. Lett.,* **58,** 549 (1978).
44. D. Möbius, *Mol. Cryst. Liq. Cryst.,* **96,** 319 (1983).
45. D. Möbius and M. Orrit, in: *Chemical Reactions in Organic and Inorganic Constrained Systems* (R. Setton, ed.), p. 315, Reidel, Dordrecht (1986).
46. U. Lehmann, *Thin Solid Films,* **160,** 257 (1988).
47. H. Bücher and H. Kuhn, *Chem. Phys. Lett.,* **6,** 183 (1970).
48. H. Bücher and H. Kuhn, *Z. Natürforsch.,* **26b,** 1323 (1970).
49. D. Möbius, *Acc. Chem. Res.,* **14,** 63 (1981).
50. T. L. Penner and D. Möbius, *J. Am. Chem. Soc.,* **104,** 7404 (1982).
51. I. Pockrand, A. Brillante and D. Möbius, *J. Chem. Phys.,* **77,** 6289 (1982).
52. D. Möbius, *J. Phys. (Paris),* **44,** C10-441 (1983).
53. M. Orrit, J. Ackermann, and D. Möbius (in preparation).
54. D. Möbius, in: *Photochemical Transformations in Non-Homogeneous Media* (M-A. Fox, ed.), ACS Symp. Ser. No. 278, Washington, D.C. (1985), p. 113.

55. D. Möbius and H. Grüniger, in: *Charge and Field Effects in Biosystems* (M. J. Allen and P. N. R. Usherwood, eds.), p. 265, Abacus Press, Tunbridge Wells (1984).
56. M. Suzuki, D. Möbius, and R. C. Ahuja, *Thin Solid Films,* **138,** 151 (1986).
57. N. J. Harrick, *Internal Reflection Spectroscopy,* Interscience, New York (1967).
58. W. Knoll, M. R. Philpott, and W. G. Golden, *J. Chem. Phys.,* **77,** 219 (1982).
59. J. F. Rabolt, F. C. Burns, N. Schlotter, and J. D. Swalen, *J. Chem. Phys.,* **78,** 946 (1983).
60. M. R. Philpott, A. Girlando, W. G. Golden, W. Knoll, and J. D. Swalen, *Mol. Cryst. Liq. Cryst.,* **96,** 335 (1983).
61. R. G. Greenler, *J. Chem. Phys.,* **44,** 310 (1966).
62. R. J. Rabolt, M. Jurich, and J. D. Swalen, *Appl. Spectrosc.,* **39,** 269 (1985).
63. J. P. Rabe, J. F. Rabolt, C. A. Brown, and J. D. Swalen, *Thin Solid Films,* **133,** 153 (1985).
64. P. Fromherz, J. Peters, H. G. Müldner, and W. Otting, *Biochim. Biophys. Aeta,* **372,** 644 (1972).
65. P-A. Chollet, *Thin Solid Films,* **52,** 343 (1978).
66. H. Nakahara and K. Fukuda, *Thin Solid Films,* **99,** 45 (1983).
67. H. Nakahara and K. Fukuda, *J. Colloid Interface Sci.,* **93,** 530 (1983).
68. F. Kimura, J. Umemura, and T. Takenaka, *Langmuir,* **2,** 96 (1986).
69. C. Naselli, J. F. Rabolt, and J. D. Swalen, *J. Chem. Phys.,* **82,** 2136 (1985).
70. D. D. Saperstein, *J. Phys. Chem.,* **90,** 1408 (1986).
71. J. D. Swalen and J. F. Rabolt, *FTIR Spectroscopy,* **4,** 283 (1985).
72. A. Barraud, A. Ruaudel-Teixier, M. Vandevyver, and P. Lesieur, *Nouv. J. Chim.,* **9,** 365 (1985).
73. J. Umemura, T. Kawai, T. Takenaka, M. Kodama, Y. Ogawa, and S. Seki, *Mol. Cryst. Liq. Cryst.,* **112,** 293 (1984).
74. E. Okamura, J. Umemura, and T. Takenaka, *Biochim. Biophys. Acta,* **81,** 139 (1985).
75. D. D. Allara and R. G. Nuzzo, *Langmuir,* **1,** 52 (1985).
76. K. Fukuda, Y. Shibasaki, and H. Nakahara, *Thin Solid Films,* **99,** 87 (1983).
77. L. Netzer and J. Sagiv, *J. Am. Chem. Soc.,* **105,** 674 (1983).
78. J. Gun, R. Isovici, and J. Sagiv, *J. Colloid Interface. Sci.,* **101,** 210 (1984).
79. W. C. Bigelow, D. L. Pickett, and W. A. Zisman, *J. Colloid Sci.,* **1,** 513 (1946).
80. D. L. Allara and R. G. Nuzzo, *Langmuir,* **1,** 45 (1985).
81. R. W. Murray, *Acc. Chem. Res.,* **13,** 135 (1980).
82. E. E. Polymeropoulos and J. Sagiv, *J. Chem. Phys.,* **69,** 1836 (1978).
83. L. Netzer, R. Iscovici, and J. Sagiv, *Thin Solid Films,* **99,** 235 (1983).
84. J. Yguerabide and M. C. Foster, in: *Membrane Spectroscopy* (E. Grell, ed.), p. 199, Springer-Verlag, Berlin (1981).
85. A. S. Waggoner and L. Stryer, *Proc. Natl. Acad. Sci. U.S.A.,* **67,** 579 (1970).
86. K. A. Zachariasse, B. Kozankiewicz, and W. Kühnle, in: *Photochemistry and Photobiology* (A. H. Zewail, ed.), Vol. 2, p. 941, Harwood, London (1983).
87. D. Möbius and H. Kuhn, *Isr. J. Chem.,* **18,** 375 (1979).
88. A. G. Tweet, G. L. Gaines, and W. D. Bellamy, *J. Chem. Phys.,* **40,** 2596 (1964).
89. D. A. Cadenhead, F. Müller-Landau, and B. M. J. Kellner, in: *Ordering in Two Dimensions* (S. K. Sinha, ed.), p. 73, Elsevier (North-Holland), New York (1980).
90. O. Gonen, H. Levanon, and L. K. Patterson, *Isr. J. Chem.,* **21,** 271 (1981).
91. D. Möbius, *Ber. Bunsenges. Phys. Chem.,* **82,** 848 (1978).
92. D. Möbius, in: *Topics in Surface Chemistry* (E. Kay and P. S. Bagus, eds.), p. 75, Plenum, New York (1978).
93. H. Heithier and H. Möhwald, *Z. Naturforsch.,* **38c,** 1003 (1983).
94. B. Kee, in: *The Chlorophylls* (L. P. Vernon and G. R. Seely, eds.), p. 253, Academic Press, New York (1966).
95. S. M. de B. Costa, J. R. Froins, J. M. Harris, R. M. Leblanc, B. H. Oger, and G. Porter, *Proc. R. Soc. London, Ser. A,* **326,** 503 (1972).
96. H. Heithier, K. Ballschmiter, and H. Möhwald, *Photochem. Photobiol.,* **37,** 201 (1983).
97. R. E. Hirsch and S. S. Brody, *Photochem. Photobiol.,* **29,** 589 (1979).
98. M. L. Agrawal, J-P. Chauvet, and L. K. Patterson, *J. Phys. Chem.,* **89,** 2979 (1985).

99. S. Vaidyanathan, L. K. Patterson, D. Möbius, and H. Grüniger, *J. Phys. Chem.*, **89**, 491 (1985).
100. J. Teissié, J. F. Tocanne, and A. Baudras, *FEBS Lett.*, **70**, 123 (1976).
101. D. A. Cadenhead, B. M. J. Kellner, K. Jacobson, and D. Papahadjopoulos, *Biochemistry*, **16**, 5386 (1977).
102. J. Teissié, J. F. Tocanne, and A. Baudras, *Eur. J. Biochem.*, **83**, 77 (1978).
103. J. Teissié, *Chem. Phys. Lipids*, **25**, 357 (1979).
104. J. Teissié, *J. Colloid Interface Sci.*, **70**, 90 (1979).
105. M. Lösche, E. Sackmann, and H. Möhwald, *Ber. Bunsenges. Phys. Chem.*, **87**, 848 (1983).
106. L. B. Å. Johansson and G. Lindblom, *Quart. Rev. Biophys.*, **13**, 1 (1980).
107. M. Lösche, J. Rabe, A. Fischer, B. U. Rucha, W. Knoll, and H. Möhwald, *Thin Solid Films*, **117**, 269 (1984).
108. M. Lösche and H. Möhwald, *Eur. Biophys. J.*, **11**, 35 (1984).
109. M. Lösche and H. Möhwald, *Rev. Sci. Instrum.*, **55**, 1968 (1984).
110. R. Peters and K. Beck, *Proc. Natl. Acad. Sci. U.S.A.*, **80**, 7183 (1983).
111. M. Lösche and H. Möhwald, *Colloids and Surfaces*, **10**, 217 (1984).
112. A. Fischer, M. Lösche, H. Möhwald, and E. Sackmann, *J. Phys. (Paris) Lett.*, **45**, L785 (1984).
113. H. M. McConnell, L. K. Tamm, and R. M. Weiss, *Proc. Natl. Acad. Sci. U.S.A.*, **81**, 3249 (1984).
114. W. M. Heckl, M. Lösche, and H. Möhwald, *Thin Solid Films*, **133**, 73 (1985).
115. A. Miller, W. Knoll, and H. Möhwald, *Phys. Rev. Lett.*, **56**, 2633 (1986).
116. A. Miller and H. Möhwald, *Europhys. Lett.* **2**, 67 (1986).
117. D. Axelrod, D. E. Koppel, J. Schlessinger, E. Elson, and W. W. Webb, *Biophys. J.*, **16**, 1055 (1976).
118. Th. Loughran, M. D. Hatlee, L. K. Patterson, and J. J. Kozak, *J. Chem. Phys.*, **72**, 5791 (1980).
119. H-J. Galla, W. Hartmann, U. Theilen, and E. Sackmann, *J. Membrane Biol.*, **48**, 215 (1979).
120. D. Vollhardt, L. Zastrow, and P. Schwarz, *Colloid Polym. Sci.*, **258**, 1176 (1980).
121. P. Fromherz and W. Arden, *J. Am. Chem. Soc.*, **102**, 6211 (1980).
122. H. Bücher, H. Kuhn, B. Mann, D. Möbius, L. v. Szentpály, and P. Tillman, *Photogr. Sci. Eng.*, **11**, 233 (1967).
123. L. v. Szentpály, D. Möbius, and H. Kuhn, *J. Chem. Phys.*, **52**, 4618 (1970).
124. R. Steiger, H. Hediger, P. Junod, H. Kuhn, and D. Möbius, *Photogr. Sci. Eng.*, **24**, 185 (1980).
125. K. H. Tews, O. Inacker, and H. Kuhn, *Nature*, **228**, 276 (1970).
126. K. H. Drexhage and M. Fleck, *Ber. Bunsenges. Phys. Chem.*, **72**, 330 (1968).
127. K. H. Drexhage, M. Fleck, H. Kuhn, F. P. Schäfer, and W. Sperling, *Ber. Bunsenges. Phys. Chem.*, **70**, 1179 (1966).
128. K. H. Drexhage, M. Fleck, and H. Kuhn, *Ber. Bunsenges. Phys. Chem.*, **71**, 915 (1967).
129. H. Kuhn, *Naturwissenschaften*, **54**, 429 (1967).
130. H. Kuhn, *J. Chem. Phys.*, **53**, 101 (1970).
131. R. R. Chance, A. Prock, and R. Silbey, in: *Advances in Chemical Physics* (I. Prigogine and S. A. Rice, eds.), p. 1, Vol. 37, Wiley, New York (1978).
132. R. E. Kunz and W. Lukosz, *Phys. Rev. B.*, **21**, 4814 (1980).
133. D. Möbius, R. C. Ahuja, and G. Debuch (in preparation).
134. I. Langmuir, *Science*, **87**, 493 (1938).
135. R. C. Ehlert, *J. Colloid Sci.*, **20**, 387 (1965).
136. E. P. Honig, *J. Colloid Interface Sci.*, **43**, 66 (1973).
137. K. Fukuda and T. Shiozawa, *Thin Solid Films*, **68**, 55 (1980).
138. Ref. 10, p. 599.
139. H. Kuhn, *Pure Appl. Chem.*, **51**, 341 (1979).
140. D. Möbius and G. Debuch (unpublished).
141. R. Steiger, *Photogr. Sci. Eng.*, **28**, 35 (1984).
142. H. Kuhn and D. Möbius, *Angew. Chem., Int. Ed. Engl.*, **10**, 620 (1971).
143. D. Möbius and G. Dreizler, *Photochem. Photobiol.*, **17**, 225 (1973).
144. O. Inacker, H. Kuhn, H. Bücher, H. Meyer, and K. H. Tews, *Chem. Phys. Lett.*, **7**, 213 (1970).

145. O. Inacker and H. Kuhn, *Chem. Phys. Lett.*, **27**, 317, 471 (1974).
146. D. Möbius and G. Debuch, *Chem. Phys. Lett.*, **28**, 17 (1974).
147. G. Wähling, D. Möbius, and H. Raether, *Z. Naturforsch.*, **33a**, 907 (1978).
148. I. Pockrand, A. Brillante, and D. Möbius, *Chem. Phys. Lett.*, **69**, 499 (1980).
149. I. Pockrand, A. Brillante, and D. Möbius, *Nuovo Cim.*, **63B**, 350 (1981).
150. C. Bubeck, B. Tieke, and G. Wegner, *Ber. Bunsenges. Phys. Chem.*, **86**, 499 (1982).
151. G. Vaubel, H. Baessler, and D. Möbius, *Chem. Phys. Lett.*, **10**, 334 (1971).
152. H. Killesreiter, *Ber. Bunsenges. Phys. Chem.*, **82**, 503, 512 (1978).
153. H. Killesreiter, *Z. Naturforsch.*, **34a**, 737 (1979).
154. P. Fromherz and W. Arden, *Ber. Bunsenges. Phys. Chem.*, **84**, 1045 (1980).
155. K. Chandrasekaran, Ch. Giannotti, K. Monserrat, J. P. Otruba, and D. G. Whitten, *J. Am. Chem. Soc.*, **104**, 6200 (1982).
156. A. F. Janzen and J. R. Bolton, *J. Am. Chem. Soc.*, **101**, 6342 (1979).
157. T. Miyasaka, T. Watanabe, A. Fujishima, and K. Honda, *Nature*, **277**, 638 (1979).
158. R. Memming and F. Schröppel, *Chem. Phys. Lett.*, **62**, 207 (1979).
159. W. R. Ware, in: *Creation and Detection of the Excited State* (A. A. Lamola, ed.), Vol. 1, Part A, p. 213, Dekker, New York (1971).
160. J. F. Rabek, *Experimental Methods in Photochemistry and Photophysics*, Wiley, New York (1982).
161. J. N. Demas, *Excited State Lifetime Measurements*, Academic Press, New York (1983).
162. R. B. Cundall and R. E. Dale (eds.), *Time-Resolved Fluorescence Spectroscopy in Biochemistry and Biology*, Plenum Press, New York (1983).
163. T. A. M. Doust and M. A. West (eds.), *Picosecond Chemistry and Biology*, Science Rev. Ltd., Northwood (1984).
164. M. Bouchy, *Deconvolution and Reconvolution of Analytical Signals*, E.N.S.I.C.–I.N.P.L., Nancy (1982).
165. E. P. Ippen and C. V. Shank, *Appl. Phys. Lett.*, **27**, 488 (1975).
166. W. Haehnel, A. R. Holzwarth, and J. Wendler, *Photochem. Photobiol.*, **37**, 435 (1983).
167. J-P. Chauvet, M. L. Agrawal, and L. K. Patterson, *Thin Solid Films*, **133**, 227 (1985).
168. P. K. J. Kinnunen, J. A. Virtanen, A. P. Tulkki, R. C. Ahuja, and D. Mobius, *Thin Solid Films*, **132**, 193 (1985).
169. A. Leitner, M. E. Lippitsch, S. Draxler, M. Riegler, and F. R. Aussenegg, *Thin Solid Films*, **132**, 55 (1985).
170. R. C. Ahuja and D. Möbius (unpublished).
171. H. Leonhardt and A. Weller, *Ber. Bunsenges. Phys. Chem.*, **67**, 791 (1963).
172. M. Gordon and W. Ware (eds.), *The Exciplex*, Academic Press, New York (1975).
173. F. D. Lewis, *Acc. Chem. Res.*, **12**, 152 (1979).
174. A. Weller, *Z. Phys. Chem., N.F.*, **133**, 93 (1982).
175. L. Laxhuber, H. Möhwald, and M. Hashmi, *Int. J. Mass Spectrom. Ion Phys.*, **51**, 93 (1983).
176. H. R. Anderson and J. D. Swalen, *J. Adhes.*, **9**, 197 (1978).
177. F. C. Burns and J. D. Swalen, *J. Phys. Chem.*, **86**, 5123 (1982).
178. J. Cunningham, E. E. Polymeropoulos, D. Möbius, and F. Baer, in: *Magnetic Resonance in Colloid and Interface Science* (J. P. Fraissard and H. A. Resing, eds.), p. 603, Reidel, Amsterdam (1980).
179. S. Kuroda, M. Sugi, and S. Iizima, *Thin Solid Films*, **133**, 189 (1985).
180. D. Marsh, in: *Spectroscopy and Dynamics of Molecular Biological Systems* (P. M. Bayley and R. E. Dale, eds.), p. 209, Academic Press, London (1985).
181. G. C. Bjorklund, D. M. Burland, and D. C. Alvarez, *J. Chem. Phys.*, **73**, 4321 (1980).
182. G. C. Bjorklund, Chr. Bräuchle, D. M. Burland, and D. C. Alvarez, *Opt. Lett.*, **6**, 159 (1981).
183. Chr. Bräuchle, D. M. Burland, and G. C. Bjorklund, *J. Am. Chem. Soc.*, **103**, 2515 (1981).
184. Chr. Bräuchle, C. M. Burland, and G. C. Bjorklund, *J. Phys. Chem.*, **85**, 123 (1981).
185. D. M. Burland, *Acc. Chem. Res.*, **16**, 218 (1983).
186. A. G. Bell, *Philos. Mag.*, **11**, 510 (1881).

187. A. Rosencwaig, *Science,* **181,** 657 (1973).
188. J. Badoz and D. Fournier (eds.), *Photoacoustic and Photothermal Spectroscopy,* in: *J. Phys. (Paris),* **44,** C6 (1983).
189. G. A. West, J. J. Barrett, D. R. Siebert, and K. V. Reddy, *Rev. Sci. Instrum.,* **54,** 797 (1983).
190. B. Rothenhäusler, J. Rabe, P. Korpiun, and W. Knoll, *Surf. Sci.,* **137,** 373 (1984).
191. H. Hediger and R. Steiger, *J. Colloid Interface Sci.,* **103,** 343 (1985).
192. H. Raether, *Phys. Thin Films,* **9,** 145 (1977).
193. I. Pockrand, J. D. Swalen, J. G. Gordon II, and M. R. Philpott, *Surf. Sci.,* **74,** 237 (1977).
194. G. Wähling, *Z. Naturforsch.,* **33a,** 536 (1978).
195. I. Pockrand, J. D. Swalen, R. Santo, A. Brillante, and M. R. Philpott, *J. Chem. Phys.,* **69,** 4001 (1978).
196. G. Wähling, *Z. Naturforsch.,* **36a,** 588 (1981).
197. W. H. Weber and C. F. Eagen, *Opt. Lett.,* **4,** 236 (1979).
198. W. Knoll, M. R. Philpott, J. D. Swalen, and A. Girlando, *J. Chem. Phys.,* **75,** 4795 (1981).
199. I. Pockrand, A. Brillante, M. R. Philpott, and J. D. Swalen, *Opt. Commun.,* **27,** 91 (1978).
200. M. Orrit, J. Bernard, J. M. Turlet, and Ph. Kottis, *J. Chem. Phys.,* **78,** 2847 (1983).
201. J. Bernard, M. Orrit, J. M. Turlet, and Ph. Kottis, *J. Chem. Phys.,* **78,** 2857 (1983).

Monolayers and Multilayers of Biomolecules

R. M. Swart

6.1. INTRODUCTION

Langmuir–Blodgett films are used for the study of ordered arrays of molecules, usually capable of forming insoluble monolayers at an air–water interface. The structure in biology which would appear to be most amenable to study by the Langmuir–Blodgett technique is the membrane. Our understanding of its composition, structure, and dynamics has to account for a wide range of its functions *in vivo*. As barriers, they are able to maintain large concentration gradients, both ionic and nonionic. Yet the proteins found embedded in the membrane are capable of dissipating and regenerating these gradients, sometimes very rapidly. Membranes are sites of contact and recognition, be it of other cells or of chemical signals and they can adopt a variety of forms. For example they can pit or fragments can "bud" off from the main structure and they can fuse with other membranes or stack as in the thylakoid membranes of plant chloroplasts (Figure 6.1). In this last array, they provide the environment for the capture of light energy to drive photosynthesis.

To carry out all these functions, an "average" membrane consists of a bilayer composed of a variety of amphiphilic molecules, mainly phospholipids and sterols (Figure 6.2). The distribution of these compounds need not be homogeneous within the plane of the membrane or symmetrical across it. Proteins are found attached to the surface of the bilayer from where they can be removed simply by a change in pH or ionic concentration. They can also be buried deeply into or across the bilayer. In

R. M. Swart • Corporate Colloid Science Laboratory, Imperial Chemical Industries plc, Runcorn, Cheshire WA7 4QE, England.

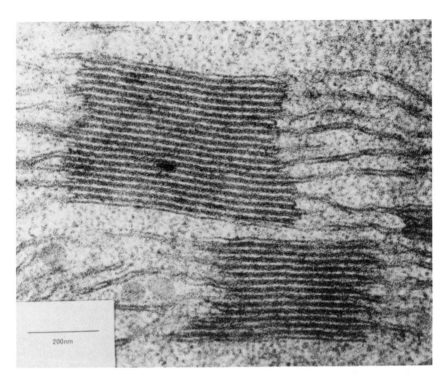

Figure 6.1. Transmission electron micrograph of a thylakoid stack found in a higher plant chloroplast.

this case, the whole structure must be dismantled before they can be isolated; even then a lipidlike environment must be provided if protein structure and function is to be maintained. Both these types of protein are called membrane proteins, but a clear distinction must be made in respect of their structural relationship to the bilayer. For a brief introduction to the amino acids as building units for proteins and a description of the important polypeptide conformations, see pp. 28–39 in Tschesche's work.[155] Carbohydrates can be found attached to the membrane, frequently as glycoproteins which may act as recognition markers for the particular surface, but also as lipid derivatives such as cerebroside or phosphatidylinositol (Figure 6.3).

In elucidating all these functions, biological membranes have been studied intact, or partly dismantled. Alternatively, model system have been built up from the synthetic or extracted components of the biomembrane. A description of the development of our understanding of membrane structure is given by Fisher and Stoeckenius[1] and the physical basis for its organization and dynamics by Sackmann.[2] A comprehensive monograph of the subject is that of Yeagle.[3] Lipid conformation in biological membranes and model systems has been well studied[4]

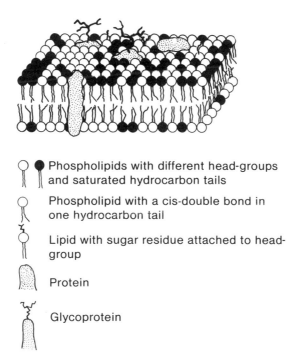

Figure 6.2. The bilayer structure of a biological membrane. Note the asymmetric distribution of lipids, the presence of surface-attached and buried membrane proteins, and the carbohydrate derivatives of both lipid and protein.

and others have considered the place in liquid crystal theory of the lipid bilayer.[5] The main gel to liquid crystal phase transition is characteristic of each lipid and important to the function of the membrane. The theory behind this property is described by Nagle.[6]

Isolation of integral membrane proteins requires considerable care and ingenuity in maintaining their structure and function. A review of the procedures developed for this and criteria for identifying carrier proteins has been written by Tanner.[7] Finally, a summary of known structural information about membrane proteins has recently appeared.[8]

The organized, planar array of a Langmuir–Blodgett film has been viewed as providing a new arrangement of the lipid membrane, suitable for probing the cooperative interactions between its constituents and the properties of the surface as a whole. It has been suggested that this "presentation" of the membrane might provide a means of using the properties of the biological system to detect small amounts of complex molecules.[9] The application of Langmuir–Blodgett films to some of these problems will be described in Section 6.4. The preparation of the

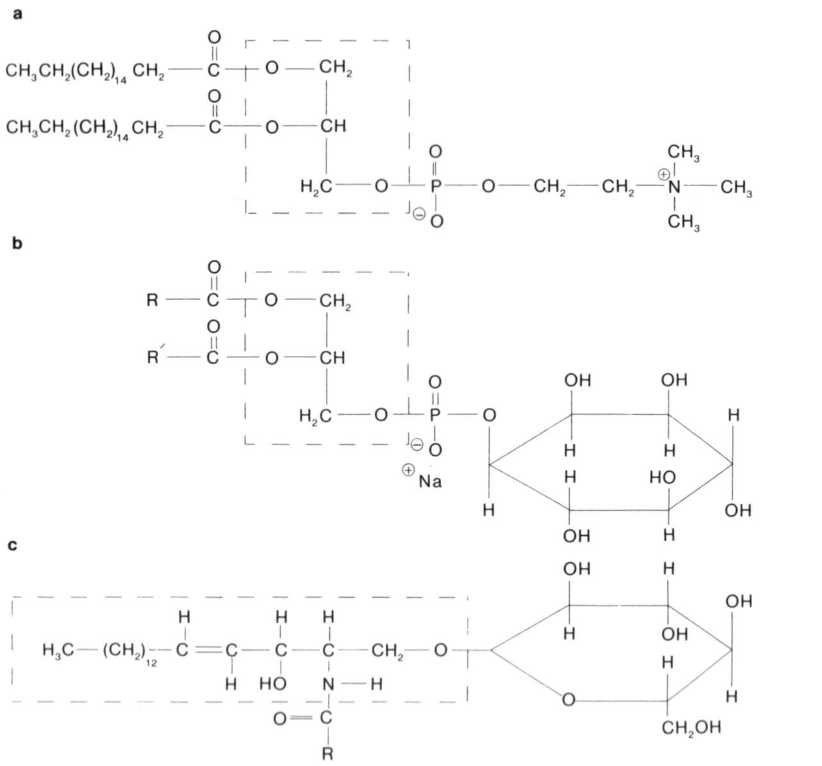

a

$CH_3CH_2(CH_2)_{14}CH_2$ — C — O — CH_2
                                    ‖
                                    O

$CH_3CH_2(CH_2)_{14}CH_2$ — C — O — CH
                                    ‖
                                    O

H_2C — O — P — O — CH_2 — CH_2 — N⊕ — CH_3
              ‖   CH₃
              O⊖ CH_3
                                                   CH_3

b

R — C — O — CH_2
     ‖
     O

R' — C — O — CH
      ‖
      O

H_2C — O — P — O
              ‖
              O⊖
              ⊕Na

c

H_3C — $(CH_2)_{12}$ — C = C — C — C — CH_2 — O
                              H   H   H
                              | | |
                              H   HO N — H
                                      |
                                  O = C
                                      |
                                      R

Figure 6.3. Structures of lipid molecules: (a) distearoylphosphatidylcholine, (b) a phosphatidylinositol, (c) a cerebroside.

films lags behind the understanding and sophistication of films prepared from nonbiological materials. In many ways, the state of understanding resembles that described by Gaines in 1980[10] for nonbiological systems. The various approaches which have been employed will be described in Section 6.3. However, since by the procedure of Langmuir and Blodgett, a monolayer at an air/water interface must first be prepared, the next section tackles the problems of producing such a monolayer from biological materials.

6.2. LANGMUIR MONOLAYERS OF BIOLOGICAL MOLECULES

Since in this volume we are interested in producing multilamellar stacks by the procedure of Langmuir and Blodgett, this section will give a brief description of the properties of monolayers of biological molecules. Apart from the usual problems of solubility in the subphase or collapse of the monolayer, there are a number of other considerations when using molecules of biological origin. While they might be chemically stable at the interface, their three-dimensional structure, in particular that of proteins, can be very sensitive to surface forces, producing changes in the protein which make the biological relevance of the subsequent supported mono- or multilayer very suspect.

6.2.1. Phospholipids and Sterols

These natural amphiphiles have been studied for some time and for earlier work Gaines[11] should be consulted. Since cholesterol and other sterols are commonly found in membranes from higher organisms, workers have also studied these molecules at the air–water surface. In general, the aim of these studies has been to elucidate the role of phospholipids and sterols in the structure of the bilayer. The relationship between monolayer, bilayer, and natural membrane is important and has been considered in a number of articles.[12,14] Since some of this work used an oil/water interface, the relevance of this configuration must also be included.[13,27] There are analogous structural changes in the monolayer and bilayer which make the former's contribution to the study of biological membranes valid.[12,29] Phase separations and clustering of components of mixed monolayers are also common and will be mentioned again below. The possibility that these structures may or may not be transferred to the surface of a solid support must be considered when removing the monolayer from the air/water interface.

The choice of spreading solvents for phospholipids can be a problem since they are only sparingly soluble in benzene or hexane. Mixtures containing chloroform and various alcohols have been used instead. However, there are problems in the use of such mixtures, especially those with a considerable proportion of alcohol (see below). The choice of a suitable solvent system was addressed by Mingins et al.[15] albeit with octadecyltrimethylammonium bromide ($C_{18}TAB$). This compound will form a monolayer from a crystal and is sufficiently soluble in water to use this as the "spreading" solvent. These monolayers were compared with those obtained using a variety of solvent mixtures and, though a good comparison was seen for part of the isotherm, discrepancies remained so that selection of one particular solvent mixture in preference to the others was not possible. The use of alcohols as spreading solvents for phospholipids causes problems[16] although the effects can vary. For example, dipalmitoyl phosphatidylethanolamine is more sensitive to ethanol in the spreading solvent than is the equivalent phosphatidylcholine.[181]

While the lowering of surface tension does cause some expansion of the monolayer at low pressures, the primary effect was film loss through increased solubility in the subphase. Alternative mixtures suggested[16] were hexane/ethanol (90:10 v/v), Freon TF/chloroform (92.8:7.2 v/v), and, for less polar lipids, Freon TF alone (Freon TF is CCl_2FCClF_2). Chloroform was also used successfully, though it should be remembered that commercially available chloroform contains a small amount of ethanol to stabilize it. These authors also pointed out that it takes some twenty minutes for all the solvent to evaporate away prior to compression. Waiting this length of time might lead to other problems such as contamination of the surface. Nonetheless, since the loss of solvent decreases exponentially with time there will be only approximately 1% left after ten minutes. It must be assumed therefore that monolayers compressed after this time will still contain some solvent.

It is not uncommon with phospholipids to see hysteresis in the pressure/area isotherm on repeated compression and expansion of a monolayer. Loss of residual spreading solvent is one cause of this, though a number of others are listed by Mingins and Taylor.[17] The effect is particularly noticeable if a phase change is traversed. Care must also be taken in purifying lipids extracted from biological material since ionic impurities can alter the properties of the monolayer.[18] For this reason, synthetic lipids are frequently used, a wide range of which are commercially available. Even here the purity should be checked by thin-layer or gas-liquid chromatography.

Hysteresis will also occur if the contact angle on the Wilhelmy plate varies during dipping. The angle should of course be zero, but this is not always the case with phospholipids. Choline head-groups seem to give rise to nonzero angles[19] while dipalmitoylphosphatidylethanolamine and fatty acids such as myristic and stearic acid give nearly zero contact angles over a surface pressure range of 0–40 mN m^{-1}.

The collapse of egg lecithin monolayers has been observed with electron microscopy.[20] The morphology of the collapsing film varies with the phase. Fluid monolayers of egg lecithin produce large, flattened oblate spheroids while solid monolayers produce narrow ribbons, flat platelets, and fiberlike structures. Dimyristoyl lecithin[21] and dimyristoyl phosphatidylglycerol[174] have been shown to form bilayers at an air-water interface. This only occurs in the presence of bulk phase liquid-crystal and at a critical temperature which is characteristic of the lipid. The formation of a hydrated bilayer has been inferred from the spreading properties of small, very pure crystals of dipalmitoylphosphatidycholine[22] (i.e., no solvent is used). Previously it was thought impossible to spread this lipid below its phase transition (gel to liquid crystalline) temperature.[23] However, pure mono-, di-, and triglycerides and the phosphatidylethanolamines will spread spontaneously from the crystal to form monomolecular films.

The range of lipids now studied at the air/water interface continues to increase. Recently galactolipids were compared with lecithin and shown to produce more stable monolayers as indicated by the increased collapse pressure.[24] Galactolipids are very common in the membranes of photosynthetic organelles, particularly the thylakoid membrane. In higher plants the constituent fatty acids of these membranes are very unsaturated, while in algae they are either monounsaturated or saturated. In this comparison with lecithin, only distearoyl galactolipids were used. Figure 6.3 (c) shows the structure of a glycolipid, cerebroside. With more complex oligosaccharide units, this structure forms the basis of a range of gangliosides which are found most notably in membranes of the nervous system. The surface behavior of these molecules, glycosphingolipids (which differ from phospholipids in that the backbone is not glycerol) and mixtures of these with phospholipids has been studied by Maggio et al.[175] The effect of carbohydrates dissolved in the subphase on the packing of a phospholipid monolayer has been studied by a number of groups. However, more recent results[25] demonstrate that most of the effects observed could be accounted for by surface active contaminants, although at higher carbohydrate levels some effect was discernible.[176] The role of charges at the interface in governing the phase behavior of the film is complex and poorly understood. Mingins, Llerenas, and Pethica[26] considered the effects of electrolyte screening, counterion type, and valency, and the presence of a second, charged long-chain species. They also compared the air/water and oil/water interfaces.

Natural phospholipids frequently have different fatty acids esterified at the glycerol 1 and 2 positions. The fatty acid can vary in chain length and degree of unsaturation. It is also normal in biological systems for there to be a mixture of phospholipids. In spreading a lipid film as a monolayer with a view to deposition onto a solid support, consideration should be given to the homogeneity of the monolayer. It is possible, for example, for phase separation to occur when one component is below its phase transition temperature[28] and a second component above it. The extent to which this occurs depends upon the fatty acid composition and its distribution; an equimolar mixture of dioleoyl and distearoyl phosphatidylcholines will not show the same surface properties over the normal range of temperatures as 1-stearoyl-2-oleoyl phosphatidylcholine. The factors affecting this behavior are discussed by Finer and Phillips[29] and the theories describing phase transitions in monolayers and bilayers are reviewed by Nagle.[6] Other components commonly found in mammalian, plant, or fungal membranes are sterols.[177] The force–area characteristics of 1-oleoyl-2-stearoyl phosphatidylcholine with 3β-, 3α-hydroxysterols and ketosteroids (Figure 6.4) have been reported.[30] For sterols the interaction depends upon a planar sterol nucleus, the presence of the 3β-hydroxygroup and an intact side-chain. (We note that, in vivo, the fatty acid at position 1 is usually more "saturated" than the residue at position 2, so that of the two mixed-chain phospholipids just mentioned, 1-stearoyl-2-oleoyl phosphatidylcholine is more likely to be found naturally.) The miscibility of cholesterol with dipalmitoyl phosphatidylcholine is discussed by Cadenhead et al.[31] who show that below the phospholipid's phase transition, the two components are immiscible at near-zero surface pressure, but that at higher pressures, similar to those seen in biological membranes, no segregation takes place. The use of an oil/water interface as a model for "half" a bilayer was mentioned earlier. Studies on phase changes and the formation of "mosaics" have been made at such interfaces.[32]

The use of fluorescent derivatives of a phospholipid mixed with the lipid itself have allowed more direct visual observation of phase changes and mixing. The phase transition of palmitoyl lecithin was studied by Teissie et al.[33] A larger study by von Tscharner and McConnell[34] indicated that in the solid condensed and the liquid-condensed/liquid-expanded regions of the isotherm, the monolayer was optically homogeneous, though the lateral diffusion of the fluorescent probe was greater in the second case. In contrast, the liquid-expanded region of the isotherm was not optically homogeneous, and flow of the surface layer was very fast. This

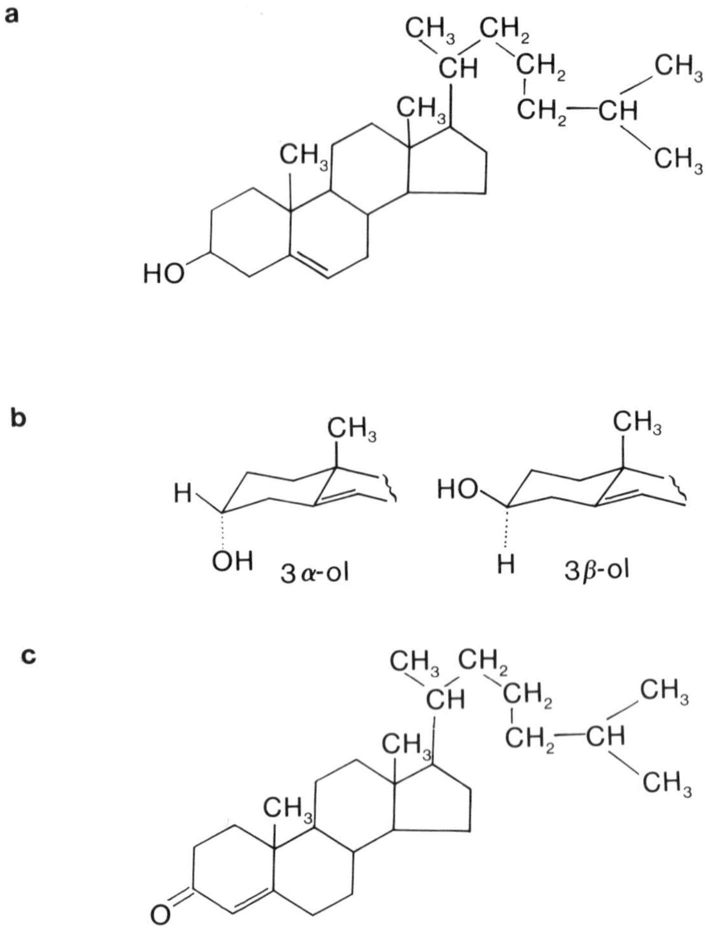

Figure 6.4. Structures of: (a) cholesterol, a 3β-sterol, (b) an α and β sterol as considered by Demel *et al.*, [30] (c) a ketosteroid, Δ4-cholesten-3-one.

study was done as a preliminary to depositing a monolayer onto a glass slide[35] and illustrates the importance of understanding the properties of the monolayer at the air/water surface prior to depositing layers onto a support. Coexisting solid and fluid phases have also been seen (Figure 6.5) using dimyristoyl phosphatidic acid with a fluorescent dye at 0.25 mol%.[36] The same type of patterns were observed using

Figure 6.5. Fluorescence micrographs (b) of a dimyristoyl phosphatidic acid (DMPA) monolayer containing 0.25% of the dye L-α-dipalmitoylinitrobenzoxadiazol phosphatidylethanolamine (pH 5.5; $T = 26.5\,°C$) at different points in the isotherm as indicated in (a). The surface pressure versus area diagram is identical to that of DMPA alone (from Losche *et al.*[36] with permission).

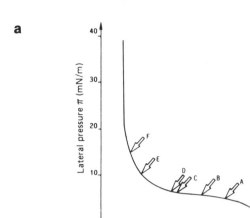

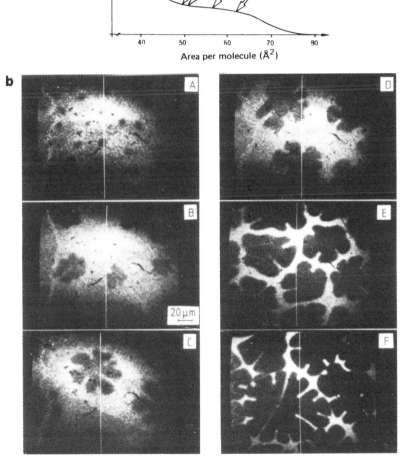

phase-contrast electron microscopy in the absence of dye, showing that the structures seen were not caused by the dye itself.

The monolayers were then transferred to a solid support where two phases remained, differing in thickness by 0.4 nm. The phase changes in the monolayer have been quantified[178] and the fluorescence studies complemented with the use synchrotron X-ray diffraction.[179] Whether patterns are observed can vary with the rate of compression and the stereochemistry of the phospholipid head-group. In a recent paper, Weis and McConnell[37] demonstrate long-range, two-dimensional order depending upon the stereochemical configuration of the head-group.

This report should be read in conjunction with that of Arnett and Gold,[38] who studied enantiomers of dipalmitoylcholines and found no way of distinguishing between the isomers by NMR, differential scanning calorimetry, or force–area diagrams. Using single-photon counting techniques, the time-resolved and steady state behavior of pyrene substituted lipids spread as a mixed monolayer have been studied.[39] The incorporation of the pyrene moiety increases the area per molecule of the lipid by about 20%, so that the presence of the fluorescent probe must perturb the lipid packing in its vicinity. The authors point out that values of the in-plane diffusion constants obtained must be viewed with caution.

6.2.2. Proteins

Two major problems arise when proteins are present at an interface:

1. Does the protein unfold?
2. Will it penetrate a monolayer, usually of phospholipid, already at the interface?

Unfolding or denaturation of the protein will result in loss of any catalytic activity it usually possesses. Recompression of the surface layer will not usually result in refolding to the original tertiary structure. The problem is that the natural conformation of the protein only has marginal stability because it is highly constrained. The hydrophobic amino acid residues are usually concentrated in the center of the structure away from the aqueous media. Unfolding to a random coil may require as little as 400 J per mol^{-1} amino acid residue since the generation of a large hydrocarbon/water interface is largely compensated for by the increase in freedom of motion.[40] Much of the early work on the adsorption of enzymes at interfaces is reviewed by James and Augenstein.[41] They point out that at some oil/water and liquid/solid interfaces, the interfacial energies are so small that unfolding does not occur. However, whether or not the tertiary structure remains intact depends as much on the original structure and flexibility of the protein. Ordered regions of the molecule, such as those joined by disulfide bridges or containing α-helices, are often retained at the surface.[40] When denaturation does take place, it is a rapid cooperative transition followed by a slower rearrangement of molecules in the film to attain the lowest free energy. This second process is a function of film compressibility.[40]

In vivo, membrane proteins either associate with the surface of the bilayer or penetrate completely across it. It is possible to inject proteins into the subphase below a preformed phospholipid monolayer and then study the interaction with, or in some cases the penetration of the protein into, the monolayer. As a test for denaturation it is possible with a segmented, circular trough to move the protein/lipid monolayer onto a fresh, protein-free subphase and measure enzyme activity when substrate is added.[42] Using this procedure, catalase associated with phospholipid was dipped onto a glass slide and its activity tested in this "Langmuir–Blodgett" configuration.

What controls the association with, or penetration into, the monolayer? Fromherz[42] reports that proteins adsorbed on a lipid film under high surface pressure retain a considerable proportion of their tertiary structure. Whether any charge associated with the phospholipid affects the interaction must depend upon the distribution of charge on the surface of the protein. If this is not especially localized, the charge and conformation of the lipid polar head-group would seem not to have a major influence on the interaction.[43] It appears that a more important factor controlling the penetration is the lateral compressibility of the lipid monolayer.[40,43–45] More protein is able to penetrate into the monolayer when it is in a liquid expanded state.[43] The compressibility of the phospholipid monolayer can be enhanced by making it a mixed monolayer in which the constituent lipids have phase transition temperatures above and below the operating temperature. This would produce regions of gel and liquid crystalline lipid, a situation which is likely to occur in biological systems. The compressibility of a lipid is also increased at its phase transition temperature.[44] It would be interesting to see if the greatest penetration occurred into a monolayer held in the region of its transition temperature. Having produced a monolayer containing lipid and protein, the miscibility of the two components must not be taken for granted. The miscibility of proteins representing the major conformational classes ("α-helix" by bovine serum albumin, "β-sheet" by β-lactoglobulin, and "disordered" by β-casein) with fatty acids, triglycerides, and phospholipids has recently been reported.[46]

Using a fluorescently labeled phospholipid, the degree of penetration of cytochrome *c* has been studied[47] with respect to the packing density of the film, the nature of the constituent lipids, and the ionic content of the subphase. The interaction was followed by measuring the loss of fluorescence caused by the haem of the protein acting as a chromophoric acceptor. This can be related to the binding and orientation of the cytochrome. The viscoelastic properties of tubulin have been studied[48] in monolayers of this protein alone and mixed with phospholipids. Surface pressure and potential measurements indicated partial unfolding of the protein at the interface, but there was evidence of stabilization in mixed monolayers. Graham and Phillips have compared dilatational[49] and shear properties[50] of β-casein, bovine serum albumin (BSA), and lysozyme at the air/water and oil/water interfaces. Viscoelastic films were formed under both conditions by the globular proteins BSA and lysozyme. Comparison of the rheological properties of the films at the two interfaces showed that the unit area of flow and activation energy of flow were less at the air/water boundary. If an adsorbed but surface denatured protein film occurred, then this could be considered as a thin protein gel layer of about 10 nm thickness.

Integral membrane proteins cannot be isolated entirely from a lipid environment without denaturation, and monolayers of these materials are produced by spreading a lipid–protein fragment or vesicle directly onto the air/water interface. In the latter case, it is possible to exchange the *in vivo* membrane lipids for others of the experimentalist's choice prior to spreading. The critical step appears to be the spreading of vesicles[51]; once spread, films can be very stable. During the spreading process, attention must be paid to the surface pressure. When this is low (i.e., a high interfacial energy exists), the membrane proteins will tend to unfold, occupying a larger surface area and producing a less compressible film. Too high a surface pressure, on the other hand, leads to multilayered films on the surface[51] (see also Figure 4 of Phillips[40]). Stable surface pressures were also found to occur only when a limited concentration range of a fragmented membrane preparation was used, sarcoplasmic reticulum in this case.[52] The membrane fragments are applied to the surface from above, either directly or by running the suspension down a wet glass rod or a wet, ground-glass slide dipping into, but at an angle to, the water surface. An alternative is to have the vesicles present in the subphase and to generate a monolayer at the surface by their spontaneous rupture. This approach has been developed by Schindler[53] for the generation of two monolayers on either side of a barrier. The monolayers are moved up the barrier and over a small hole punched therein, where they appose producing a bilayer spanning the hole. The size and concentration of the vesicles, together with the mole percent of protein, determines the stability of the monolayer and will regulate the surface pressure as the area of the monolayer increases or, from another point of view, as the monolayer is removed from the surface. This might be a profitable approach to the generation of protein-containing monolayers for the production of Langmuir–Blodgett films. The exchange of lipid between vesicles in the subphase and a monolayer has recently been considered by Jähnig.[54] A monolayer of dioleoyl phosphatidylcholine produced from a vesicle has been shown[180] to be very similar to that produced by spreading the lipid in an organic solvent. This study used ellipsometry to follow compression of the monolayer, a method which might be applicable to monolayers of protein and lipid where, apart from the use of fluorescence probes to study the structure of the mixed monolayer,[47] electron microscopy has also been employed.[55]

6.2.3. Pigments

Studies of the properties of chlorophyll monolayers mixed with stearic acid[56,57] and phospholipid[58] have been made. A report of the photoelectrochemical properties of chlorophyll *a* (Figure 6.6) deposited onto a semitransparent platinum electrode is given by Villar.[59] The monolayer characteristics and production of Langmuir–Blodgett multilayers of simple, esterified prophyrins are reported by Jones *et al.*[60]

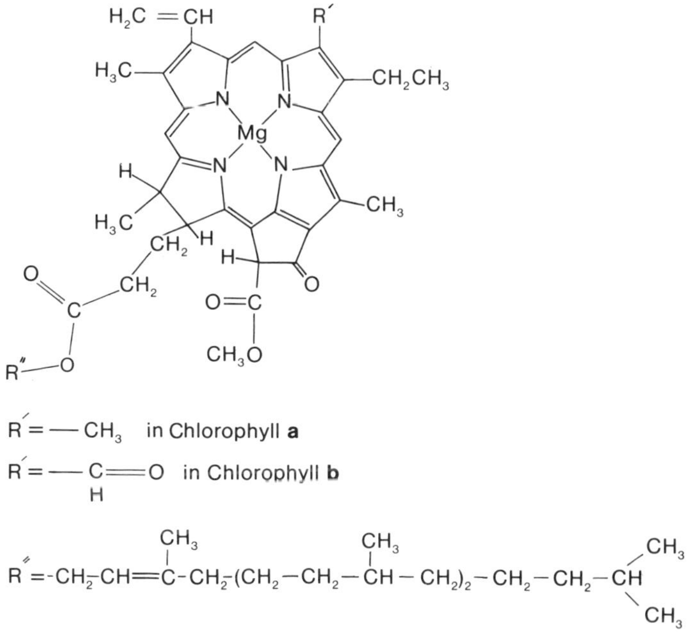

Figure 6.6. Structures of chlorophyll *a* and *b*.

The photophysical properties (fluorescence intensities, lifetimes, self-quenching, and energy transfer) of chlorophyll *b* (Figure 6.6) have been studied[61] in mixed monolayers with dioleoylphosphatidylcholine. The self-quenching process for chlorophyll *b* seems to be more complex than that for chlorophyll *a*. In addition to the more usual measure of surface pressure and surface potential, Ducharme *et al.*[62] have used ellipsometry to study components of rod outer segment (from retinal rod cells of the human eye responsible for vision in dim light). Mixed monolayers of phospholipid and rhodopsin were used as well as 11-*cis*-retinal, all-*trans*-retinal (Figure 6.7), rhodopsin, and discal membranes. Rhodopsin consists of the protein opsin and 11-*cis*-retinal and is a transmembrane protein. Its orientation in the monolayer with lipid was interpreted as changing from the horizontal to the vertical on compression of the monolayer.

6.2.4. Fatty Acids and Polymerizable Phospholipids

Free fatty acids are rarely found in membranes *in vivo*. However, much work has been done on monolayers composed of these materials and reference will be made to a few articles where a comparison with phospholipids is useful. The use of

a

b

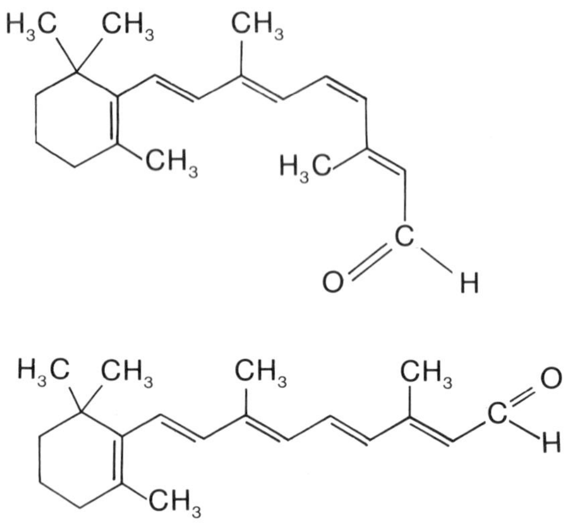

Figure 6.7. Structures of: (a) 11-*cis*-retinal, (b) all-*trans*-retinal.

polymerizable lipids for producing "stable" bilayers and perhaps more biocompatible surfaces requires that some mention be made of their monolayer properties.

The different shapes of pressure/area diagrams for fatty acids are considered by Baret *et al.*[63] They used compounds with an even number of carbons from C_{14}–C_{24} and proposed that over a sufficiently large temperature range any insoluble, fatty acid would produce eight different types of pressure–area diagram. When only two phase transitions are seen, they propose an explanation which involves only the interaction between the polar head-groups and suggest that their model may be applicable to phospholipid monolayers. This paper can be considered with the experimental observations of phospholipid monolayers made by von Tscharner and McConnell[34] where the lateral diffusion of the monolayer and its flow were considered at different points of the isotherm. In the next section the deposition of an underlying fatty acid layer will be described as a preliminary to depositing phospholipids. Earlier work on fatty acids at interfaces is described by Gaines.[11] The effect of cis or trans double bonds in the constituent fatty acids of mixed monolayers of saturated and unsaturated analogues was considered by Feher *et al.*[64] It should be reiterated that phospholipids do not usually have the same two fatty acids esterified to the 1 and 2 glycerol hydroxyls. They vary frequently not only in chain length but also in degree of unsaturation and the cis isomers are almost invariably found.

A large variety of surfactant and phospholipid analogues have now been synthesized containing polymerizable groups. Vesicles or films are prepared from these molecules, which are then crosslinked photochemically producing more stable surfaces. The preparation of Langmuir–Blodgett layers using phospholipid analogues will be considered in the next section, but studies have been made of these compounds in monolayers. The effect of head-group volume and charge has been determined[65] using a range of surfactants containing diacetylenic groups in the hydrocarbon chain and increasing in complexity up to diglycerides and phospholipids. The reaction of the conjugated triple bonds to form a poly-double/triple bonded conjugated system is "topochemically" controlled. It will only occur in the condensed phase and not in the melt, solution, or liquid-crystalline phases. The production of the polymerized surface can be followed spectroscopically since the products are usually highly colored. Details of this approach for phospholipids, together with information on monolayer properties and the effect of one or both of the fatty acids containing a diacetylenic group, can be found in the paper by Johnston et al.[66] The osmotic properties and permeability of polymerized vesicles prepared from these materials is also considered by these authors, as well as the mobility of spin labels in the bilayer center.

The properties of mixed monolayers and vesicles containing some of the diacetylenic phospholipid dispersed in saturated lipid have been considered[67] as a means of avoiding the greatly diminished mobility of completely polymerized model membranes. The resistance of the polymerized lipid to enzymatic hydrolysis was proposed as a means of selectively opening such mixed lipid vesicles by solubilization of the monomeric component. Studies of the monolayer properties and multilayer formation of diacetylenic compounds and 4-hydroxy-cinnamic acid derivatives have been made by Tieke et al.[68] Long-chain fatty acids are further considered by Day et al.[69] and the polymerization of micelle forming monomers is considered by Martin et al.[70] Finally, it is possible to form stable bilayers from dialkylammonium and methyldialkyl-sulfonium salts.[71] These display many properties associated with phospholipid bilayers, and have been cast as bilayers on glass slides.[72]

6.3. PREPARATION OF SUPPORTED MULTILAYERS

Having prepared a monolayer at an air/water interface and considered its properties, what procedures have been employed to deposit a mono- or multilayer onto a solid support? What special problems arise with the use of biological material? These questions will be considered in this section. There have recently been some reports of the preparation of a mono- or bilayer of phospholipid using a variation of the standard Langmuir–Blodgett technique and these will be outlined in terms of the problems arising from earlier methods. Finally, the preparation of multilayers by adsorption onto the surface directly from a solution or suspension will be described. Although well removed from the traditional Langmuir–Blodgett procedure, direct adsorption can prove a useful and complementary approach.

6.3.1. Dipping Phospholipids onto a Vertical Support

The usual orientation of the solid support is to have the plane of its largest surface vertical with respect to the monolayer at the air/water interface. Using this approach Honig et al.[73] published a report of Langmuir–Blodgett deposition ratios obtained from a variety of fatty acid esters and mono-, di-, and triglycerides. Problems were encountered with all the materials, but by varying the treatment of the glass, or depositing several layers of cadmium arachidate first, the fatty acid esters and triglycerides could be dipped. However, the mono- and diglycerides proved difficult. As might be expected, using hydrophilic glass no material was deposited on the first immersion. A monolayer was deposited on the upward trip, but came away from the surface on subsequent immersion. Pretreatment of the glass with $(CH_3)_2SiCl_2$, making it hydrophobic, resulted in no deposition. Glass made hydrophobic by the deposition of cadmium arachidate picked up a monolayer on the first dip but lost it on the upward movement.

A similar problem was encountered[74] when dipping phosphatidylethanolamine containing saturated fatty acids onto a hydrophilic surface. Having picked up monolayers on the first upward trip, the subsequent immersion and upward trip, the third monolayer was lost in the next immersion. This was repeated with each of the following cycles. Phosphatidic acid could be deposited providing calcium was present in the subphase (10^{-4} M) and the pH was between 4 and 9, though transfer could be obtained less successfully outside this range. Mixtures of phosphatidic acid and saturated phosphatidylethanolamine (55:45 and 40:60) could be dipped but ratios with less phosphatidic acid would not work. The authors also remarked about the suitability of the Wilhelmy plate method for following surface pressure. While caution must be exercised,[19] this method has been used elsewhere with success.[35] Corrections can be made for the meniscus curvature at the trough periphery[75] which will alter the area of the monolayer very slightly. The effect would only be important for very accurate work or if the trough were small.

Saturated phosphatidylethanolamine is reported to have been dipped successfully.[181] Green et al.[76] found that it was possible to produce regular multilayers of this lipid providing the monolayer was condensed. Polytetrafluoroethylene (PTFE) or glass were used as substrates and the subphase was pure water (pH 5.5). The 2 cm glass plate was dropped through the monolayer in about 1 s. This would seem very fast, yet deposition only occurred during immersion (X-type). This procedure did not yield uniform multilayers in the hands of Hasmonay et al.,[74] who considered the rate of immersion far too quick. However, the X-ray spacing reported by Green et al. do suggest a regular, bilayer arrangement of the lipid. Very rapid immersion has been used more recently[77] with acetylenic phospholipids. In this case, deposition only occurs on the upward stroke; the rapid immersion is to avoid loss of the last deposited monolayer.

6.3.2. Dipping Phospholipids onto a Horizontal Support

While it would appear to be a departure from the Langmuir–Blodgett procedure, the placing of the support *flat* onto the monolayer surface was first used by Langmuir[78] to deposit protein onto hydrophobic surfaces. This was pointed out by von Tscharner and McConnell in their preparation of supported lipid monolayers.[35] In order to obtain an optically homogeneous covering of lipid, it was important to leave the monolayer on the air/water surface for 30 min in the liquid expanded region of the isotherm. The monolayer was then compressed to 40 mN m^{-1} and left for a further 2 h before being dipped. This procedure would certainly avoid problems with entrapped spreading solvent,[16] but some care must have been taken to avoid surface contamination. The hydrophobic glass slide was then placed horizontally against the monolayer in one short movement and 10–15 s later pushed through the surface. Providing no air bubbles had been trapped and the slide was kept under water, the monolayer was reported to be quite stable. A fluorescent lipid probe was incorporated into the lipid layer to measure lateral diffusion ($<10^{-10}$ cm^2 s^{-1} if the lipid was in the solid-condensed phase prior to dipping; $>10^{-8}$ cm^2 s^{-1} if the lipid was in the liquid condensed/liquid expanded region of its isotherm) and to show the homogeneity of the layer.

In an extension of this procedure, a recent report[79] from the same group demonstrated the formation of a supported bilayer of dipalmitoyl, dimyristoyl, or dioleoylphosphatidycholine. The substrate used was p-type silicon wafer with an approximately 800 μm thick oxide layer. The wafer was dipped through the layer in the conventional manner (i.e., vertically); no deposition occurred. On withdrawing rapidly, but slowly enough to allow water to drain from the surface, a monolayer was picked up. The second monolayer was deposited by "horizontal" immersion through the recompressed lipid on the water surface. Glass and quartz have also been used as substrates using a conventionally spread lipid monolayer and by fusion of phospholipid vesicles when in contact with the hydrophilic surface.[182] Incorporation of a fluorescently labeled lipid probe indicated very little diffusion in the first monolayer (this was deposited polar head-group down onto the hydrophilic support) but diffusion in the bilayer above its phase transition temperature was very similar to that seen in biological membranes. This is an important piece of evidence for proposing the supported bilayer as a reasonable model of a biomembrane. The bilayer had to be handled carefully, since dipalmitoylphosphatidylcholine deposited below its phase transition generated a variety of liposomes on heating the bilayer above 41 °C and a supported bilayer of dioleoyl phosphatidylcholine developed holes as it was cooled. These different approaches and some of the uses to which they have been put are summarized in a review by McConnell.[182]

A similar approach to the production of one bilayer had been reported earlier,[117] in which a porous disk (cellulose acetate, 0.13 mm thick with a 0.2 μm pore size) was first withdrawn through a compressed monolayer. The disk was then "touched" horizontally onto the recompressed monolayer and then withdrawn, i.e., it was not pushed through into the subphase. This procedure was used to prepare a supported bilayer composed of cholesterol and a variety of phospholipids. The visual pigment rhodopsin was also spread with phospholipid, picked up onto the disk, and the supported bilayer used for studying light-induced ion diffusion.

A careful observation of the formation of Langmuir–Blodgett films using dimyristoyl phosphatidylcholine has been reported[36] in which either the traditional procedure or one more akin to that just discussed was used. This involved rapid withdrawal of the support through the monolayer, such that a thin film of water (about 50 μm) was taken with the lipid; this "drained" away slowly to a thickness of less than 0.5 nm. The authors report that this method allows the transfer of monolayers which are very expanded. Using a fluorescent lipid probe, various structures were seen at the air–water interface (Figure 6.5). The shape of these varied with speed of compression and are similar to those reported elsewhere[37] and related to stereochemical configuration of the polar head-group. These structures could be transferred onto a solid support.

6.3.3. Dipping Phospholipids: General Considerations

It is apparent that dipping phospholipids is not a straightforward application of the Langmuir–Blodgett technique. Apart from a few reports of success with phosphatidylethanolamine derivatives, no more than three layers are usually deposited; however, we note again the work by Chapman's group[77] and the remarkable effect of basic proteins attached to the lipid surface during dipping (Section 6.3.6). The use of "horizontal" dipping has resulted in very useful studies of lipid monolayers and bilayers, and it is a valid question to ask why any more layers need be dipped if the intention is to use this configuration only as a model bilayer membrane. The problem is that, for a number of spectroscopic techniques, one or two layers may not present a long enough path length through the material to obtain a reasonable signal. However, the usefulness of fluorescent probes has already been demonstrated and if the layers are deposited onto, for example, a germanium substrate, attenuated total reflectance spectroscopy in the infrared is feasible. With the advent of Fourier-Transform infra-red spectrometers, it is now possible to obtain a spectrum by reflection-absorption spectroscopy from a very few layers.[216]

The reason for the instability of multiple (>3) layers of phospholipids may reside in the poor interaction between adjacent, neutral head-groups. Explanations for the observed transfer behavior of nonbiological monolayers have certainly been based upon dipole–dipole interactions between adjacent layers.[80] Phosphatidic acid multilayers prepared with calcium in the subphase, and hence between the supported layers, were stable.[74] However, zwitterionic phosphatidylcholines are not greatly stabilized by subphase cations, though sometimes it is impossible to deposit even one to three layers in their absence. An approach to the formation of Langmuir–Blodgett films of defined crystallinity and a regular array of domains is described by Lösche et al.[81] They studied the electrostatic properties of the monolayer by varying the ionic composition of the subphase, in particular the type and concentration of divalent ions. The electrostatic forces were calculated using the Gouy–Chapman–Stern theory.

Another part of the problem may be rearrangement of the layers. This was recognized by Honig[82,183] as an explanation for X- or Y-type deposition. A study of long-chain esters[83] showed that Y-type deposition was favored by a high surface pressure of the monolayer, low temperature (presumably also giving a condensed phase), and rapid withdrawal. Y-type transfer of dipalmitoylphosphatidylcholine was obtained with cations in the subphase,[84] but the lecithin peeled off on subsequent immersion. The transfer-work on immersion or exit from the subphase was calculated in this paper. It was shown that there was no interaction between the monolayer and substrate and that the transfer-work was only that of wetting. During the first immersion and exit the values for the transfer-work were different from those measured during subsequent cycles. This difference was also underlined and explained in a study of ω-tricosenoic acid[85] where molecular reorganization was shown to occur. The rate-limiting step for this reorganization appears to be the loss of the last few molecules of water from between the layers.

While these reports concern nonbiological material, they suggest a reason for the success of rapidly withdrawing the slide through the monolayer. In addition, it might be beneficial to leave a period before reimmersing the trilayer coated substrate. This would allow for more complete "drainage" and any reorientation of the lipids. Some preliminary results do show this approach to help in dipping five layers of dioleoylphosphatidylcholine onto a glass slide.[86]

6.3.4. Dipping Polymerizable Derivatives of Phospholipids

The preparation of polymeric surfaces with lipid head-groups by dipping acetylenic phospholipids (Figure 6.8a) has been studied particularly by Ringsdorf's group[65,87] and that of Chapman.[77,88] At dipping speeds of $0.3-5$ cm min^{-1} only three layers of the diacetylenic phosphatidylcholine could be deposited,[77] as had been found for the saturated lipid. However, by increasing the rate of immersion to $10-14$ cm min^{-1} while maintaining the upward stroke at 0.5 cm min^{-1} or less, it was found that a layer was not lost on immersion, but that one was gained on each upward stroke. By this procedure they were able to produce up to 43 layers. PTFE was used as a hydrophobic substrate; glass, quartz, steel, and Perspex were used as hydrophilic surfaces. In a subsequent paper[89] a monolayer of diacetylenic phosphatidylcholine was deposited by dipping at only 1 mm min^{-1} over a subphase containing cadmium chloride. There do not appear to be further reports of the rapid dipping procedure for these or saturated phospholipids.

If the polymerizable unit is attached to the lipid headgroup (Figure 6.8b) via a hydrophilic spacer, successful Y-type deposition is possible[90] with a dipping speed of 5 cm min^{-1} downward and 0.5 cm min^{-1} upward. The slide was left in air for 5 min before commencing the next dip, otherwise the last deposited monolayer was lost (cf. comments in Section 6.3.3). It was also possible to dip the polymerized monolayer because the fluidity of the hydrocarbon chains was maintained; the pressure/area isotherms still exhibited fluid phases. The pressure/area isotherms of molecules polymerized in the monolayer and those of the prepolymerized spread monolayer were very similar. The polymerization on the surface was achieved by

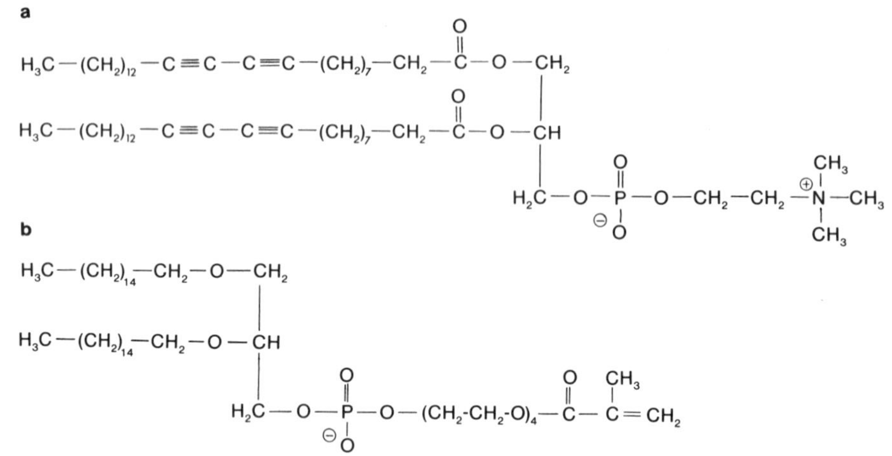

a

$$H_3C-(CH_2)_{12}-C\equiv C-C\equiv C-(CH_2)_7-CH_2-\overset{\overset{\displaystyle O}{\|}}{C}-O-CH_2$$

$$H_3C-(CH_2)_{12}-C\equiv C-C\equiv C-(CH_2)_7-CH_2-\overset{\overset{\displaystyle O}{\|}}{C}-O-CH$$

$$H_2C-O-\overset{\overset{\displaystyle O}{\|}}{\underset{\underset{\displaystyle O}{\ominus}}{P}}-O-CH_2-CH_2-\overset{\overset{\displaystyle CH_3}{|}}{\underset{\underset{\displaystyle CH_3}{|}}{\overset{\oplus}{N}}}-CH_3$$

b

$$H_3C-(CH_2)_{14}-CH_2-O-CH_2$$

$$H_3C-(CH_2)_{14}-CH_2-O-CH$$

$$H_2C-O-\overset{\overset{\displaystyle O}{\|}}{\underset{\underset{\displaystyle O}{\ominus}}{P}}-O-(CH_2\text{-}CH_2\text{-}O)_4-\overset{\overset{\displaystyle O}{\|}}{C}-\overset{\overset{\displaystyle CH_3}{|}}{C}=CH_2$$

Figure 6.8. Structures of: (a) a diacetylene phosphatidylcholine (see Hupfer and Ringsdorf[65]), (b) 2,3-bis(hexadecyloxy)propyl 12-methacryloyl-3,6,9,12-tetraoxadodecylsuccinate (see Elbert et al.[90]).

UV irradiation and in contrast to topochemically controlled reactions (see Section 6.2.4 and other work[65,68], complete conversion could be achieved.

6.3.5. Oriented Multilayers by Other Methods

There are a variety of methods for producing ordered multilayers on a solid support which do not use a preformed monolayer. These techniques will be mentioned briefly here, because in the main they produce a large number of multilayers and can be considered complementary to the Langmuir–Blodgett technique. For example, layers giving a final thickness of 500 μm over a 1 cm² area could be formed from phosphatidylcholine by annealing the lipids at elevated temperatures.[91] The multilayers were characterized using polarized and dark-field microscopy. That these multilayers were comprised of a large monodomain has been shown[92] by optical birefringence, light scattering, and X-ray diffraction. Further optical studies of such macroscopic domain samples and those containing cholesterol, chlorophyll, and a variety of antibiotics have been reported briefly.[93,94] Approximately, 6 to 15 layers of phosphatidylcholine were deposited by dialyzing a dispersion of the lipid and deoxycholate.[95] Lateral diffusion was measured using a fluorescent lipid probe and gave the expected values depending on whether the lipid was in the gel or liquid crystalline phase.

Polarized Raman spectra have been obtained from oriented monodomains of dipalmitoylphosphatidylcholine.[96] These were prepared by allowing a drop of a suspension of small, unilamellar vesicles[97] to dry slowly on a clean, glass slide. The conformation of phospholipid head-groups has been studied using Raman spectroscopy with solid samples of the lipid or solutions in alcohol[98] and a study of the oriented multilayers has also been completed.[99]

Unsaturated lipid multilayers have been prepared[100] by allowing a water/alcohol/lipid mixture to equilibrate under a humidified nitrogen atmosphere for 5 h. They were then macroscopically oriented by gently rubbing the hydrated lipid between two microscope cover slips. The alignment was checked by optical microscopy.[91] The satisfactory samples were sealed round the edges of the cover slips with epoxy resin to maintain the hydration. The samples were 12 to 40 µm thick and were used for studying the angle-resolved fluorescent polarization of a probe (1,6-diphenyl-1,3,5-hexatriene) incorporated into the lipid layers.

6.3.6. Proteins and Pigments

In a paper referred to earlier,[76] the preparation of Langmuir–Blodgett films of synthetic polypeptides was considered. X-ray diffraction patterns were obtained and indicated that long helical sections are stable at the air–water interface and in multilayers; it seems that shorter regions of α-helix adjacent to random-coil portions of the protein unfold easily. This results in a two-dimensional array of amino-acid residues. Using the same two materials poly (γ-methyl-L-glutamate) and poly (γ-benzyl-L-glutamate) though of lesser molecular weight, Winter and Tredgold[101] demonstrated the formation of bilayers on the surface of the trough. Both mono-layers and bilayers could be dipped (Z type) and good straight-line regions were obtained when reciprocal capacitance was plotted against the number of layers. The slopes were very similar, assuming that on each upstroke through material in the bilayer region of the pressure/area isotherm two layers were deposited. Baking the material at 100 °C for 24 h did not change the dielectric constant of 8.0 for the methyl derivative, suggesting that "bound" water was present in the α-helical structure. The authors suggest that the molecular order and large dipole moment associated with the α-helical structure might be used in improving the in-plane order of the deposited films.

"Horizontal" dipping of pure, synthetic polypeptides has been demonstrated.[102] The structure appeared to be amorphous, unless the scattering caused by the electron density fluctuations in such a film was too small to observe. Preparing mixed monolayers of these materials with stearic acid produced very stable monolayers, which produced Y-type deposition on "vertical" dipping. The bilayer thickness was similar to that of stearic acid alone, though the crystallinity of the film normal to the surface was not as good. If the lipid is acidic and there are basic proteins in the subphase, some of the protein will be transferred with the lipid onto a support. Using quite rapid immersion (9.1 mm s^{-1}) and very slow withdrawal (0.08 mm s^{-1}), MacNaughton et al.[103] were able to deposit up to 50 bilayers onto aluminum foil! They note that in the absence of protein it was much more difficult to deposit lipid bilayers even with divalent cations such as calcium in the subphase, as has been noted before (Section 6.3.3).

Mixed monolayers of natural protein and lipid can be prepared more successfully[104,105] from vesicles and, while these have been deposited onto a PTFE septum with a small hole (thus producing a bilayer across the hole), this approach does not seem to have been used widely for producing multilayers, although

McConnell's group have produced deposited ilayers from vesicles.[182] Earlier work by Stoeckenius's group[106] with purple membrane fragments from *Halobacterium halobium* demonstrated the production of monolayers. These were used to produce multilayers and this approach has since been shown to produce films on lipid and paraffin impregnated filters.[107,108] Mixed monolayers of purple membrane fragments and phospholipids are miscible.[109] Deposition of a single layer onto a hydrophilic surface is straightforward, but subsequent deposition of a monolayer of phospholipid to complete the supported "bilayer" proved difficult. *In vivo*, the protein acts a light-driven proton pump, so that to span the bilayer at least two hydrophilic domains must be present. In the monolayer, therefore, there are two possible orientations at the interface. In the work by Stoeckenius's group[106] 85% of the bacteriorhodopsin was oriented in the same direction. However, when deposited onto the lipid impregnated filters[108] the effect of ionophores on the photopotential showed that a major rearrangement of the film had occurred on the filter. Workers using Langmuir–Blodgett films to probe the orientation of proteins incorporated in the lipid bilayers should always be aware of this possibility.

In the development of model systems for photosynthesis, chlorophyll and other pigments have been prepared as multilayers to examine their spectral properties and interaction with other components of the photosystems. Porter's group used this approach,[111] but in preparing the multilayers they found it necessary to deposit four layers of cadmium arachidate before depositing the pigments. Chlorophyll *a* multilayers have also been prepared on glass slides by Iriyama,[112] and on an optically transparent tin oxide (SnO_2) electrode.[113] In this second paper, the pickup was almost completely Z type (on upward stroke only) particularly for the first five cycles. Sixteen cycles were completed. Preparation of multilayers of chlorophyll *a* mixed with phosphatidylcholine has also been achieved.[114] The photostability of the pigment was enhanced over that of pure mono/multilayers of chlorophyll which were themselves much less labile than solutions of chlorophyll *a* in organic solvents. Strong adhesion between layers of porphyrin resulted in good multilayer formation.[115] These were anionic and cationic derivatives. Simple porphyrin esters have been studied[60] and gave reasonably stable multilayers. The reaction centre and light harvesting chlorophyll protein from the photosynthetic bacterium *Rhodopseudomonas sphaeroides* have been reconstituted into phospholipid monolayers and transferred onto a glass slide. Using fluorescently labelled lipids or the intrinsic fluorescence of the light harvesting protein, Heckl and coworkers[110] were able to determine whether the distribution was homogeneous or not. Protein concentrations of 10^{11}–10^{12} cm^{-2} were achieved and they demonstrated that the protein would still function normally for several days. They note however that these proteins are rather stable and well characterized and that control of surface concentration and aggregation require significant further work.

While not a pigment in the sense of its physiological activity being dependent upon its spectral qualities, haemoglobin has been studied[116] as a monolayer and deposited onto supports. It proved impossible to obtain pinhole-free layers on glass coated with an evaporated film of silver or gold. However, it was possible to evaporate a silver electrode onto a film deposited onto an aluminum base electrode. The electrical properties of the film (up to 15 layers, deposition on withdrawal only) were measured and found to depend to some extent on the water content.

6.4. STUDIES EMPLOYING SUPPORTED MOLECULAR LAYERS

The application to which mono- and multilamellar films have been put fall into a number of categories.

1. *Samples for Spectroscopy.* Most of these measurements have examined the structure of the multilayer and the properties of spectroscopic probes used in membrane studies. It is important to be aware of the relationship between the LB structure and that adopted by the same materials in nature, particularly if, for example the degree of hydration is low in the LB film.
2. *Membrane Studies.* Some of these reports also involve spectroscopic techniques; however, the aim here has been to discover more about the function of the lipid bilayer as it relates to biological membranes.
3. *Proteins.* A small number of papers have considered the interaction of protein with the supported mono- or multilayer. These will be mentioned in particular, because they demonstrate measurements which are less effective using a vesicle or a monolayer at an air/water interface.
4. *Pigments and Photosynthesis.* Particular interest has focused on the use of supported layers of the photosynthetic or visual pigments and proteins because of their equivalent ordering *in vivo.* In addition, the experiments of Kuhn's group[118-120] on the transfer of energy between nonbiological donors and acceptors, precisely separated by the preparation of Langmuir–Blodgett films with "spacers," have presented a guide for studies with biological systems.
5. *Electrical Interactions.* The dielectric properties of the bilayer, the arrangement and orientation of surface dipoles, and the role of water molecules at the interface have resulted in a number of studies on these properties of built-up multilayers. Here again the relevance to biological systems has sometimes been tenuous.
6. *Polymeric Surfaces.* The aim has been to produce biologically compatible surfaces based on polymerized phospholipids.

6.4.1. Spectroscopic Studies

The spectroscopic methods developed for multilayers of nonbiological materials are discussed elsewhere in this volume, but many are applicable to biological materials. In particular, the review by Kuhn, Möbius, and Bücher[121] of the spectroscopy of monolayer assemblies should be consulted.

Fluorescence depolarization has been used to study the fluidity of biological

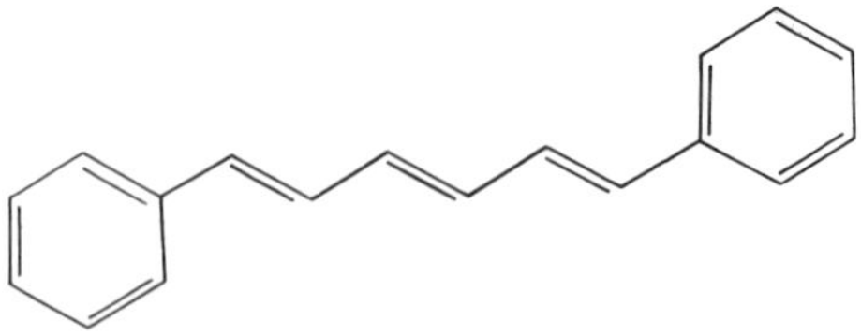

Figure 6.9. Structure of 1,6-diphenyl-1,3,5-hexatriene (DPH).

membranes. One of the most popular probes employed has been 1,6-diphenyl-1,3,5-hexatriene (DPH) (Figure 6.9). It is important to know the properties of the probe and especially the angle between its absorption and emission moments, usually *assumed* to be zero, i.e., the moments are parallel. Using a macroscopically ordered array of lipids (prepared by adsorption and alignment), the value of the fluorescence anisotropy at time zero (r_o) was determined[122] for a variety of lipids. This showed an increasing proportion of the probe was aligned perpendicular to the lipid chains (and not parallel) as the temperature increased. The measurements were carried out using angle-resolved fluorescence depolarization and, in a further paper,[100] problems with the use of DPH in unsaturated lipid systems were reported. The absorption and emission moments diverge and the probe no longer appears to be cylindrically symmetric. This becomes apparent in unsaturated systems because the probe movement appears to be slower. The samples were prepared with 30% (by weight) of water, allowed to equilibriate in a humidified nitrogen atmosphere and then sealed. Under these circumstances, a lipid bilayer would form. The phase diagram for dipalmitoyl phosphatidylcholine (Figure 12.7 in work by Sackmann[22]) shows that, near the phase transition temperature, the maximal water absorption from vapor is 22% and from liquid water, 37%. The percent weight of water absorbed on submersing a sample of the same lipid in water at 80 °C has been determined by Powers and Pershan[123] in their study of macroscopically aligned lamellar phases of dipalmitoylphosphatidylcholine.

Phospholipids containing the pyrene fluorophor have been dipped[124] onto arachidic acid coated quartz glass slides. Of the different lipids employed, only phosphatidylethanolamine proved difficult to transfer. The lipid head-group altered the preferred orientation of the pyrene, though the exact orientation of the excitation dipole could not be unambiguously assigned. Excimers are formed by the collision of an excited state pyrene with a ground state molecule, giving rise to an excited dimer. Excimer decay was measured in this study with respect to temperature. These papers show the advantage of using a supported multilayer system to define the properties of a fluorescent probe normally used in vesicular or cell membranes.

The percent by weight of water is important in determining the structure of the multilayer and hence in any comparison to be made with the biological system. X-ray diffraction studies[125] of a supported phospholipid monolayer showed that equilibration in a water-saturated helium atmosphere only produced approximately 5% (by weight) of water in the layer which was equivalent to about two water molecules per lipid head-group. This low degree of hydration produced an increased gel to liquid-crystalline phase transition temperature of 65 °C. This had been observed previously[126] in samples of similar water content. Small and wide-angle X-ray scattering of lecithin multilayers at 23% (by weight) of water were compared[127] and the domain spacing was found to be larger than reported by others. This increase was explained by the presence of interlamellar water.

The effects of temperature, hydration, and surface pressure of the monolayer have been studied by electron diffraction.[128] The use of a new ultrasonic technique, called laser-induced phonon spectroscopy, to study the mechanical properties of aligned lipids has recently been reported.[129] Lipid area compressibility and viscosity in the presence of 2–20% (by weight) water was determined. Calcium present at 100 mM was found to have no effect on these parameters at a 20% water content. The multilayers were prepared by gentle shearing and compression as described previously (Section 6.3.5 and work by Asher and Pershan[91]).

A series of papers have appeared reporting the use of Raman spectroscopy for studying hydrocarbon chain configuration[130,99] and head-group conformation.[98] These papers used multilayers prepared by adsorption or from lipid vesicles. Techniques for obtaining the Raman spectrum from single monolayers of lipid prepared by the Langmuir–Blodgett technique have also appeared.[131–133] The angular distribution of porphyrins within a supported film has been determined using polarized resonance Raman spectroscopy.[134] Langmuir–Blodgett films of some synthetic dyes and photosynthetic pigments deposited on metal surfaces have been used to define the configuration (metal–film thickness/metal–molecule separation) required for surface-enhanced Raman spectroscopy.[135]. These techniques, are capable of detecting very low concentrations of chemical species at certain metal surfaces (such as silver, gold, and copper).

The rearrangement of Langmuir–Blodgett layers of tripalmitin was determined by electron microscopy and attenuated total reflectance (ATR) infrared spectroscopy.[136] The experiment lasted several days however, and the sample in the spectrophotometer was flushed with dry air. A very interesting and recent study[137] using Fourier transform infrared ATR has been able to link the information obtained from the spectrum to a description of the monolayer at the air/water interface. The presence of surface islands or micelles was postulated and these were related to different sections of the pressure/area isotherm. Dipalmitoylphosphatidylcholine was the lipid used. A similar study[34] has already been referred to in Section 6.2.1 in which the properties of the monolayer at the air/water interface were better understood through fluorescence measurements of the supported layer.

6.4.2. MEMBRANE PROPERTIES

The surfaces of biological membranes are often charged and, depending on the concentration of charged species and the ionic strength of the surrounding media, a double layer of counterions will be generated. The pH at the surface will not necessarily be that of the bulk media. A method for determining this difference has

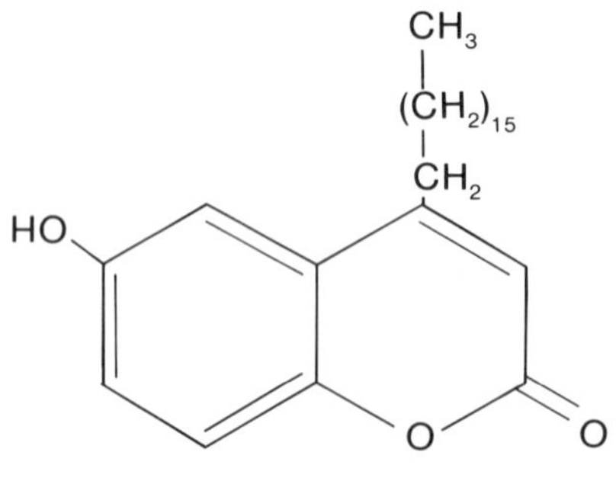

Figure 6.10.   Structure of 4-heptadecylumbelliferone.

been developed by Fromherz[138] in which 4-heptadecylumbelliferone (Figure 6.10) was spread as a mixed monolayer with a lipid and then dipped onto a glass "slide," which becomes the fourth wall of a cuvette. The monolayer was, of course, deposited on both sides of the glass but that on the outside of the cuvette was removed. The probe has a high fluorescence quantum yield in the dissociated state, but a very low one when undissociated. Hence, the pH and ionic strength of the cuvette contents could be altered and the pH at the membrane interface measured.

In a lipid arrangement far removed from the usual Langmuir–Blodgett film, but of relevance to the type of studies that might be done, dimyristoyl phosphatidylcholine was adsorbed onto the surface of hydrated potassium bromide.[139] From the infrared spectrum, the orientation of the lipid head-group at high and low surface densities was determined as well as the distribution of water molecules. A germanium crystal could be used and the deposited monolayers studied by infrared ATR.

In studies of cell surface recognition by the various components of the immune system, reconstituted vesicular bilayers have been used containing specific antigens or lipid haptens (a hapten is a small molecule, foreign to the immune system, which will not elicit a response by itself, but can do so if attached to a macromolecule). One problem in interpreting this response was the contribution made by the curvature of the membrane. Hafeman et al.[140] deposited a monolayer of lipid, a nitroxide lipid hapten, and a fluorescently labeled lipid onto an alkylated glass slide. Using fluorescence microscopy they were able to follow the interaction with fluorescently labeled rabbit anti-nitroxide antibody. (We note that the lipid hapten was used for preparing the monolayer; the antibodies were raised using the same hapten attached to a much larger macromolecule.[141]) This approach has been applied to the more general question of how antigens are presented to the immune system using deposited planar layers.[182,184]

The interaction between lipid surfaces as they are brought very close together (about 1 nm apart) was the subject of a paper by Marra and Israelachvili.[142] Using atomically smooth mica surfaces, bilayers can be deposited and the interaction between the two surfaces measured using a force balance. Phosphatidylethanolamine and phosphatidylcholine were used and the interbilayer forces measured as the ionic strength and composition of the aqueous media between them was altered. Attractive van der Waals, repulsive electrostatic (double layer) and, at short range, repulsive steric hydration forces were all measured. Further details of the deposition of mono- and bilayers onto the mica surfaces are provided by Marra,[143] together with direct force measurements between galactolipid bilayers.

The order and alignment of absorption dipoles of fluorescently labeled lipids have been measured[144] by exciting the fluorescence with polarized evanescent radiation. This is generated by a laser beam internally reflected at the interface between glass and water. It is hoped that this technique will be applicable to protein orientation on or in membranes prepared on surfaces and more generally to biological systems at interfaces.[185]

The use of Raman spectroscopy has been mentioned above with relation to lipids. This approach has also been used for determining the penetration of the antibiotic amphotericin into a thin film containing cholesterol.[145]

Using stacks of lipid bilayers prepared by allowing solvent to slowly evaporate from a lipid solution, Franks has reported a number of studies on the permeability of small molecules across membranes[146] and anaesthesis.[147,148] These experiments were done using X-ray and neutron diffraction. The application of these procedures to stacks of protein in lipid bilayers appeared more recently.[103] The permeability of deuterated molecules was determined and the technique has the advantage of avoiding the problem associated with permeability studies, that of the unstirred water layer adjacent to the membrane. A new procedure for determining water transport through bilayers has been reported.[149] The lipids were deposited onto a cylinder of agar-gel, which was then transferred to a vacuum chamber and the evaporation rate measured under reduced pressure. The results compare favorably with previous work employing octadecanol. Distearoyl and dipalmitoyl-phosphatidylcholines were included in the new procedure. As might be expected intuitively, the resistance to water transport increases as the surface pressure of the dipped monolayer approaches 40 mN m^{-1}.

6.4.3. Proteins

In the earlier section (6.2.2) discussing the spreading of proteins at interfaces, the procedure developed by Fromherz[42] for moving monolayers from one subphase to another was described. Suppose the first subphase contains an enzyme which adheres to the spread monolayer; the enzyme can be transported with the monolayer onto a second subphase containing the enzyme's substrate. Alternatively, it can be moved over a subphase of water or buffer, effectively washing away unbound enzyme, and then dipped onto a solid support. With trypsin, it was found[150] that the enzyme's activity was maintained for 3 h and there was an increase in its effective substrate concentration of 170-fold compared to the enzyme activity in the bulk. This enzyme is only bound to the surface of the monolayer; it does not penetrate into it. Similarly, malate dehydrogenase is only adsorbed onto the surface. The effect of the phospholipid composition and ionic composition of the subphase have been studied by Fromherz,[151] though in this case the monolayer was not subsequently deposited onto a support. If this were done for an adsorbed protein and the supported monolayer then placed in an environment of markedly different ionic strength or pH, it is quite likely that the protein would desorb and/or denature.

The distribution and conformation of three basic proteins (myelin, cytochrome c, and polylysine) at the surface of acidic, charged lipids have been studied[103] by X-ray diffraction. Polylysine was found to attach to the surface as a fully extended chain, while cytochrome c maintained its native structure. The myelin basic protein covers an area of the membrane of about 25 nm^2; it is intimately associated with the surface but does not significantly penetrate it. The use of the Langmuir–Blodgett film has enabled a much more direct measurement of these proteins at an interface and, as the authors point out, "should provide a framework for interpreting the results obtained using similar lipid bilayer systems and other spectroscopic techniques."

To incorporate intrinsic membrane proteins (those which are embedded in the membrane) into a Langmuir–Blodgett film, it is necessary to spread the protein as part of the monolayer with phospholipid. For example, the work with purple membrane[106] and rhodopsin[117] has been mentioned. The effect of an embedded decapeptide on a supported phospholipid monolayer was recently reported.[152] The peptide was covalently linked to a phospholipid, which was itself fluorescently labeled. The peptide lipid diffused laterally at 1.5×10^{-9} cm^2 s^{-1}, which was much less than that of lipids in fluid membranes (0.75×10^{-7} cm^2 s^{-1} for dimyristoylphosphatidylcholine at 30 °C[153]). The use[90] of lipids linked through their head-groups by hydrophilic spacers (Figure 6.8b) may be applicable to the incorporation of proteins into stable lipid matrices. The hydrocarbon chains maintain their mobility and deposition of the prepolymerized monolayer is successful.

Proteins are composed of amino acids and polymerization of derivatized amino acids deposited as Langmuir–Blodgett films has recently been reported.[154] Monolayers of glycine octadecyl ester (condensed) and dl- or l-alanine octadecyl ester (somewhat expanded) gave Y-type deposition onto glass previously coated with one layer of iron stearate. Interbilayer polymerization yielded a helical structure from the l-form or a random coil from the dl-form. The rate of reaction was much faster than in bulk powder or the melt. If the layers of polymerizable material were alternated with nonpolymerizable layers, the polymerization occurred in the plane of the layer leading to a β-sheet structure. (For a brief introduction to the amino acids as building units for proteins and a description of the important polypeptide conformations, see pp. 28–39 of Tschesche's work.[155])

6.4.4. Pigments and Photosynthesis

Procedures for dipping chlorophyll described by Costa *et al.*[111] were referred to in an earlier section. By incorporating cadmium arachidate spacers into these assemblies, they were able to look at the effect of quinone quenchers on chlorophyll fluorescence.[156] A quinone layer adjacent to one of chlorophyll enhanced the rate of photoreduction by threefold. However, photoreduction was reduced if the quencher was incorporated into the chlorophyll layer. To prevent self-quenching the chlorophyll layer was diluted by calcium stearate and oleate. It also appeared possible for the quinone to diffuse through the arachidate layers.

Reorganization of chlorophyll molecules was mentioned in Costa's papers and noted again more recently.[157] In this paper by Leblanc, the aggregation of molecules was reported to vary between the mono- and multilayer arrangement; the role of water in determining the aggregation state was noted. This work has been updated recently[158] by the use of electronic and photoacoustic spectroscopy in mono- and multilayers of chlorophyll *a*. There was good qualitative agreement between the two spectral techniques and the results were compared with a number of models to try and understand the aggregation of chlorophyll *a*.

Models of photosynthetic systems prepared as multilayers from fragments of purple membrane have been described[107,108] with an eye to the possibilities of designing artificial photosynthetic systems. Once again, rearrangement of the lipid films was observed.[108] Photoelectrical effects of chlorophyll *a* on a semitransparent platinum electrode[59] have been measured. The action spectrum has been obtained using a similar arrangement[159] and the role of acceptor molecules, ubiquinone, and plastoquinone studied.[160]

Under less physiological conditions a range of physical measurements of chlorophyll multilayers have been made. Photovoltaic effects,[161] photoconduction,[162] and surface potential measurements[163] have all been reported.

The assumption that mono-/multilayers of chlorophyll mimic to some degree the organization and properties of the pigment *in vivo* is not unreasonable. Spectroscopic studies of these model systems may well explain some of the spectral properties of the real system. However, the development of artificial photosynthetic systems based on the same supported multilayers of natural pigments would still appear to be a long way off.

6.4.5. Electrical Properties

Many of the studies to be described in this section were conducted with dry films, sometimes *in vacuo*. However, the measurements relate these substances to others of a nonbiological origin which have been better characterized and whose electrical properties are understood and described by theory. As a result of this comparison it may be reasonable to consider and look for biochemical properties, which at first sight one might not expect it to possess in its *native* environment. The mistake is to put forward explanations of biological phenomena based *solely* on the properties of a biological molecule removed from its physiological environment, especially as hydration is known to cause major changes in electrical properties.

The electrical properties of dipalmitoyl and distearoyl phosphatidylcholines deposited as a Langmuir–Blodgett film of 1 to 5 layers between two electrodes (Al/Al_2O_3) are reported by Procarione and Kauffman.[164] They observed that the structures themselves generated an EMF which might originate from permanent dipoles embedded in the structure, though movement of ions within the insulating layer or electrode asymmetry might also provide an explanation. Curiously, they observed changes in conductance and capacitance at temperatures associated with the phase transition of the lipid when fully hydrated. The multilayers in this experiment were dry. For the thinner samples, electron tunneling appeared to be the main conduction mechanism.

Tunneling was also suggested for the steady-state conduction of dipalmitoylphosphatidylcholine.[165] The samples were kept for long periods (several days) under electrical stress *in vacuo*, removing all the water. The tunneling distance was reported as 2.7 nm. The electrical properties of ethanolamine lipids are described by the same group.[166,167] Tunneling effects in monolayers, the application of an understanding of this to photosynthesis, and the role of lipids was considered earlier by Kuhn.[168] Theory and calculations of the electrical interactions in phospholipid bilayers[169] highlight the importance of the phosphatidylcholine head-group dipole. It is shown that these dipoles are likely to form arrays in Langmuir films with a net dipole moment. By considering dipole–dipole interactions and the effects of pH and double layers, it is possible to calculate the pressure–area isotherms of Langmuir monolayers and predict phase transitions.

The measurement of electrical properties of Langmuir–Blodgett films of biological molecules is not only restricted to phospholipids and chlorophyll. Dielectric measurements were recently reported of built-up multilayers of hemoglobin[170] and electrical transport through Langmuir–Blodgett films of haemoglobin has been studied.[116]

6.4.6. Polymeric Surfaces

The preparation of stable, polymeric surfaces composed of polar phospholipid head-groups is described by McLean et al.[89] Diacetylenic fatty acids and phosphatidylcholine analogs were studied. The supported monolayer is stabilized by photochemically crosslinking the diacetylenic groups while still under water. Alternatively, after immersion, the surface monolayer is replaced by one of stearic acid and the coated slide raised through this before being exposed to UV radiation to cause polymerization. Applications of these surfaces to the manufacture of molecular sensors or the preparation of blood-compatible surfaces were discussed.

Polymerization of a dispersion of diacetylenic phospholipids did not alter the mobility of spin labels.[66] One explanation could be lateral diffusion and phase separation of the polymerized lipids, with the spin probe remaining in the unpolymerized region. Gaub et al.[171] demonstrated that this occurs in vesicles prepared from dimyristoylphosphatidylcholine and a butadiene containing lipid. However, if vesicles prepared only from the butadiene lipid and fluorescent probe were polymerized, the lateral diffusion coefficient only decreased fourfold from 3×10^{-7} cm^2 s^{-1} to 8×10^{-8} cm^2 s^{-1}. The authors suggest that this occurred because of the limited extent (~10 nm) of the cross-linking resulting in small aggregates.

Although there are reports of lateral diffusion measurements in monolayers on supports, these are restricted to nonpolymerized systems. It would be interesting to obtain such figures for a Langmuir–Blodgett film of polymerized lipid cross-linked to different degrees and dispersed in saturated phospholipid.

6.5. CONCLUDING REMARKS

The preparation of Langmuir–Blodgett films from biological molecules has not proved easy if the usual procedure of repeated dipping cycles is employed. Variations of this technique employing "touching" or horizontal placement of one dipped layer against a monolayer at the air/water interface, have proved successful at generating bilayers with properties akin to biological membranes. The advent of more sophisticated and sensitive spectroscopic techniques (e.g., Fourier–Transform, Infra-Red ATR, Raman microprobe, fluorescent bleaching, and laser-induced phonon spectroscopy) enable a single monolayer or bilayer to be studied. However, if a greater "path length" is required, ordered multilayers can be prepared, but not always by the procedure of Langmuir and Blodgett.

As research in biomembranes moves on from a consideration of the basic structure to studies of surface properties and, in particular, the role of glycoproteins as recognition markers for different cell membranes, deposited lipid bilayers incorporating such proteins will be a very useful tool in studying the type of interactions occurring. Similarly, the preparation of ordered arrays of antigen-labelled lipids and proteins will be important in understanding the forces involved in antibody/antigen binding. The recent studies of surface charges associated with thylakoid stacking,[172] the control of energy transfer between photosystems, and the forces between lipid surfaces[142,143] indicate an increase in the application of the principles of colloid and surface science to biological systems, which might be studied very effectively using supported mono- and bilayers.

The use of more stable surfaces prepared from polymerized phospholipids as immobilization media for proteins or antibodies remains an intriguing possibility. However, significant progress in producing sensing devices based on these materials has not been reported. Arya et al.[173] illustrate the difficulties involved. In general, the incorporation of proteins (surface or embedded) has not been achieved on a wide scale. An exception to this is the deposition of up to fifty lipid bilayers with surface-associated protein by MacNaughton *et al.*[103] The protein appeared to stabilize the deposited layers of phospholipid. The preparation of supported lipid bilayers polymerized at the head-group via a hydrophilic spacer may provide a medium for the incorporation of transmembrane proteins, since the hydrocarbon region remains fluid.[90] The formation of stable lipid/protein monolayers would still seem to require effort. This type of monolayer has been prepared[53,104] for the generation of model bilayers across small (100 μm diameter) holes in septa, so perhaps the techniques employed here might be beneficially applied to the manufacture of Langmuir–Blodgett films containing proteins.

A range of physical measurements on supported biological monolayers is still being applied under nonphysiological conditions. This type of study is sometimes difficult to relate to the biological role of the compound and caution must be shown in interpreting results in this way. However, these measurements can be worthwhile in comparing the physical chemistry of the compound with nonbiological materials and could suggest uses and properties of the biochemical which had not previously been appreciated.

6.6. APPENDIX

Since this chapter was first written, a certain amount of new information has been published. This will now be summarized and the comments will, in general, follow the subject order used in the main text.

The phase changes observed on compressing a phospholipid monolayer have been further characterized by fluorescence microscopy and X-ray diffraction.[186,187] Using carefully obtained experimental data, Pallas and Pethica[188] have shown that for monolayers of *n*-pentadecanoic acid, *n*-hexadecanoic acid, and dihexadecanoylphosphatidylcholine, the so called "liquid-expanded to liquid-condensed" transition is first order and that reported "higher order" transitions are a consequence of poor technique or purity of the monolayer substrate or subphase. They proposed that the "liquid-expanded to liquid-condensed" transition would be better considered as a liquid to close packed or simply as a liquid to solid transition. The phase behavior of phospholipid monolayers has also been the subject of a number of publications[189,190,191] attempting to describe the observed pressure/area isotherms by a variety of theoretical methods.

Several papers have considered the viscoelastic properties of phospholipid monolayers. The surface elasticity and viscosity of dipalmitoylphosphatidylcholine monolayers have been studied[192] by surface light scattering techniques at the air/water and oil/water interfaces (the 'water' phase contained 0.01 M sodium chloride). The response to shear of monolayers of the same phospholipid has also been studied[193] at the air/water interface as surface density and pH were varied. By observing the wavelength change of capillary waves propagated on the surface, the elasticity and viscosity of monolayers of dimyristoylphosphatidylserine and bovine brain phosphatidylserine have been followed[194] as a function of pH and surface pressure. The dimyristoyl analogue had a measureable viscosity which was sensitive to surface pressure and subphase pH. The authors attributed this to hydrogen bonding between the serine headgroup and water. However, the large number of cis double bonds in the analogue from bovine brain disrupted such interactions either at the headgroup or hydrocarbon chain and the monolayer showed low surface viscosity which was independent of surface density. Finally, monolayer viscosities have been monitored by fluorescence microscopy as the gel phase monolayer was forced to move by the application of an inhomogeneous electric field.[195] While the authors stated that the viscosity values obtained were subject to rather large uncertainties, they were obtained without touching the film and the importance of microscopic observation during viscoelastic measurements was demonstrated.

The contribution to the surface potential of monolayers, made by the various chemical structures at the interface, has been described by Vogel and Möbius.[196] By comparing the values obtained from a variety of structures, they were able to assign effective dipole moments to a range of polar groups. They included dipalmitoylphosphatidylcholine and ethanolamine in their study. Using a new procedure to measure surface potentials with a lateral resolution of 40 μm, combined with fluorescence microscopy, Heckl et al.[197] demonstrated that the surface potential of the fluid phase of dimyristoyl phosphatidylethanolamine is 90 ± 10 mV lower than that in the solid phase. After allowing for changes in molecular densities, this figure was used to calculate a change in tilt angle of the hydrocarbon chains on going through the phase transition.

As can be seen so far, there has been an increase in the number of studies undertaken using the more common phospholipids. The range of lipids studied has also increased. Omega-cyclohexyl fatty acids are found in the lipids of certain bacterial membranes. In particular, they comprise at least 70 mol % of membrane lipids of Bacillus acidocaldarius which will tolerate high temperatures (65°C) and acid conditions (~pH 3). The cyclohexyl group creates considerable disturbance in the packing of hydrocarbon chains. In their paper, Asgharian et al.[198] demonstrate that while the lipid occupies a greater area/molecule, the chain–chain interactions are strong leading to expanded states of the monolayer which are stable at high temperatures. This observation is noteworthy in view of the abundance of such lipids in the thermophilic organism. In contrast, polar substituents in the hydrocarbon chains of some phosphatidylcholines produced[199] molecules with two points of surface contact at large areas per molecule. On compression the fatty acid derivatives adopted a more vertical orientation. However, a phospholipid derivative with a keto group at the 12 position in each chain produced an area per molecule consistent with one chain vertical in the air, and the other in the water.

Moving on to protein monolayers, human plasma fibronectin, a glycoprotein involved in, among other things, cell adhesion and differentiation, has been studied at the air/solution interface.[200] Thermodynamic considerations suggested that a number of folding and packing configurations were possible depending upon the charge of the glycoprotein. The interaction between the carbohydrate binding protein concanavolin A and phospholipid monolayers has been investigated by fluorescence and electron microscopy.[201] Cross-linking by polysaccharide was demonstrated and the authors showed that it was possible to transfer the monolayers with the cross-linked protein onto a solid support.

The fluorescence polarization of chlorophyll a and chlorophyll b spread in monolayers of dioleoylphosphatidylcholine has been determined.[202] The polarization of chlorophyll b was 3–4 times greater than that of chlorophyll a and varied with surface pressure, whereas the polarization of chlorophyll a was only slightly affected. The photosynthetic unit is a complex mixture of pigments, proteins, and electron transfer agents and the interaction between chlorophyll a and some of the quinones normally found has been monitored in a study[203] of mixed monolayers. The methods for studying mixtures are well known (see, for example, Gaines[11]) but their application, in particular to monolayers of biological components, is very important in ascertaining the miscibility of mixtures and understanding their interaction.

Moving on to the transfer of monolayer from the air/water surface to a solid support, successful deposition of multilayers of dipalmitoylphosphatidycholine has now be reported[204] using mica as a substrate and with 1.1×10^{-5} M uranyl acetate in the subphase at pH 4.4. On withdrawal, the substrate plus deposited layer appeared wet and had to be allowed to drain. This took about ten minutes. Up to 50 layers were deposited by Y-type transfer. An X-ray study suggested that the uranyl ions were not attached to either side of the bilayer but remained uniformly distributed in the residual water layer. The importance of water entrainment has been demonstrated[205] using a capacitance technique with monolayers of arachidic acid and n-docosyl derivatives of two cyanine dyes. Only poor deposition was seen on the downstroke if insufficient time had been allowed for drainage. On the other hand, too long a period spent below the surface,[183] together with the effects of dipping speed and subphase ion concentration,[206] can give rise to a change from Y-type to X-type transfer.

As part of a general review of molecular monolayers and films, Swalen and co-authors[207] considered the production of thin films of proteins and noted that the manipulation of proteins at interfaces remains difficult to control. The problems increase if an integral membrane protein is to be used since a hydrophobic/hydrophilic environment must be maintained throughout the isolation procedure from original membrane to purified vesicle suspension and subsequent spreading at an air/water interface with accompanying lipid. The use of Langmuir–Blodgett films as chemical sensors (including enzyme biosensors and immunosensors) is further discussed by Mariizumi.[208] Work from this group using the surface associated enzyme, glucose oxidase, demonstrated[209] the importance of electrostatic interactions between the protein and lipid monolayer and the manipulation possible by expansion and recompression. The film was deposited onto a hydrogen peroxide sensor and the sensitivity monitored as a function of monolayer composition and surface pressure at which deposition occurred. Another mixture of glucose oxidase and lipid has been deposited onto a platinum electrode.[210] The amperometric response to a range of glucose concentrations (0.05–1.2 mM glucose) was linear.

In a novel attempt[211] to mimic the biological processes of taste and smell, the fluorescence quenching of deposited films of anthroyloxystearate/stearate and perylene/arachidate by mixtures of amino acids was measured. The natural processes rely upon the response to a range of chemical stimuli which are then processed by the nervous system. In the model system, the amino acids used could be divided into five groups according to the different responses of the fluorescent molecules.

A variety of cyclic polypeptides are capable of acting as ionophores in conveying cations across lipid bilayers. Valinomycin consists of twelve amino acids and is selective for potassium ions. Configured as a Langmuir–Blodgett film, it has been considered as the basis of solid phase potassium selective sensor. However, monolayers of the pure material at an air/water interface appear to adopt the same conformation regardless of subphase composition, be it buffer alone or buffer with either sodium or potassium.[212] Valinomycin has been deposited onto silver, glass with three layers of lead stearate, polymethylmethacrylate,[213] single crystal silicon and quartz.[214] The last two substrates also required a few layers of lead stearate, but interestingly it was show that valinomycin could be deposited onto either the polar or nonpolar surface of the underlying lead stearate. Up to sixty layers could be transferred. Interaction between a Langmuir–Blodgett film of valinomycin and potassium was recently demonstrated.[215] It was necessary to have mixed monolayers of the ionophore with arachidic acid. On its own, a deposited film of valinomycin showed no interaction with potassium.

In Section 6.5 the lack of progress in sensor development was noted. Clearly progress has been made in this field and a more fundamental understanding of the interactions between the "sensing" molecules and supporting monolayer is beginning to develop. The application of a variety of spectroscopic techniques is proving very important in monitoring structural changes in the film before, during, and after deposition, and the further complexities of mixed monolayers are beginning to be addressed. This process could prove very important if both more complex and labile integral membrane proteins are to be incorporated and more realistic mimicry of photosynthetic processes is to be achieved. The difficulty of depositing certain phospholipids may be overcome by the judicious use of subphase ions. However, by understanding the viscoelastic properties of the monolayer during compression and deposition together with better control of wetting/drainage, a more systematic approach to the deposition of these materials should develop.

REFERENCES

1. K. A. Fisher and W. Stoeckenius, in: *Biophysics* (W. Hoppe, W. Lohmann, H. Markl, and H. Zeigler, eds.), pp. 413–425, Springer-Verlag, Berlin (1983).
2. E. Sackmann, in: *Biophysics* (W. Hoppe, W. Lohmann, H. Markl, and H. Zeigler, eds.), pp. 425–460, Springer-Verlag, Berlin (1983).
3. P. Yeagle, *The Membranes of Cells,* Academic Press (1987).
4. J. Seelig and A. Seelig, Lipid conformation in model membranes and biological membranes, *Q. Rev. Biophys.,* **13,** 19–61 (1980).

5. D. Chapman, in: *Liquid Crystals: Fourth State of Matter*, Conference Proceedings, pp. 305–334, Dekker, New York (1979).

6. J. F. Nagle, Theory of the main lipid bilayer phase transition, *Annu. Rev. Phys. Chem.*, **31**, 157–195 (1980).

7. M. J. Tanner, Isolation of integral membrane proteins and criteria for identifying carrier proteins, *Top. Membranes Transp.*, **12**, 1–51 (1979).

8. D. R. Nelson and N. C. Robinson, Membrane proteins: a summary of known structural information, *Methods Enzymol.*, **97**, 571–618 (1983).

9. G. G. Roberts, Langmuir–Blodgett films, *Contemp. Phys.*, **25**, 109–128 (1984).

10. G. L. Gaines, Jr., From monolayer to multilayer: some unanswered questions, *Thin Solid Films*, **68**, 1–5 (1980).

11. G. L. Gaines, *Insoluble Monolayers at Gas–Liquid Interfaces*, Interscience, New York (1966).

12. M. C. Phillips, Physical state of phospholipids and cholesterol in monolayers, bilayers and membranes, *Progr. Surf. Membr. Sci.*, **5**, 139–121 (1972).

13. S. Ohki and C. B. Ohki, Monolayers at the oil/water interface as a proper model for bilayer membranes, *J. Theor. Biol.*, **62**, 389–407 (1976).

14. D. Papahadjopoulos, Phospholipids as model membranes. Monolayers, bilayers and vesicles, *Biophys. Biochim. Acta Library*, **3**, 143–169 (1973).

15. J. Mingins, N. F. Owens, and D. H. Iles, Properties of monolayers at the air–water interface. 1. The effect of spreading solvent on the surface pressure of octadecyltrimethylammonium bromide, *J. Phys. Chem.*, **73**, 2118–2126 (1969).

16. D. A. Cadenhead and B. M. Kellner, Monolayer spreading solvents with special reference to phospholipid monolayers, *J. Colloid Interface Sci.*, **49**, 143–145 (1974).

17. J. Mingins and J. A. G. Taylor, Physicochemical properties of phospholipid monomolecular layers, *Proc. R. Soc. Med.*, **66**, 383–385 (1973).

18. G. Collacicco, Lipid monolayers: ionic impurities and their influence on the surface potentials of neutral phospholipids, *Chem. Phys. Lipids*, **10**, 66–72 (1973).

19. S. Sato and H. Kishimoto, The contact angle of phospholipid monolayer on a Wilhelmy plate, *J. Colloid Interface Sci.*, **69**, 188–191 (1979).

20. H. E. Ries Jr., G. Albrecht and, L. Ter-Minassian-Saraga, Collapsed monolayers of egg lecithin, *Langmuir*, **1**, 135–137 (1985).

21. N. L. Gershfeld and K. Tajima, Spontaneous formation of lecithin bilayers at the air–water surface, *Nature*, **279**, 708–709 (1979).

22. A. G. Bois and N. Albon, Equilibrium spreading pressure of L-alpha-dipalmitoyl lecithin below the main bilayer transition temperature: can it be measured? *J. Colloid Interface Sci.*, **104**, 579–582 (1985).

23. M. C. Phillips and H. Hauser, Spreading of solid glycerides and phospholipids at the air–water interface, *J. Colloid Interface Sci.*, **49**, 31–39 (1974).

24. M. Tomoaia-Cotisel, E. Chifu, A. Sen, and P. J. Quinn, Galactolipid and lecithin monolayers at the air/water interface, *Dev. Plant Biol.*, **8**, 393–396 (1982).

25. E. M. Arnett, N. Harvey, E. A. Johnson, D. S. Johnston, and D. Chapman, No phospholipid monolayer–sugar interactions, *Biochemistry*, **25**, 5239–5242 (1986).

26. J. Mingins, E. Llerenas, and B. A. Pethica, The role of interfacial charges in the phase behaviour of lipid monolayers and bilayers, *Colston Papers No.29* (D. H. Everett and Vincent, eds.), pp. 41–68, Scientechnica (1978).

27. S. Ohki, C. B. Ohki, and N. Duzgunes, Monolayer at the air/water interface vs. oil/water interface as a bilayer membrane model, *Colloid Interface Science* (Proc. International Conf.), **5**, 271–284 (1976).

28. M. C. Phillips, H. Hauser, and F. Paltauf, The inter- and intra-molecular mixing of hydrocarbon chains in lecithin/water systems, *Chem. Phys. Lipids*, **8**, 127–133 (1972).

29. E. G. Finer and M. C. Phillips, Factors affecting molecular packing in mixed lipid monolayers and bilayers, *Chem. Phys. Lipids*, **10**, 237–252 (1973).

30. R. A. Demel, K. R. Bruckdorfer, and L. L. M. Van Deenen, Structural requirements of sterols for the interaction with lecithin at the air–water interface, *Biochim. Biophys. Acta,* **225,** 311–320 (1972).

31. D. A. Cadenhead, B. M. J. Kellner, and M. C. Phillips, The miscibility of dipalmitoyl phosphatidylcholine and cholesterol in monolayers, *J. Colloid Interface Sci.,* **57,** 224–227 (1976).

32. J. A. G. Taylor, J. Mingins, B. A. Pethica, Beatrice Y. J. Tan, and C. M. Jackson, Phase changes and mosaic formation in single and mixed phospholipid monolayers at the oil–water interface, *Biochim. Biophys. Acta,* **323,** 157–160 (1973).

33. J. Teissie, J. F. Tocanne, and A. Baudras, Phase transitions in phospholipid monolayers at the air–water interface: a fluorescence study, *FEBS Lett.,* **70,** 123–126 (1976).

34. V. Von Tscharner and H. M. McConnell, An alternative view of phospholipid phase behaviour at the air-water interface. Microscope and film balance studies, *Biophys. J.,* **36,** 409–419 (1981).

35. V. Von Tscharner and H. M. McConnell, Physical properties of lipid monolayers on alkylated planar glass surfaces, *Biophys. J.,* **36,** 421–427 (1981).

36. M. Lösche, J. Rabe, B. Fischer, U. Rucha, W. Knoll, and H. Möhwald, Microscopically observed preparation of Langmuir–Blodgett Films, *Thin Solid Films,* **117,** 269–280 (1984).

37. R. M. Weis and H. M. McConnell, Two dimensional chiral crystals of phospholipid, *Nature,* **310,** 47–49 (1984).

38. E. M. Arnett and J. M. Gold, Chiral aggregation phenomena. 4. A search for stereospecific interactions between highly purified enantiomeric and racemic dipalmitoyl phosphatidylcholines and other chiral surfactants in monolayers, vesicles and gels, *J. Am. Chem. Soc.,* **104,** 636–639 (1982).

39. R. Subramanian and L. K. Paterson, Paper presented at the Second International Conference on Langmuir–Blodgett Films, Schenectady, NY, U.S.A., July 1–4 (1985).

40. M. C. Phillips, The conformation and properties of proteins at liquid interfaces, *Chem. Ind.,* 170–176 (1977).

41. L. K. James and L. G. Augenstein, Adsorption of enzymes at interfaces: film formation and the effect on activity, *Adv. Enzymol.,* **28,** 1–40 (1966).

42. P. Fromherz, A new technique for investigating lipid protein films, *Biochim. Biophys. Acta,* **225,** 382–387 (1971).

43. M. C. Phillips, H. Hauser, R. B. Leslie, and D. Oldani, Comparison of the interfacial interactions of the apoprotein from high density lipoprotein and beta-casein with phospholipids, *Biochim. Biophys. Acta,* **406,** 402–414 (1975).

44. M. C. Phillips, D. E. Graham, and H. Hauser, Lateral compressibility and penetration into phospholipid and bilayer membranes, *Nature,* **254,** 154–156 (1975).

45. M. C. Phillips, M. T. A. Evans, and H. Hauser, Interaction of proteins with phospholipid monolayers, *Adv. Chem. Ser.* **144** (*Monolayers Membr. Syn.*), 217–230 (1975).

46. D. G. Cornell and R. J. Carroll, Miscibility in lipid–protein monolayers, *J. Colloid Interface Sci.,* **108,** 226–233 (1985).

47. J. Teissie, Interaction of cytochrome *c* with phospholipid monolayers. Orientation and penetration of protein as functions of the packing density of film, nature of the phospholipids and ionic content of the aqueous phase, *Biochemistry,* **20,** 1554–1560 (1981).

48. P. M. Vassilev, S. Taneva, I. Panaiotov, and G. Georgiev, Dilatational viscoelastic properties of tubulin and mixed tubulin–lipid monolayers, *J. Colloid Interface Sci.,* **84,** 169–174 (1981).

49. D. E. Graham and M. C. Phillips, Proteins at liquid interfaces. 4. Dilatational properties, *J. Colloid Interface Sci.,* **76,** 227–239 (1980).

50. D. E. Graham and M. C. Phillips, Proteins at liquid interfaces. 5. Shear properties, *J. Colloid Interface Sci.,* **76,** 240–250 (1980).

51. R. Verger and F. Pattus, Spreading of membranes at the air/water interface, *Chem. Phys. Lipids,* **16,** 285–291 (1976).

52. T. Kanno, M. Setaka, T. Hongo, and T. Kwan, Spontaneous formation of a monolayer membrane from sarcoplasmic reticulum at an air–water interface, *J. Biochem. (Tokyo),* **94,** 473–477 (1983).

53. H. Schindler, Exchange and interactions between lipid layers at the surface of a liposome solution, *Biochim. Biophys. Acta,* **555,** 316–336 (1979).

54. F. Jähnig, Lipid exchange between membranes, *Biophys, J.,* **46,** 687–694 (1984).

55. D. G. Cornell and R. J. Carroll, Electron microscopy of lipid-protein monolayers, *Colloids Surf.,* **6,** 385–393 (1983).

56. T. Miyasaka, T. Watanabe, A. Fujishima, and K. Honda, Light energy conversion with chlorophyll monolayer electrodes. In vitro electrochemical simulation of photosynthetic primary processes, *J. Am. Chem. Soc.,* **100,** 6657–6665 (1978).

57. T. Watanabe, T. Miyasaka, A. Fujishima, and K. Honda, Photoelectrochemical study of chlorophyll monolayer electrodes, *Chem. Lett.,* 443–446 (1978).

58. T. Miyasaka, T. Watanabe, A. Fujishima, and K. Honda, Highly efficient quantum conversion at chlorophyll *a* -lecithin mixed monolayer coated electrodes, *Nature,* **277,** 638–640 (1979).

59. J-G. Villar, Etude des effets photoélectrochimiques de films multimoléculaires de chlorophylle *a* formés sur une électrode de platine semi-transparante, *C.R. Acad. Sci. Paris, Ser. D.,* **275,** 861–864 (1972).

60. R. Jones, R. H. Tredgold, and P. Hodge, Langmuir–Blodgett films of simple esterified porphyrins, *Thin Solid Films,* **99,** 25–32 (1983).

61. J-P. Chauvet, M. L. Agrawal, G. L. Hug, and L. K. Patterson, Effects of molecular organization on photophysical behaviour: steady-state and real-time behaviour of chlorophyll *b* fluorescence in spread monolayers of dioleoylphosphatidylcholine, *Thin Solid Films,* **133,** 227–234 (1985).

62. D. Ducharme, C. Salesse, and R. M. Leblanc, Ellipsometric studies of rod outer segment phospholipids at the nitrogen–water interface, *Thin Solid Films,* **132,** 83–90 (1985).

63. J. F. Baret, H. Hasmonay, J. L. Firpo, J. J. Dupin, and M. Dupeyrat, The different types of isotherm exhibited by insoluble fatty acid monolayers. A theoretical interpretation of phase transitions in the condensed state, *Chem. Phys. Lipids,* **30,** 177–187 (1982).

64. A. I. Feher, F. D. Collins, and T. W. Healy, Mixed monolayers of simple saturated and unsaturated fatty acids, *Aust. J. Chem.,* **30,** 511–519 (1977).

65. B. Hupfer and H. Ringsdorf, Spreading and polymerization behaviour of diacetylene phospholipids at the gas–water interface, *Chem. Phys. Lipids,* **33,** 263–282 (1983).

66. D. S. Johnston, L. R. McLean, M. A. Whittam, A. D. Clark, and D. Chapman, Spectra and physical properties of liposomes and monolayers of polymerizable phospholipids containing diacetylene groups in one or both chains, *Biochemistry,* **22,** 3194–3202 (1983).

67. R. Büschl, B. Hupfer, and H. Ringsdorf, Polyreactions in oriented systems. 30. Mixed monolayers and liposomes from natural and polymerizable lipids, *Macromol. Chem. Rapid Commun.,* **3,** 589–596 (1982).

68. B. Tieke, V. Enkelmann, H. Kapp, G. Lieser, and G. Wegner, Topochemical reactions in Langmuir–Blodgett multilayers, *J. Macromol. Sci., Chemistry A,* **15,** 1045–1058 (1981).

69. D. R. Day, H. Ringsdorf, and J. B. Lando, Polymerization of surface active diacetylene monomers at the gas–water interface, *Polym. Prepr., Am. Chem. Soc., Div. Polym. Chem.,* **19,** 176–178 (1978).

70. V. Martin, H. Ringsdorf, and D. Thunig, Polymerization of micelle forming monomers, *Midl. Macromol. Monogr.,* **3,** 175–188 (1977).

71. T. Kunitake and Y. Okahata, Bilayer membranes prepared from modified dialkylammonium salts and methyldialkylsulphonium salts, *Chem. Lett.,* 1337–1340 (1977).

72. N. Nakashima, R. Ando, and T. Kunitake, Casting of synthetic bilayer membranes on glass and spectral variation of membrane-bound cyanine and merocyanine dyes, *Chem. Lett.,* 1577–1580 (1983).

73. E. P. Honig, Th. J. H. Hengst, and D. den Engelsen, Langmuir–Blodgett deposition ratios, *J. Colloid Interface Sci.,* **45,** 92–102 (1973).

74. H. Hasmonay, M. Caillaud, and M. Dupeyrat, Langmuir–Blodgett multilayers of phosphatidic acid and mixed phospholipids, *Biochim. Biophys. Res. Commun.,* **89,** 338–344 (1979).

75. G. L. Gaines Jr., Interface curvature corrections for film balance measurements, *J. Colloid Interface Sci.,* **98,** 272–273 (1984).

76. J. P. Green, M. C. Phillips, and G. G. Shipley, Structural investigations of lipid, polypeptide and protein multilayers, *Biochim. Biophys. Acta*, **330**, 243–253 (1973).

77. O. Albrecht, D. S. Johnston, C. Villaverde, and D. Chapman, Stable biomembrane surfaces formed by phospholipid polymers, *Biochim. Biophys. Acta*, **687**, 165–169 (1982).

78. I. V. Langmuir and V. J. Schaefer, Activities of urease and pepsin monolayers, *J. Am. Chem. Soc.*, **60**, 1351–1360 (1938).

79. L. K. Tamm and H. M. McConnell, Supported phospholipid bilayers, *Biophys. J.*, **47**, 105–113 (1985).

80. M. F. Daniel, O. C. Lettington, and S. M. Small, Langmuir–Blodgett films of amphiphiles with cyano headgroups, *Mol. Cryst. Liq. Cryst.*, **96**, 373–385 (1983).

81. M. Lösche, C. Helm, H. D. Mattes, and H. Möhwald, Formation of Langmuir–Blodgett films via electrostatic control of the lipid/water interface, *Thin Solid Films*, **133**, 51–64 (1985).

82. E. P. Honig, Molecular constitution of X- and Y-type Langmuir–Blodgett films, *J. Colloid Interface Sci.*, **43**, 66–72 (1973).

83. K. Fukuda and T. Shiozawa, Conditions for formation and structural characterization of X-type and Y-type multilayers of long-chain esters, *Thin Solid Films*, **68**, 55–66 (1980).

84. M. Saint-Pierre and M. Dupeyrat, Measurement and meaning of the transfer process energy in the building up of Langmuir–Blodgett multilayers, *Thin Solid Films*, **99**, 205–213 (1983).

85. I. R. Peterson, G. J. Russell, and G. G. Roberts, A new model for the deposition of omega-tricosenoic acid Langmuir–Blodgett film layers, *Thin Solid Films*, **109**, 371–378 (1983).

86. B. L. Eyres and R. M. Swart, unpublished results.

87. B. Hupfer and H. Ringsdorf, Polymeric monolayers and liposomes as models for biomembranes and cells, *ACS Symp. Ser.*, **175** (*Polym. Sci. Overview*), 209–232 (1981).

88. D. S. Johnston, S. Sanghera, A. Manjon-Rubio, and D. Chapman, The formation of polymeric model biomembranes from diacetylenic fatty acids and phospholipids, *Biochim. Biophys. Acta*, **602**, 213–216 (1980).

89. L. R. McLean, A. A. Durrani, M. A. Whittam, D. S. Johnston, and D. Chapman, Preparation of stable polar surfaces using polymerizable long-chain diacetylenic molecules, *Thin Solid Films*, **99**, 127–131 (1983).

90. R. Elbert, A. Laschewsky, and H. Ringsdorf, Hydrophilic spacer groups in polymerizable lipids: formation of biomembrane models from bulk polymerized lipids, *J. Am. Chem. Soc.*, **107**, 4134–4141 (1985).

91. S. A. Asher and P. S. Pershan, Alignment and defect structures in oriented phosphatidycholine multilayers, *Biophys. J.*, **27**, 393–422 (1979).

92. L. Powers and N. A. Clark, Preparation of large monodomain phospholipid bilayer smectic liquid crystals, *Proc. Natl. Acad. Sci. U.S.A.*, **72**, 840–843 (1975).

93. L. Powers, J. P. LePesant, and P. S. Pershan, Optical studies of monodomain phospholipid bilayers, *Biophys. J.*, **16**, 138a (1976), abstract.

94. L. Powers and P. S. Pershan, Optical studies of monodomain phospholipid bilayers containing various biological membrane components, *Biophys. J.*, **16**, 138a (1976), abstract.

95. L. Huang and H. M. McConnell, Formation of lipid multilayer on alkylated glass surface, *Biophys. J.*, **41**, 115a (1983), abstract.

96. W. L. Peticolas, M. Harrand and R. Dupeyrat, Polarized Raman spectra of oriented monodomains of phospholipid monolayers, *J. Raman Spectrosc.*, **12**, 130–132 (1982).

97. B. P. Gaber and W. L. Peticolas, On the quantitative interpretation of biomembrane structure by Raman spectroscopy, *Biochim. Biophys. Acta*, **465**, 260–274 (1977).

98. E. Bicknell-Brown, K. G. Brown, and W. B. Person, Conformation dependent Raman bands of phospholipid surfaces, *J. Raman Spectrosc.*, **12**, 180–189 (1982).

99. M. Harrand, Polarized Raman spectra of oriented dipalmitoylphosphatidylcholine (DPPC). 1. Scattering activities of skeletal stretching and methylene vibrations of hydrocarbon chains, *J. Chem. Phys.*, **79**, 5639–5651 (1983).

100. M. J. M. Van de Ven and Y. K. Levine, Angle-resolved fluorescence depolarization of macroscopically ordered bilayers of unsaturated lipids, *Biochim. Biophys. Acta*, **777**, 283–296 (1984).

101. C. S. Winter and R. H. Tredgold, Langmuir–Blodgett multilayers of polypeptides, *Thin Solid Films*, **123**, L1–L3 (1985).

102. T. Furuno, H. Sasabe, R. Nagata, and T. Akaike, Studies of Langmuir–Blodgett films of poly(1-benzyl-L-histidine) stearic acid mixtures, *Thin Solid Films*, **133**, 141–152 (1985).

103. W. MacNaughton, K. A. Snook, E. Caspi, and N. P. Franks, An X-ray diffraction analysis of oriented lipid multilayers containing basic proteins, *Biochim. Biophys. Acta*, **818**, 132–148 (1985).

104. H. Schindler and U. Quast, Functional acetylcholine receptor from *Torpedo marmorata* in planar membranes, *Proc. Natl. Acad. Sci. U.S.A.*, **77**, 3052–3056 (1980).

105. N. Nelson, R. Anholt, J. Lindstrom, and M. Montal, Reconstitution of purified acetylcholine receptors with functional ion channels in planar lipid bilayers, *Proc. Natl. Acad. Sci. U.S.A.*, **77**, 3057–3061 (1980).

106. S. B. Hwang, J. I. Korenbrot, and W. Stoeckenius, Structural and spectroscopic characteristics of bacteriorhodopsin in air–water interface films, *J. Membr. Biol.*, **36**, 115–136 (1977).

107. J. R. Brocklehurst and M. T. Flanagan, Multilayer models of photosynthetic membranes, *Comm. Eur. Communities Report (EUR 7688)*, 1–135 (1982).

108. M. T. Flanagan, The deposition of Langmuir–Blodgett films containing purple membrane on lipid and paraffin impregnated filters, *Thin Solid Films*, **99**, 133–138 (1983).

109. J. Schildkraut and A. Lewis, Purple membrane and purple membrane–phospholipid Langmuir–Blodgett films, *Thin Solid Films*, **134**, 13–26 (1985).

110. W. M. Heckl, M. Lösche, and H. Möhwald, Langmuir–Blodgett films containing proteins of the photosynthetic process, *Thin Solid Films*, **133**, 73–81 (1985).

111. S. M. de B. Costa, J. R. Froines, J. M. Harris, R. M. Leblanc, B. H. Orger, and G. Porter, Model systems for photosynthesis. 3. Primary photo-processes of chloroplast pigments in monomolecular arrays at solid surfaces, *Proc. R. Soc. London, Ser. A.*, **326**, 503–519 (1972).

112. K. Iriyama, Methods for preparing chlorophyll *a* multilayers on glass plates, *Photochem. Photobiol.*, **29**, 633–636 (1979).

113. K. Iriyama, M. Yoshiura, and F. Mizutani, Deposition of chlorophyll *a* Langmuir–Blodgett films onto a SnO_2 optically transparent electrode, *Thin Solid Films*, **68**, 47–54 (1980).

114. K. Iriyama, Preparation of multilayers containing chlorophyll *a* and/or phosphatidylcholine and the chemical stability of chlorophyll *a* molecules in the multilayers, *J. Membr. Biol.*, **52**, 115–120 (1980).

115. A. Ruaudel-Teixier, A. Barraud, B. Belbeoch, and M. Roulliay, Langmuir–Blodgett films of pure porphyrins, *Thin Solid Films*, **99**, 33–40 (1983).

116. J. B. Hasted, A. K. Ko, Y. Al-Baker, S. Kadifachi, and D. Rosen, Electrical transport in haemoglobin Langmuir–Blodgett films, *J. Chem. Soc., Faraday Trans.*, **81**, 463–472 (1985).

117. C. N. Kossi and R. M. Leblanc, Rhodopsin in a new model bilayer membrane, *J. Colloid Interface Sci.*, **80**, 426–436 (1981).

118. H. Kuhn, Functionalized monolayer assembly manipulation, *Thin Solid Films*, **99**, 1–16 (1983).

119. H. Kuhn, Synthetic molecular organizates, *J. Photochem.*, **10**, 111–132 (1979).

120. H. Kuhn and D. Möbius, Systems of monomolecular layers. Assembly and physico-chemical behaviour, *Angew, Chem., Int. Ed. Engl.*, **10**, 620–637 (1971).

121. H. Kuhn, D. Möbius, and H. Bücher, in: *Physical Methods of Chemistry* (A. Weissberger and B. W. Rossiter, eds.), **Part 3B**, pp. 577–702, Wiley-Interscience, New York (1972).

122. M. H. Vos, R. P. H. Kooyman, and Y. K. Levine, Angle resolved fluorescence depolarisation experiments on oriented lipid membrane systems, *Biochem. Biophys. Res. Commun.*, **116**, 462–468 (1983).

123. L. Powers and P. S. Pershan, Monodomain samples of dipalmitoyl phosphatidylcholine with varying concentrations of water and other ingredients, *Biophys. J.*, **20**, 137–152 (1977).

124. P. K. J. Kinnunen, J. A. Virtanen, A. P. Tulkki, R. C. Ahiya, and D. Möbius, Pyrene-fatty acid-containing phospholipid analogues: characterisation of monolayers and Langmuir–Blodgett assemblies, *Thin Solid Films*, **132**, 193–203 (1985).

125. M. Seul, P. Eisenburger, and H. McConnell, X-ray diffraction by phospholipid monolayers on single-crystal surfaces, *Proc. Natl. Acad. Sci. U.S.A.*, **80**, 5795–5797 (1983).

126. J. B. Stamatoff, W. F. Graddick, L. Powers, and D. E. Moncton, Direct observation of the hydrocarbon chain tilt angle in phospholipid bilayers, *Biophys. J.*, **25**, 253–261 (1979).

127. M. Hentschel and R. Hoseman, Small and wide angle X-ray scattering of oriented lecithin multilayers, *Mol. Cryst. Liq. Cryst.*, **94**, 291–316 (1983).

128. S. W. Hui, M. Cowden, D. Papahadjopoulos, and D. F. Parsons, Electron diffraction study of hydrated phospholipid single bilayers. Effects of temperature, hydration and surface pressure of the "precursor" monolayer, *Biochim. Biophys, Acta*, **382**, 265–275 (1975).

129. G. Eyring and M. D. Fayer, A laser-induced ultrasonic probe of the mechanical properties of aligned lipid multilayers, *Biophys. J.*, **47**, 37–42 (1985).

130. L. J. Lis, S. C. Goheen, and J. W. Kauffman, Raman spectroscopy of fatty acid Blodgett–Langmuir multilayer assemblies, *Biochem. Biophys, Res. Commun.*, **78**, 492–497 (1977).

131. M. Delhaye, M. Dupeyrat, R. Dupeyrat, and Y. Lévy, An improvement in the Raman spectroscopy of very thin films, *J. Raman Spectrosc.*, **8**, 351–352 (1979).

132. A. Aurengo, M. Masson, R. Dupeyrat, Y. Lévy, H. Hasmonay, and J. Barbillot, Technical device for obtaining Raman spectra of ultrathin films of phospholipids, *Biochem. Biophys, Res. Commun.*, **82**, 559–564 (1979).

133. R. Dupeyrat and M. Masson, in: *Microbeam Analysis* (K. F. J. Heinrich, ed.), pp. 286–288, San Francisco Press (1982).

134. M. Vandevyver, A Ruaudel Teixier, I. Richamet, and M. Lutz, Polarized resonance Raman spectroscopy of Langmuir–Blodgett films, *Thin Solid Films*, **99**, 41–44 (1983).

135. R. A. Uphaus, T. M. Cotton, and D. Möbius, Surface-enhanced resonance Raman spectroscopy of synthetic dyes and photosynthetic pigments in monolayer and multilayer assemblies, *Thin Solid Films*, **132**, 173–184 (1985).

136. F. Kopp, U. P. Fringeli, K. Mühlethaler, and H. Hs. Güthard, Spontaneous rearrangement in Langmuir–Blodgett layers of tripalmitin studied by means of ATR infrared spectroscopy and electron microscopy, *Z. Naturforsch.*, **30c**, 711–717 (1975).

137. E. Okamura, J. Umemura, and T. Takenaka, Fourier transform infrared attenuated total reflection spectra of dipalmitoyphosphatidylcholine monomolecular films, *Biochim. Biophys. Acta*, **812**, 139–146 (1985).

138. P. Fromherz, A new method for investigation of lipid assemblies with a lipid pH indicator in monomolecular films, *Biochim. Biophys. Acta*, **323**, 326–334 (1973).

139. A. Mellier, O. Auge, and P. Crouigneau, Molecular interactions at the phospholipid–water interface. Infrared spectrum of dimyristoyl-L-alpha-lecithin adsorbed on hydrated potassium bromide, *Colloids Surf.*, **7**, 325–337 (1983).

140. D. G. Hafeman, V. van Tscharner, and H. M. McConnell, Specific antibody dependent interactions between macrophages and lipid haptens in planar lipid monolayers, *Proc. Natl. Acad. Sci. U.S.A.*, **78**, 4552–4556 (1981).

141. G. M. Humphries and H. M. McConnell, Antibodies against spin labels, *Biophys. J.*, **16**, 275–277 (1976).

142. J. Marra and J. N. Israelachvili, Direct measurements of forces between phosphatidylcholine and phosphatidylethanolamine bilayers in aqueous electrolyte solutions, *Biochemistry*, **24**, 4608–4618 (1985).

143. J. Marra, Controlled deposition of lipid monolayers and bilayers onto mica and direct force measurements between galactolipid bilayers in aqueous solutions, *J. Colloid Interface Sci.*, **107**, 446–458 (1985).

144. N. L. Thompson, H. M. McConnell, and T. P. Burghardt, Order in supported phospholipid monolayers detected by the dichroism of fluorescence excited polarised evanescent radiation, *Biophys, J.*, **46**, 739–747 (1984).

145. M. Masson and R. Dupeyrat, in: *Microbeam Analysis* (K. F. J. Heinrich, ed.), pp. 289–293, San Francisco Press (1982).

146. N. P. Franks and W. R. Lieb, Rapid movement of molecules across membranes. Measurement of the permeability coefficient of water using neutron diffraction, *J. Mol. Biol.*, **141**, 43–61 (1980).

147. N. P. Franks and W. R. Lieb, Where do general anaesthetics act? *Nature*, **274**, 339–342 (1978).

148. N. P. Franks and W. R. Lieb, The structure of lipid bilayers and the effects of general anaesthetics. An X-ray and neutron diffraction study, *J. Mol. Biol.*, **133**, 469–500 (1979).

149. R. J. Vanderveen and G. T. Barnes, Water permeation through Langmuir–Blodgett monolayers, *Thin Solid Films*, **134**, 227–236 (1985).

150. P. Fromherz and D. Marcheva, Enzyme kinetics at a lipid protein monolayer, induced substrate inhibition of trypsin, *FEBS Lett.*, **49**, 329–333 (1975).

151. J. Peters and P. Fromherz, Interaction of electrically charged monolayers with malate dehydrogenase, *Biochim. Biophys. Acta*, **394**, 111–119 (1975).

152. N. L. Thompson, A. A. Brian, and H. M. McConnell, Covalent linkage of a synthetic peptide to a fluorescent phospholipid and its incorporation into supported phospholipid monolayers, *Biochim. Biophys. Acta*, **772**, 10–19 (1984).

153. J. L. R. Rubenstein, B. A. Smith, and H. M. McConnell, Lateral diffusion in binary mixtures of cholesterol and phosphatidylcholines, *Proc. Natl. Acad. Sci. U.S.A.*, **76**, 15–18 (1979).

154. K. Fukuda, Y. Shibasaki, and H. Nakahara, Molecular arrangement and polymerizability of amino-acid derivatives and dienoic acid in Langmuir–Blodgett films, *Thin Solid Films*, **133**, 39–49 (1985).

155. H. Tschesche, in: *Biophysics* (W. Hoppe, W. Lohmann, H. Markl, and H. Zeigler, eds.), pp. 20–41, Springer-Verlag, Berlin (1983).

156. S. M. de B. Costa and G. Porter, Model systems for photosynthesis. 4. Photosensitization by chlorophyll *a* monolayers at a lipid/water interface, *Proc. R. Soc. London, Ser. A.* **341**, 167–176 (1974).

157. R. Leblanc, Optical properties of biological pigments in monolayer and multilayer arrays, *Thin Solid Films*, **99**, 140 (1983), abstract.

158. A. Désormeaux and R. M. Leblanc, Electronic and photoacoustic spectroscopies of chlorophyll *a* in monolayer and multilayer arrays, *Thin Solid Films*, **132**, 91–99 (1985).

159. J-G. Villar, Photoelectrochemical effects in the electrolyte–pigment–metal system. 3. Chlorophyll films short-circuit photocurrent transients light energy conversion efficiency, *J. Bioenerg. Biomembr.*, **8**, 199–208 (1976).

160. A. F. Janzen and J. R. Bolton, Photochemical electron transfer in monolayer assemblies. 2. Photoelectric behaviour in chlorophyll *a*-acceptor systems, *J. Am. Chem. Soc.*, **101**, 6342–6348 (1979).

161. R. Jones, R. H. Tredgold, and J. E. O'Mullane, Photoconductivity and photovoltaic effects in Langmuir–Blodgett films of chlorophyll *a*, *Photochem. Photobiol.*, **32**, 223–232 (1980).

162. K. J. McCree, Photoconduction and photosynthesis. 1. The photoconductivity of chlorophyll monolayers, *Biochim. Biophys. Acta*, **102**, 90–95 (1965).

163. R. H. Tredgold and G. W. Smith, Surface potential studies on Langmuir–Blodgett multilayers and adsorbed monolayers, *Thin Solid Films*, **99**, 215–220 (1983).

164. W. L. Procarione and J. W. Kauffman, The electrical properties of phospholipid bilayer Langmuir films, *Chem. Phys. Lipids*, **12**, 251–260 (1974).

165. D. M. Taylor and M. G. B. Mahboubian-Jones, Steady state conduction in synthetic phospholipid films, *Thin Solid Films*, **99**, 149–156 (1983).

166. M. G. B. Mahboubian-Jones and D. M. Taylor, Electrical properties of thin biological films grown by the Langmuir–Blodgett technique, *J. Electrostat.*, **13**, 1–8 (1982).

167. D. M. Taylor and M. G. B. Mahboubian-Jones, The electrical properties of synthetic phospholipid Langmuir–Blodgett films, *Thin Solid Films*, **87**, 167–179 (1982).

168. H. Kuhn, Electron tunneling effect in monolayer assemblies, *Chem. Phys. Lipids*, **8**, 401–404 (1972).

169. P. J. Bowen and T. J. Lewis, Electrical interactions in phospholipid layers, *Thin Solid Films*, **99**, 157–163 (1983).

170. A. Y. Ko, D. Rosen, and J. B. Hasted, Dielectric properties of Langmuir–Blodgett films of haemoglobin, *Proc. First Int. Conf. Conduct. & Breakdown of Solid Dielectrics*, pp. 175–179 (1983).

171. H. Gaub, E. Sackmann, R. Büschl, and H. Ringsdorf, Lateral diffusion and phase separation in

two dimensional solutions of polymerized butadiene lipid in dimyristoylphosphatidylcholine bilayers. A photobleaching and freeze fracture study, *Biophys. J.,* **45,** 725–731 (1984).

172. J. Barber, Influence of surface charges on thylakoid structure and function, *Annu. Rev. Plant Physiol.,* **33,** 261–295 (1982).

173. A. Arya, U. J. Krull, Michael Thompson, and H. E. Wong, Langmuir–Blodgett deposition of lipid films on hydrogel as a basis for biosensor development, *Anal. Chim. Acta,* **173,** 331–336 (1985).

174. N. L. Gershfeld, W. F. Stevens, Jr., and R. J. Nossal, Equilibrium studies of phospholipid bilayer assembly, *Faraday Discuss. Chem. Soc.,* **81,** 19–28 (1986).

175. B. Maggio, F. A. Cumar, and R. Caputto, Molecular behaviour of glycosphingolipids in interfaces. Possible participation in some nerve membranes, *Biochim. Biophys. Acta.,* **650,** 69–87 (1981).

176. J. H. Crowe, L. M. Crowe, J. F. Carpenter, A. S. Rudolf, C. A. Wistrom, B. J. Spargo, and T. J. Anchordoguy, Interactions of sugars with membranes, *Biochim. Biophys. Acta.,* **947,** 367–384 (1988).

177. R. A. Demol and B. DeKruyff, The function of sterols in membranes, *Biochim. Biophys. Acta.,* **457,** 109–132 (1976).

178. M. Lösche, H-P. Duwe, and H. Möhwald, Quantitative analysis of surface textures in phospholipid monolayer phase transitions, *J. Colloid Interface Sci.,* **126,** 432–444 (1988).

179. C. A. Helm, H. Möhwald, K. Kjaer, and J. Als-Nielsen, Phospholipid monolayers between fluid and solid states, *Biophys. J.,* **52,** 381–390 (1987).

180. C. Salesse, D. Ducharme, and R. M. Leblanc, Direct evidence for the formation of a monolayer from a bilayer—An ellipsometric study at the nitrogen–water interface, *Biophys. J.* **52,** 351–352 (1987).

181. H. Nakahara, K. Fukuda, H. Akutsu, and Y. Kyogoku, Monolayers and multilayers of phosphatidylethanolamine: effects of spreading solvent, monovalent cations and substrate pH, *J. Colloid Interface Sci.,* **65,** 517–526 (1978).

182. H. M. McConnell, T. H. Watts, R. M. Weiss, and A. A. Brian, Supported planar membranes in studies of cell–cell recognition in the immune system, *Biochim. Biophys. Acta.,* **864,** 95–106 (1986).

183. E. P. Honig, The transition from Y- to X- type Langmuir–Blodgett films, *Langmuir,* **5,** 882–883 (1989).

184. T. H. Watts and H. M. McConnell, in: *Processing and Presentation of Antigens* (B. Pernis, S. C. Silverstein and H. J. Vogel, eds.), pp. 143–155, Academic Press (1988).

185. D. Axelrod, T. P. Burghardt, and N. L. Thompson, Total internal reflection fluorescence, *Ann. Rev. Biophys. Bioeng.,* **13,** 247–268 (1984).

186. H. Möhwald, Direct characterization of monolayers at the air–water interface, *Thin Solid Films,* **159,** 1–15 (1988).

187. K. Kjaer, J. Als-Nielsen, C. A. Helm, P. Tippmann-Krayer, and H. Möhwald, An X-ray scattering study of lipid monolayers at the air-water interface and on solid supports, *Thin Solid Films,* **159,** 17–28 (1988).

188. N. R. Pallas and B. A. Pethica, Liquid-expanded to liquid-condensed transitions in lipid monolayers at the air–water interface, *Langmuir,* **1,** 509–513 (1985).

189. R. S. Cantor and K. A. Dill, Theory for the equation of state of phospholipid monolayers, *Langmuir,* **2,** 331–337 (1986).

190. D. Stigter and K. A. Dill, Lateral interactions among phospholipid head groups at the heptane/water interface, *Langmuir,* **4,** 200–209 (1988).

191. J. A. Poulis, A. A. H. Boonman, P. Gieles, and C. H. Massen, Influence of clustering on the behavior of insoluble monolayers on water: a theoretical approach, *Langmuir,* **3,** 725–729 (1987).

192. B. B. Sauer, Y-L. Chen, G. Zografi, and H. Yu, Surface light-scattering studies of dipalmitoyl phosphatidylcholine monolayers at the air/water and heptane/water interfaces, *Langmuir,* **4,** 111–117 (1988).

193. B. M. Abraham and J. B. Ketterson, Dipalmitoyllecithin monoalyers at the air/water interface: measurements of the response to shear as a function of surface density and pH, *Langmuir,* **1,** 708–713 (1985).

194. B. M. Abraham and J. B. Ketterson, Viscoelastic measurements at the air/water interface on monolayers of dimyristoylphosphatidylserine (DMPS) and of bovine brain phosphatidylserine (BBPS), *Langmuir*, **2**, 801–805 (1986).

195. W. M. Heckl, A. Miller, and H. Möhwald, Electric-field induced domain movement in phospholipid monolayers, *Thin Solid Films*, **159**, 125–132 (1988).

196. V. Vogel and D. Möbius, Hydrated polar groups in lipid monolayers: effective local dipole moments and dielectric properties, *Thin Solid Films*, **159**, 73–81 (1988).

197. W. M. Heckl, H. Baumgärtner, and H. Möhwald, Lateral surface potential distribution of a phospholipid monolayer, *Thin Solid Films*, **173**, 269–278 (1989).

198. B. Asgharian, D. K. Rice, D. A. Cadenhaed, R. N. A. H. Lewis, and R. N. McElhaney, Monomolecular film behaviour of a homologous series of 1,2-bis(ω-cyclohexyacyl) phosphatidylcholines at the air/water interface, *Langmuir*, **5**, 30–34 (1989).

199. F. M. Menger, S. D. Richardson, M. G. Wood, Jr., and M. J. Sherrod, Chain-substituted lipids in monomolecular films. Effect of polar substituents on molecular packing. *Langmuir*, **5**, 833–838 (1989).

200. N. F. Zhou and B. A. Pethica, Monolayers of human plasma fibronectin at the air/water interface, *Langmuir*, **2**, 47–50 (1986).

201. W. M. Heckl, M. Thompson, and H. Möhwald, Fluorescence and electron microscopic study of lectin-polysaccharide and immunochemical aggregation at phospholipid Langmuir–Blodgett monolayers, *Langmuir*, **5**, 390–394 (1989).

202. J-P. Chauvet and L. K. Patterson, Measurement of fluorescence depolarization for chlorophyll-a and chlorophyll-b in spread monolayers at the nitrogen-water interface, *Thin Solid Films*, **159**, 149–157 (1988).

203. D. Guay and R. M. Leblanc, Excess free energies of interaction of chlorophyll-a with α-tocopherylquinone and plastoquinone 3 and 9. A mixed-monolayer study, *Langmuir*, **3**, 575–580 (1987).

204. J. B. Peng, M. Prakash, R. Macdonald, P. Dutta, and J. B. Ketterson, Fornation of multilayers of dipalmitoyl phosphatidylcholine using the Langmuir–Blodgett technique, *Langmuir*, **3**, 1096–1097 (1987).

205. M. P. Srinivasan, B. G. Higgins, P. Stroeve, and S. T. Kowel, Entrainment of aqueous subphase in Langmuir–Blodgett films, *Thin Solid Films*, **159**, 191–205 (1988).

206. J. B. Peng, J. B. Ketterson, and P. Dutta, A study of the transition from Y- to X-type transfer during deposition of lead stearate and cadmium stearate Langmuir–Blodgett films, *Langmuir*, **4**, 1198–1202 (1988).

207. J. D. Swalen, D. L. Allara, J. D. Andrade, E. A. Chandross, S. Garoff, J. Israelachvili, T. J. McCarthy, R. Murray, R. F. Pease, J. B. Rabolt, K. J. Wynne, and H. Yu, Molecular monolayers and films, *Langmuir*, **3**, 932–950 (1987).

208. T. Moriizumi, Langmuir–Blodgett films as chemical sensors, *Thin Solid Films*, **160**, 413–429 (1988).

209. M. Sriyudthsak, H. Yamagishi, and T. Moriizumi, Enzyme-immobilized Langmuir–Blodgett film for a biosensor, *Thin Solid Films*, **160**, 463–469 (1988).

210. Y. Okahata, T. Tsuruta, K. Ijiro, and K. Ariga, Langmuir–Blodgett films of an enzyme-lipid complex for sensor membranes, *Langmuir*, **4**, 1373–1375 (1988).

211. M. Aizawa, M. Matsuzawa, and H. Shinohara, An optical chemical sensor using a fluorophor-embedded Langmuir–Blodgett film, *Thin Solid Films*, **160**, 477–481 (1988).

212. B. M. Abraham and J. B. Ketterson, Determination of the viscosity of valinomycin monolayers as a function of surface density and a comment on conformation, *Langmuir*, **1**, 461–464 (1985).

213. J. B. Peng, B. M. Abraham, P. Dutta, J. B. Ketterson, and H. F. Gibbard, Langmuir–Blodgett deposition of a ring-shaped molecule (valinomycin), *Langmuir*, **3**, 104–106 (1987).

214. V. A. Howarth, M. C. Petty, G. H. Davies, and J. Yarwood, The deposition and characterization of multilayers of the ionophore valinomycin, *Thin Solid Films*, **160**, 483–489 (1988).

215. V. A. Howarth, M. C. Petty, G. H. Davies, and J. Yarwood, Infrared studies of valinomycin containing Langmuir–Blodgett films, *Langmuir*, **5**, 330–332 (1989).

216. I. D. Swalen, Linear and nonlinear optical and spectroscopic properties of Langmuir–Blodgett films, *Thin Solid Films*, **160**, 197–208 (1988).

Potential Applications of Langmuir–Blodgett Films

G. G. ROBERTS

7.1. INTRODUCTION

The considerable surge of activity in Langmuir–Blodgett film research occurred approximately forty years after the first detailed report of sequential monolayer transfer.[1] Some of the motivation and enthusiasm arose from the results of fundamental work[2,3] on energy transfer in monomolecular assemblies. However, the resurgence of activity in academic and industrial laboratories, and increased funding for the subject, can mainly be attributed to British and French groups highlighting their possible applications in the field of electronics.[4–6] Several applications-oriented review articles[7–10] have been written but no commercial device is yet available.

7.1.1. Molecular Electronics

Before reviewing future prospects for the exploitation of LB films, it is instructive to consider how they might be fitted into the requirements of the electronics industry. It is accepted that the microelectronics and optoelectronics industries will continue to grow vigorously well into the 21st century. Until now, they have relied largely on inorganic materials such as silicon and lithium niobate in single crystal form. However, as the perceived limitations restrict the realization of more complex system designs, more attention is being focused on the organic solid state. The

G. G. ROBERTS • Department of Engineering Science, University of Oxford, Oxford OX1 3PJ, England, and Thorn EMI plc, Hayes, Middlesex UB3 1HH, England.

richness of the variety of organic molecular materials offers enormous potential compared with the relative paucity of structures achievable with inorganic compounds, even when due allowance is made for the recent exciting developments in inorganic quantum well semiconductors.[11]

The ability to enlist the assistance of synthetic organic chemists to produce organic materials with tailored properties has, of course, been used to advantage in several applications. The best known is that of liquid crystals and their use in displays and digital thermometers. New phenomena and types of molecule are still being discovered and seem likely to lead to successful large-area displays for high-definition television and to high-density information stores. Other examples are the use of piezoelectric polymers in very sensitive hydrophones for submarine detection, photoconducting polymers for electrocopying, and photochromic molecules for reversible high-density optical storage and signal processing. Biosensors and chemical sensors for converting specific biochemical or chemical solute or gas interactions into electrical signals for use in industrial or medical diagnostics are also examples of "Molecular Electronics"; that is, they are fields in which organic molecular materials perform an active function in the processing of information and its transmission and storage. This definition does not embrace their use in possible roles such as insulation, adhesion, or encapsulation. Thus, molecular electronics is interpreted broadly and is not limited to phenomena concerning the movement of electrons only. Electromagnetic radiation and polarization phenomena, and various forms of electromechanical and electrochemical energy transfer are also included in the definition. A common feature of all the examples cited and of the area in general is that progress is achieved via "molecular engineering," that is, using that ability to manipulate the architecture of a material to optimize a specific physical parameter.

7.1.2. Supermolecular Electronics

An alternative definition exists for molecular electronics; this is formulated in terms of switching on a molecular scale and is aimed more at the long-term problem of fabricating molecular electronic devices suitable for assembly into a computer.[12] It is interesting to note that only a modest diminution in the size of electronic circuit components is required before the scale of individual molecules is reached. An illustration of the rapid evolution of silicon based microelectronics may be gained by plotting the linear feature size of commercial electronic circuits versus time using a log-linear scale. If the good straight line signifying the rapid rate of reduction in feature size is extrapolated into the future, then device geometries with nanometer dimensions are predicted in approximately thirty years' time. The requirements of reliability and testing of complex structures suggest a system approach rather than the traditional one which uses the properties of individual circuit elements. It appears likely that sequential designs, because of their vulnerability, will be abandoned in favor of supermolecular arrays acting as concurrent processor networks. For this reason, and to differentiate it from "molecular electronics," signal transport and control in nanometer scale assemblies is referred to here as "supermolecular electronics."

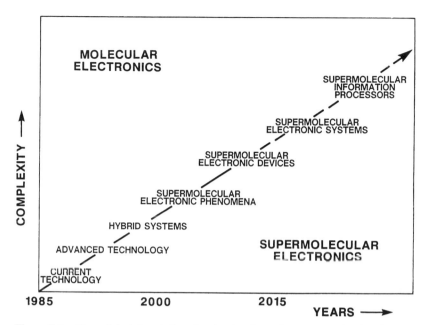

Figure 7.1. The anticipated evolution of molecular electronics to supermolecular electronics.

Three-dimensional integration is extremely difficult using silicon technology while, chemically, nonspecific methods such as molecular beam epitaxy are relatively crude. A self-assembly technique is a far more attractive alternative if regular, three-dimensional, ordered structures are required. This involves the construction of unique assemblies whose architectures depend on the shapes and charge distributions of the units from which they are built; as distinct from the methods used to assemble them. There is a considerable degree of self-assembly associated with organic monomolecular films deposited using the Langmuir trough technique. It seems likely that an understanding of their physical properties and their utilization in molecular-based devices will assist generally in the transition from molecular to supermolecular electronics. The likely pattern for this evolutionary process is given in Figure 7.1. The plan is speculative and envisages three main stages before the advent of applied, complex supermolecular systems. In the near term it predicts that current research and development will result in organic materials ousting inorganic materials in certain existing applications, such as optoelectronics. Hybrid technologies, comprising a novel device based on conventional solid-state materials and on organic compounds, could possibly follow in the mid-term, say 5 or 8 years. Thereafter, at some stage dictated by the emergence of reliable, stable supermolecular assemblies, a true watershed will occur. The materials and process technologies associated with conventional solid-state devices will be superseded by radically new types based on a different class of device. This era will be equivalent to that witnessed about forty years ago when inorganic semiconductor materials were developed. Just as then, interesting new effects should be discovered and these in turn will lead to the fabrication of special devices which can be integrated into novel systems.

7.1.3. Possible Niches for Langmuir–Blodgett Films

Outlined above is a key reason for sustaining a well-coordinated research activity in LB films, embracing basic as well as applied research. That is, in the longer term, investigations of switching processes and rectification effects may well prove to be the stimulus for the design of novel supermolecular information processors. However, before then, current research and development is likely to lead to organic materials produced in LB film form replacing inorganic materials in more conventional situations. Most of this chapter is devoted to this topic but some attention is given to Langmuir and Blodgett's early thoughts on how to apply LB films and their usefulness in fundamental research.

Organic materials are already being used successfully in high-technology applications despite their inherent lack of stability compared with most inorganic materials. Having established that organic films can be superior to their inorganic counterparts, the onus is then on the LB film enthusiasts to demonstrate the special advantages to be gained using their technique. More often than not this will require attention to be placed on a figure of merit for a device rather than a single physical parameter of the material. For commercial applications, the monomolecular and multilayer films will need to play an essential role before a company is prepared to invest in a new technology. That is, one must capitalize on their special features such as the degree of control over molecular architecture, their ultrathinness, and the selective way in which they might react with the environment. These considerations, taken together with the requirements of mechanical and thermal stability, will certainly restrict the niches likely to be occupied by LB films to a relatively small number.

Despite the increased activity during the past ten years or so, no important commercial venture has yet been started based on LB films. However, it should be remembered that the level of financial investment, even in the general field of organic molecular solids, is still a small fraction of the sum being devoted to inorganic materials. When an institution decides to invest in an expensive items such as molecular beam epitaxial growth equipment, it invariably ensures that a reasonable number of workers are associated with the project. Langmuir troughs are relatively inexpensive and normally much smaller-size teams are involved. This factor, coupled with the difficulty of organizing interdisciplinary teams, probably accounts for the current scenario. It is pertinent to draw a comparison between the present situation for LB films and that which existed fifteen years ago for liquid crystals. At that time, when the display application was clearly appreciated, there was a large background knowledge of structure–property relationships. This enabled applied scientists to replace deficient materials by more desirable, tailored structures which had already been thoroughly investigated by chemists. Very few correlations of this kind exist for LB films, and hence it is unreasonable to expect technical progress to take place at the same dramatic rate as occurred for liquid crystals. Now that there is a genuine commitment to the subject, with higher levels of funding and manpower being made available, it is likely that at least one innovative application will emerge. When this happens there will be a need to produce a specially designed trough, for example, one capable of coating a moving belt or multiple wafers of silicon. No difficulties are envisaged in constructing continuous fabrication equipment and substantial progress has already been made in this direction.

7.2. PATENTS FILED BY LANGMUIR AND BLODGETT

Even in the early stage of development of the subject, some of the motivation for floating monolayer LB film work lay in their possible applications. For example, the first attempt[13] to control evaporation and thus prevent the loss of fluids from storage reservoirs was carried out in 1924. This subject has since formed the basis of a special conference,[14] the general conclusion being that the best evaporation retardants are straight-chain compounds such as fatty alcohols.

Langmuir and Blodgett filed several patents based on their discoveries[15–20] but none appears to have led to any significant commercial advantage for their employers, the General Electric Company. Nevertheless, it is interesting to reflect on their initial thoughts regarding how to exploit monolayer and multilayer films. They relate mainly to the use of skeletonized films described below.

7.2.1. Image Reproduction

Elsewhere[15] Blodgett states:

An important aspect of my present invention consists in the provision of stratified films in which there are numerous voids of molecular dimensions in the various film strata. Such film structures may be prepared by forming successive monolayers comprising two or more separable constituents and thereafter selectively abstracting one of such constituents to leave a molecular skeleton film or lattice.

A factor which is significant in respect to the utility of skeleton films of the character referred to consists in the ability of such a film to absorb into its structural voids other substances. From an optical standpoint the phenomenon just referred to may be utilized by heating various portions of a skeleton film permeated with a volatile impregnating medium to cause selective vaporization of the medium. Such vaporization will be accompanied by corresponding color changes in the selectively heated portions of the films as a result of local variations in refractive index.

Practically, this feature may be employed in various ways in image reproduction devices. Thus, if the heating means comprises, for example, an invisible infra-red light image projected on the film, a visible or photographically reproducible facsimile of such image will be produced on the film surface. Other media than infra-red light, such as cathode ray producing means, may also be used to stimulate the film. For this reason, the invention has an obvious application in the field of television.

7.2.2. Optical Devices

Katharine Blodgett's earliest patent[15] also alludes to the optical properties of LB films which "may be built up to have a thickness which is controllable within extraordinarily narrow limits, for example, within one ten millionth of an inch." However, more details are to be found in subsequent patents[16,17] and papers[21,22] dealing with low reflectance glass and the reduction of surface reflection. She claimed that:

> Because of the accuracy with which their thickness may be controlled, it is practical to use the films as colour standards, as standards for minute thickness measurements, and as devices for the detection of very minute quantities of material substances. These uses depend primarily upon the optical properties of the films, although other properties may also be utilized. A further important aspect of my invention consists in the provision and application of certain types of reflection-reducing coatings, and especially in the provision of coating substances suitable for use with transparent bodies having indices of refraction on the order of that of glass. Such a substance should have a refractive index between that of air and that of glass and preferably between the values of about 1.2 and about 1.3. The substance should also be of such character as neither to absorb nor scatter an appreciable amount of light. The best value of refractive index for a material to be applied to ordinary glass (index 1.52) is substantially below those which characterize known solid substances. However, a refractive index within the desired range may be realized by the formation of a so-called skeleton film.
>
> In the case of cadmium arachidate, a film of about forty layers thickness is appropriate. After its formation the cadmium arachidate film may be skeletonized by subjecting the film to a suitable solvent. For example, a cadmium arachidate film which has been built from the surface of a body of water containing a dissolved cadmium salt in a concentration 10^{-4} molar and having a pH = 5.7 comprises about 50% cadmium arachidate and 50% arachidic acid. When a film of this type is soaked for about one to five minutes in a solvent such as alcohol or acetone and is then withdrawn, the character of the film is substantially altered. This is due to the solution from the film of the arachidate being left as a skeleton with air filling the spaces previously occupied by the arachidic acid. A film subjected to a procedure such as is outlined in the foregoing is found to have a refractive index of approximately 1.25, this being relatively close to the geometric mean between the refractive index of air and that of a glass of index 1.42. Such a film is also highly uniform as to thickness and is characterized by negligible absorption and scattering of light.
>
> In reading a meter, the observer is frequently troubled by glare produced by reflection from the surface of the viewing window. Such glare makes it difficult to read the meter by dazzling or obscuring the observer's vision. This difficulty may be substantially eliminated through my invention by applying to the inner and outer surfaces of the viewing window, coatings of the character described above. It may be applied with equal advantage in any situation or arrangement involving the use of a viewing window. Particular examples which may be enumerated comprise watch dials, show windows, showcase windows, picture frame glass, automobile windshields, spectacle lenses, and the like. It has equivalent utility in connection with many other optical devices such as telescopes, binoculars, microscopes, spectrographic apparatus and other laboratory instruments. In many of these devices the occurrence of reflection imposes the main limitation on sensitivity. Consequently, the removal of this factor by utilization of my invention will make possible a general advance in the optical arts.

Further work[23] on the optical properties of skeletonized films has been published confirming that they remain optically clear even after solvent treatment has reduced the amount of acid molecules to less than 60% of the original. It has also been confirmed that the change in color is due to a change in the refractive index and not a change in thickness. Using a microscope glass slide of refractive index 1.5 coated on both sides with multilayers of $\lambda/4$ optical thickness and refractive index 1.245, a reflection coefficient as low as 0.25% is found for monochromatic light in the visible region. Using simple fatty acid type molecules, the refractive index can be adjusted between 1.20 and 1.50 using different pH values. These results suggest that there may be some merit in forming $\lambda/4, \lambda/2$ sequences aimed at producing tailor-made band-pass filters as discussed by Smith.[24] The reversibility of the process and the extraordinary rigidity of the lattice is confirmed by experiments such as those carried out by Race and Reynolds[25] showing how the optical thickness is restored by filling the spaces with some polar oil.

7.2.3. Step Gauge

Skeletonized film also formed the basis of the step gauge sold by the General Electric Company which is illustrated in Figure 7.2. In her patent and associated paper[26] Blodgett states:

> In carrying out my invention, a base member of polished X-ray shield glass treated with hot hydrogen is provided with a coating of barium stearate separated into gradu-

Figure 7.2. A step gauge built up of barium stearate layers, patented by K. Blodgett and sold by the General Electric Company.

ated steps of known thicknesses to form a gauge. When it is required to determine the thickness of a film, the gauge and the film are exposed to light from a single source, and the color reflected from the film and the matching or complementary color from a step of the gauge are compared. The thickness of the film is proportional to the index thickness of the step from which the compared color is reflected.

7.2.4. Mechanical Filters

Langmuir and Blodgett also anticipated one of the applications discussed later in this chapter, namely that of ultrafine filtration. In her first patent[15] Blodgett states:

> Skeleton films of the class referred to also have many other uses which are independent of the optical properties. For example, such films may be employed as sieves or filters for segregation of previously non-filterable substances of molecular magnitudes. It is only necessary for such filtration that the particles or molecules to be filtered be of a size greater than the dimensions of the molecular voids formed in the film structure.

7.2.5. Sensors

One of the important topics discussed later is that of sensors. Langmuir published many papers on built-up layers of proteins and filed a patent[18] entitled "Method of substance detection." In it he described how

> One may prepare a monolayer which is conditioned selectively to absorb the suspected substance and no other likely to be present. The prepared surface may then be exposed to the carrier medium, the occurrence or non-occurrence of a change in the properties of the surface being indicative of the presence or absence of the suspected substance. Various properties of monolayers may be employed as indicia of the adsorption of a foreign substance desired to be detected. For example, one may observe variations in thickness, refractive index, solubility, volatility, contact angle against various liquids, adsorbing power for other substances etc.
>
> Selective absorbency may be obtained by employing as a surface monolayer a material which is specifically reactive with respect to the suspected substance. Specific reactions are numerous in biological and toxicological fields, so that the method just outlined is capable of application in such fields. Many proteins, including disease toxins and viruses, are also capable of specific reactions, so that the invention may be used in the study and control of biological reactions involved in the diagnosis and treatment of disease.

7.2.6. Summary

Despite the intense activity in recent years involving sophisticated troughs, improved deposition techniques, and specially designed molecularly engineered materials, there is a surprising amount of useful information to be found by studying the early papers and patents on the subject. Gaines's book[27] is an excellent source of useful hints and data for the newcomer to the field. Some of the very early papers describe results which others were later able to exploit. For example, Porter and

Wyman[28] discovered that potentials of 70 mV per layer were produced using X-type deposition. These large local fields, which arise due to bipolar orientation, have relevance to work on semiconductor surface barriers and pyroelectric detectors. Germer and Storks[29] were the first to characterize LB films using electron diffraction techniques. Based on this structural information Henke[30] struck on the idea of using fatty acid monolayers as large two-dimensional spacing analyzers for use in the ultrasoft X-ray region. In the range 1.5 nm to 15 nm, resolving powers in the range 50 to 150 were obtainable using lead stearate. In fact, the concept was commercialized and several companies marketed lead stearate as part of light element analysis kits. Another interesting paper, and one which heralded more sophisticated and controlled studies of electron energy ranges, is that by Widdowson and Gregg[31] who showed how the range of recoil atoms could be determined using radioactive LB films.

7.3. MODEL SYSTEMS IN BASIC RESEARCH

Whatever the commercial prospects for LB films, there is no doubting the applicability of the Langmuir trough in fundamental research. There are many areas of science that can benefit from investigations of model systems based on monomolecular assemblies.

7.3.1. Energy Transfer in Complex Monolayers

The Langmuir trough technique provides a method of constructing simple artificial systems of cooperating molecules on a substrate. The pioneer in this field has been Kuhn[2,3]; the elegance of his and his colleagues' work is evident in their reviews of the subject. These describe the use of LB films to investigate intermolecular interactions and various photophysical and photochemical processes. Their supermolecular structures normally involved long-chain fatty acids as matrices for appropriate synthetic dyes; these were ingeniously designed to clarify the different interactions that can occur between molecules via photon, electron, and proton transfer. An example of their research, designed to investigate the Forster type of energy transfer from a sensitizing molecule S to an acceptor molecule A is given in Figures 7.3 and 7.4. If S is a compound which absorbs in the ultraviolet part of the spectrum and fluoresces in the blue, while A absorbs in the blue and fluoresces in the yellow, then interesting effects are observed when the system is irradiated with ultraviolet light. If there is a sufficient distance between S and A, as in Figure 7.3a, the fluorescence of S appears since A does not absorb UV radiation. However, below a certain threshold distance, as in Figure 7.3b, the excitation energy of S is transferred to A and the yellow fluorescence of A is expected. Similar experiments based on fluorescence quenching indicate that the rate constant of the electron transfer decreases exponentially with increasing barrier thickness separating a donor chromophore and an electron acceptor. In the example shown in Figure

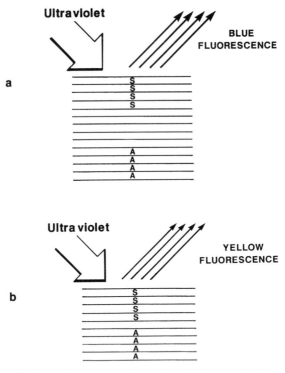

Figure 7.3. Schematic diagram showing monolayers of a sensitizing dye S and an acceptor molecule A separated by spacer monolayers of fatty acids. In a no interaction is possible but in b the spacing is close enough for quantum mechanical tunneling and energy transfer to occur between A- and S-type layers.

7.4, a cyanine dye has been used in conjunction with a viologen acceptor layer to observe the steady fluorescence intensities of the cyanine dye monolayer in the absence (I_o) and presence (I) of the acceptor layer. The quantity, $[(I_o/I) -1]$ is proportional to the rate constant of the electron transfer; its linear dependence with d, the distance between the chromophores, is evidence of electron tunneling. In a similar series of experiments it has been possible to investigate the energy transfer mechanism responsible for spectral sensitization.[2]

7.3.2. Magnetic Monolayers

The addition of divalent ions into the liquid subphase in a Langmuir trough sometimes increases both the shear resistance and cohesion of the monolayer.[27] Consequently, much of the early work on LB films was carried out on fatty acid salts of metals such as Cd, Ba, and Ca. By adjusting the pH of the subphase it is then possible to build up multilayers containing metal ions separated by the width of

FLUORESCENCE
QUENCHING

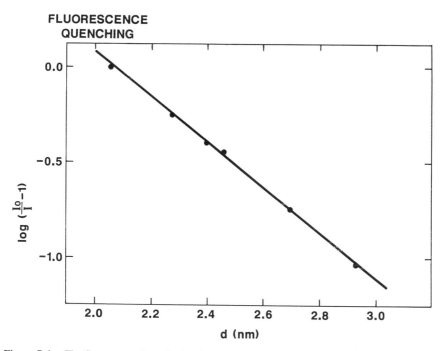

Figure 7.4. The fluorescence intensity I_o of a donor dye is reduced to value I in the presence of an acceptor dye. The logarithm of $[(I_o/I) - 1]$ is shown as a function of d, the spacing between the donor and acceptor planes. (See Möbius.[32])

an integral number of monolayers. This facility has been capitalized upon in two types of fundamental investigations. The first to be discussed here is magnetism.

Theoretical considerations show that, in three dimensions, magnetic ordering is always possible but in one dimension it is impossible. However, theoretical predictions concerning its existence in two dimensions are model-dependent. Pomerantz[35] and Haseda et al.[34] were the first to recognize the importance of using LB films to study this interesting space. Their two-dimensional magnetism experiments were carried out using fatty acid salts of manganese. Using electron spin resonance, Pomerantz demonstrated that at temperatures near 2 K the resonance field and lineshapes were affected, thus indicating the rapid development of a large internal magnetic field. He claimed that the nature of the ordered state was likely to be predominantly antiferromagnetic, but with a weak ferromagnetic component. Owing to the sensitivity of low-dimensional structures to external perturbations it is naturally more desirable to examine these magnetic transitions in a zero magnetic field, for example, carry out an experiment to detect the magnetization of a single magnetic monolayer using a SQUID magnetometer. If it provided evidence of magnetic order this might stimulate more applied research, for example, that aimed at magnetic control of superconducting memory devices.

One of the most interesting challenges in the field of organic molecular materials is the design of molecular ferromagnets.[35] Using polymetallic systems, one is thwarted by symmetry requirements from achieving ferromagnetic interactions between nearest-neighbor ions and therefore an alternative approach has been tried involving complexes of metals with local spins of $\frac{5}{2}$ and $\frac{1}{2}$. One of the most exciting developments has been the observation of magnetic bistability involving a spin transition phenomenon. The hexacoordinated iron(II) complexes displaying this effect show a transition from an $s = 2$ ground state to an $s = 0$ state as the temperature is lowered through 174 K. A color change accompanies the change of spin. The LB films which have been deposited based on the same basic molecule. $Fe(OP_3)_2NCS_2$, where OP is $1-10$ phenanthrolene substituted with three aliphatic chains, show similar behavior[36] but the transition is less abrupt. Using a 400-monolayer-thick sample, the midpoint of the temperature range over which the transition takes place is approximately 250 K. Possibly, due to the disorder present, both spin states are present throughout the range; but at a local level the transition is reasonably sharp. Clearly, there are applied prospects for this work in the field of information storage.

7.3.3. Determination of Electron Energy Ranges

Another example of the usefulness of the Langmuir trough technique in being able to control the concentration of metal ions incorporated in an organic matrix is to be found in electron energy range determination.

In the area of surface science, X-ray photoemission (XPS) has been used extensively to study polymer surfaces. One of the most important aspects of this procedure is a knowledge of electron mean free paths as a function of kinetic energy. The first definitive study of this kind in LB films was carried out by Brundle *et al.*[37] on a fatty acid salt system; they reported a mean free path of approximately 3.5 nm for electrons of kinetic energy 1 keV. Since then Clark *et al.*[38] have extended the work to diacetylene polymer and anthracene, and have confirmed that the mean free paths for ordered multilayers are significantly longer than those for conventional polymers.

Another interesting use of LB films involving a measurement of the range of electrons has been carried out by Mori *et al.*[39]; they studied radioactive stearate monolayers in which some of the hydrogen atoms had been replaced by nuclides such as ^{51}Cr, ^{54}Mn, ^{55}Fe, ^{57}Co, ^{65}Zn, and ^{109}Cd. They showed that it is possible to deposit a dilute radioactive source using the Langmuir trough technique and then use standard monolayers as absorbers of well-controlled dimensions. This approach avoids the customary difficulties due to self-absorption in measuring the range of L-shell Auger electrons. Results show that Auger electrons from the L shell with an energy of approximately 0.5 keV are almost completely absorbed by 15 monolayers of barium stearate. On the other hand, Auger electrons from the K shell, with energy one order of magnitude higher, are barely affected in this distance. By studying the intensity ratio of L to K electrons, Mori *et al.*[39] showed that it is possible to avoid fluctuations in the radioactive source strength. By labeling the overlayer molecules with ^{14}C and examining the autoradiographs, information concerning the uniformity of the deposition process can be obtained. Experiments of this kind to determine the range of electrons with energy <1 keV are of importance in fields such as medical physics and upper atmosphere science. A further development of the principle discussed here has been the report of an ultrathin film tritium source for calibration purposes with low-energy beta emitters.[40] Similar work has been carried out by Holmes *et al.*[41]; they used tritiated palmitic acid to examine the tritium spectrum and thus place improved limits on the mass of the electron antineutrino.

7.3.4. Biological and Permeable Membranes

Biochemists have long been aware of the usefulness of the Langmuir trough technique to assemble biological materials of interest. Langmuir, for example, was the first to report on multilayers of chlorophyll *a*, the green pigment in higher plants. Other examples of nature's materials that readily form LB multilayers are given in Chapter 6.

The physical structure and chemical nature of classical LB films give them a close resemblance to naturally occurring biological membranes. For example, because the two ends of a lipid molecule have incompatible solubilities, they spontaneously organize themselves in the form of a bilayer, or essentially a two-layer LB film. Scientists have suggested that they might provide a suitable model of the lipid membrane for probing the cooperative interactions between its constituents. However, caution must be exercised in assessing the biological relevance of this type of work and associated studies aimed at incorporating ionophores into phospholipid layers.

Much of this research is targeted at novel integrated solid-state devices incorporating biological molecules such as enzymes; this topic is discussed later in this chapter. However, in some cases, LB films have been useful to facilitate physical studies of biological molecules, for example, to measure the ionic permeability of reconstituted membranes.[42] Supermolecular structures have also been designed to mimic the primary process in photosynthesis and for achieving an efficient photoinduced charge separation by appropriate modeling of potential profiles.

As an alternative to conventional solar cells made from inorganic semiconductors, it is possible to envisage incorporating an optically active biological material into a model membrane. The efficiency of such a battery would be small but the low fabrication costs in energy terms would be attractive. A natural example is bacteriorhodopsin whose molecules, when illuminated, undergo a number of physical and chemical changes which result in protons being transported from one side of the membrane to another. An attempt to reproduce this process artificially has been reported.[43]

This type of energy conversion mechanism could also be adapted for use in other areas of biotechnology involving the permeability characteristics of mem-

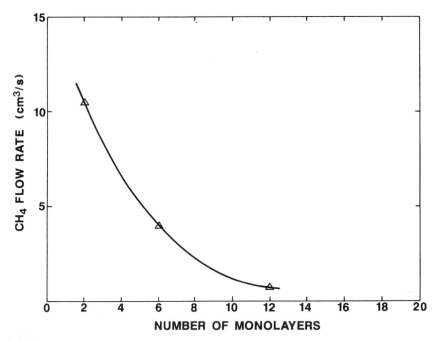

Figure 7.5. Flow of CH_4 through polypropylene substrate coated with different number of monolayers of the cadmium salt of unpolymerized hexacosa-10, 12-diynoic acid. (See Albrecht *et al.*[46])

branes, for example, LB films as synthetic membranes for ultrafiltration, gas separation, and reverse osmosis. Existing developments on a commercial scale are hindered by relatively poor separation efficiencies and, partly because of their amorphous nature, the short lifetimes of the materials used. Although membranes can be made from several types of material, most work has been carried out on polymers. Membrane-based gas separators are proving to be very useful in the recovery of gases such as hydrogen and the removal of corrosive gases such as water and carbon dioxide.[44] Rose and Quinn[45] were the first to fabricate composite membranes using the LB technique, but more successful results have been obtained by Albrecht *et al.*[46] The general principle is that molecular-size pores can impart high selectivity to the membrane while the ultrathinness of the LB film permits high mass transfer rates. Figure 7.5 shows the effect of depositing multilayers of a diacetylene lipid onto a porous polypropylene support. Eighteen monolayers of hexacosa-10,12-diynoic acid reduce the gas flow by a factor of thirty compared to an uncoated substrate. In this case, stabilizing the film by polymerization caused no apparent change in the permeation characteristics. Promising data have also been obtained by Kajiyama and co-workers[47] showing how LB multilayers can control the permeability of K^+ ions and O_2/N_2 gases. Heckmann *et*

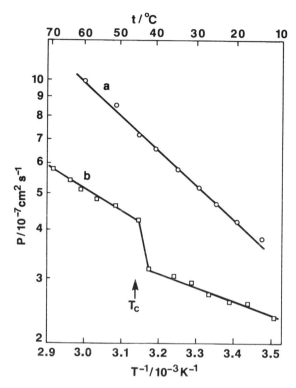

Figure 7.6. Arrhenius plots of NaCl permeation through (a) a glass plate and (b) an identical substrate coated with a polymeric $2C_{18}-Si$ monolayer. T_c represents a phase transition temperature. (See Okahata et al.[49])

al.[48] have alluded to the use of biological membranes for hyperfiltration. An excellent practical demonstration has been given by Okahata et al.[49] These authors covalently bonded dialkylsilane monolayers onto a porous glass plate. This polymeric $2C_{18}-Si$ molecular layer exhibits a sharp endothermic peak at 44 °C, probably reflecting a phase transition from a solid to a disordered liquid crystalline state. Figure 7.6 shows results obtained when the monolayer immobilized porous plate is placed between an aqueous solution of NaCl and distilled water. The permeation of NaCl through the membrane was established by monitoring the increased conductance of the distilled water at different temperatures. The original porous glass plate is semipermeable to NaCl and gives a straightforward Arrhenius plot. The trace for the monolayer coated substrate shows a very definite kink at the phase-transition temperature which confirms the expected reduction in permeability. This approach of using a monolayer membrane as a permeation valve could find application in separating larger water-soluble species such as glucose.

7.4. PASSIVE THIN-FILM APPLICATIONS

Langmuir–Blodgett films may find application in many applied areas of traditional interest to the industrial chemist such as adhesion, encapsulation, catalysis, and the prevention of corrosion. However, little of substance appears to have been published to add to early reports[50,51] of improved results using LB layers. We simply note here that single monolayers can be used to reduce rusting and improve the adhesion between metals and polymers.

Recently, several chemical companies have started paying more attention to the requirements of the electronics industries. There is a major requirement for high-purity chemical precursors for organometallic chemical vapor deposition, lithographic polymer resists and encapsulants, etc. The term "chemtronics" is often used to describe relevant *passive* uses of organic materials. Some of those are mentioned below.

7.4.1. Electron Beam Microlithography

In integrated circuit technology the demand for faster and increased memory storage capacity has led to a continuous refinement of microlithographic techniques for producing ever-smaller circuit elements. As has been mentioned in Section 7.1, submicron resolution is now required and this has necessitated a move away from conventional photolithography to techniques involving X rays, and electron or ion beams.[52] Further developments in photosensitive etching resists will have a marked impact on the geometries of submicron silicon/GaAs integrated circuitry, on integrated optics, and on surface acoustic wave devices. Currently VLSI component dimensions fall in the $1-2$ μm range, close to the limit of optical lithography. State-of-the-art components are already demanding electron-beam lithography in direct write mode to achieve tens of gegahertz operating frequencies. Scanned electron beams are an attractive possibility owing to the excellent depth of focus and the high accuracy of registration that may be achieved; the pattern data can also be processed by computer. The main disadvantage of electron-beam systems lies with their electron scattering characteristics. Even though it is possible to focus the electron beam to a diameter of less than 10 nm, a halo of much larger dimension is formed around the point of impact because of scattering in the resin material on the substrate. The resolution attainable is improved only by using very thin resists. Unfortunately, below a thickness of 1 μm, conventional spin-coated polymer films display unacceptably large pinhole densities and variations of thickness. The LB technique thus provides a novel method of depositing suitable materials onto a substrate to aid in microstructure fabrication.

There are good examples of both positive and negative resists.[53–59] Several types of polymerizable material have been reported including diacetylenes, synthetic glycerides, and phospholipids, and long chain derivatives of vinyl alcohol, acrylic acid, oxiran carboxylic acid, and pentadienoic acid.

Broers and Pomerantz[60] have demonstrated the remarkable resolution (better

Table 7.1. Parameters for a Resist
in the Direct Fabrication of Devices

Parameter	Desirable[63]	Achievable with ωTA[64]
Contrast	>1.0	1.6
Film thickness	<0.9 μm	87 nm
Resolution	>1.0 μm	60 nm

than 10 nm) that is possible using multilayers of simple fatty acid salts; similar results have been observed by Zingsheim.[61] These molecules disappear when irradiated due to sublimation, and therefore the film is regarded as a positive resist. Higher sensitivities are more easily achieved with a negative resist material where an electron initiates the growth of a long polymer chain. The most studied of these is ω-tricosenoic acid.[56-58] This molecule can be deposited onto suitable substrates at the fast rate of 0.5 cms^{-1}, so that the film preparation time is similar to that required for spin coated resists.[62]

Thompson et al.[63] have stated the desirable parameters for a resist to be used in the direct fabrication of devices. These are listed in Table 7.1 alongside the best values obtained by Barraud and his co-workers for ω-tricosenoic acid. In purified form, ωTA has adequate sensitivity (2.5 mC m^{-2} at 5 kV) and appears to meet all the requirements. The main drawback appears to be its lack of resistance to rigorous plasma etching; this suggests that films with lower levels of defect densities are required. One of the features of using this LB film negative resist is the use of short-development times to improve the contrast and the sensitivity. Figure 7.7 shows the exposure curves of a 90 nm thick ωTA film as a function of the development time in ethanol.[64]

It has been suggested[65] that protein layers might be useful in microlithography. The method involves first depositing a synthetic protein such as polylysine on the surface of a semiconductor wafer and then covering it with a layer of conventional resist material. Chemical etching following polymerization with an electron beam then leaves the desired pattern in the form of an exposed polylysine layer. This may then be physically developed using a solution of $AgNO_3$; that is, the Ag$^+$ ions are absorbed by the NH_2 groups of the protein and converted to metallic silver to form the electrode pattern.

In summary, none of the LB film materials studied to date presents a totally acceptable solution as either a positive or negative resist. The use of trilayer resists involving inorganic compounds is now well established and appears to serve the immediate needs of the electronics industry. However, the results achieved to date do demonstrate the potential importance of LB film resists when feature sizes close to 0.1 μm are required in the semiconductor industry and for specialist applications involving nanolithography.

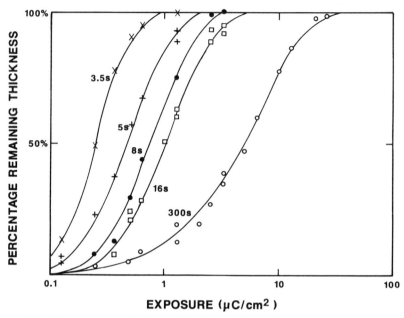

Figure 7.7. Exposure curve of a 90-nm-thick film of ωTA as a function of the development time in ethanol.

7.4.2. Lubrication

Lord Rayleigh showed that a single monolayer deposited on glass or porcelain was sufficient to drastically reduce static friction. This topic was largely responsible for rekindling Langmuir's interest in monolayer research in 1933 when his company encountered problems with friction in meter jeweled bearings. Many years earlier he had demonstrated how work on insoluble monolayers could provide information about the attachment of oily substances in solids important in lubrication.[66] Bowden, Tabor, and co-workers[67-69] have carried out elegant research in this difficult area. For example, they have studied the sliding friction between two oriented monolayers using molecularly smooth mica substrates in the form of contacting cylinders. Their results show that the magnitude of the interface shear strength and associated parameters are relatively insensitive to the chain length of the alkyl group, but partial fluorination of the chain produces a marked increase in friction. The presence of a single monolayer on each surface acts as an efficient lubricant; the coefficient of friction can be reduced by a factor of 20 and the rate of wear by more than 10,000. The mechanisms controlling the friction and responsible for maintaining the monolayer's integrity during sliding have been discussed by Briscoe and Evans.[70]

Lubrication and encapsulation are two very important areas of research in the magnetic recording industry. A good permeability barrier protects the oxidized

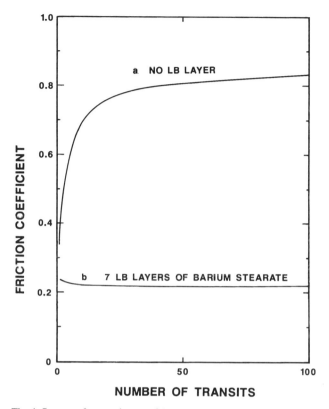

Figure 7.8. The influence of seven layers of barium stearate (curve b) in reducing the frictional coefficient between a cobalt magnetic tape and a recording head. Curve a is for an unprotected metal surface. (See Seto *et al.*[72])

magnetic media and a good lubricant avoids wear and tear during transits of the tape across the magnetic head. The challenge is to find a coating that will meet both requirements. For use in high-density magnetic recording systems it is essential that the spacing between the magnetic layer and the recording head be very thin, typically no more than 10 nm. Various surface coatings have been tried, including fluorine-containing polymers,[71] but only two reports have been published describing the use of LB films. The first was by Seto *et al.*[72] Some of their data are reproduced in Figure 7.8. This shows the significant effect on the frictional coefficient of coating an evaporated cobalt tape with an LB film of barium stearate. For a bare metal layer, the coefficient increases to a value of 0.85 after a hundred transits of the tape past the head. With a surface coating of seven monolayers of barium stearate the coefficient remains unchanged at the relatively low figure of 0.22 during operation. The effect is larger using an odd number of layers when the hydrophobic end of the molecule is exposed at the surface. Further experiments to confirm the beneficial effects of the LB film were obtained by analyzing the playback signal of a video tape recorder. A dramatic order-of-magnitude improvement in durability was observed with just a single monolayer coating. Suzuki *et al.*[73] have reported similar intriguing data but as yet only for glass substrates. They find that extremely low values of frictional coefficient (about 0.1) can be achieved using a bilayer structure, where the monolayer is used as a binder between a polymer and the substrate. Clearly there is a strong relationship between the adhesive properties of monolayers and their effectiveness as lubricants. A closely packed LB multilayer will automatically limit moisture penetration and serve as an encapsulant at the same time.

The magnetic tape industry is huge, approximately five times that of silicon, and therefore would present a marvellous opportunity to establish LB films in technology. Optical disks too could benefit from specialized coatings of this kind. However, economic factors are such that the process involved would have to be extremely cheap; it seems unlikely that companies will be prepared to make the necessary investment in a special production unit for coating LB films onto rapidly moving belts of magnetic tape.

7.4.3. Enhanced Device Processing

There may well be some interesting niches for LB films as passive layers in devices. Their role as good insulators on semiconducting surfaces will be discussed in Section 7.8. Some close-packed materials such as ω-tricosenoic acid can effectively seal the surface of etched silicon against the atmosphere and thereby greatly retard the development of interface states.[74] Related work[75] also proves that even a single monolayer is sufficient to increase the breakdown strength of a leaky "oxide" film. To illustrate how the controlled deposition of an ordered monolayer film can optimize device performance, we have selected two contrasting but significant situations of practical importance:

7.4.3.1. Surface Acoustic Waves

Surface acoustic wave (SAW) devices[76] are commonly used in the field of signal processing, especially as filters and delay lines. Their use as microgravimetric sensors is discussed in Section 7.9. Most SAW devices are based on quartz or lithium niobate and contain complex interdigital electrodes whose form is defined using lithography. These perform the conversion between electrical and acoustic energy on the piezoelectric substrate. The operating frequency of a typical device with bandwidth 1MHz is 100 MHz. For specialized applications, the substrate also contains other two-dimensional structures consisting of geometric shapes or arrays of shapes either made of metal film or grooves in the surface. The accuracy of a device clearly depends on the uniformity of the piezoelectric substrate and the precision with which the patterns have been defined using lithography. When exacting specifications are demanded, normally in military applications, it is desirable to fine-tune the characteristics of the device after it has been fabricated and tested. This

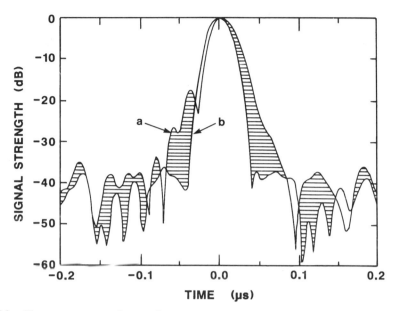

Figure 7.9. The resonance curve for a surface acoustic wave device. Curve (a) is for the imperfect device; curve (b) shows the dramatic improvement in terms of sharpening the resonance main peak when the device is coated with an appropriate number of LB monolayers.

costly procedure normally requires an additional photolithographic step and a new mask to correct each device. The LB film technique can provide a more convenient solution. In this case, the propagation characteristics of the device can be adjusted by depositing specified thicknesses of monolayers on different regions of the substrate. The advantage of this method is that the corrective coatings can be deposited in well-controlled bands, each of precisely known thickness, to provide desired amounts of phase or amplitude correction. A further advantage of the use of an insulating film is that it does not interfere with the operation of the metal layers. Figure 7.9 shows a compression pulse for an in-line reflective array compressor device before and after deposition of an LB film,[77] The reduction in the signal strength of the sidelobes, the main aim of the phase correction process, is from 17.5 dB to 36.2 dB below that of the main pulse. Examples of other SAW devices which can be improved in the same way include delay lines, oscillators, filters, and convolvers. However, it should be acknowledged that there is only a very limited demand for these specialized devices and that LB films are unlikely to be adopted as the standard method for improving their performance.

7.4.3.2. Liquid Crystal Alignment

Liquid crystal displays have been mentioned as a classical example of molecular electronics. Considerable research is still in progress aimed at applications such as high-definition television, computer terminals, and information storage systems. The types of molecule used in devices have been refined during the past twenty years. Conventional displays are based on smectic or nematic liquid crystals, but exciting research is currently aimed at materials with an intrinsic polarization. The advantage of these ferroelectric molecules is that they do not require addressing by a thin-film transistor in television applications. Instead they simply require flat plates of glass coated with a relatively straightforward array of conductors. When a pulsed voltage is applied, the liquid crystal molecules are switched and locked into a specific state, which can be reversed when the voltage is changed in sign. Material stability, switching characteristics, and color considerations must all be taken into account and fully optimized before large-screen thin-film panels based on this technology can be commercialized.

Another important factor is the alignment of the liquid crystal molecules on the substrate. One method for obtaining uniform orientational alignment is to treat the surface with materials such as lecithin[78] or alkoxysilanes,[79] which induce some appropriate physical and chemical interactions between the substrate and the liquid crystal. Another is to adopt the technique of rubbing a polyimide film. The objective in both cases is to produce a high pretilt angle, ideally 10° or more. In the first case, there is a problem in producing controlled layers of organic molecules using conventional dipping or spin-coating techniques. The rubbing technique introduces low yields for sophisticated device processing as it causes damage to the switching elements and unwanted electrostatic effects. Clearly, LB films with their controlled architecture and large-area capability are possible contenders for liquid-crystal alignment layers. The first report was given by Hiltrop and Stegemyer,[80] who used lecithin films. A number of different molecules, used by Saunders et al.,[81] included cadmium arachidate, ω-tricosenoic acid, and merocyanine dyes. Their most interesting results were obtained with ωTA; the 4° tilt angle observed was influenced by electron beam polymerization of the film. Very recently Japanese scientists[82,83] have succeeded in fabricating working displays of various types based on LB films of polyimide.[84,85] The maximum birefringence was observed using five monolayers. In all cases the performance was as good as that obtained using rubbed polyimide films prepared conventionally. However, the excellent flatness of the cells containing the LB layer afforded a control that was lacking in the case of films spread using the normal technique. Thus the production yield of active, matrix driven, twisted nematic liquid-crystal displays was improved. Another tangible benefit of incorporating the LB layer was discovered using the ferroelectric display. Figure 7.10 shows that it is possible to obtain almost perfect bistability characteristics with excellent reproducibility in comparison with those obtained using rubbed polyimide films. It may well be that empirical approaches may be discovered which are easier to adopt than LB technology. However, it would appear that a slight improvement in tilt angle using a molecule that was strongly bonded to the underlying substrate would force the liquid-crystal display community to seriously entertain using monomolecular films in their devices.

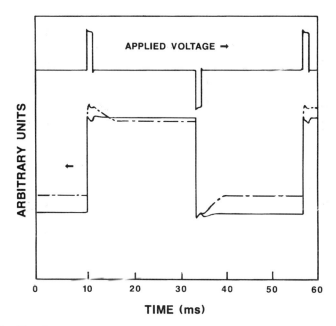

Figure 7.10. The electrooptic response of a ferroelectric liquid crystal display to a voltage pulse of approximately ± 20 V. The solid and dashed curves are for polyimide alignment layers produced using LB film and rubbing techniques, respectively. (See Suzuki *et al.*[84])

7.5. PIEZOELECTRIC AND PYROELECTRIC ORGANIC FILMS

There is evidence that many organic molecules possess very high nonlinear coefficients. If these molecular properties can be maintained on a microscopic or macroscopic level in LB film form, there is considerable scope for novel devices. Good quality monomolecular assemblies are invariably Y-type; these structures are centrosymmetric and are therefore unlikely to yield the required effects. However, the less perfect X- and Z-type structures should display noncentrosymmetric effects. There is some evidence that many X- or Z-type layers undergo a subsequent transformation to the symmetrical Y-type form. However, some molecules deposit naturally and permanently in one or other of these ways so that film transfer is achieved only when the substrate is immersed in or withdrawn from the liquid subphase. For example, polybutadiene[86] terminated by a quaternary ammonium salt deposits X-type. Crosslinking of the material between each monolayer transfer step results in the formation of a stable structure.

Figure 7.11a shows how a simple Langmuir trough can be modified to deposit a Z-type film such as phthalocyanine onto a moving fiber belt. An alternative approach to producing noncentrosymmetric structures is to use alternate layers of

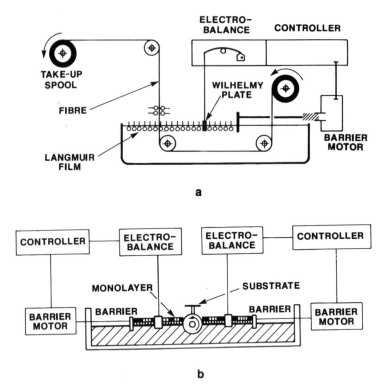

Figure 7.11. The upper schematic diagram shows a moving belt or fiber being coated with a mono-layer. This Z-type deposition can be repeated to form a noncentrosymmetric film. The lower diagram shows an alternative approach. In this case an organic superlattice is formed by rotating a substrate between suphase surfaces containing two different molecules.

two different materials where the contributions of adjacent molecules do not cancel each other. Styles of Langmuir trough which achieve interleaved layer deposition are described in Chapter 3. The designs basically fall into three categories, two dipping arms, the rotating dipper shown in Figure 7.11b, and the canal lock. In the first of these,[87] an auxiliary arm detaches the substrate from one arm, pushes it under the barrier separating the two compartments, and transfers it to the other arm, which is then withdrawn from the subphase and eventually moved to the original location so that the process can be repeated. The second method[88,89] uses a fixed beam and a revolving center section and thus enables a rotating dipper containing the substrate to rotate so that the surface to be coated leaves the subphase on the upstroke in one compartment and enters the second compartment on the down-stroke. In the canal-lock method,[90,91] the substrate is translated from one compart-ment to another through an interlock; sometimes a third compartment is used to help reduce contamination between layers.

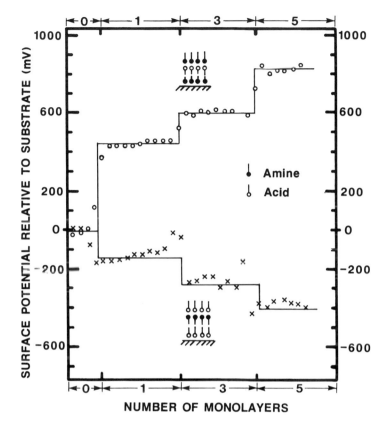

Figure 7.12. Surface potential of ω-tricosenoic acid/diocosylamine alternate layers as a function of the number of deposited layers for two complementary orientations of the polar axis. (See Christie et al.[127])

One of the quickest techniques to establish that X, Z, or alternate layer Y-type deposition has occurred is to measure the surface potential.[86,92–95] Figure 7.12 shows this parameter as a function of the number of monolayers for two different organic superlattices, both comprising acid and amine molecules deposited alternately.[95] The upper section corresponds to the case where the amine layer is deposited first; it may be seen that the surface potential steps are of different sign when the acid layer is deposited first.

There is good evidence of short-range order in LB monolayers produced using both conventional and alternate layer troughs. However, with the exception of polymerized materials such as diacetylenes, neither the translational nor orientational order extends to macroscopic dimensions. Macroscopic ordering by electric and magnetic fields is a well-established technique in the fabrication of liquid crystal cells.[96] The first publication describing attempts to induce long-range order

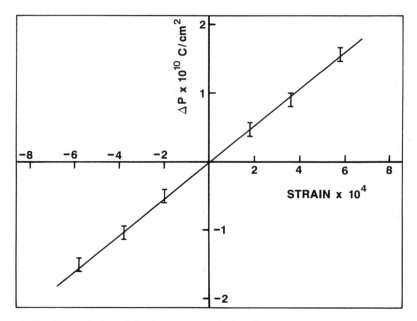

Figure 7.13. Change in the polarization component P of an X-type azobenzene multilayer film versus the magnitude of the strain along the X axis. (See Novak and Myagkov.[107])

with electric fields in order to improve device performance was by Grunfeld and Pitt.[97] More comprehensive data have since been obtained by Eldering *et al.*[98] Pressure changes were examined on monolayers of hemicyanine dye subjected to electric fields of 10^6 V cm^{-1}. These experiments showed that there was a permanent change induced in the floating film. Although forces introduced by magnetic fields are weaker than those possible with electric fields, they have the advantage that only orientational order is involved. The technique is commonly used to good effect with liquid crystalline materials. However, no positive results have yet been achieved with LB films. In a series of careful experiments on floating monolayers of porphyrins and a synthetic polypeptide, Tredgold and Jones[99] have observed nil effects using magnetic fields of the order 1 Tesla.

The prospect for obtaining highly anisotropic films that display interesting nonlinear physics phenomena are excellent. This section concentrates on the piezoelectric and pyroelectric LB film work reported to date. The impressive progress made on noncentrosymmetric films suitable for nonlinear optics is described in the next section.

7.5.1. Piezoelectric Properties

The piezoelectric effect in quartz was discovered over a century ago and the first transducer applications were suggested by Langevin in 1918. Single crystals of this material are used extensively as simple resonators and as substrates for surface acoustic wave devices. However, following the discovery in 1947 that barium titanate can be made piezoelectric by poling, an extensive series of ceramics is now available for many applications. Nevertheless, an important sector of the electronics industry is based on polymeric piezoelectric films, particularly those based on polyvinylidene fluoride (PVF_2). The properties of this important semipolycrystalline polymer, in which piezoelectric activity was first reported by Kawai,[100] are now utilized in underwater membrane hydrophones, scanning sonar arrays, and pulse-echo transducers, for medical applications and nondestructive testing. Its possible use as a pyroelectric element is discussed in Section 7.5.2.

PVF_2 consists of long chain molecules with the repeat unit CF_2CH_2; its piezoelectric properties are well documented[101–103] and have been shown to depend on crystal form, stretch ratio, and poling conditions. Although its piezoelectric "d" constants are an order of magnitude lower than those for an optimized ceramic such as PZT, the electric field strength, that can be applied to the polymer without causing depolarization, is approximately two orders of magnitude larger. The maximum strain achievable is thus ten times larger than for ceramic systems, a key advantage in electroacoustic and electromechanical transducers. On the other hand, ceramic materials have the edge regarding long-term stability and operating temperature range. There have been several efforts made to improve existing materials. Promising approaches involve the use of piezoelectric ceramic fillers distributed in a polymer matrix[104] and the development of copolymers, in particular difluoroethylene (VF_2) with related fluorinated vinyl monomers such as vinyl fluoride, trifluoroethylene (VF_3), and tetrafluoroethylene. The combination of VF_2 and VF_3 is particularly interesting because it exhibits a Curie temperature and ferroelectric behavior.[105,106] Typically a 70/30 mix shows enhanced piezoelectric activity compared with PVF_2 and a Curie temperature near 100 °C.

The chemistry and processing of copolymers have not yet been fully explored and further improvements may be possible. However, a singular disadvantage of this approach is that poling is required in high electric fields (20 MV m^{-1} or more). Experience has shown that partial depolarization due to mechanical relaxation occurs with time. The opportunities offered by the Langmuir trough technique, especially using the alternate layer configuration, are obvious and yet relatively little research has been carried out to date.

The only publication describing the piezoelectric properties of LB films is that by Novak and Myagkov.[107] These authors described results on 4-nitro-4'-N-octadecylazobenzene prepared in thin-film form using the Langmuir/Schaefer method. By extracting the substrate through a clean water surface they obtained X-type layers, typically 30 monolayers thick. They evaluated the piezoelectricity by measuring the charge developed on the plates of a capacitor upon extension of the LB film in the longitudinal direction as a stress was applied in the transverse direction. The mechanical stress was achieved by bending the substrate in a controlled way using micrometer control. The polarization P was observed to vary linearly with strain. An example of Novak and Myagkov's data is shown in Figure 7.13. They showed that the piezoelectric moduli d_{31} and d_{33} are opposite in sign; the latter has a magnitude of 1.5 pC N^{-1}, approximately one order of magnitude less than the figure for PVF$_2$.

There is considerable scope for designing noncentrosymmetric piezoelectric LB films. For example, electric dipoles based on polarized π-electron systems could be grafted onto a polymer backbone or alternatively molecules could be synthesized containing some of the features of PVF$_2$. Most practical applications require thin films about 1 μm thick; this, together with the fact that the compliance of the substrate must be accounted for, possibly restricts the usefulness of LB films in practical applications. Nevertheless, there is considerable justification in carrying out more basic measurements in this field. To date only a few fundamental investigations have been reported. O'Brien et al.[109] and Abraham et al.[108] have studied the shear and compressional moduli of floating monolayers. The most interesting data for LB films are those of Zanoni et al.[110] Using Brillouin scattering spectroscopy they have confirmed that the elastic properties of cadmium arachidate are highly anisotropic; their compressional constants are comparable to those of oriented polymer films, but the shear strains are much smaller and similar to those reported for smectic B liquid crystals.

7.5.2. Pyroelectric Properties

A traditional night-vision device is based on the narrow bandgap semiconductor cadmium mercury telluride. In this type of quantum detector, an infrared photon excites a carrier to produce an electrical signal. Although reliable and high-resolution images are possible using this method, an enormous drawback is the requirement that the device be operated using liquid refrigerants. Naturally, this restricts the range of applications especially in the civil sector. An alternative approach is to use a thermal detector. In this case, incident radiation is absorbed, resulting in a change of temperature. This in turn changes a physical parameter and produces an electrical signal. An example of this kind of detector relies on the pyroelectric activity of a material,[111] that is, its temperature-dependent spontaneous polarization. Devices based on this principle respond to a rate of change of temperature rather than to change of temperature, as in semiconducting quantum detectors. Therefore, by modulating the incoming infrared radiation they are capable of operation at ambient temperature. Despite this inherent advantage their full potential has yet to be realized. For applications[112,113] where both high speed and sensitivity are required, conventional materials have been unsuccessful. This can be attributed partly to the fact that they have not been available in very thin film form. For device applications, a thin-film geometry is preferred because one can then form large-area imaging devices directly on microelectronic amplifying circuits. These so-called "staring arrays" use as many detectors as picture points; typically a mosaic array of 250 × 400 pixels is necessary to obtain the required resolution.

7.5.2.1. Theoretical Consideration in a Thermal Imager

The pyroelectric coefficient p is a useful parameter with which to compare different materials. If the thin film acts as a dielectric in a capacitor and an external resistance is connected between the electrodes, a pyroelectric current I flows in the circuit; this can be expressed as

$$I = pA \, dT/dt$$

where dT/dt is the rate of change of temperature, and A is the cross-sectional area of the device. The pyroelectric coefficient is a measure of the current generated by a specific rate of change of temperature. However, the induced voltage V is a more useful parameter in an infrared detection system. An incremental change in temperature, dT, generates a charge $dQ = pA dT$. Therefore, $dQ = CdV$ gives

$$dV/dT = pd/\epsilon$$

where ϵ is the permittivity and d is the film thickness. The quantity p/ϵ is a useful parameter for a pyroelectric material. Table 7.2 lists typical data for a number of different materials. They are only approximate, for the values are dependent on preparation conditions, ambient, proximity to Curie point, etc. The single-crystal materials are not suitable in that they are only available in bulk form. The ceramic materials have a high value of p but this is offset by their large values of relative permittivity.[114,115] The polymeric materials are of some interest; they fall into the two categories of plastic polymer membranes such as polyvinylidene fluoride

Table 7.2. Comparison of Pyroelectric and Dielectric Properties of Materials in Non-LB Film Form

Material	p (μC m^{-2} K^{-1})	ϵ	p/ϵ (μC m^{-2} K^{-1})	tan δ	Reference
Single crystal					
Li-Ta-O$_3$	190	46	4.1	0.003	120
TGS	300	50	6.0	0.008	120
Ceramic					
Sr$_{0.6}$Ba$_{0.4}$Nb$_2$O$_6$	850	607	1.4	0.020	120
PLZT	1700	3800	0.5	0.020	115
PZFNTU[a]	380	290	1.3	0.002	121
Polymer					
PVF$_2$	30	10	3.0	0.020	120
VF$_2$/VF$_3$	10	11	0.9	0.010	117
Biphenyl/ester[b]	4	3.6	1.1	0.015	120

[a]PZFNTU: PbrO$_3$–PbTiO$_3$–PbFe$_{1/2}$Nb$_{1/2}$O$_3$ + UO$_3$.
[b]Properties measured 5 °C from Curie point.

$(PVF_2)^{[116-118]}$ discussed previously and ferroelectric liquid-crystal polymers.[119,120]

The figure of merit must also include the material aspects of detector noise. The analysis is complex but the dominant contribution is likely to be due to dielectric loss. Thus it is important to measure the tan δ ($1/\omega CR$) of a pyroelectric material. Approximate values of this parameter for different materials are given in Table 7.2.[121]

In a practical thermal imager, many other considerations must be borne in mind when designing a pyroelectric detector array capable of resolving a temperature difference in the scene temperature of 0.1 K. A useful indicator of performance is the voltage responsivity V_R of the device, which is defined as the amplitude of the voltage response divided by the total modulated power incident on the detector. Whatmore[111] has shown that

$$V_R \propto \frac{p\omega}{[(G^2 + \omega^2 H^2)(R^{-2} + \omega^2 C^2)]^{1/2}}$$

where ω is the modulation frequency, G is the thermal conductance per unit area of the detector array, H is the heat capacity of the detector, R is the electrical resistance, and C is the capacitance of the detector element and amplifier. The values of R^{-1} and C are inversely proportional to pyroelectric film thickness but H increases with thickness. Thus there is an optimum thickness d_o where V_R is a maximum. For a typical set of values for ceramic materials, $d_o \sim 5$ μm, but existing thermal imagers based on these materials use thicknesses several times larger than this.

The pyroelectric properties of LB film materials are discussed in the following subsection. However, it is worth mentioning at this stage that their optimum thickness is less than those for ceramics. Using figures for an acid–amine alternate layer structure, theoretical modeling provides the information shown in Figure 7.14. The voltage responsivity is seen to reach a maximum when $d_o \sim 0.03$ μm, corresponding to ten monolayers or so if the preferred modulating frequency 25 Hz is used.

In order to capitalize on the advantages of organic films a suitable supporting structure must be designed. This must provide a means for extracting the electrical signal while at the same time providing good thermal isolation of the individual sensors of the array. These conflicting demands can be satisfied by combining the planar techniques of microelectronics with silicon micromachining. Starting with a silicon wafer, it is possible to prepare a supporting substrate for the pyroelectric material which possesses all the desired properties. The method of fabrication depends on the nonisotropic etching characteristics of silicon. The advantage of using this material is that both the supporting structure and the electronic amplifiers can be produced on the same substrate. The sketch shown in Figure 7.14 assumes a value of $G = 10$ W m^{-2} K^{-1}, a value thought to be achievable using silicon

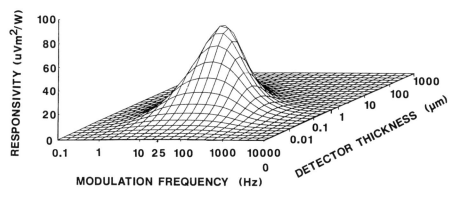

Figure 7.14. Theoretical modeling results for an acid–amine LB film deposited onto a substrate with $G = 10$ W m^{-2} K^{-1}. For a convenient modulating frequency $\omega = 25$ Hz, the optimum detector thickness is approximately 0.03 μm.

microengineering. If the thermal conductance is two orders of magnitude greater, the optimum thickness is closer to 0.2 μm.

The pyroelectric effect in a material is made up of two components,[122] labeled primary and secondary. The primary effect is a result of the temperature dependence of the dipole moments of the mechanical properties of the material. The secondary effect arises due to the fact that every pyroelectric is also piezoelectric. Therefore, any mechanical stress which results due to thermal expansion or contraction leads to an additional change in the polarization. Zook and Liu[123] have discussed how mechanical clamping due to the substrate influences the pyroelectric activity of thin-film materials. In the case of LB films, the molecular assembly is free to expand in the direction perpendicular to the substrate plane, but expansion within this plane is restricted by the substrate. The magnitude of the secondary contribution is thus influenced by the thermal expansion coefficient of the substrate.

The above considerations emphasize the importance of evaluating more than just the thin-film material parameters in a device application such as a thermal imager. In this case the properties and design of the substrate are equally critical.

7.5.2.2. Pyroelectric Langmuir–Blodgett Films

The pyroelectric effect arises as a result of the change with temperature of the spontaneous polarization. In LB films, the temperature dependence is likely to involve either ionic interaction between molecular headgroups or a realignment of polarized molecules.

The first report of pyroelectricity in multilayer films was by Blinov et al.[124]; they reported the effect in X-type layers of a series of amphiphilic azoxy compounds. This work was later extended[125] to investigate the dependence of the

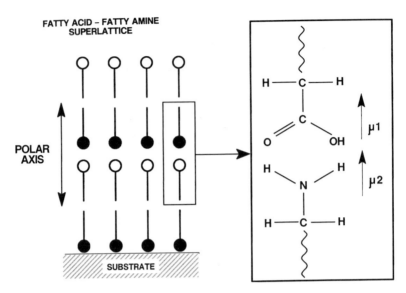

Figure 7.15. The left-hand diagram shows an organic superlattice with a unique polar axis. The two molecules involved are a fatty acid and a fatty amine. The insert is designed to show that these two materials have dipole moments in opposite senses with respect to the hydrophobic chain. Thus the Y-type film has a resultant dipole moment. (See Christie *et al.*[127])

pyroelectric coefficient on thickness. Using a similar approach Sakuhara *et al.*[126] observed pyroelectric activity in monolayer assemblies of 4-cyano-4″-pentyl-*p*-ter-phenyl. The Z-type films of this liquid-crystal-type compound were not of sufficient quality by themselves but when diluted in a 2 : 1 mixture with cadmium arachidate they gave reasonable results. The alternative approach to producing noncentrosym-metric structures is to use alternate layers of two different materials where the contributions of adjacent molecules do not cancel each other. The additional scope available using this method has been successfully used by Christie *et al.*,[95–127] Smith *et al.*,[128,129] and Jones *et al.*[130–132] using various acid– amine combina-tions. The schematic diagram shown in Figure 7.15 illustrates the principle of a superlattice comprising acid and amine molecules whose dipole moments are in opposite senses but when deposited in Y-type LB film form are aligned in the same direction. Infrared studies[132] have indicated that adjacent layers are effectively ionically bonded as a result of proton transfer from the acid to the amine. The temperature dependences of the pyroelectric coefficients of acid/amine and acid/aniline structures are shown in Figure 7.16. These data[133] are complicated by the presence of both primary and secondary effects; above approximately 240 K the substrate thermal expansion coefficient becomes appreciable. This tendency for the secondary effect to have an opposite sign to the primary one and thus decrease the overall pyroelectric activity has also been reported in organic crystals.[134]

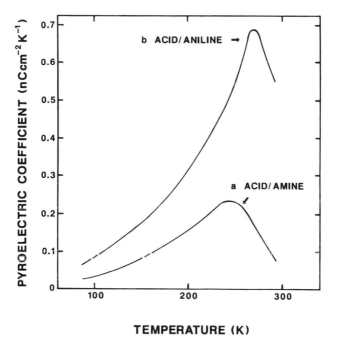

Figure 7.16. Temperature dependence of the pyroelectric coefficient of an acid/aniline (curve b) and an acid/amine (curve a) LB film superlattice. (See Davies *et al.*[133])

Ruthenium compounds complexed with a range of liquid-crystal-type molecules have been investigated in LB film form.[135] These molecules, which are based on ruthenium (η^5 − cyclopentadienyl)(bistriphenylphosphine) are described further in Section 7.5.6. They generally deposit Z-type and exhibit pyroelectric activity. However, improved structure and higher values of p are observed (1) when the molecules are mixed with a fatty acid (typically 30% molar percent) and (2) when alternate layers are formed with a fatty acid.[135] The improvement can be quite dramatic, presumably as a result of less aggregation and better order in the superlattice. Figure 7.17 shows a typical current trace observed when the pyroelectric test structure is subjected to a temperature ramp. The thermal mass of the device substrate (in this case it is made of aluminized glass) accounts for why an ideal square wave current is not observed.

The origin of the pyroelectric activity in this class of compounds[136] is not clear. However, preliminary experiments involving a variety of similar molecules have indicated that the cause is principally molecular realignment rather than ionic interactions between headgroups. The molecular dipole moment responsible for the effect arises due to a distortion of the normal distribution of delocalized electrons in the conjugated system. If one assumes that the molecular dipole moment is normal to the substrate, then simple estimates suggest that the tilt angle due to a unit change

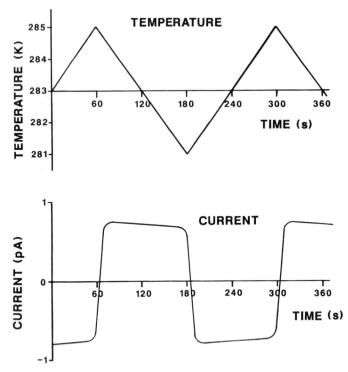

Figure 7.17. The upper diagram shows the form of the temperature cycle applied to the pyroelectric test structure. The lower trace is the corresponding current observed. In this case the pyroelectric material is an alternate-layer LB film comprising behenic acid and a ruthenium liquid-crystal complex. (See Tien and Ulrich.[136])

of temperature is typically 1°. It may well be that a compromise has to be reached in designing pyroelectric molecules of this kind, for intuitively one might anticipate that the tan δ will increase as more freedom is given to the dipoles to tilt more in the presence of a temperature change.

The values of p reported in the literature for a range of LB films are summarized in Table 7.3. Too much emphasis should not be placed on absolute values and their relative magnitudes for several reasons. First, the measurements are thickness dependent and are influenced by the substrate material. Second, the size of the coefficient depends critically on the degree of order in the thin film. Suffice to say that the p/ϵ values observed for LB films are comparable with the best reported data for carefully optimized inorganic materials and for organic films produced using other means (see Table 7.2).

The rationale for using LB films in a thermal imaging application can thus be summarized as follows:

Table 7.3. Comparison of Pyroelectric and Dielectric Properties of LB Film Materials

Material	p (μC m^{-2} K^{-1})	ϵ	p/ϵ (μC m^{-2} K^{-1})	tan σ	Reference
Azo Compound					
X-type multilayer	3.0	2.6	1.2	0.008	124
Liquid Crystal					
Z-type multilayer	1.0	5.8	0.2	—	126
Alternate Layer					
ωTA–docosylamine	1.9	2.7	0.7	—	127
Carboxylic acid/ stearylamine[a]	1.1	2.9	0.4	0.09	130
ωTA–octadecylaniline	6.5	2.9	2.2	—	133
ωTA–ruthenium	2.7	—	—	—	136

[a]A factor of two increase was observed when a proprietary dye was incorporated in the stearylamine.

1. Their values of p/ϵ are as good, if not better, than those reported for other compounds; using molecular engineering principles, there is every prospect of improving substantially on existing values.
2. The dielectric loss in LB films is very low; thus, low-noise structures can be envisaged.
3. Using carefully engineered, low thermal capacity substrates, the optimum thickness of a pyroelectric film is typically 30 nm. The LB film technique is particularly well suited to providing well-ordered films of this order.
4. Other high-performance pyroelectric materials are ferroelectric and require poling. They are operated close to their Curie point to obtain high values of p, but the permittivity and tan δ values also increase rapidly near this temperature. By contrast, the polarization of an organic LB film superlattice is solidly built into the structure and is only removed if fields in excess of the dielectric breakdown field are applied.

7.6. APPLIED OPTICS USING LANGMUIR–BLODGETT FILMS

Molecular optics defines that area of science where organic and polymeric materials are designed for application in optical and electrooptical devices which store and process optical signals. Interest in the subject has grown steadily during the past decade primarily due to the requirements of the telecommunications industry for high-bandwidth optical switching and processing devices and the recording industry's desire for optical rather than magnetic information-storage systems. There have been sufficient advances in the fiber and laser technologies required to meet present objectives, but less than adequate progress in producing and developing the active nonlinear materials. Inorganic materials are currently being used in commercial situations, but it is generally thought that organic materials offer a viable alternative because of their diversity, relative ease of fabrication, potential low cost and, most importantly, their high but controlled nonlinearity.

7.6.1. Waveguiding in Langmuir–Blodgett Films

Some of the elementary optical applications of organic thin films have been mentioned in Section 7.2. Blodgett and others measured the optical constants of fatty acid monolayers using conventional reflection and transmission techniques. Others have used more sophisticated methods utilizing waveguide structures. In the simplest waveguide, the guiding film is required to have a higher refractive index than those of the materials lying adjacent to it. However, the guided wave fields extend into the bounding media and therefore it is possible to conceive of organic materials being used as (1) the guiding film, (2) the substrate onto which the guiding film is deposited, or (3) a thin overcoating deposited on top of the guiding layer. Coupling is usually achieved via a prism or a grating. Because of an organic materials lack of mechanical strength, grating couplers are normally preferred.

Swalen[137] has shown that the three-layer (substrate–thin film–air) structure can be modeled successfully and is useful for studying both the in-plane and perpendicular refractive indices of films in the thickness range 0.3 μm to 4 μm; the desirable thickness is approximately 2 μm. On the other hand, using a four-layer system, the optimum thickness of the overcoating layer is less than the wavelength of light, typically 0.1 μm. This represents a far more convenient thickness as far as LB films are concerned. (When one of the surfaces involved is a metal, surface electromagnetic modes can be observed; these so-called surface plasmon effects are described in Section 7.9.3.) Because light incident from within the film onto either of its boundaries encounters a material of lower refractive index, total internal reflection will occur at both interfaces. If the guiding film is thick enough for transverse constructive interference to occur, then light can be propagated over distances limited only by absorption of the field or by scattering due to defects, imperfections, etc. Loss due to scattering is a very important parameter and should for practical purposes be less than 1 dB cm^{-1}. It should be remembered that by varying the refractive index and the film thickness, it is possible to select which media carry most of the energy. Thus, because attenuation levels vary from one material to another, the measured value for a signal propagating along a four-level waveguide is very dependent on thickness.

The first successful waveguide measurements in LB films were reported by Pitt and Walpita[138,139] and by Columbini and Yip.[140] Both sets of experiments used fatty acid salts illuminated by helium–neon lasers. Swalen et al.[141] and Novak[142] have extended these investigations and reported the effects of subphase pH, film thickness, and wavelength. Using optimum conditions Novak reported an attenuation of 3 dB cm^{-1} for films 300 nm thick but a value one order of magnitude higher for films 50% thicker. The most impressive results have been obtained using waveguide structures containing LB films of preformed polymers.[143] In this case the measured attenuation was only 11 dB cm^{-1} using relatively thick films. Waveguiding measurements displaying enormous losses due to scattering have been reported in LB films of polydiacetylcne.[144,145]

The incorporation of thin-layer organic materials in waveguides has been of particular interest to those working in the field of nonlinear optics. The advent of fiber optic communications has generated the prospect of compact signal processing units relying on nonlinear optical effects such as the variation of refractive index with high electric or optical fields. Some background theory is described in the following section, so that the results obtained to date for LB films can be placed in a proper perspective.

7.6.2. Nonlinear Optics: Relevant Theory

There are several review articles[146–149] and books describing the role of organic materials in optical components for use in the field of optoelectronics. All emphasize that a high possibility exists that they will oust inorganic materials for certain applications, provided stability problems can be overcome.

Guided wave optics is currently dominated by lithium niobate (LiNbO$_3$), but considerable research effort is underway to find alternative materials which are less prone to radiation damage and amenable to integration with silicon or gallium arsenide structures. The major stimulus for using organic materials is that "molecular engineering" enables the nonlinear optical properties to be accurately controlled and optimized; the important parameters are thus locked into the molecular structure rather than being dependent on the processing technique.

Some of the basic equations relating to nonlinear optics have already been referred to in Chapter 4, but are repeated here for convenience.

The microscopic polarization P of a solid comprising many individual molecules may be expressed as

$$P = P_0 + \chi_1(E) + \chi_2(E^2) + \chi_3(E^3) + \cdots$$

where P_0 is a constant, E is the electric field, and χ_n is the nth order susceptibility tensor. In this equation χ_2 and χ_3, respectively, are responsible for higher-order effects such as second- and third-harmonic generation. When an individual molecule is involved, it is customary to express the polarizing effect in terms of an induced dipole moment μ and a series of polarizabilities, namely

$$\mu = \alpha E + \beta E^2 + \gamma E^3 + \cdots$$

where the coefficients α, β, and γ are tensor quantities.

The basic molecular properties which give rise to high values of the hyperpolarizabilities such as β and γ are reasonably well understood in organic solids. For quadratic molecular effects, conjugation and intramolecular charge transfer are important, while cubic nonlinearity related effects are enhanced in one-dimensional conjugated structures, but no charge transfer induced asymmetry is required. In both cases, of course, there is the scope to tailor the molecular unit to meet a specific requirement. A second crucial step in engineering a molecular structure for nonlinear applications is to optimize the crystal structure. For second-order effects, a noncentrosymmetric geometry is essential. Anisotropic features such as parallel conjugated chains are also useful for third-order effects. An important factor in the optimization process is to shape the material for a specific device, so as to enhance the nonlinear efficiency of a given structure. A thin-film geometry is normally preferred, for then it is possible to integrate nonlinear interactions, linear filtering and transmission functions into one precision monolithic structure.

7.6.2.1. Second-Order Effects

Molecules containing conjugated π-electronic systems with charge asymmetry exhibit extremely large values of β. In order to assist the chemist in designing organic molecules with optimal properties, Oudar and Chemla[150] separated the contribution to β from charge transfer (β_{CT}, the dominant term) and the component due to the induced asymmetry in the charge distribution. They showed that

$$\beta_{CT} = \omega f \Delta \mu / (\omega - 2\hbar\omega)^2 (\omega - \hbar\omega)^2$$

where $\hbar\omega$ is the energy gap, f is the oscillator strength of the charge transfer transition in the molecule, and $\Delta\mu$ is the change in the dipole moment involved in that transition. Thus, in order for a material to be of practical use for second-harmonic generation, it must be closely matched to the laser it is to double; the absorption edge should be near the wavelength of the second harmonic but must not include it. The equation also shows that the molecule must possess a large transition moment and exhibit a large change in the dipole moment on excitation. In order to select the best combination of donors and acceptors, one must consider the relative strengths of the range of available groups. An approximate measure of the donor strength is given by the first ionization potential of a simple nonconjugated molecule containing the relevant group, since this represents the ease with which it releases electrons. The effectiveness of acceptor groups tends to be more variable in that the nature of the rest of the chromophore can have an important influence.

Generally speaking, the two-level model has been successful in accounting for the nonlinear optical features of various classes of molecule. For example, the prediction that highly polar organic molecules, such as the cyanine and merocyanine dyes, should possess large hyperpolarizabilities has been verified experimentally.[151,152] It has also been possible to relate the polarizability to the length L of the conjugated system. Theory predicts that $\beta \propto L^3$.[153]

In organic molecular solids each molecule largely retains its electronic integrity and, if suitably aligned, contributes its second-order nonlinear polarizability β to the overall macroscopic quantity χ_2. Zyss and Oudar[154,155] have developed the required algebraic theory to relate the macroscopic second-order nonlinear response of a molecular crystal to the microscopic response of the constituent molecules.

The equation expressing the intensity $I_{2\omega}$ of the second-harmonic radiation due to a plane wave of intensity I_{ω} propagating through a nonlinear material is given by

$$I_{2\omega} = I_\omega^2 \chi_2^2 l^2 \left(\frac{2\omega^2}{\epsilon_0 c^3 n_{2\omega}^2 n_\omega} \right) \text{Sinc}^2 \left(\frac{\Delta k l}{2} \right)$$

where l is the path length, c is the velocity of light, and n signifies the refractive index at the fundamental or second-harmonic frequency. The sinc function ($\sin x/x$) is the phase-matching condition.

Phase matching is an essential prerequisite for the efficient operation of nonlinear materials in parametric interactions. In its absence, a fraction of the crystal length smaller than the coherence length effectively contributes to the outgoing emission. When phase modeling occurs, $\Delta k = 0$ or $n_{2\omega} - n_\omega = 0$; thus the sinc function equals unity and the conversion efficiency is seen to depend only on the product $\chi_2^2 l^2$. This strong functional dependence on χ_2 thus provides an excellent opportunity for chemists to design organic materials which can compete effectively with inorganic compounds. Moreover, organic materials have the added significant advantage that both the linear and nonlinear properties can ultimately be tuned more or less independently. Another factor worth noting in the above equation is that the second-harmonic signal strength is proportional to the square of the path length, or thickness of the nonlinear material. This relationship is illustrated graphically later in this section.

For any practical application, it is necessary to establish an accurate value of χ_2 for a specific material. This requires a whole sequence of experiments in transmission and reflection modes, with all possible combinations of input and output polarizations. However, in order to assist scientists to quickly screen a large number of different compounds and assess their relative importance, several approximate techniques have been developed. One of the most convenient is the Kurtz[156] powder method which avoids the need to grow good quality crystals or film. It detects a convolution of all the tensor components of χ_2 but nevertheless enables general trends to be observed. Another popular technique is electric field induced second harmonic (EFISH) generation.[157] Here, a strong dc electric field is applied to a liquid or a solution of the molecules of interest.

The value of χ_2 for lithium niobate is approximately 1.4×10^{-8} esu, while the corresponding figure for potassium dihydrogen phosphate (KDP) is approximately 2×10^{-9} esu. Many organic materials have exhibited higher nonlinear coefficients than these.[158] For example, methyl (2,4-dinitrophenyl) aminopropanoate (MAP) is phase matchable over its entire transparency range (0.5 μm–0.2 μm) and has a χ_2 value 2.8 times that of $LiNbO_3$. Sometimes the parameter χ_2^2/n^3 is used as a figure of merit, in which case MAP is fifteen times better than the best inorganic material. The equivalent figures for 2-methyl-4-nitroaniline (MNA)[159] are 4.8 and 4.5, respectively.

The first reported demonstration of harmonic doubling in an organic planar waveguide was that of Hewig and Jain[160] using an evaporated film of p-chlorophenylurea. Second-harmonic generation has also been observed in an MNA-coated tapered waveguide.[161]

Another method of measuring the second-order nonlinearity in a material is to carry out a Pockels effect experiment. The principal conceptual difference between the electrooptic technique and a conventional second-harmonic generation measurement is that only one of the fields oscillates at optical frequencies. The second is normally a dc electric field. Effectively, what is observed is the modification of the relative permittivity or refractive index of a material by a slowly varying applied electric field.

7.6.2.2. Third-Order Effects

Third-order phenomena in inorganic semiconductors are relatively large and therefore, as long as good transparency is not a problem, there is every likelihood that they will be improved further by the advent of multiquantum well techniques. Nevertheless, there may well prove to be a niche for organic films provided the third-order coefficients are reasonable. Most of the projected applications relate to input and output beams of the same frequency and rely on the ability of a material to display an intensity-dependent refractive index.

The basic nonlinear optical quantity of interest is χ_3, which is associated with a modification of the relative permittivity of the material. In the isotropic situation

$$\epsilon = \epsilon_1 + 4\pi\chi_3 E^2$$

where ϵ_1 is the linear permittivity and E is the optical electric field in the material. This leads to the approximate expression

$$n = n_1 + n_2 I$$

where n_1 is the linear index of refraction, I is the optical intensity, and $n_2 = 16\pi^2\chi_3/c\epsilon_1$. Thus the change in refractive index is proportional to both χ_3 and I. This is known as the Kerr effect.

A favorable organic structure for the enhancement of cubic susceptibility is that of a one-dimensional conjugated periodic system. For example, bulk crystals of polydiacelytene[162] have a value of $\chi_3 = 0.7 \times 10^{-10}$ esu, approximately one order of magnitude greater than that of GaAs in the 111 direction. The large nonlinearity in this material is attributed to the delocalized π-electron wave functions along the one-dimensional carbon backbone yielding semiconductor-like states. More generally, in materials of this kind theory predicts that $\gamma\alpha L^6$ where L is the conjugated system length.[163]

7.6.3. Optoelectronic Applications of Organic Films

It is beyond the scope of this article to discuss the myriad of nonlinear effects associated with χ_2, χ_3, and combinations of optical frequencies and electric fields. Table 7.4 summarizes a few of the main physical phenomena and their possible applications.

The nomenclature $\chi_n(-\omega_3; \omega_2, \omega_1)$ signifies that a susceptibility of the nth order is used to produce a resultant field of frequency ω_3 for input frequencies ω_1 and ω_2. For second-harmonic generation the expression is $\chi_2(-2\omega; \omega, \omega)$. Many of the application areas are discussed at length in the review paper by Stegeman et al.[164]; Kowel et al.[10] and Potember et al.[165] also provide useful insights into novel structures incorporating organic films that display nonlinear optical behavior. One of the applications not mentioned in Table 7.4 is that group associated with photorefractive behavior. This requires materials which exhibit high nonlinearity and good photoconductive properties. Novel organic compounds are required that will compete effectively with inorganics such as lithium niobate and bismuth silicon oxides for use in holography and wavefront conjugation. Other proposed special devices using organic films include picosecond transient digitizers,[166] nonlinear coherent directional couplers,[167] and induced birefringence electrooptic modulators.[10] These integrated optoelectronic devices use electrical inputs to gate the switching of an optical signal and are therefore limited by capacitance and electron transit time effects. Great excitement has been generated by the possibility of all optical switches that are controlled by the incident intensity of the optical beam. In particular, it is believed that third-order nonlinear materials with large χ_3 can be used for optical bistable devices that are up to 1000 times faster than corresponding electrically controlled devices.

This subsection has mentioned several device areas where organic films, because of their high nonlinearities, could compete effectively with their inorganic counterparts. In all cases it is important that the materials can sustain reasonable

Table 7.4. Electronic Susceptibility Functions χ_2 and χ_3 for Various Types of Interacting Field Components; The Associated Physical Phenomenon and Application (See Williams[146])

Susceptibility	Physical effect	Application
$\chi_2(-2\omega;\omega,\omega)$	Frequency doubling	Second-harmonic generation
$\chi_2(-\omega;\omega,0)$	Electrooptic (Pockels effect)	Modulation phase retardaton
$\chi_2(-\omega;\omega_a,\omega_b)$	Frequency mixing	Parametric amplification
$\chi_3(-3\omega;\omega,\omega,\omega)$	Frequency tripling	Deep UV conversion
$\chi_3(-\omega;\omega,\omega,-\omega)$	Electrooptic (Kerr effect)	Optical bistability phase con-
	Degenerate four-wave mixing	jugation
$\chi_3(-\omega;\omega,0,0)$	Electrooptic	High speed optical gating

power densities. For example, frequency doublers may need to withstand 1 GW m^{-2}. This will demand high levels of purity; otherwise operation will be restricted to the picosecond regime. A clear advantage of organics is that their nonlinear molecular properties can be engineered for specific technical applications. They also have the edge in situations where the speed of the nonlinearity is involved; that is, the time it takes for the optically induced change in the refractive index to disappear. All measurements on electronic nonlinearities in organic materials indicate that subpicosecond relaxation times are attainable, even in regions of strong absorption. When it comes to debating the relative strengths of LB films in comparison with organics produced in other ways, there are several positive features to mention. First, the anisotropy is likely to lead to second-order coefficients of larger magnitude. Second, they are more convenient to use in systems capitalizing on the tetrahertz bandwidth capacity of optical fibers. Their main benefit would appear to be their ability to integrate many different signal processing operations into one precision monolithic structure.

7.6.4. Nonlinear Optics: Langmuir–Blodgett Films

The previous sections have described some of the relevant background theory and possible ways to exploit films in nonlinear optics. Many researchers have recognized the potential benefits of LB film technology and their results are described here.

7.6.4.1. Second-Order Effects

There are several published reports of second-harmonic generation (SHG) in floating[168] and single monolayer LB films. Aktsipetrov et al.[169] observed the effect in nitro-octadecylazobenzene and Girling et al.[170] in a substituted merocyanine. In neither case was the nonlinear coefficient particularly large and it must be remembered that a one-layer film has to be noncentrosymmetric. Monolayers have also been used to enhance SHG from surface plasmons generated on silver surfaces.[171–173] These results on single monolayers were later extended to multilayers[174,175] deposited in X- or Z-type fashion. More recently a χ_2 value of 2.5×10^{-20} C^3 J^{-2} has been calculated from SHG results obtained for Z-type films of 4-[4-(N-n-dodecyl-N-methylamino) phenylazo]-3-nitrobenzoic acid (DFNA),[176] irradiated with a 1.06 μm (Nd^{3+}:YAG) laser. However, as with all other reports of nonlinear behavior in X- or Z-type films, the signal strength did not show the expected square-law dependence on number of layers, probably due to structural defects. By varying the angle of incidence of the laser beam it is possible to deduce the orientation of the molecular dipoles responsible for the nonlinearity. Figure 7.18 confirms that the optic axis of a particular azobenzene molecule is normal to the plane of the film.[174]

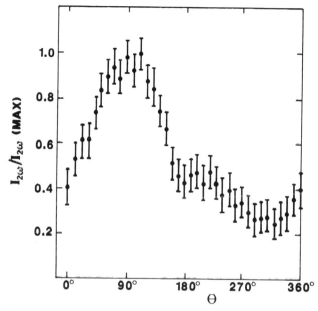

Figure 7.18. Second-harmonic intensity versus angle θ for a 4-nitro-4'-octadecylazobenzene LB film illuminated with radiation pulses from a Q-switched Nd^{3+}:YAG laser of wavelength 1.06 μm. The peak shows that the preferred orientation of the microcrystallites is normal to the plane of the film. (See Aktsipetrov et al.[174])

No attempt is made in this review of nonlinear effects to compare values of β or χ$_2$ for different materials. This is because the magnitudes of these coefficients are a strong function of the amount of order, level of aggregation, degree of polymerization, etc., in the film. The clearest evidence for this has come from investigating the nonlinear optical properties of a hemicyanine dye diluted with a fatty acid.[177,178] In this case the SHG efficiency of the diluted material is superior to that of a film of pure dye. For the hemicyanine in arachidic acid film investigated, the efficiency increases monotonically with hemicyanine concentration up to a value of 50%. The explanation[179] is that in a pure dye film the dye is almost completely in the form of H (blue shifted) aggregates; in a mixed dye–arachidic acid film, the dye is mostly monomeric. Stroeve et al.[180] have successfully reduced the level of aggregation in merocyanine dyes using poly(methylmethacrylate) and observed SHG.

There have been several reports of large second-order nonlinear coefficients produced using Y-type deposition. In one unusual case,[181] the noncentrosymmetry results from a unique alignment of the chromophores in a single direction. This unusual herringbone morphology has been reported for multilayers of 2-doc-osylamino-5-nitropyridine. However, the remainder have relied on the principle, described in the previous section, of depositing alternate layers of two different materials.[182,183] Particular interest attaches to the work of Neal et al.[184,185] who showed that the nonlinear response of a hemicyanine/nitrostilbene layer (see Figure

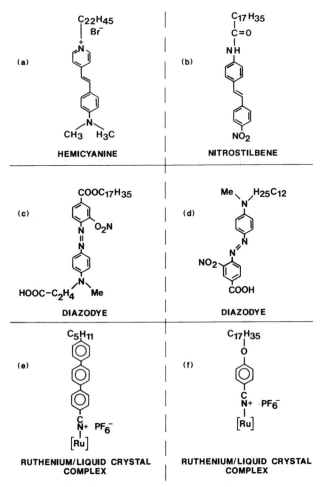

Figure 7.19. Molecules which have displayed useful second-order effects when used in an alternate-layer geometry. The pairs of molecules (a), and (b), and (c) and (d) can be used to form an organic superlattice displaying a high coefficient for second-harmonic generation. (See Neal *et al.*[184] and Neal *et al.*,[183] respectively.) Figures (e) and (f) are examples of novel ruthenium–liquid crystal complexes which can be used to form Z-type films or be part of an alternate-layer structure.

7.19) is greater than that expected from the simple addition of contributions arising from the individual (separated) monolayers. The coefficient for second-harmonic generation, β, of the alternate layer structure is approximately five times the average value of the same parameter measured for the hemicyanine and nitrostilbene molecules shown in Figures 7.19a and b; this superadditive effect is best explained in terms of improved film structure with adjacent molecules influencing each other to orient more vertically with respect to the substrate; however, there may be a synergistic interaction between the dipoles due to the unique crystal structure involved.

This interesting result, when translated into macroscopic terms, gives a χ_2 value fifty times greater than that of lithium niobate. It is also satisfying to note that the second-harmonic power increases quadratically with the number of deposited bilayers; this dependence on the areal density of the dye is to be expected for a film which is thin compared with the incident wavelength, but is rarely observed in practice.

In order to meet the stringent thermal, mechanical, and chemical stability requirements necessary in the field of optoelectronics, a serious attempt has been made to develop a novel group of noncentrosymmetric compounds based on organotransition metal complexes.[186] Figures 7.19e and f show two examples of specially engineered molecules which have displayed both interesting SHG and pyroelectric data. The complexing of an organoruthenium group to a cyanophenyl molecule results in excellent monolayer and multilayer formation. The molecular hyperpolarizabilities of both compounds illustrated in Figure 7.19 are high; the SHG signal strength is also a strong function of the film quality. Figure 7.20 shows the expected square-root dependence of the second-harmonic intensity with number of deposited monolayers.

One of the undoubted advantages of LB films in the field of nonlinear optics is

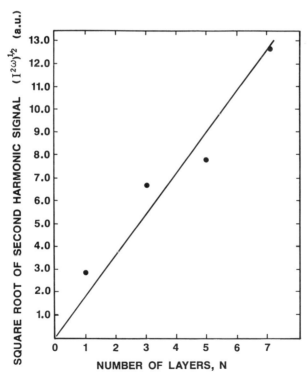

Figure 7.20. Thickness dependence of the second-harmonic generation signal in a Z-type film of N-[ruthenium(cyclopentadienyl)-(bis triphenylphosphine)]-4-cyano-4"n-p-terphenyl. (See Davies *et al.*[187])

that they can be coated readily onto length of fiber. Selfridge *et al.*[188] have reported interesting second-harmonic data; they first removed the protective outerlayer and cladding of standard fibers and dipped them, ten at a time, into a Langmuir trough. Hemicyanine coatings were placed on 5 cm lengths of 600 μm diameter optical fiber core. The measured conversion efficiency (second-harmonic power divided by input power) was about 10^{-6} but improvements are expected using smaller-diameter fibers and materials with higher values of χ_2.

7.6.4.2. Electrooptic Effects

Polar LB film materials such as those shown in Figure 7.19 have large values of β and χ_2. One might expect therefore that they should also exhibit sizeable electrooptic effects. There are only a couple of publications describing the Pockels effect in LB films. The first report showing the modulation of the output of a He:Ne laser using a single monolayer was by Cross *et al.*[189] The cross section of the structure used to apply electric fields ~ 1 MV m^{-1} across a hemicyanine layer is shown in Figure 7.21. The differential reflectivity as a fraction of the total laser output is shown versus angle of incidence in Figure 7.22. Using these results it is possible to calculate the dc induced charges in the real and imaginary parts of the dielectric constant to be approximately 2.0×10^{-4} and 4.8×10^{-5}, respectively. Similar data have recently been reported for *N*-stilbazene[190] and the azobenzene (DPNA)[191] discussed earlier in this section.

7.6.4.3. Third-Harmonic Effects

The fundamental properties of a high χ_3 material may be very different from those required for second-harmonic generation. A favorable structure for the enhancement of cubic susceptibilities is a one-dimensional conjugated periodic system. A good example is polydiacetylene with a diacetylene monomer: bis (*p*-toluene

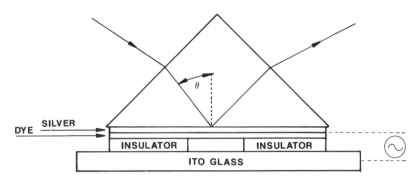

Figure 7.21. Cross section of the electrooptic cell used to study the Pockels effect in an LB film. (See Cross *et al.*[189])

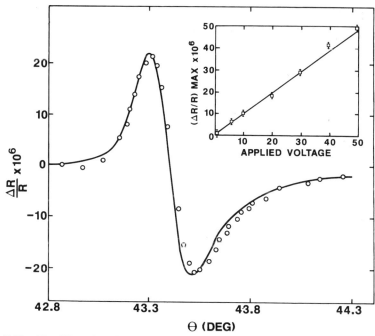

Figure 7.22. The differential reflectivity measured for an applied voltage of 20 V; the estimated value of electric field across the single hemicyanine monolayer is 1.7×10^6 FNI55w1. The inset confirms the linear dependence of $(\Delta R/R)_{max}$ with applied voltage. (See Cross *et al.*[189])

sulfonate) 2,4-hexadiyne-1,6-diol, abbreviated PTS/PDA. The cubic susceptibility parallel to the polymer chains in bulk single crystals of this material is approximately thirteen times the value of χ_3 for gallium arsenide. The tensor coefficients perpendicular to the polymer chain are similar to those in the monomer crystal, thus reflecting the high anisotropy of the structure. These encouraging results have prompted several research groups to study third-harmonic generation in diacetylene-based LB films. Early measurements indicated χ_3 values approximately two orders of magnitude lower than those reported for single-crystal PTS/PDA.[192,193] More recent data[194] are shown in Figure 7.23, which shows the cubic susceptibility as a function of wavelength. Of special interest is the resonance observed at $\lambda = 1.35$ μm, as this can be used for high-powered laser conversion. At this wavelength the measured value of the cubic susceptibility is only a factor of six lower than that observed for single crystals of PTS/PDA and is larger than that of gallium arsenide. However, the attenuation losses reported by Carter *et al.*[195] would appear to dampen any hopes of exploiting this material in waveguiding devices.

Optical third-harmonic generation has also been observed in dye-based LB film systems. For example, Kajzar *et al.*[196,197] have reported detailed studies of χ_3 in merocyanine and stilbazolium dyes. They have shown that the thickness precision

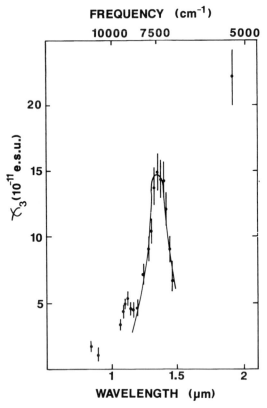

Figure 7.23. Cubic susceptibility measured as a function of wavelength for 78 monolayers of poly-diacetylene. (See Kajzar and Messier.[194])

available with LB films enables the phase of the third-order susceptibility and its dispersion to be measured over a wide range of wavelengths. Interesting applications-oriented results have also been obtained in polydiacetylene.

Kajzar et al.[198] have described a very large enhancement of the Kerr susceptibility and response times faster than 3 ps in this material. Carter et al.[199] have demonstrated the intensity dependence of the refractive index in the same material. Some of their reflectivity data are shown in Figure 7.24. To obtain these results laser holographic techniques were used to etch a grating in the surface of a silicon wafer. A silver film was then deposited to form a low-index substrate before the planar waveguide structure was completed with a 500-nm-thick polymer coating. Using this technique it is possible to observe refractive index changes which occur as fast as those in the optical intensity.

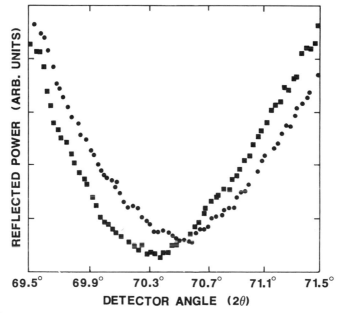

Figure 7.24. Intensity-dependent reflected power versus detector angle (twice the angle of incidence) for a polydiacetylene waveguide. When the input pulse energy is increased from 4 μJ to 14 μJ, there is a distinct shift of the reflectivity minimum to higher angles. (See Carter *et al.*[199])

7.6.4.4. Summary

There is thus abundant evidence that LB films have values of χ_2 and χ_3, well in excess of those of inorganic materials currently in use. As in the case of pyroelectric applications it is necessary to then ask the question: is the figure of merit also attractive? Many factors are involved including refractive index, phase matching and transparency, etc. There is every indication, especially in the case of second-order processes, that LB films can be tailored to compete effectively provided (1) stability requirements can be satisfied, (2) devices are designed to capitalize on the fine control and thickness of the films. Otherwise, organic materials produced using alternative methods may be equally suitable. Clearly, more research is required to improve on the existing molecules synthesized for third-order nonlinear optical effects. It is perhaps in optical logic systems where high χ_3 materials will find their niche.

7.6.5. Optical Information Storage

Optical information storage and retrieval is now a commercial reality; an essential part of this technology is the recording medium. Most effort is being devoted to write-once, read-only media used in conjunction with solid-state GaAlAs semiconductor diode lasers. This type of laser has effectively defined the wavelength region for recording, reading, and erasing plus the sensitivity and other parameters required of the media. For an optical system operating at a wavelength close to 800 nm with a numerical aperture of about 0.5, the storage density capability is approximately 10^8 bits cm^{-2}. The laser beam is typically focused to a diffraction-limited stop 1 μm in diameter; this produces a thermally induced physical transformation of the media. In ablative recording there is material displacement to create some form of pit; however, it is also possible to obtain optical contrast by introducing some optically detectable transformation in the film, such as a phase change or aggregation–deaggregation change. The performance criteria required for an erasable medium are similar, but there is the additional need for the medium to undergo rapid, laser-induced reversibility from its marked state to its original state, many times over. Practical devices require the erasure to occur in the micro- to nanosecond time scale, and be reproducible for up to 1 million erase–record cycles.

In ablative write-once systems, marking occurs when the hole-forming medium is heated above its melting point. This thermal process thus dictates that low melting point, low vaporization, low flow temperature materials with low thermal conductivity are desirable. Organic materials are ideally suited for this mode of recording although tellurium, a low melting point metal, has often been used as the active layer. A disadvantage of using metals for archival applications is that they are susceptible to hydrolytic and oxidative degradation.

A variety of organic–polymeric based materials have been studied for optical recording. A good review of the subject has been given by Pearson.[200] In all cases the light absorption function is provided by a dye or metal with the polymer serving as a binder. Several classes of dye, including phthalocyanine, absorb as required in the near-infrared part of the spectrum. In systems of this type it is important to prevent the amorphous material undergoing crystallization, which will subsequently degrade the recording quality.[201]

Dautartas et al.[202] have obtained very promising data using metal-free phthalocyanine and cite its additional benefits over other media as low thermal conductance, cheapness and lack of toxicity, and thermal stability. It is possible to define a set of design criteria for such dye materials and also calculate the optimum thickness required. Typically, for film thicknesses greater than 100 nm, the ablative recording sensitivity starts to decrease owing to the amount of material that must be displaced to form the pit. The primary function of the dye is to absorb the incident laser energy while the associated polymer binder influences the final dimensions of the pit.

Apart from the benefits associated with uniformity of thickness there is little to suggest that LB films should be used in the write-once situation. It should be noted, however, that vacuum sublimed films of metal-free phthalocyanine, in the thickness range 5–7 nm, have been successfully incorporated into a tellurium-based medium and shown to significantly enhance the writing contrast.[202]

There are a few examples of organic–polymer-type reversible recording media.

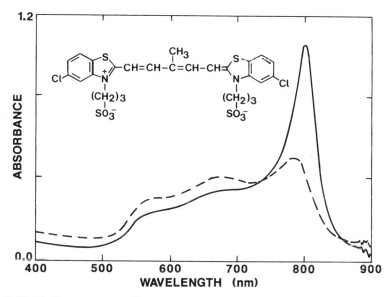

Figure 7.25. Reflectance spectra of a cyanine dye medium coated on an aluminum surface before (——) and after (----) irradiation with a laser diode of wavelength 780 nm.

Some of these[203] are of the ablative type and depend on the use of low melt viscosity, low softening temperature polymers as the binding matrix. Such materials spread and refill the pits during the erase cycle. Mey[204] has described a system which operates via an aggregation–disaggregation mechanism. The active layer in this case consists of tiny dye/polymer complex particles dispersed in a continuous phase of the same polymer. Exposing this aggregated film to a laser pulse results in disaggregation with an attendant shift in the absorption spectrum. Reaggregation of the dye/polymer complex is achieved using a thermal process. This method is not all that different to that described in the interesting paper by Ishimoto *et al.*[205] These authors have capitalized on the narrow absorption bands of J (red-shifted) aggregates of organic dyes deposited in LB film form.

An example of a typical reflectance spectrum for 25 monolayers of a cyanine dye is shown in Figure 7.25. The dye monomer absorbs strongly near 660 nm. The increased absorbance near 780 nm arises because of electronic interactions in regularly aligned J aggregates formed in a mixture of diotadecyldimethyl ammonium chloride and methyl stearate. Laser light disorders the structure of the aggregates so that the J aggregate band is diminished (the lower curve in Figure 7.25). Ishimoto *et al.*[205] studied the influence of laser radiation of various power and pulse lengths. They reported that a reflectivity change of 20% was obtained using an energy density of 0.03 J cm^{-2}. The reflectance spectrum before recording could be recovered by placing the medium in an environment of 100% humidity at 30 °C. They also extended their measurements to a system incorporating two different dyes deposited on an aluminum reflective layer. This enabled them to restrict the structural change to within a single recording layer. There is much to be learnt about the physical processes occurring in J aggregates. Electrochromism as well as photochromism in LB films appears to be strongly influenced by such molecular structures.[206,207]

Another facet of optical information storage which has been discussed in connection with LB films is that of photochemical hole burning. A good review of its prospects has been given by Moerner.[208] It is a technology which, if successful, would permit a thousand times more bits of information to be stored in the volume irradiated by a focused laser beam. The disadvantages of this frequency domain approach to optical storage is that the medium must be cooled to liquid helium temperatures. Until this limitation is removed, there appears to be little chance of commercializing the technology, with or without LB films.

7.7. ELECTRICAL PROPERTIES OF MONOLAYERS AND MULTILAYERS

Langmuir–Blodgett films of many materials show good insulating behavior and low dielectric loss. These attractive "passive" attributes suggest that they may find application, for example, as the dielectric layer in a simple capacitor. More recently, there has been considerable progress in designing molecules suitable for conducting LB films. If a stable material with adequate conductivity were found, then it could well find a role as an interconnect in microelectronic circuits. This section begins with a brief review of the salient features of conduction in monolayers (the reader is referred to Chapter 4 of this book for more details) before touching briefly on a few tentative possibilities showing where their insulating or conducting properties may be useful. The important application of LB films to control and modify the surfaces and interfaces associated with semiconductors is discussed in the next section.

7.7.1. Electrical Conductivity in Langmuir–Blodgett Films

Most of the electronic conductivity data for LB films are suspect in that they have been obtained for multilayers deposited onto metals that are invariably coated with semiinsulating natural oxides.[7] The first monolayer provides a uniform coating on a relatively rough surface, and protects the weak spots, thus increasing the breakdown strength. Agarwal and Srivastava [209,210] found that the breakdown voltages tended to become independent of LB film thicknesses above a certain number of layers. The unusual dc electrical characteristic first observed by Roberts et al.,[211] but since confirmed many times by other researchers,[212–214] is also best explained in terms of the "oxide" sustaining most of the electric field. They reported that the current through an LB film sandwich structure varied with voltage according to the relation $\log J \propto V^n$ with $n = \frac{1}{4}$. This is the result one would expect[215] if exceptionally large fields were present, sufficiently high to influence the image forces controlling charge flow through the Schottky barrier at the metal–oxide interface. Sometimes, the expected Poole–Frenkel law ($n = \frac{1}{2}$) is observed and the data can be analyzed in a conventional way.

There is no doubt that many monomolecular assemblies contain a large number of defects. Peterson[216] has suggested that those responsible for influencing the conductivity are analogous to the disclinations observed in liquid crystals. In any event, the complications arising from imperfect electrodes and defects means that conduction mechanisms in LB films are not well understood.

As a result of their unique structure, there is every reason to expect the electrical conductivity of LB films to be highly anisotropic. The first in-plane studies[217,218] were made using anthracene, which had been appropriately substituted with *short* chains (see Chapter 2). In this interesting series of molecules the aliphatic regions no longer totally dilute the electrical properties of the π-electron systems. The resistivity in the plane is higher by eight orders of magnitude than that in the film normal direction. There have also been reports[219] of large anisotropies in other materials, particularly the charge transfer complexes discussed below. Provided reasonable stability can be introduced into the specially designed molecules, there is a possibility that some practical use may be made of the inherent anisotropy present in such systems.

For many years there has been considerable research activity aimed at producing highly conducting organic materials. A variety of different approaches has been used, the most popular involving polymers such as polydiacetylene and charge transfer complexes such as those based on tetracyanoquinodimethane (TCNQ) and tetrathiofulvalene (TTF). There has also been interesting work on intercalated compounds consisting of alternating layers of organic and inorganic (such as dichalcogenides) materials. In these structures it is the inorganic plane which is the conductor.

As a first step to obtaining metallic-like characteristics, LB film enthusiasts have used the same basic molecules but have modified them to suit the trough technique. The first success was achieved by Barraud and his co-workers.[220–224] In their initial studies, they built a nonconducting precursor film from a N-docosylpyridinium$^+$-TCNQ$^-$ salt and then exposed it to iodine vapor. They have since shown that this last stage can be avoided by using mixed valence conducting salts such as n-stearyl pyridinium TCNQ. It is also possible to avoid iodine doping by using the charge-transfer complex approach first suggested by Vincett and Roberts.[5] The majority of the TCNQ compounds studied exhibit conductivities in the range of 10^{-2} to 10 S cm^{-1} in the plane of the film. Measurement techniques such as thermoelectric power[225] and surface acoustic-wave propagation[226] are best suited in these instances because they are less sensitive to grain boundary effects. Preliminary Hall effect data[227] suggest that the majority carriers are holes with a mobility of about 0.5 cm^{-2} V^{-1} s^{-1}.

The first experiments based on TTF/TCNQ complexes were carried out by Kawabata and his collaborators.[228–230] Others have extended the research to novel LB films based on different materials, such as polypyrrole and ferrocene.[231–233]

There is plenty of evidence to suggest that all conducting films studied to date are relatively imperfect. More molecular engineering is required to achieve the high conductivities and anisotropy potentially available using the Langmuir trough technique.

7.7.2. Langmuir–Blodgett Films as Insulators

One of the most common characterization experiments for LB layers involves measuring the capacitance of a staircase-type structure. The linear dependence of C^{-1} on the number of monolayers demonstrates clearly the repeatability of the dielectric thickness of each monolayer. Impedance data[234–236] show that the relative permittivity is nearly independent of frequency up to 1 MHz. Since the films can be made so thin, it is natural to consider their use as low voltage, low loss capacitors.[237] Simple calculations based on applying a few volts across a hundred angstroms of organic material show the enormous charge storage capability of metal–oxide–LB film–metal systems. The relative permittivity of fatty acid materials is relatively low, typically 2.5. However, it is possible to design molecules with large inbuilt polarizations; a good example of this kind is polybutadiene.[86]

Polymerized multilayers, with their improved mechanical strength and stability, are more likely to be used in capacitor-type applications. Indeed, because of their sensitivity to water vapor, it has been proposed that fatty acid structures could be used as hygrometers, thus exploiting the large, fast, and linear variation in capacitance with ambient humidity.

One interesting approach[238–240] to achieving thermally stable layers is to remove the long alkyl chains at the air–water interface or after LB film deposition. In this way it has been possible to obtain wholly aromatic layers at the air–water interface and ultrathin layers of polyimide.[83,84] In the latter case, polymer precursors in the form of a long-alkyl amine salt were used for LB film preparation. The amine is then extracted via a solution treatment after deposition and the film is converted to polyimide.

Interesting effects, such as large internal voltages and differential negative resistance, have been observed in capacitor-like structures fitted with asymmetrical electrodes.[241–253] For example, Leger et al.[242] connected 226 junctions in series to obtain a battery of 50 V with a short-term internal impedance of 10^{10} Ω, a capacity of about 1 μC, and a volume of 0.1 mm^3. The mechanisms responsible for these observations are not well understood but are likely to originate from interfacial barriers and defects in the multilayers.

The fine control of thickness available with monomolecular films, even to dimensions as low as 1 nm, has already been utilized to great advantage in fundamental research, and is also likely to find application in device areas involving quantum-mechanical tunneling. The first claim of tunneling currents in LB films was by Miles and McMahon[254] about twenty-five years ago. Since then numerous workers have inferred from the shapes of their observed characteristics that this type of conduction mechanism does occur. However, it is unfortunate that much of the evidence for tunneling at low temperatures and low bias values, e.g., when superconducting electrodes are used, must be viewed with caution on the grounds that another dielectric (normally an oxide film) is present in the structure.

Although there remains a question concerning the stability of LB films at elevated temperatures, there is abundant evidence that such layers can be cycled to liquid helium temperatures without damage. Using polymerized films of vinyl stearate and diacetylene, Larkins et al.[255] have observed the hysteretic dc features of Josephson effect devices at 4.2 K. They have also conducted magnetic-field threshold reduction tests to help support their claim of superconductivity. These interesting results highlight the considerable benefits of the trough technique in terms of producing tailored materials which might control the critical current, switching speeds, and energy gap parameters in low temperature devices such as SQUIDs. The low thermal expansion coefficient of LB films is an added advantage in this connection.

Inorganic materials can fulfil most requirements in the application areas mentioned above and therefore it seems unlikely that LB films will be used. However, insulating monolayers should find a role in pyroelectric thermal imaging systems. Their vital role in modifying the surface properties of semiconductors is discussed in Section 7.8.

7.7.3. Photoelectronic Behavior of Langmuir–Blodgett Films on Metals

The photoelectrical transport properties of LB films have been discussed in an early review by Vincett and Roberts.[5] This paper includes a summary of early work on photoconductivity, photovoltaic and photomagnetic effects in a wide range of materials. However, none of these results seems likely to be exploited in the future. Their main benefit lies in helping to improve our basic understanding of intermolecular interactions and photophysical and photochemical processes in organic materials.

A considerable volume of work is available describing the properties of dye LB films. Some of these relate to their use in sensitizing or desensitizing silver halide. Kuhn and others[2] have used a variety of long-chain substituted dyes to modify the sensitivity of photographic films beyond the limitation imposed by the bandgap of the silver halide. In a related way the photopolymerization of diacetylene films can be sensitized to visible light by surface active dyes embedded in the multilayers.[256] Sugi and his co-workers have carried out extensive studies on the three main families of dye sensitizers: mercocyanine, cyanine, and oxonol. Their initial investigations described photoconductivity measurements[257–261]; they showed that the data were anisotropic and influenced by postdeposition treatment and molecular coordination. Aggregates, in particular the J type, which lead to a red shift of the absorption curve, always seem to play a key role. Penner[262] has shown that the efficiency of energy transfer between layers of J aggregates separated by 10 nm can be greater than 90%. He has speculated on the possibility of building an efficient light-collecting array leading to photoinduced electron transfer. Photosensitization is also used in electrophotography. Here, the phenomena of electronic transport, photosensitization, and triboelectricity are all brought into play. There is no doubt that several industries are studying LB films with this application in mind. Sugi and his colleagues[263–267] have fabricated several types of heterojunction cells and Schottky-type diodes based on mixed monolayers of dyes and fatty acids. Using n- and p-type dyes with different redox potentials they have observed rectification effects, voltage-dependent capacitances, and photovoltages up to 0.7 V. However, the energy conversion efficiency of these p–n junction devices is very small. The exact mechanisms are not well understood and they may indeed be more a function of the metal–oxide–dye layer interface than the p–n junction. This is also true of the simpler photovoltaic cells based on phthalocyanine,[268] In common with organic films prepared using other techniques[269–271] a major breakthrough is required for the commercial prospects to be attractive.

Some of the aggregated cyanine dye multilayers exhibit electroluminescence. The first report [272,273] of light emission from LB films involved the model aromatic compound, anthracene, deposited as multilayers with only short aliphatic side chains instead of the very long chains that were previously believed to be necessary for handling all conjugated compounds. In the direction normal to the film plane, multilayers of anthracene substituted with a C_4H_9 hydrophobic chain (and a C_2H_4 COOH hydrophobic chain on the other side of the molecule) showed ohmic behavior up to 10^6 V cm^{-1} and thereafter space-charge limited conduction. Under some circumstances, and always when the gold electrode voltage was positive, quite strong blue electroluminescence was observed which was visible under normal room illumination. The effect was probably the result of double injection, that is, hole injection from the gold electrode and weak electron injection from the aluminum electrode. A typical current–electric field characteristic is shown in Figure 7.26 for both an LB and evaporated film of the short-chain substituted anthracene. The use of an electric field rather than voltage axis disguises the fact that the voltages employed to obtain the electroluminescence were typically 6 V and 200 V, respectively. The insert shows that the light emission decays with time, probably due to degradation of the anthracene (possibly due to the formation of anthraquinone).

Despite these observations ten years ago proving that light emission can be obtained at much lower applied voltages than previously, there appears to have been no equivalent study of electroluminescence in aromatic compounds. Clearly, two primary problems remain those of finding a suitable electron injecting contact so that more electron–hole recombination will occur, and searching for a molecule with less tendency to photochemical degradation, e.g., perylene. There is no sound reason why thin-film electroluminescent devices based on LB film technology should not be a practical proposition.

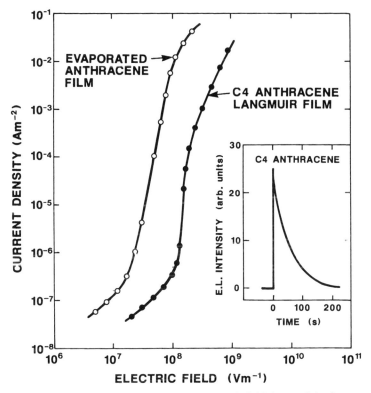

Figure 7.26. A comparison of the current density versus electric field characteristics for an evaporated film with that measured for an LB film of 69 layers of C4 anthracene, deposited onto an Al/Al_2O_3 substrate; a positive bias was applied to the top gold electrode. The inset shows the decay of electroluminescence when a dc field is applied to the structure, (See Roberts *et al.*[272])

7.7.4. Molecular Rectification and Cooperation

This section adopts a rather general definition of "rectification" in that it describes work on molecular pumps as well as more ambitious endeavors aimed at constructing molecular systems capable of performing straightforward electronic-charge rectification.

7.7.4.1. Molecular Pumping

The light-driven molecular pump involves the formation in a multilayer of cooperative groups of molecules which act in an analogous way to cooperative molecular units in biological systems.[274–277] The object is to produce electron–hole pairs which are then made to separate for long enough for efficient photochemical reactions to occur. The principle is to harvest energy via a large number of absorbing molecules and to channel it to a small number of exciton trapping species incorporated in the film. Naturally, it is essential that the charges remain separated for a significant time despite the tendency for back-tunneling to occur. These charge motion processes should be possible in a system where the donor is separated from the acceptor by a wide, low tunneling barrier and from the source by a high, fairly narrow barrier. To obtain the high barrier a hydrocarbon chain can be used; there are various thoughts as to how the low barrier may be achieved. Given the advances in LB film work during the ten years since these pioneering experiments by Kuhn and his co-workers, it will be surprising if novel supermolecular structures do not emerge improving on these attempts to combine suitable dyes with molecular wires. They will clearly be significant in helping our understanding of biological photosynthetic systems but may also have relevance in designing photochemical conversion devices. However, as with many other areas of possible application for LB films, it should be remembered that light-driven molecular pumps might also be fabricated using related techniques.[278]

7.7.4.2. Superconductivity

The type of organized supermolecular functional units described above might also form the basis of organic superconductors. Ginzburg and Little[279–282] have prophesied that room temperature effects may be possible if the crucial attractive interaction between pairs of electrons could be made to occur via an excitonic mechanism rather than the customary phonon-mediated situation. The basic idea is that electrons in a conducting or semiconducting spine or layer might experience such an interaction if they were in very close proximity to (but separated from) a molecular system with a rich excitonic spectrum such as suitable dye molecules. The observation of warm superconductivity in inorganic ceramics has clearly reduced the incentive to test out these controversial ideas for organic superconductors. The proposed mechanisms have never been confirmed or definitely tested, principally because of the very great difficulty in constructing them. It is clear that, because of the fine thickness control available, the LB film technique provides the best hope of overcoming this problem. A different approach to achieving the same objective might be to form LB films of such materials as the platinum complex of phenanthroline.[282] Here, the platinum atoms which are located at the center of an organic ring system may stack to form a linear metallic chains surrounded by polarizable compounds.

7.7.4.3. Organic Rectification

The last decade has witnessed dramatic progress in submicron configured electronic devices and integrated circuits. Using X-ray lithography it is possible to obtain silicon logic circuits with 0.1 μm channel lengths and novel devices, such as those relying on ballistic electronics with active dimensions close to 0.01 μm. Electronic processes and device operation at these levels are nonclassical and do not generally scale down from their bulk equivalents. However, it is apparent that computational power can be increased further by reducing the size of the gates, increasing the charge mobility, and providing methods for parallel operations. As we have mentioned at the start of this chapter, to overcome the limitations of inorganic semiconductor devices, significant research effort is now being directed toward developing optical and molecular electronic signal processors. These have the inherent advantages of increasing the speed of computer operations and decreasing the amount of heat which must be dissipated. The reader is referred to the proceedings of the three international workshops on molecular electronics[283] for a comprehensive view of the various approaches being studied. It is evident that substantial problems still remain concerning how to make the individual switches and how to address and interrogate molecular electronic devices.

The early contributions of Aviram and Ratner[284,285] in the field are always generously acknowledged even though it has proved difficult to synthesize the molecular systems patented by these authors. Their model of organic rectification is based on knowledge of molecules such as ferrocene, quinones, or pyridinium. What they have proposed is that metal sandwich structures, containing LB multilayers of type D–σ–A, may act as rectifiers of electric current. Here D is a good one-electron donor, A is a good one-electron acceptor, and σ is a covalent bridge that keeps the molecular orbitals of D separate from those of A. Several attempts to synthesize such molecules have met with limited success.[286,287] For example, LB films have been made with D = bishexylaminophenyl, σ = carbonate, and A = hydroxymethoxyan traquinodimethan, but no rectification has been observed. More complex mixed valence systems containing double-well potentials with a spirocycloalkane bridge have also been proposed.

Aviram and co-workers[288] have reported results indicating that switching and rectification effects are present in a hemiquinone compound which contains on either end catechol and *o*-quinone rings.

In this way they believe that an asymmetrical double-well potential for two protons has been formed. However, their observations were based on the scanning tunneling microscope technique and it is believed that the gold substrate may be influencing the results.

The reader is referred to other reviews for an account of other speculative approaches toward the ambitious goal of organic switching elements.[288–291]

7.8. LANGMUIR–BLODGETT FILMS ON SEMICONDUCTORS

We have already discussed the critical role played by surface layers in applications such as adhesion and friction. Frequently, a single monolayer containing appropriate functional groups can alter dramatically the properties of materials. This is particularly true of semiconductors whose relatively low charge-carrier concentrations are strongly influenced by the presence of another species. This section is devoted to possible practical uses of this effect.

To date, surface LB film coatings have not been designed with specific substrates in mind. Ideally, substrate parameters such as the geometry of the bonding sites and the strength of the substrate–head group bond would be taken into account. Nevertheless, even without optimization, it is possible to observe significant effects when an oriented LB film is deposited onto most semiconductors. In all cases, it should be borne in mind that a double dielectric interface is involved because the inorganic semiconductor already possess a nascent "oxide" prior to organic film deposition.

7.8.1. Metal–Insulator–Semiconductor Diodes

The surface characteristics of semiconductors may be investigated by measuring the electrical properties of metal–insulator–semiconductor (MIS) structures. However, the lack of suitable insulating layers has confined this study almost exclusively to silicon. The first MIS measurements using LB films were those of Tanguy et al.[292] Capacitance data as a function of bias voltage [$C(V)$ measurements] were obtained, mainly for multilayers of orthophenanthroline substituted with long aliphatic chains. Accumulation and inversion regions were observed, with a large capacitance modulation as expected from thin insulating layers and high resistivity semiconductors. The capacitance in the accumulation region was consistent with that measured between metal electrodes. Both "normal" and "abnormal" hysteresis were observed in the $C(V)$ curves, the former being due to the motion of small numbers of ions; this effect was also detected in thermally stimulated current measurements. Similar results for multilayers of fatty acids on silicon have been reported by Lundström et al.[293–295] The injection of charges into the insulator and their trapping were investigated by obtaining $C(V)$ measurements after applying dc voltages to the structures for various intervals, and after waiting at zero voltage for different times. It was concluded[293] that injection into the LB layers (which were regarded as a model of biological membranes) and accumulation of charge could occur at fields low enough to arise in physiological situations. The voltage dependence[294] of the injection currents implied a Schottky-like process. Modification of the layer by substitution of the end-group methyl hydrogens by fluorine changed the charge trapping, as did exposure to fatty-acid-soluble substances; mixing some unsaturated acid into the layers gave significantly increased injection currents. All these effects demonstrated that the trapping actually occurred within the multilayer, most probably at the first "methyl gap" between successive monolayers. The sensitivity of the silicon surface to free and bound charge can be used to advantage to monitor the amount of charge trapped within an LB film. Evans et al.[296] have reported results for ω-tricosenoic acid multilayers deposited onto single-crystal silicon. They showed that the density of charge trapped in a bilayer is significantly less than the surface state density at the silicon surface. It is also interesting to note that compact LB films can be used to reduce the formation of interface states on silicon. Roberts et al.[74] have used the electron spin resonance technique to monitor the formation of defect centers on silicon with and without a protective LB layer.

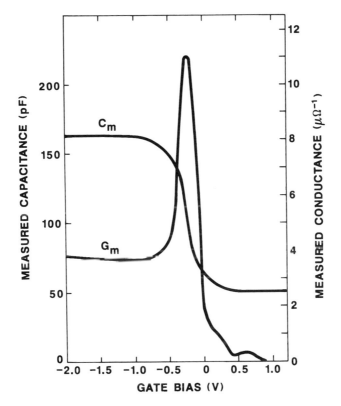

Figure 7.27. Measured values of capacitance and conductance at 100 kHz for a p-type CdTe/CdSt$_2$ capacitor. The dielectric thickness is 80 nm and the device area is 5.5×10^{-7} m^2. (See Evans *et al* [296])

It is harder to obtain classical C/V and conductance–voltage (G/V) plots using semiconductors other than silicon because of the absence of a naturally good insulating oxide layer. However, data[297,298] for cadmium telluride, using cadmium stearate as the LB film, show accumulation, depletion, and inversion characteristics, similar to those reported for many Si–SiO$_2$ devices at high frequencies. Some of these are reproduced in Figure 7.27. These curves display some of the classical characteristics expected using a metal–insulator–p-type-semiconductor structure. That is, at negative values of gate bias in the accumulation region, the capacitance is equal to the insulator capacitance. The value of the accumulation capacitance is therefore in good agreement with that expected from the known values of top electrode area, relative permittivity, and thickness of the LB film. However, when the gate is biased positively, the capacitance decreases due to the creation of a depletion layer in the cadmium telluride. A minimum capacitance value is then reached resulting from a fixed depletion layer width and the formation of inversion charge at the interface. The conductance peak in the depletion region is thought to be due to losses at interface states. By analyzing the frequency dependence of the

G/V data, it is possible to estimate the density of surface states as approximately 10^{12} cm^{-2} eV^{-1} at flat band condition.

The first attempts to passivate the surface of the important semiconductor gallium arsenide, GaAs, are described in a doctorate thesis.[299] More recently, using ω-tricosenoic acid, Tabib-Azar et al.[300] have reported accumulation, depletion, and deep depletion characteristics for this semiconductor. Interface trap densities in this case were calculated to be about 10^{11} cm^{-2} eV^{-1}.

Considerable effort has been expended in finding a suitable technology to passivate the indium phosphide (InP) surface,[301] Fermi-level pinning presents a problem in this important semiconductor. However, by preparing a suitable semiconductor surface precoating using a bromine-based etch,[302] it has been possible to invert both n- and p-type InP using LB films.[303−305]

During the past decade there have been dramatic advances in inorganic semiconductor growth, especially using techniques such as molecular beam epitaxy (MBE) and metal-organic chemical vapor deposition (MOCVD). Transistor performance has improved due to increased control of physical parameters, e.g., semiconducting alloys can be formed whose dimensions are perfectly matched to the lattice of the underlying substrate and higher transconductances can be obtained by forming adjacent layers of high- and low-bandgap materials. These developments have increased the importance of ternaries such as InGaAs and InAlAs, which have high electron mobility and great potential for high-frequency transistor action. One of the drawbacks of these materials is that they commonly form low barrier heights. Thomas[306] has shown, however, that it is possible to increase the barrier height by depositing one or more monolayers on the semiconductor surface. For example, using InGaAs coated with ω-tricosenoic acid, the barrier height is more than double its 0.2 eV normal value using a gold electrode.

There are, of course, many methods other than the LB film technique for depositing organic and inorganic films on semiconductors. However, experience has shown that when an energetic process such as evaporation, sputtering, or growth from a plasma is used to deposit a thin film onto a semiconductor, a surface-damaged layer is produced which invariably dominates the electrical characteristics of the junction so formed. The Langmuir trough technique, being a low-temperature deposition process, provides a means of circumventing this particular difficulty. On the other hand, it does mean the preparation of the substrate before dipping is of considerable importance in determining the quality of the interface produced. That is, the nascent "oxide" layer formed during the etching procedure remains relatively undisturbed and this can play a vital role even after it has been coated with an LB film.

Organic films prepared using alternative methods do appear to suffer badly from hysteresis effects[307] and, in this regard, LB films do appear to have the edge. However, experience to date suggests that physicists and engineers engaged in this field tend to learn from experiments with organic materials. Rather than planning to use them in commercial devices, they are stimulated to try harder to achieve similar or better data with more stable inorganic materials. Thus, even though the surface of

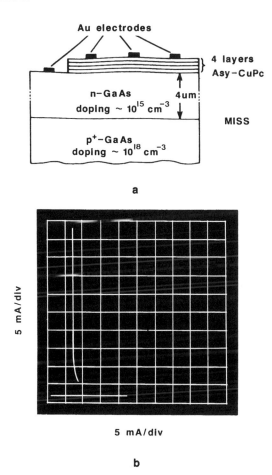

Figure 7.28. (a) The structure of a GaAs MISS device containing four monolayers of an asymmetrically substituted copper phthalocyanine molecule. The trace in (b) shows a typical device characteristic using a series resistance of 1 kΩ; the diode area is approximately 2×10^{-3} cm². (See Thomas *et al.*[310])

n-type InP was first inverted using an LB film,[303] alternative methods based on complex inorganic insulators are now being successfully applied in the industry.

The MIS structure is, of course, an integral part of insulated-gate field-effect transistors and other field-effect devices. These topics are discussed later in this section.

Bistable switching characteristics have been observed in metal–insulator–n–p⁺ (or equivalent p–n⁺) structures.[308,309] Potential applications include memories, shift registers, and light-sensitive or gas-sensitive switches. An essential requirement in such devices is an ultrathin high-quality insulator of tunneling dimensions. For silicon it is possible to use oxide layers approximately 3 nm thick, even though they are difficult to grow uniformly and reproducibly. As has already been indicated, for high mobility group III–V semiconducting compounds like GaAs, the problem is more severe. Thomas *et al.*[310,311] have demonstrated the usefulness of LB films in this connection. Figure 7.28 reproduces some of their data for a GaAs/phthalocyanine MISS switching device. A punch-through mechanism has been postulated to account for the dramatic rise in current at a particular threshold voltage.

7.8.2. Schottky Barrier Modification

The organic/inorganic semiconductor interface is not well understood; the matching of energy levels based on narrow bandgaps, low charge-carrier mobilities in organic materials, and the presence of a poorly characterized and unwanted interfacial layer, all complicate the situation. However, a number of applications have emerged for metal–organic insulator–semiconductor structures.

One of these is the MIS solar cell[312]; the insertion of an insulator between a metal and a semiconductor increases the open circuit photovoltage of the cell over that of the Schottky barrier device. If the insulator is of tunneling thickness, no reduction in photocurrent is observed. Another MIS device exploits the enhanced minority carrier injection ratios that are possible with these diodes.[313] Many efficient luminescent materials are not available in both p- and n-type forms (e.g., some II–VI compounds). In these cases the MIS structure offers the possibility of producing useful electroluminescent (EL) displays. Theoretical predictions emphasize the crucial role for the insulating layer in determining the operation of both types of MIS diode. In particular, the insulator thickness and nature of the insulator/semiconductor interface must be optimized to obtain efficient devices. The poor control of insulator thickness (in the 1–10 nm range) provided by many thin-film deposition methods, together with the questionable quality of the resulting films, has suggested the use of LB films as the organic media.

7.8.2.1. Photovoltaic Cells

There have been many attempts to optimize the efficiency of solar cells by incorporating a thin insulating layer between the photoconductor and the top transparent electrode.[314–316] The exact role of the interfacial layer depends on many factors, such as its thickness and the amount of fixed charge it introduces into the heterojunction structure. Under illumination an excess of minority carriers is produced at the surface of the semiconductor. If the solar cell device is short-circuited, some of these carriers will tunnel through the insulator to the metal producing a short-circuit current density, J_{sc}. If P is the solar input power per unit area, V_{oc} the open circuit voltage, and F the fill-factor, then the efficiency of a photovoltaic solar cell may be written in the form

$$\eta = V_{oc} J_{sc} F/P$$

Therefore, an increase in V_{oc} produces an overall improvement in efficiency, provided the introduction of the interfacial layer does not reduce the values of J_{sc} and F.

The value of the short-circuit current is determined by the diffusion of minority carriers from the neutral region of the semiconductor into the depletion region, and by minority carrier production in the depletion region. The tunnel resistance of very thin insulating layers incorporated in Schottky-barrier solar cells has been shown to be much larger for majority carriers than for the photogenerated minority carriers.[317] Thus the effect of the interfacial layer will be to reduce the solar-cell dark current without affecting the short-circuit photocurrent. However, beyond a certain critical insulator thickness, the additional resistance introduced will lead to a reduction in the value of J_{sc}. Thus an optimum thickness of insulator is required in order to obtain maximum efficiency from an MIS solar cell.

Figure 7.29 reproduces some data from the first report of MIS solar cells incorporating LB films.[318] The addition of a single monolayer of cadmium stearate between the surface of single-crystal cadmium telluride and a gold electrode is seen to increase the value of V_{oc}, but not degrade J_{sc} or the fill factor. However, a

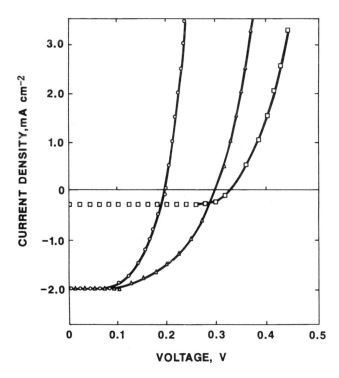

Figure 7.29. Current–voltage characteristics measured under AM1 conditions for a Au/CdTe solar cell incorporating no LB film, monolayer of $CdSt_2$, and 3 monolayers of $CdSt_2$. (See Dharmadasa et al.[318])

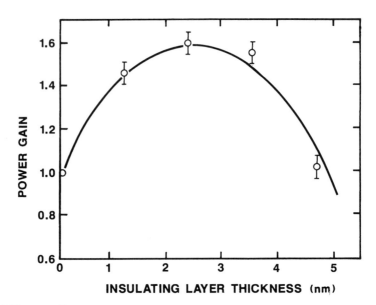

Figure 7.30.   The efficiency of an MIS cadmium telluride solar cell (normalized to its value without an LB film) as a function of insulator thickness. The points refer to 1, 2, 3, and 4 monolayers of C4 anthracene. (See Roberts *et al.*[319])

further two monolayers reduces the current density. In order to obtain finer control over insulator thickness, the measurements were repeated[319] using the short-chain substituted anthracene material mentioned in Section 7.7.3. The data have been replotted in Figure 7.30 and show that the solar cell efficiency is optimum near 2.1 nm. Above this thickness the slight improvement in open circuit voltage is more than offset by a sharp reduction in short-circuit current density. These two papers on CdTe also provided clear evidence for the first time that the introduction of an LB film is capable of increasing the effective Schottky-barrier height on a semiconductor.

Tredgold and Jones[320] studied the effect of LB films of stearic acid and other amphiphilic materials on GaP photodiodes. They found that a single layer of stearic acid more than doubled the solar cell efficiency and proceeded to repeat the effect with more stable amphiphilic polymers.[321] The same authors[322,323] demonstrated that the density of carboxylic acid head groups affected the degree of band bending and thus the efficiency. Some of their data are reproduced in Figure 7.31. A complete theoretical analysis is difficult in that it involves many physical processes including quantum-mechanical tunneling across and through a complex interfacial layer containing trapped charge and highly oriented surface dipoles. More recently Tredgold and El-Badawy[324,325] have studied the benefits of including LB films in GaAs solar diodes and reported that three monolayers of a suitable polymer lead to a substantial improvement in diode efficiency. The results shown in Figure 7.32 were

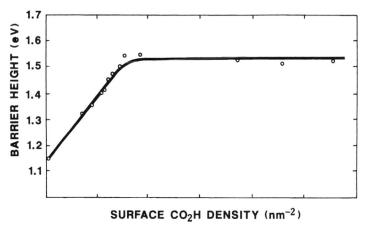

Figure 7.31. Schottky barrier height at the interface between GaP and derivatives of polysty-rene/maleic anhydride LB films. The parameter is shown as a function of the number of interfacial carboxylic acid groups per unit area. (See Winter *et al.*[323])

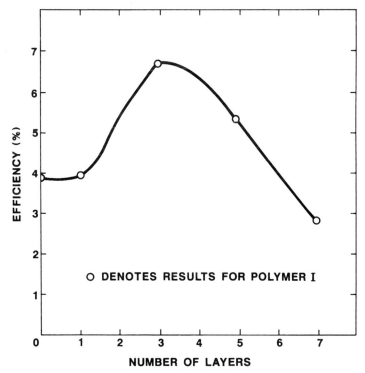

Figure 7.32. The efficiency of GaAs photodiodes as a function of the number of preformed polymer monolayers. (See Tredgold.[326])

obtained without using antireflection coatings or other means to improve the efficiency.

7.8.2.2. Electroluminescent Diodes

The use of LB films as the active layer in an electroluminescent device has been mentioned in Section 7.7.3. Traditional light-emitting diodes rely on a rather different principle, i.e., a p–n homojunction is used to stimulate radiative recombination. An alternative method must be sought for semiconductors which do not exhibit n- and p-type conductivities. One possibility is to use a metal–semiconductor Schottky barrier structure. A critical parameter is the minority carrier injection ratio for such a device. Its value can be enhanced by incorporating a thin insulator between the semiconductor and metal electrode. For example, Card and Smith[327] have demonstrated that the optimum insulator thickness for electroluminescence (EL) output is approximately 4 nm for a $Au/SiO_2/n$-type GaP system. However, controlling the interface presents similar difficulties to those encountered with MIS solar cells and thus there may be some benefit in using the good insulating and controlled thickness properties of LB films for this purpose.

The only published work on electroluminescent MIS diodes incorporating LB films is by Roberts, Petty, and co-workers. Their early papers described results for GaP.[328–331] Following a systematic series of experiments designed to provide conclusive proof that oriented monolayers were producing a change in the effective barrier height, they showed that an optimum thickness of LB film could improve the electroluminescent efficiency by several orders of magnitude. Although all the initial results were obtained for $CdSt_2$ and ω-tricosenoic acid, the later devices relied on more robust monolayers such as phthalocyanine. A typical curve is shown in Figure 7.33. The maximum in efficiency occurs using a thickness of approximately 21 nm, exactly the value at which the short-circuit photocurrent reduces in a solar cell based on the same materials.[331] Both effects are dependent on the ability of minority carriers to cross the semi-insulating film. The thickness involved is well above that expected for conventional tunneling; more research is required to establish the transport mechanism involved.

The success achieved with the model GaP system described above has led to further work on II–VI compounds which cannot be fabricated as p–n junctions. Using the same principle, blue light emission has been observed for the first time at room temperature from an MIS structure based on MOCVD ZnSe and a substituted phthalocyanine LB film.[332–334] The devices were completely stable using pulsed conditions. Another desirable feature of coating a semiconductor with a compact monolayer is that it smoothes out any potential variations on the surface and thus ensures uniform brightness.

7.8.3. Field-Effect Transistors

The success of integrated electronic devices on silicon is due largely to the impressive qualities of its native oxide. By carefully thermally oxidizing the surface

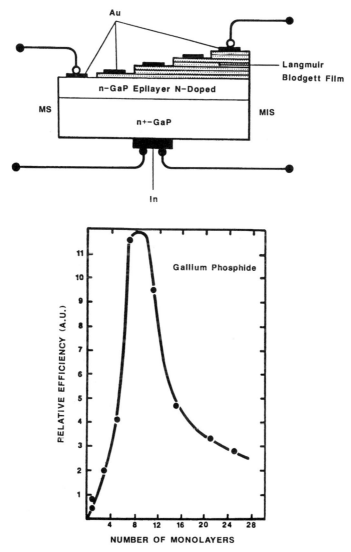

Figure 7.33. The electroluminescent efficiency versus number of ωTA monolayers for the metal–LB film–semiconductor structure shown in the upper section of the diagram.

of this semiconductor, it is possible to produce a predictable interface that is characterized by an extremely low surface-state density. It is this feature that makes possible the development of high-performance silicon MOS transistors and other integrated electronic devices. Considerable effort has been devoted to finding a suitable technology for the formation of compatible insulating films on other semiconductors that do not possess a natural oxide layer of the required quality. The III–V compounds, because of their relatively high charge-carrier mobilities, have attracted a great deal of attention in this respect.

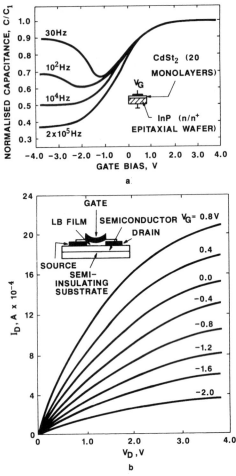

Figure 7.34. (a) Normalized capacitance–voltage characteristics of an n-type InP/cadmium stearate MIS structure measured in the dark at different frequencies. (b) Source–drain characteristics of an InP MISFET incorporating a 30 monolayer cadmium stearate LB film. Source–drain spacing is 70 μm. (See Barlow et al.[335])

7.8.3.1. Group III–V Semiconductors

The first transistor incorporating LB multilayers as the insulator was reported more than a decade ago[4,335]; it followed successful preliminary work aimed at inverting the surface of n-type indium phosphide with cadmium stearate as the insulator in an MIS structure.[303] Figure 7.34a shows the C/V data at various frequencies. As the frequency is decreased to values where the minority carriers are able to follow the variation in the measurement signal, an increase in capacitance is observed. Only minor hysteresis was present using a voltage scanning rate of 10 mV s^{-1}. The corresponding field-effect transistor characteristics are shown in Figure

7.34b. It can be seen that the channel conductivity between the source and drain is modulated by the action of a gate electrode.

The commercial benefit of realizing high-speed InP transistors has prompted other researchers to develop this work further,[336] In particular, it is important to mention the work of Fowler *et al.*[337] who have described a depletion-mode InP MISFET using standard planar integrated-circuit technology. These authors used iron-doped semi-insulating material with an n-channel layer formed by silicon ion implantation and a titanium gate electrode. Their novel molecular material was spread as a monomer but polymerized on the water surface to form poly[1,4-phenylene-5,5' (6,6')-dibenzimidazole-2,2'-diyl], which is stable up to 400 °C. Following the coating process the LB film was selectively etched to expose the source–drain electrodes, by patterning with about 1 μm of photoresist followed by oxygen reactive ion etching. A careful analysis was made of the interface; the surface state charge density distribution showed a minimum value near 10^{11} cm^{-2} eV, comparable with those obtained by Roberts *et al.*[335] using cadmium stearate as the LB layer. The source–drain characteristics of the MISFET fabricated by Fowler *et al.* using twelve monolayers of their polymeric insulator is shown in Figure 7.35.[337] It is unlikely that a commercial application will be based on this depletion-mode InP device. However, it is important to note that the LB film approach was successful while other popular techniques, such as chemical vapor deposition of silicon oxide, have not shown promise. High-speed and low-noise GaAs field-effect transistor technology is successful largely due to the high Schottky barrier

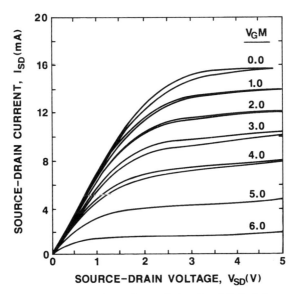

Figure 7.35. Source–drain current–voltage characteristics of an InP depletion-mode MISFET (gate length 1.2 μm, gate width 200 μm) incorporating 12 monolayers of a polybenzimidazole LB film as gate insulator. (See Fowler *et al.*[337])

height (typically 0.85 eV) achievable with standard metal electrodes. This is not the case for other III–V compound semiconductors; for example, with InP, depending on the surface etching procedure, the barrier height lies in the range 0.36 eV to 0.55 eV.

In Section 7.8.1 we indicated that important ternary compounds such as GaIn-As present even more of a challenge, but that improvements are possible with LB surface coatings.[306] An interesting development of this idea has been reported by Chan et al.[338] who took cognisance of the results of Meiners et al.[339] showing that unusually high Schottky barrier heights could be obtained on InP using electroplated cadmium. Their approach was to deposit several monolayers of cadmium stearate onto the surface of the alloy $In_{0.53}Ga_{0.47}As$, and then remove the organic components with an oxygen plasma. The barrier height increased as a result from 0.20 eV to 0.52 eV and yielded a modulation-doped field-effect transistor with excellent dc and microwave performance (transconductance 170 mS mm^{-1} and cutoff frequency 19 GHz). It would be interesting to examine carefully the chemical composition of the interface following the oxygen plasma treatment and also to extend the principle to other semiconductors. Other metal atoms can be incorporated into fatty acid LB layers and therefore the technique could clearly be adapted as an alternative to vacuum deposition. It is compatible with standard III–V processing technology and well suited for the small dimensions required for high-frequency transistors.

7.8.3.2. Silicon

The difficulties associated with making a silicon transistor incorporating an LB film are nowhere near as acute as those encountered with groups II–VI and III–V semiconductors. The protective insulating oxide layer ensures that the effect of any pinholes or defects in the LB film can be reduced. Nevertheless, it is important to report the progress made to date because of its possible relevance to chemical sensors, which are discussed in the following section.

Fung and Larkins[340] have obtained characteristics similar to those observed with conventional insulated gate FETs using silicon oxide as the gate insulator. The n-channel devices used phosphorous-diffused sources and drains on a boron-doped p-type Si substrate. The planar fabrication process resulted in 11 transistors and associated test structures being formed on a single chip. The gate insulator used was ten monolayers of 16-8 diacetylene, a material rugged enough to "survive" the metal gate patterning process.

Hydrogenated amorphous silicon ($\propto$Si;:H) is an important semiconductor for solar cells and for thin-film transistors used to address liquid-crystal displays. Lloyd et al.[341,342] have used LB films to form transistors on this material. Their results show clearly that the source–drain current can be altered by more than two orders of magnitude by applying low voltages to the gate electrode. However, their results are complicated by the presence of a surface coating on the amorphous silicon. In any event, inorganic thin-film insulators such as amorphous silicon nitride meet the device requirements adequately and therefore there seems little purpose in pursuing the LB film approach in applications based on this material.

In summary, the good insulating properties of LB films suggest their possible use in field-effect devices. Because of the huge investment made in integrated circuit processing technologies based on inorganic materials, especially silicon, it is unlikely that organic films will play a major role. A possible niche exists for them as a result of their ability to change the effective barrier height at a semiconductor surface. However, one is more likely to capitalize on the advantages of being able to incorporate an organic layer at the heart of a transistor device in the field of sensors, the subject of the next section of this chapter.

7.9. CHEMICAL/BIOLOGICAL SENSORS AND TRANSDUCERS

There is an increasing requirement for improved analytical techniques in the pharmaceutical and biomedical industries and for process and environmental control. It is difficult to estimate the share of the large market that organic-based devices may occupy. However, the enormous scope available with organic molecular systems to produce carefully designed, optimized, and engineered materials that are specific (ability to recognize) and selective (ability to discriminate) should in due course lead to important commercial developments. In many of the situations described earlier in this chapter, it is possible to encapsulate the active layer and thus utilize a "fragile" organic film. For obvious reasons, it is not normally possible to protect a sensitive and selective organic coating designed to monitor interactions with a gas or a fluid.

From a device perspective the operation of chemically and biochemically sensitive sensors can be divided into three stages. The first involves recognition; in an ideal device there will be a lock–key mechanism governed by a specific interaction between a molecule and the sensing surface. Second, the binding process must result in a detectable and systematic change in a physical parameter. Finally, a suitable transducer is required to convert the change into an observable signal. There are many types of device that can be used in conjunction with a suitable sensing element. It is fashionable to employ a transistor-type structure containing a number of different sensor layers so that a multiplicity of chemical species can be detected. Examples of this type and those based on other principles, such as acoustic and optical wave propagation, are discussed in this section. For a practical system many other requirements must be satisfied, not least of which is low cost and reliability. From a convenience viewpoint it is important that the sensing surface show self-regeneration and recalibration following exposure to the gas or fluid. Past response and recovery times are desirable. By using a single monomolecular layer one can avoid complex formation and bulk diffusion processes likely to lengthen interaction times. LB films have a distinct advantage in this context because of their high ratio of surface area to bulk volume.

There are many reviews emphasizing developments in LB film-based chemical[343] and biosensors.[344,345] To a large extent the emphasis has been on proving that monomolecular layers can be incorporated in a large range of different systems, sometimes to advantage compared with films deposited by other means. Insufficient attention has been placed to date on the more difficult task of designing stable, specific molecules tailor-made to detect certain gases or fluids. Some interest has been shown in using LB films as a matrix for immobilization of a recognition molecule which may prove particularly suitable for antibody–antigenic-type reactions. Biosensors of this type are discussed separately at the end of this section following a brief review of work to date on LB film devices designed to monitor gas and vapors.

7.9.1. Conductivity Devices

Owing to the ease of construction, simplicity, and low cost, most gas and vapor sensors have relied on measurement of the dc conductivity across the surface of a material fitted with appropriately positioned planar electrodes. Additional charge carriers are introduced as a result of the interaction between the organic compound and the reaction gas. Two of the most popular systems are the phthalocyanine and porphyrins; both have been examined in LB film form. In general their quality is relatively imperfect compared with those of the classic film-forming materials, but their significant advantage lies in their thermal and mechanical stabilities.[346–352] Like their single-crystal and vacuum-deposited film counterparts, they interact strongly with powerful electron acceptors such as N_2O_4, Cl_2, Br_2, and I_2. In fact, the problem with all ligand gas sensors of this type lies not with their sensitivity but their selectivity, for porphyrins and phthalocyanine molecules bind several gases.

The first chemiresistor gas sensor incorporating an LB film of phthalocyanine is described by Baker et al.[353] Figure 7.36 shows some of their data for an asymmetrically substituted copper phthalocyanine exposed to a few volume parts per million of NO_2. Similar research was reported by Wohltjen et al.,[354] who studied the effects of ammonia and NO_2 on copper tetracumylphenoxy phthalocyanine. Tredgold et al.[355] have reported conductivity changes in complexes of mesoporphyrin IX and tetraarylporphyrin upon exposure to NO_2, H_2S, and CO. The copper diol complex was sensitive to NO_2 but insensitive to H_2S and CO. The conductivity changed by several orders of magnitude but long response times were observed, probably due to defects within the 25 monolayer thick film. It is well known that acceptor gases produce a partial oxidation of the tetracyanoquinodimethane (TCNQ) charge transfer salt to TCNQ° with a corresponding increase in conductivity. Henrjon et al.[356] have reported such effects in LB films of this compound and have shown that they are completely reversible when the device is exposed to ammonia.

7.9.2. Field-Effect Devices

Many semiconductor field-effect devices incorporating LB films were described in Section 7.8. These included MIS and MISS diodes and transistors, but

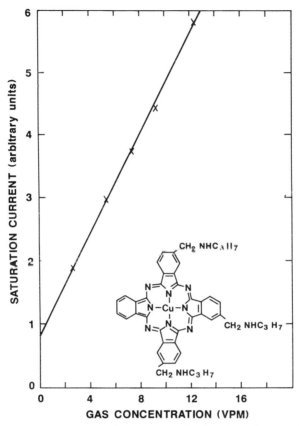

Figure 7.36. Saturation current versus NO_2 gas concentration for a device incorporating 8 monolayers of the asymmetrically substituted phthalocyanine molecule shown in the insert. (See Baker *et al.*[353])

there are, of course, numerous other structures as well that rely on modifying the charge distribution at a semiconductor surface. Minute changes in the dielectric properties of the interfacial region can have profound effects on the characteristics of a device. For this reason, it is sometimes preferable to use a field-effect structure as a sensor because of the "built in" amplification involved, even though it is only essentially a conductivity change that takes place.

It is well known that the surface state distribution at a silicon–silicon oxide interface can be modified by the ambient. In particular, if palladium is used as the metal gate electrode, an excellent hydrogen sensing device is obtained.[357,358] Recently Evans *et al.*[359] have shown that the incorporation of multilayer films of ω-tricosenoic acid can increase the density of interface states and thus enhance the sensitivity. There are a couple of recent examples of field-effect devices incorporating phthalocyanine films.[360,361] If this approach is to be used commercially, it will be essential to add some intelligence feature to improve their discrimination capabilities, for example, multiplexing using a number of different devices.

7.9.3. Optical Sensors

Optical waveguides and optical fibers can be used in a variety of sensors. In many ways they are simpler than FETs as sensors in that normally a critical resonance or threshold effect is involved. Most attention has been given to the surface plasmon resonance (SPR) technique, first introduced by Lundstrom and his workers.[362,363]

Surface plasma waves are electromagnetic waves at the interface of a metal and a dielectric. Their quanta are called surface plasmons. Any changes in the properties of the dielectric layer in the immediate vicinity of the interface influence the excitation of the surface plasmons. The properties of LB films are ideally suited to a systematic examination of surface plasmon resonance. Very elegant basic work on the subject has been carried out by workers at the IBM San Jose Laboratory.[364–367] A comprehensive account of relevant applications has been given by Raether.[368]

Surface plasmons may be excited by several methods but the most popular utilizes the photon reflection technique. The resonance can be observed by changing the angle of incidence of the light and monitoring the amount of light reflected from the metal surface. At the resonance condition there is considerable reduction in light throughput because of plasmon absorption. The simplest and most convenient dielectric to use is glass and the metal which is reported to show the strongest resonance is silver.

The angle at which the resonance occurs is dependent on the wavelength of light used and the refractive index of the glass, but is typically in the region of 43°, which is above the critical angle. A consequence of this is that a rectangular-section glass slab cannot be used to create the resonance condition; an inclined incident surface is required, and the solution most commonly employed is the use of a prism.

The thickness of the metal layer is of great importance; if it is too thin then a true resonance does not occur, if it is too thick then insufficient energy can be coupled through the absorbing metal layers and the resonance is only very shallow. Once the optimum metal thickness has been established the next step is to optimize the thickness of the organic layer. Its presence shifts the resonance to a higher angle of incidence and also broadens the peak. The final stage is to expose the overall coated prism unit to the required vapor and monitor any changes in the resonance condition accompanying surface adsorption of the vapor. When this occurs, the incident angle for peak resonance increases. This shift can be noted either by mapping out the overall curve, or, more conveniently, by previously setting the angle of incidence to be at the steepest part of the resonance curve and monitoring the intensity change. This affords the greatest sensitivity and the presence of the vapor is indicated simply by an intensity change and does not involve mechanical adjustment of the apparatus.

A good illustration of the sensitivity of the shape and position of a SPR reflectivity curve is given in Figure 7.37.[367] Similar data have been reported by Lloyd et al.[369] using a substituted phthalocyanine LB film. As expected, the

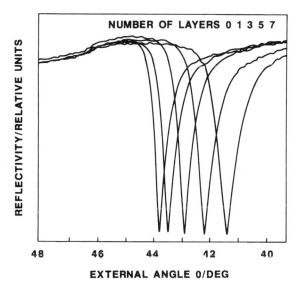

Figure 7.37. Reflectivity minima due to surface plasmon resonance in a 50-nm-thick film of silver coated with different numbers of monolayers of cadmium dimethylarachidate. The minima from left to right are for films of bare silver and silver with 1, 3, 5, and 7 monolayers. (See Brown *et al.*[367])

results are sensitive to the presence of acceptor gases. Figure 7.38 shows the reflectivity change measured at a fixed angle of incidence for various concentrations of NO_x. These data are very similar to those first reported by Liedberg *et al.*[362] where a silicon glycol polymer was used to detect the anaesthetic gas halothane.

The SPR technique provides an extremely simple and elegant method to

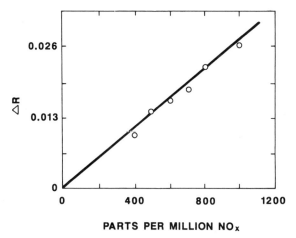

Figure 7.38. Reflectance change after 10 m exposure of phthalocyanine LB film to various concentrations of NO_x. (See Lloyd *et al.*[369])

monitor both gases and fluids. It appears to offer considerable advantages compared with those discussed earlier in this section. Developments are underway to miniaturize the equipment; prospects for exploiting the effect in medical diagnostic areas, perhaps using LB films, are excellent.

7.9.4. Acoustoelectric Sensors

When materials are added or removed from a vibrating body, its resonant frequency is changed. This phenomenon has been used for mass determination, normally with a resonator made of piezoelectric quartz cut with a specific crystallographic orientation.[370-372] A more sophisticated acoustoelectric microgravimetric sensor is a surface acoustic wave (SAW) structure. Such devices have had a great impact in the field of signal processing, especially as filters and delay lines.[373] Absorption of a gas introduces minute changes in the mass of a sensing layer. Thus both bulk quartz oscillators[374] and SAW devices[375,376] can form the basis of gas detectors.

7.9.4.1. Quartz Oscillator

Traditionally, the most common method to check the reproducibility of LB film monolayer deposition has been the capacitance method, where the aim is to produce good linear plots of inverse capacitance versus number of layers. However, this technique involves evaporating metal contacts onto the organic film and possibly damaging the surface region. The use of simple quartz oscillators circumvents this difficulty.[377] Typical results for piezoelectric crystals loaded with LB films of pyridinium TCNQ are shown in Figure 7.39. The reproducibility of the dipping process is confirmed by the linear relationship between the change in the resonant frequency (Δf) and N. The resolution of the measurement system used was 0.1 Hz, thus enabling single monolayer films to be detected with ease.[378]

Ross and Roberts[377] have discussed the responses to ammonia and hydrogen sulfide of quartz crystals coated with LB films of ω-tricosenoic acid and a substituted phthalocyanine. In all cases, the coatings responded reversibly to the presence of gases at room temperature; the lowest detection limit using the single-crystal resonators was approximately 1 ppm. These results demonstrate that Δf varies approximately linearly with gas concentration in the range 5 ppm to 100 ppm. The magnitude of the responses for both gases was the same for both three- and five-layer films suggesting that the coated crystals are responding principally to surface adsorption. It is interesting to note the large difference in selectivity between the two types of LB film. The ωTA is clearly more responsive to hydrogen sulfide than ammonia, but the opposite effect is observed for the substituted phthalocyanine coating. These results were obtained for LB film materials developed for applications other than sensors. Nevertheless, they demonstrate the enormous potential of the technique using specially tailored receptor molecules for specific target gases. In a practical device, it would be preferable to use a differential sensing system rather

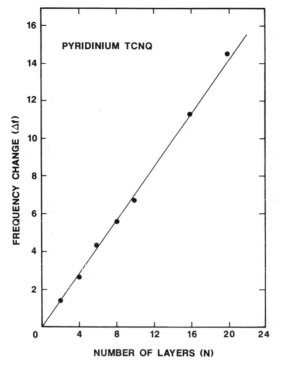

Figure 7.39. Change of resonant frequency of 18 MHz piezoelectric quartz crystals coated with LB films of pyridinium TCNQ. (See King.[371])

than one using a single-crystal oscillator as described here. That is, the gas-sensitive coated crystal would be run in conjunction with an identical reference uncoated crystal.

7.9.4.2. Surface Acoustic Wave Oscillator

The most common SAW devices are based on quartz or lithium niobate. In due course it may be possible to use piezoelectric LB films to launch the surface waves. To date, however, the reported values of the piezoelectric coefficients in monolayers are too small. In SAW devices, input and output interdigitated electrodes are defined using lithography. These perform the conversion between electric and acoustic energy. The single-crystal surface region between the transducers serves as a propagation path for acoustic waves and thus forms a delay line. Figure 7.40 shows a dual delay-line configuration specially designed for sensing purposes. Basically, the device comprises two identical SAW oscillators positioned alongside each other.[379,380] The hatched regions are earth shields to minimize reflections and cross talk between the two oscillators. The selective coating is placed in the propagation path of one of the oscillators thus affecting the delay time; the relative shift in

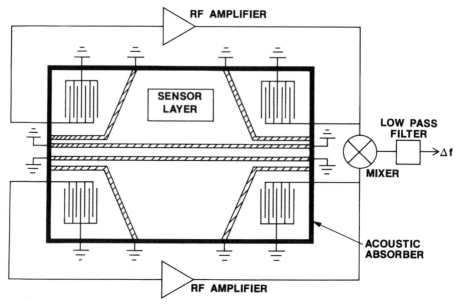

Figure 7.40. A surface acoustic wave dual delay-line oscillator designed for application as a gas sensor. (See Holcroft *et al.*[379])

frequency between the two oscillators is measured using a mixer circuit to obtain the differences frequency and then passing the resulting signal through a low-pass filter. The change in frequency (Δf) between the two channels is then directly attributable to the sensor layer; other extraneous effects, such as those due to temperature changes, are eliminated or very much reduced. Alternative geometries, such as a surface acoustic wave resonator, are also possible.

Using this delay line, results have been obtained for both insulating and semi-conducting LB films. Figure 7.41 shows data for a substituted phthalocyanine molecule deposited onto a lithium niobate device. As expected, the velocity change Δv scales almost perfectly with the number of deposited layers. Similar data for insulating LB films also yield straight-line graphs, thus confirming that the change is governed purely by simple microgravimetric effects. Thus a SAW device provides essentially the same information as can be obtained using single-crystal resonators. However, they do offer increased sensitivity, for even with single monolayers frequency changes in excess of 1 kHz are observed.

Contrasting results are obtained when a more conducting LB film[381] is deposited onto the SAW device. Additional effects are introduced as a result of interactions between the surface acoustic wave and mobile charges in the organic layer. By subtracting the component due to mass-loading effects, it is possible to measure the conductivity of the surface film. For films of different thickness, a conductivity of approximately 0.5 Ω^{-1} cm^{-1} is obtained. Such an experiment is difficult to

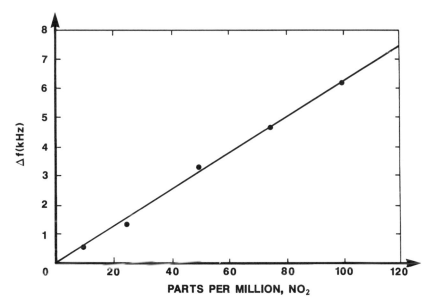

Figure 7.41. Change in resonant frequency of a 98.6 MHz SAW oscillator exposed to NO_2. The chemically sensitive coating is pyridinium TCNQ. (See Aizawa.[382])

achieve by means other than the Langmuir trough; this is because, for films thicker than a few tens of nanometres, mass loading effects dominate those due to the electric field. The ability to separate electric-field and mass-loading effects when the devices are exposed to gases means that much more selectivity is possible.

Acoustoelectric sensors have no inherent specificity and therefore, as for optical sensors, there is a requirement for tailored materials to bind specific molecules. Multiplexed sensing can be used to alleviate problems of response selectivity. Using SAW devices it is relatively straightforward to envisage an array of sensors on the same substrate. A patterned electrode system on the piezoelectric material can be used to deflect the sound wave to separate areas of the substrate coated with different molecules. Such devices have the potential of being mass produced at low cost and thus used as disposable sensors. Sensitivity is clearly not a problem in that mass changes as low as femtograms per square centimeter can be detected. However, for use with fluids rather than gases, it will be necessary to use nonconventional surface wave propagation modes.[344]

7.9.5. Biosensors

We have learnt in Chapter 5 that scientists have long been aware of the usefulness of the Langmuir trough to form monolayers of biological materials, such as chlorophyll or cholesterol, on a water surface. They have also known that monomolecular films bear a strong resemblance to naturally occurring biological membranes. The membrane typically consists of long-chain compounds called phospholipids in a bilayer arrangement with globular proteins inserted within the layer. The phospholipid molecules have both fatty and polar regions and thus the bilayer is essentially a two-layer LB film. As well as holding a cell together, a biological membrane is a highly selective barrier whose permeability characteristics are intimately involved.

The ordered structure of LB films thus creates an inert matrix in which biologically specific binding agents can be dispersed. Many research groups are excited by the prospect of constructing highly specific biosensor devices aimed at antibody–antigenic-type lock–key interactions. Excellent reviews of this topic have been given by Aizawa,[382] Reichert et al.,[345] and Nylander.[383] A competitor to LB films as the host matrix is black lipid membranes.[384] Thompson et al.[385,386] have described various ways of supporting these fragile membranes, including use of the LB technique.[387] The lipid used was egg-derived phosphatidyl choline and cholesterol. Others have also adopted this method to produce stable bilayers.[388,389]

The electrical properties of phospholipid membranes in LB film form have been studied and shown to possess the desired resistive properties for use as an inert matrix.[390–393] Stability requirements are more difficult to meet and it is unlikely that chemically reactive phospholipid analogs will be used in any practical device. Some attention has already been given to diacetylene and other polymers[394–397]; an advantage of this approach is that cross-linking could be used to help bind the receptor molecules within the matrix.

The biological systems most suitable for chemically specific sensor applications are enzyme–substrate, antibody–antigen, receptor–transmitter, etc. In the first of these the topography and surface charge distribution of a chosen enzyme provides binding sites for a specific molecule, the substrate; enzymatic activity catalyzes the conversion of the substrate to a product. For example, the enzyme glucose oxidase in LB film form has been used as the basis of a solid-state glucose sensor.[344,398] Antibodies are proteins typified by having a Y-shaped structure whose two arms contain the active regions that bind only to a specific antigen. If an antigen is bound to a membrane and free antibody is specifically adsorbed on it, the potential across the membrane will alter. Receptor–transmitter systems are distinctive in that the binding event induces a conformational change in the receptor which indicates the transduction mechanism. The incorporation of receptors, ionophores, and other biologically interesting species into bilayer structures has been reported by many authors.[399–402] In many cases the work has progressed as far as producing LB films and confirming that the active molecular species is present. For example, the integral membrane protein, cytochrome B5, has been incorporated into phospholipid monolayers at the air–water interface to test the practicability of the immobilization technique for amphipathic proteins.[403] Increasing the fatty acid chain length led to increased cytochrome penetration; this was confirmed by labeling the protein with radioactive iodine and assaying the mixed monomolecular assembly.

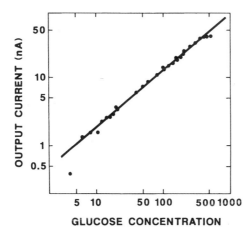

Figure 7.42. Calibration curve for a glucose sensor; the device contains two monolayers of glucose oxidase–cadmium arachidate. (See Moriizumi.[344])

There have been several efforts to develop biosensors based on LB films. All the different device configurations mentioned earlier in this section, namely amperometric, field effect, optical, and acoustical, could in principle be used in conjunction with suitably designed, biochemical layers. Moriizumi and co-workers[404,398] have extended their early work using photosensitive polyvinylalcohol for enzyme immobilization. Molecules of the enzyme glucose oxidase (GOD) were absorbed onto the hydrophilic surface of a spread monolayer of cadmium arachidate. The deposition process used was that introduced by Fromhertz.[405] Further details of the device geometry, including the presilanization of the chip surface, are to be found in the original paper. Their calibration curve for a device incorporating two monolayers is reproduced in Figure 7.42. The advantages of this glucose sensor were perceived to be that no output saturation occurs at high glucose concentrations and that control is possible of the dynamic range by appropriate selection of the number of GOD/cadmium arachidate layers.

Ion-sensitive field-effect transistors (ISFETS) are formed by using an ion-selective membrane in place of the metal gate normally found in a field-effect transistor. Any resultant change in membrane potential results in a change in the source–drain current. When used in conjunction with materials displaying biological selectivity, other acronyms describe the devices, for example, IMFET (immunosensitive field-effect transistors) and ENFET (enzyme field-effect transistors). Progress towards the goal of producing reliable LB film biosensors of this type has been reported. For example, Katsube et al.[406] have adsorbed immunoglobin G (IgG) antigen onto stearic acid monolayers, deposited the layers onto conducting InO_x electrodes, and used a simple electronic circuit to detect 5×10^{-8} M concentration of IgG antibody.

We have already mentioned the potential advantages of using optical processes to monitor gases and fluids. LB films can readily be deposited onto fibers or planar waveguiding structures. It should also be remembered that no large electric fields are involved and therefore the integrity of the organic layer need not be as good as that required in a field-effect device. The majority of the optical sensors featuring biorecognition use a fluorescently labeled antibody; antigen concentration is detected by the observed fluorescence. Very little work of this kind has been reported to date using LB films but Aizawa and Furuki,[407] Ives *et al.*,[408] and Krull *et al.*[409] have reported relevant data.

The possible application of LB films as molecular sieves and filters was mentioned in Section 7.3.4. This area is very germane to the sensor field in that suitable layers could be used to enhance the specificity of a device. Recent examples of LB films displaying interesting permeation features have been given by Kawabata *et al.*[410] and Higashi *et al.*[411] The filter itself can alternatively be used as the sensing element. Yoshikawa *et al.*[412] have deposited dioleyl lecithin LB films onto microporous filters and observed switching and negative resistance effects when the filters were positioned between NaCl and KCl aqueous solutions.

The biometric approach in research on sensors should, in the long term, pay dividends. At the present time scientists are taking readily available biological materials and combining them with traditional transduction devices. These hybrid structures should be viewed as only the first stage toward the design of novel devices which mimic sensing functions used in nature. Whether LB films are used as multimolecular templates or not, there can be no question of their importance as providers of model structures.

7.10. SUMMARY

This chapter has attempted to summarize most of the applications-oriented articles and papers written to date on LB films. In the majority of cases, the monolayers have simply been used as substitutes for inorganic materials. Moreover, in all but a few instances, minimal efforts have been made to optimize the properties of the molecules used. Published papers do not dwell on many of the unresolved problems in the field and rarely devote sufficient attention to reproducibility of results, stability of materials, and characterization of multilayers, etc. Understandably, more emphasis is placed on the excitement and promise of working in this interdisciplinary field in order to encourage researchers to participate and attract sponsorship. In the United Kingdom, there has been a government initiative in LB films since 1980 and many other countries have now arranged earmarked financial support for the subject. It is interesting to note in this context that the British Prime Minister Mrs. Margaret Thatcher once researched the properties of Langmuir films.[413]

Organic thin films are normally more fragile than their inorganic counterparts and therefore a determined effort must be made to improve their mechanical, thermal, and chemical stability if they are to have a realistic chance of being included in a commercial venture. Various approaches must be explored including cross-linking and removal of fragile side groups after deposition. In some cases it will be appropriate to use a combination of deposition techniques involving chemical modification of the transferred film. Organic materials are already being used successfully in high-technology applications, but only in situations where they offer a definite advantage compared with inorganic materials. Realistically, this will probably restrict opportunities to those that rely on collective phenomena in LB films, that is, to monomolecular assemblies with cooperative properties that are different from those of the individual molecular components. The design of these optimum structures will clearly involve a multidisciplinary approach and require contributions from physicists, organic and physical chemists, surface and colloid scientists, biologists, and engineers. New characterization techniques, such as scanning tunneling microscopy, will be necessary to extend our present analytical capabilities. The analyses of single monolayers, depth profiles in assemblies of molecules, and subtle features of alternate-layer organic superlattices present a considerable challenge, which must be overcome if specific architectures are to be designed and fabricated.

There are many different techniques available for producing thin organic films and therefore the LB film enthusiast should always be aware of recent research developments in associated fields.[414] For example, in many application areas where only a single ordered monolayer is required, the self-assembly technique[415–420] may be equally appropriate. Alternative methods are also available to form bilayer films.[421,422] However, taking our present knowledge into consideration and bearing in mind that the collective properties of molecules in organized structures are those most likely to be exploited in LB films, experience would suggest that the applied areas most worthy of attention are those mentioned in Sections 7.5.2, 7.6.4, and 7.9.4. Surely though, the future of LB films lies in designing structures that perform new functions, such as novel phase transitions. There is a need to sustain the present level of research effort, because good fundamental science will also emerge from studies of floating monolayers and LB films. It is interesting to reflect on the words of Irving Langmuir in a film he made in 1937 to commemorate his award of the Nobel Prize in Chemistry:

> The experiments with oil films on water were started with the purpose of finding out something about the forces that hold molecules on surfaces. They soon led to the discovery of the cause of the spreading of oils on surfaces and to accurate methods of measuring the sizes and shapes of molecules. This all illustrates the fact that experiments started perhaps just for the fun of it, or to satisfy scientific curiosity, often lead to unexpected useful results in fields that could not have been predicted.

In conclusion, it is also worth stating an old Japanese proverb: "Beauty is only one layer."

REFERENCES

1. K. B. Blodgett, *J. Am. Chem. Soc.*, **57**, 1007 (1935).
2. H. Kuhn, D. Möbius, and H. Bücher, in: *Physical Methods of Chemistry* (A. Weissberger and B. Rossiter, eds.), Vol. 1, New York, John Wiley, p. 577 (1972).
3. H. Kuhn, *Thin Solid Films*, **99**, 1 (1983).

4. G. G. Roberts, K. P. Pande, and W. A. Barlow, *Proc. IEEE, Part 1,* **2,** 169 (1978).
5. P. S. Vincett and G. G. Roberts, *Thin Solid Films,* **68,** 135 (1980).
6. A. Barraud, C. Rosilio, and A. Ruaudel Teixier, *Thin Solid Films,* **68,** 91 (1980).
7. G. G. Roberts, *Adv. Phys.,* **34,** 475 (1985).
8. M. Sugi, *J. Mol. Electron.,* **1,** 3 (1985).
9. J. D. Swalen, *J. Mol. Electron.,* **2,** 155 (1986).
10. S. Kowel, *Thin Solid Films,* **152,** 377 (1987).
11. M. J. Kelly and C. Weisbach, *The Physics and Fabrication of Microstructures and Microdevices,* Springer-Verlag, Berlin (1986).
12. F. L. Carter, *Superlattices and Microstructures,* **2,** 113 (1986).
13. G. Hedestrand, *J. Phys. Chem.,* **28,** 1244 (1924).
14. V. K. LaMer, *Retardation of Evaporation by Monolayers,* Academic Press, New York (1962).
15. K. B. Blodgett, Film structure and method of preparation, U.S. Patent 2, 220, 860 (1940).
16. K. B. Blodgett, Reduction of surface reflection, U.S. Patent 2, 220, 861 (1940).
17. K. B. Blodgett, Low-reflectance glass, U.S. Patent 2, 220, 862 (1940).
18. I. Langmuir, Method of substance detection, U.S. Patent 2, 232, 539 (1941).
19. K. B. Blodgett, Step gauge for measuring thickness by means of interference, U.S. Patent 2 587 282 (1952).
20. K. B. Blodgett, *Science,* **89,** 60 (1939).
21. K. B. Blodgett and I. Langmuir, *Phys. Rev.,* **51,** 964 (1937).
22. K. B. Blodgett, *Phys. Rev.,* **55,** 391 (1939).
23. H. Hasmonay, M. Dupeyrat, and R. Dupeyrat, *Opt. Acta,* **23,** 665 (1976).
24. S. D. Smith, private communication.
25. H. H. Race and S. I. Reynolds, *J. Am. Chem. Soc.,* **61,** 1425 (1939).
26. K. B. Blodgett, *Rev. Sci. Instrum.,* **12,** 10 (1941).
27. G. L. Gaines, Jr., *Insoluble Monolayers at Liquid Gas Interfaces,* Interscience, New York (1966).
28. E. E. Porter and J. Wyman, *J. Am. Chem. Soc.,* **59,** 2746 (1937).
29. L. H. Germer and K. H. Storks, *J. Chem. Phys.,* **6,** 280 (1938).
30. B. L. Henke, *Adv. X-Ray Anal.,* **8,** 269 (1985).
31. E. E. Widdowson and S. J. Gregg, *Nature,* **143,** 769 (1939).
32. D. Möbius, *Acc. Chem. Res.,* **14,** 63 (1981).
33. M. Pomerantz, *Phase Transitions in Surface Films,* Plenum Press, New York (1980).
34. T. Haseda, H. Yamakawa, M. Ishizuka, Y. Okuda, T. Kubota, M. Hata, and K. Amaya, *Solid State Commun.,* **24,** 599 (1979).
35. T. Sugawara, S. Bandow, K. Kimura, H. Iwamura, and K. Itoh, *J. Am. Chem. Soc.,* **108,** 368 (1986).
36. A. Ruaudel-Teixier, A. Barraud, P. Coronel, and O. Kahn, *Thin Solid Films,* **160,** 107 (1988).
37. C. R. Brundle, H. Hopster, and J. D. Swalen, *J. Chem. Phys.,* **70,** 5190 (1979).
38. D. T. Clark, Y. C. T. Fok, and G. G. Roberts, *J. Electron Spectrosc. Relat. Phenom.,* **22,** 173 (1981).
39. C. Mori, H. Noguchi, M. Mizuno, and T. Watanabe, Jap. *J. Appl. Phys.,* **19,** 725 (1980).
40. P. J. Ballard, M. F. Daniel, and G. D. Whitlock, Proc. 6th Congr. Int. Radiat. Prot. Assoc., p. 1083 (1984).
41. P. A. Holmes, N. A. Jelley, S. S. Williams, and R. Alizadeh (to appear).
42. N. P. Franks and K. A. Snook, *Thin Solid Films,* **99,** 139 (1983).
43. M. T. Flanagan, *Thin Solid Films,* **99,** 133 (1983).
44. T. R. Chern, W. J. Koros, H. B. Hopfenberg, and V. T. Starrett, *Materials Science of Synthetic Membranes,* ACS Symp. Ser., **269,** Washington, p. 25 (1985).
45. G. D. Rose and J. A. Quinn, *Science,* **159,** 636 (1967).
46. O. Albrecht, A. Laschewsky, and H. Ringsdorf, *J. Membrane Sci.,* **22,** 187 (1985).
47. T. Kajiyama, A. Kumano, M. Takayanagi, and T. Kunitake, *Chem. Lett.,* 915 (1984).
48. K. Heckmann, C. Stobl, and S. Bauer, *Thin Solid Films,* **99,** 265 (1983).
49. Y. Okahata, K. Ariga, H. Nakahara, and K. Fukuda, *Chem. Commun.,* 1069 (1986).
50. H. E. Ries, H. D. Cook, and C. M. Luane, *ASTM, Special Tech. Publ.,* **211,** 55 (1957).

51. W. D. Harkins, *The Physical Chemistry of Surface Films*, Reinhold, New York (1952).
52. A. N. Broers and T. H. P. Chang, in: *Microcircuit Engineering* (H. Ahmed and W. C. Nixon, eds.), p. 1, Cambridge University Press (1980).
53. A. Banerjie and J. B. Lando, *Thin Solid Films*, **68**, 67 (1980).
54. G. Fariss, J. Lando, and S. Rickert, *Thin Solid Films*, **99**, 305 (1983).
55. G. Fariss, J. Lando, and S. Rickert, *J. Mater. Sci.*, **18**, 2603 (1983).
56. A. Barraud, C. Rosilio, and A. Ruaudel-Teixier, *J. Colloid Interface Sci.*, **62**, 509 (1977).
57. A. Barraud, C. Rosilio, and A. Ruaudel-Teixier, *Thin Solid Films*, **68**, 91 (1980).
58. A. Barraud, *Thin Solid Films*, **99**, 317 (1983).
59. P. A. Delaney, R. A. W. Johnstone, B. L. Eyres, R. A. Hann, I. McGrath, and A. Ledwith, *Thin Solid Films*, **123**, 353 (1985).
60. A. N. Broers and M. Pomerantz, *Thin Solid Films*, **99**, 323 (1983).
61. H. P. Zingsheim, *Scanning Electron Microsc.*, **1**, 357 (1977).
62. I. R. Peterson, G. J. Russell, and G. G. Roberts, *Thin Solid Films*, **109**, 371 (1983).
63. L. F. Thompson, L. E. Shillwagon, and E. M. Duerries, *J. Vac. Sci. Technol.*, **15**, 938 (1978).
64. I. R. Peterson, *IEE Proc., Part 1*, **130**, 252 (1983).
65. J. H. McAlear and I. M. Wehrung, *Molecular Electronic Devices*, Dekker, New York (1982).
66. I. Langmuir, *Trans. Faraday Soc.*, **15**, 62 (1920).
67. F. B. Bowden and L. Leben, *Proc. R. Soc. London, Ser. A*, **169**, 371 (1939).
68. E. Rabinowicz and D. Tabor, *Proc. R. Soc. London, Ser. A*, **208**, 455 (1951).
69. F. P. Bowden and D. Tabor, *The Friction and Lubrication of Solids*, Clarendon Press, Oxford (1986).
70. B. J. Briscoe and D. C. B. Evans, *Proc. R. Soc. London, Ser. A*, **380**, 389 (1982).
71. C. M. Pooly and D. Tabor, *Proc. R. Soc. London, Ser. A*, **329**, 251 (1972).
72. J. Seto, T. Nagai, C. Ishimoto, and H. Watanabe, *Thin Solid Films*, **134**, 101 (1985).
73. M. Suzuki, Y. Saotome, and M. Yanagisawa, *Thin Solid Films*, **160**, 453 (1988).
74. G. G. Roberts, M. C. Petty, P. J. Caplan, and E. H. Poindexter, in: *Insulating Films on Semiconductors* (J. Verwey and D. R. Wolters, eds.), p. 20, Elsevier, North-Holland, Amsterdam (1983).
75. B. Holcroft, D.Phil. thesis, University of Oxford (1988).
76. J. D. Maines and E. G. S. Paige, *IEE Rev.*, **120**, 1078 (1973).
77. F. Huang, UK Patent Application No. 8729310 (1987).
78. K. Hilltrop and H. Stegemeyer, *Mol. Cryst. Liq. Cryst.*, **49**, 61 (1978).
79. F. J. Kahn, *Appl. Phys. Lett.*, **22**, 386 (1975).
80. K. Hiltrop and H. Stegemyer, *Liquid Crystals and Ordered Fluids*, Vol. 4, p. 515, Plenum Press, New York (1983).
81. F. C. Saunders, J. Staromlynska, G. W. Smith, and M. F. Daniel, *Mol. Cryst. Liq. Cryst.*, **122**, 297 (1985).
82. H. Ikeno, A. Oh-saki, N. Ozaki, M. Nitta, K. Nakaya, and S. Kobayashi, Proc. Int. Conf. Soc. for Information Display, Anaheim (1988).
83. M. Kakimoto, M. Suzuki, T. Konishi, Y. Imai, M. Iwamoto, and T. Hino, *Jpn. Chem. Soc.*, 823 (1986).
84. M. Suzuki, M. Kakimoto, T. Konishi, Y. Imai, M. Iwamoto, and T. Hino, *Chem. Lett.*, 395 (1986).
85. Y. Nishikata, M. Kakimoto, A. Morikawa, and Y. Imai, *Thin Solid Films*, **160**, 15 (1988).
86. P. Christie, M. C. Petty, G. G. Roberts, D. H. Richards, D. Service, and M. J. Stewart, *Thin Solid Films*, **134**, 75 (1985).
87. A. Barraud, J. Leloup, A. Gouzerh, and S. Palacin, *Thin Solid Films*, **133**, 117 (1985).
88. I. R. Girling and D. R. J. Milverton, *Thin Solid Films*, **115**, 85 (1984).
89. B. Holcroft, M. C. Petty, G. G. Roberts, and G. J. Russell, *Thin Solid Films*, **134**, 83 (1985).
90. M. F. Daniel, J. C. Dolphin, A. J. Grant, K. E. N. Kerr, and G. W. Smith, *Thin Solid Films*, **133**, 235 (1985).
91. S. T. Kowel, R. Selfridge, C. Eldering, N. Matloff, P. Stroeve, B. G. Higgins, M. P. Srinivasan, and L. B. Coleman, *Thin Solid Films*, **152**, 377 (1987).
92. R. H. Tredgold and G. W. Smith, *J. Phys. D*, **14**, L193 (1981).

93. R. H. Tredgold and G. W. Smith, *Thin Solid Films*, **99**, 215 (1983).
94. R. Jones, R. H. Tredgold, and P. Hodge, *Thin Solid Films*, **123**, 307 (1985).
95. P. Christie, G. G. Roberts, and M. C. Petty, *Appl. Phys. Lett.*, **48**, 1101 (1986).
96. G. W. Gray and P. A. Winsor (eds.), *Liquid Crystals and Plastic Crystals*, Ellis Harwood, Chichester (1974).
97. F. Grunfeld and C. W. Pitt, *Thin Solid Films*, **99**, 249 (1983).
98. C. A. Eldering, S. T. Kowel, A. Mortazavi, M. P. Srinivasan, B. G. Higgins, and P. Stroeve, *Thin Solid Films* (to appear).
99. R. H. Tredgold and R. Jones, private communication.
100. H. Kawai, *Jpn. J. Appl. Phys.*, **8**, 975 (1969).
101. G. R. Davies, *Physics of Dielectric Solids*, Institute of Physics Conf. Series No. 58, p. 50 (1980).
102. P. Pantelis, *Phys. Technol.*, **15**, 239 (1984).
103. G. M. Sessler, *J. Acoust. Soc. Am.*, **70**, 1596 (1981).
104. J. R. Giniewicz, R. E. Newnham, A. Safari, and D. Moffatt, *Ferroelectrics*, **73**, 405 (1987).
105. T. Yamada and T. Kitayama, *J. Appl. Phys.*, **52**, 6859 (1981).
106. A. J. Lovinger, *Science*, **220**, 4602 (1983).
107. V. R. Novak and I. V. Myagkov, *Sov. Tech. Phys. Lett.*, **11**, 159 (1985).
108. B. M. Abraham, K. Miyano, S. Q. Xu, and J. B. Ketterson, *Phys. Rev. Lett.*, **49**, 1643 (1982).
109. K. C. O'Brien, J. Long, and J. B. Lando, *Langmuir*, **1**, 414 (1985).
110. R. Zanoni, C. Naselli, J. Bell, G. I. Stegeman, and C. T. Seaton, *Phys. Rev. Lett.*, **57**, 2838 (1986).
111. R. W. Whatmore, *Rev. Prog. Phys.*, **49**, 1335 (1986).
112. M. E. Lines and M. Glass, *Principles and Applications of Ferroelectrics*, p. 561, Clarendon Press, Oxford (1977).
113. A. A. Turnbull, *Infrared Phys.*, **22**, 299 (1982).
114. Y. Yamashita, K. Yokoyama, H. Honda, and T. Takahashi, *Jpn. J. Appl. Phys.*, **20**, 183 (1981).
115. L. Eyrand, P. Eyrand, and F. Bauer, *Adv. Ceramic Materials*, **1**, 223 (1986).
116. U. Korn, Z. Rav-Noy, and S. Shtrikman, *Appl. Opt.*, **20**, 1980 (1981).
117. E. Yamaka, *Ferroelectrics*, **57**, 337 (1984).
118. E. L. Nix, J. Nanayakkara, G. R. Davies, and I. M. Ward, *J. Polym. Sci.*, **26**, 127 (1988).
119. K. Yoshino, M. Ozaki, T. Sakurai, K. Sakamoto, and M. Honma, *Jpn. J. Appl. Phys.*, **23**, L175 (1984).
120. M. Glass, J. S. Patel, J. W. Goodby, D. H. Olson, and J. M. Geary, *J. Appl. Phys.*, **60**, 2778 (1986).
121. R. Watton (RSRE, Malvern), private communication.
122. S. B. Lang, *Sourcebook of Pyroelectricity*, Gordon and Breach, New York (1974).
123. J. D. Zook and S. T. Liu, *J. Appl. Phys.*, **49**, 4604 (1978).
124. L. M. Blinov, N. N. Davydova, V. V. Lazarev, and S. G. Yudin, *Sov. Phys. Solid State*, **24**, 1523 (1982).
125. L. M. Blinov, L. V. Mikhnev, E. G. Sokolova, and S. G. Yudin, *Sov. Tech. Phys. Lett.*, **9**, 640 (1983).
126. T. Sakuhara, H. Nakahara, and K. Fukuda, *Thin Solid Films*, **159**, 345 (1988).
127. P. Christie, C. A. Jones, M. C. Petty, and G. G. Roberts, *J. Phys. D*, **19**, L167 (1986).
128. G. W. Smith, M. F. Daniel, J. W. Barton, and N. M. Ratcliffe, *Thin Solid Films*, **132**, 125 (1985).
129. G. W. Smith, N. M. Ratcliffe, S. J. Roser, and M. F. Daniel, *Thin Solid Films*, **151**, 9 (1987).
130. C. A. Jones, M. C. Petty, and G. G. Roberts, Proc. IEEE Conf. Applications of Ferroelectrics, p. 195, Lehigh (1980).
131. C. A. Jones, M. C. Petty, G. G. Roberts, G. H. Davies, J. Yarwood, N. M. Ratcliffe, and J. W. Barton, *Thin Solid Films*, **155**, 187 (1987).
132. C. A. Jones, M. C. Petty, and G. G. Roberts, *Thin Solid Films*, **160**, 117 (1988).
133. G. H. Davies, J. Yarwood, M. C. Petty, and C. A. Jones, *Thin Solid Films*, **159**, 461 (1988).
134. J. Giermanska, R. Nowak, and J. Sworakowski, *J. Mater. Sci.*, **10**, 77 (1984).
135. R. Colbrook, G. G. Roberts, and B. Holcroft, Proc. Zurich Conference on Applications of Polar Solids, *Ferroelectrics*, **91** (1989).

136. P. K. Tien and R. Ulrich, *J. Opt. Soc. Am.,* **60,** 1325 (1970).
137. J. D. Swalen, *J. Mol. Electron.,* **2,** 155 (1986).
138. C. W. Pitt and L. M. Walpita, *Electron. Lett.,* **12,** 479 (1977).
139. L. M. Walpita and C. W. Pitt, *Electron. Lett.,* **13,** 210 (1977).
140. E. Columbini and G. I. Yip, *Trans. IEEE (Jpn.),* **61,** 154 (1978).
141. J. D. Swalen, K. E. Rieckhoff, and M. Tacke, *Opt. Commun.,* **24,** 146 (1978).
142. V. R. Novak, *Mikroelectronika,* **12,** 181 (1983).
143. R. H. Tredgold, M. C. J. Young, P. Hodge, and E. Khoshdel, *Thin Solid Films,* **151,** 441 (1987).
144. F. Grunfeld and C. W. Pitt, *Thin Solid Films,* **99,** 249 (1983).
145. Y. C. Chen, S. K. Tripathy, and G. M. Carter, *Mol. Cryst. Liq. Cryst.,* **106,** 403 (1984).
146. D. J. Williams, *Angew. Chem., Int. Ed. Engl.,* **23,** 690 (1984).
147. J. Zyss, *J. Mol. Electron.,* **1,** 25 (1985).
148. J. D. Swalen, *Thin Solid Films,* **160,** 197 (1988).
149. D. S. Chemla and J. Zyss, *Nonlinear Optical Properties of Organic Molecules and Crystal,* Vols. 1 and 2, Academic Press, New York (1987).
150. J. L. Oudar and D. S. Chemla, *J. Chem. Phys.,* **66,** 2664 (1977).
151. L. G. S. Brooker, G. H. Keyes, and D. W. Heseltine *J. Am. Chem. Soc.,* **73,** 5350 (1951).
152. A. Dulcic and C. Flytzanis, *Opt. Commun.,* **25,** 402 (1978).
153. J. L. Oudar and D. S. Chemla, *Opt. Commun.,* **13,** 164 (1975).
154. J. Zyss and J. L. Oudar, *Phys. Rev. A,* **26,** 2028 (1982).
155. J. L. Oudar and J. Zyss, *Phys. Rev. A,* **26,** 2016 (1982).
156. S. K. Kurtz and T. T. Perry, *J. Appl. Phys.,* **39,** 3978 (1968).
157. B. F. Levine and C. G. Bethea, *J. Chem. Phys.,* **63,** 2666 (1975).
158. J. L. Oudar and R. Hierle, *J. Appl. Phys.,* **48,** 2699 (1977).
159. B. F. Levine, C. G. Bethea, C. D. Thurmond, R. T. Lynch, and J. L. Bernstein, *J. Appl. Phys.,* **50,** 2523 (1979).
160. G. H. Hewig and K. Jain, *Opt. Commun.,* **47,** 347 (1983).
161. K. Sasaki, T. Kinoshita, and N. Karasawa, *Appl. Phys. Lett.,* **45,** 333 (1984).
162. C. Sauteret, J. P. Hermann, R. Frey, F. Pradère, J. Ducuing, R. R. Chance, and R. H. Baughman, *Phys. Rev. Lett.,* **36,** 956 (1976).
163. K. C. Rustagi and J. Ducuing, *Opt. Commun.,* **10,** 258 (1974).
164. G. I. Stegeman, C. T. Seaton, and R. Zanoni, *Thin Solid Films,* **152,** 231 (1987).
165. R. S. Potember, R. C. Hoffman, S. H. Kim, K. R. Speck, and K. A. Stetyick, *J. Mol. Electron.,* **4,** 5 (1988).
166. C. Liau, P. Bundman, and G. I. Stegeman, *J. Appl. Phys.,* **54,** 6213 (1983).
167. S. M. Jenson, *IEEE J. Quantum Electron.,* **18,** 1580 (1982).
168. T. Rasing, Y. R. Shen, M. W. Kim, P. Valint, and I. Block, *Phys. Rev. A,* **31,** 537 (1985).
169. O. A. Aktsipetrov, N. N. Akhmedier, E. D. Mishgina, and V. R. Novak, *JETP Lett.,* **37,** 209 (1983).
170. I. R. Girling, N. A. Cade, P. V. Kolinsky, and C. M. Montgomery, *Electron. Lett.,* **21,** 169 (1985).
171. I. Pockrand, A. Brillante, and D. Möbius, *J. Chem. Phys.,* **77,** 6289 (1982).
172. Z. Chen, W. Chen, J. Zheng, W. Wang, and Z. Zhang, *Springer Ser. Opt. Sci. Laser Spectroscopy,* **49,** 324 (1985).
173. I. R. Girling, N. A. Cade, P. V. Kolinsky, G. H. Cross, and I. R. Peterson, *J. Phys. D,* **19,** 2065 (1986).
174. O. A. Aktsipetrov, N. N. Akhmediev, I. M. Baranova, E. D. Mishina, and V. R. Novak, *Sov. Phys. JETP,* **62,** 524 (1985).
175. I. R. Girling, N. A. Cade, P. V. Kolinsky, J. D. Earls, G. H. Cross, and I. R. Peterson, *Thin Solid Films,* **132,** 101 (1985).
176. I. Ledoux, D. Joise, P. Vidakovik, J. Zyss, R. A. Hann, P. F. Gordon, B. D. Bothwell, S. K. Gupta, S. Allen, P. Robin, E. Chastaing, and J. C. Dubors, *Europhys. Lett.,* **3,** 803 (1987).
177. I. R. Girling, B. A. Cade, P. V. Kolinsky, R. J. Jones, I. R. Peterson, M. M. Ahmad, D. B. Neal, M. C. Petty, G. G. Roberts, and W. J. Feast, *J. Opt. Soc. Am.,* **4,** 950 (1987).

178. M. M. Ahmad, W. J. Feast, D. B. Neal, M. C. Petty, and G. G. Roberts, *J. Mol. Electron.*, **2**, 129 (1986).
179. J. S. Schildkraut, T. L. Penner, C. S. Willand, and A. Ulman, *Opt. Lett.* (to appear).
180. P. Stroeve, M. P. Srinivasan, B. G. Higgins, and S. T. Kowel, *Thin Solid Films*, **146**, 209 (1987).
181. G. Decher, B. Tieke, C. Bosshard, and P. Gunter (to appear).
182. I. R. Girling, P. V. Kolinsky, N. A. Cade, J. D. Earls, and I. R. Peterson, *Opt. Commun.*, **55**, 289 (1985).
183. I. Ledoux, D. Joise, P. Frenaux, J. P. Piel, G. Post, J. Zyss, T. McLean, R. A. Hann, P. F. Gordon, and S. Allen, *Thin Solid Films*, **160**, 217 (1988).
184. D. B. Neal, M. C. Petty, G. G. Roberts, M. M. Ahmad, W. J. Feast, I. R. Girling, N. A. Cade, P. V. Kolinsky, and I. R. Peterson, *Electron. Lett.*, **22**, 460 (1986).
185. D. B. Neal, M. C. Petty, G. G. Roberts, M. M. Ahmad, I. R. Girling, N. A. Cade, P. V. Kolinsky, and I. R. Peterson, Proc. 6th IEEE Conf. on Applications of Ferroelectrics, p. 89 (1986).
186. T. Richardson, G. G. Roberts, M. E. C. Polywka, and S. G. Davies, *Thin Solid Films*, **160**, 231 (1988).
187. S. G. Davies, M. E. C. Polywka, T. Richardson, and G. G. Roberts, UK Patent No. 8717566 (1988).
188. R. H. Selfridge, S. T. Kowel, P. Stroeve, J. Y. S. Lam, and B. G. Higgins, *Thin Solid Films*, **160**, 471 (1988).
189. G. H. Cross, I. R. Girling, I. R. Peterson, and N. A. Cade, *Electron. Lett.*, **22**, 1111 (1986).
190. G. H. Cross, I. R. Peterson, I. R. Girling, N. A. Cade, M. J. Goodwin, N. Carr, R. S. Sethi, R. Marsden, G. W. Gray, D. Lacey, A. M. McRoberts, R. M. Schrowston, and K. J. Toyne, *Thin Solid Films*, **156**, 39 (1988).
191. J. C. Loulergue, M. Dumont, Y. Levy, P. Robin, J. P. Pocholle, and M. Papuchon, *Thin Solid Films*, **160**, 399 (1988).
192. F. Kajzar, J. Messier, J. Zyss, and I. Ledoux, *Opt. Commun.*, **45**, 13 (1983).
193. G. M. Carter, Y. J. Chen, and S. K. Tripathy, *ACS Symp. Ser.*, **233**, 213 (1983).
194. F. Kajzar and J. Messier, *Thin Solid Films*, **132**, 11 (1985).
195. G. M. Carter, M. K. Thakar, Y. J. Chen, and J. V. Hryniewicz, *Appl. Phys. Lett.*, **47**, 457 (1985).
196. F. Kajzar, J. Messier, I. R. Girling, and I. R. Peterson, *Electron. Lett.*, **22**, 1231 (1986).
197. F. Kajzar, I. R. Girling, and I. R. Peterson, *Thin Solid Films*, **160**, 209 (1988).
198. F. Kajzar, L. Rothberg, S. Etemad, P. A. Chollet, D. Grec, and T. Jedju, *Thin Solid Films*, **160**, 373 (1988).
199. G. M. Carter, Y. J. Chen, and S. K. Tripathy, *Opt. Eng.*, **24**, 609 (1985).
200. J. M. Pearson, *ACS Symp. Ser.* (to appear).
201. P. Kivitz, R. de Bont, and J. van der Veen, *J. Appl. Phys.*, **A26**, 101 (1981).
202. N. F. Dautartas, S. Y. Suh, S. R. Forrest, M. L. Kaplan, A. J. Lovinger, and P. H. Schmidt, *Appl. Phys.*, **A36**, 71 (1985).
203. M. C. Gupta and F. C. Strome, *J. Appl. Phys.*, **60**, 2932 (1986).
204. W. Mey, U.S. Patent 4 513 071 (1985).
205. C. Ishimoto, H. Tomimoro, and J. Seto, *Appl. Phys. Lett.* **49**, 1677 (1986).
206. S. Imazeki, M. Takeda, Y. Tomioka, A. Kakuta, A. Mukoh, and T. Narahara, *Thin Solid Films*, **134**, 27 (1985).
207. H. Yamamoto, T. Sugiyama, and M. Tanaka, *Jpn. J. Appl. Phys.*, **24**, L305 (1985).
208. W. E. Moerner, *J. Mol. Electron.*, **1**, 55 (1985).
209. V. K. Agarwal and V. K. Srivastava, *J. Appl. Phys.*, **44**, 2900 (1973).
210. V. K. Agarwal and V. K. Srivastava, *Thin Solid Films*, **27**, 49 (1975).
211. G. G. Roberts, P. S. Vincett, and W. A. Barlow, *J. Phys. C*, **11**, 2077 (1978).
212. R. Jones, R. H. Tredgold, A. Hoorfar, and P. Hodge, *Thin Solid Films*, **113**, 115 (1984).
213. T. M. Ginnai, D. P. Oxley, and R. G. Pritchard, *Thin Solid Films*, **68**, 241 (1980).
214. S. Kuniyoshi, C. Okazaki, and K. Tanaka, *Thin Solid Films* (to appear).
215. S. M. Sze, *Physics of Semiconductor Devices*, Wiley, New York (1969).
216. I. R. Peterson, *J. Mol. Electron.*, **2**, 95 (1986).

217. G. G. Roberts, T. M. McGinnity, W. A. Barlow, and P. S. Vincett, *Solid State Commun.*, **32**, 683 (1979).

218. G. G. Roberts, T. M. McGinnity, W. A. Barlow, and P. S. Vincett, *Thin Solid Films*, **68**, 223 (1979).

219. R. Jones, R. H. Tredgold, and P. Hodge, *Thin Solid Films*, **123**, 307 (1985).

220. A. Ruaudel-Teixier, M. Vandevyver, and A. Barraud, *Mol. Cryst. Liq. Cryst.*, **120**, 319 (1985).

221. M. Vandevyver, A. Barraud, P. Lesieur, J. Richard, and A. Ruaudel-Teixier, *J. Chim. Phys.*, **83**, 559 (1986).

222. A. Barraud, P. Lesieur, A. Ruaudel-Teixier, and M. Vandevyver, *Thin Solid Films*, **134**, 195 (1985).

223. A. Barraud, P. Lesieur, J. Richard, A. Ruaudel-Teixier, and M. Vandevyver, *Thin Solid Films*, **133**, 125 (1985).

224. J. Richard, A. Barraud, M. Vandevyver, and A. Ruaudel-Teixier, *Thin Solid Films*, **160**, 81 (1988).

225. J. Richard, M. Vandevyver, P. Lesieur, A. Barraud, and K. Holczer, *J. Phys. D*, **19**, 2421 (1986).

226. G. G. Roberts, B. Holcroft, A. Barraud, and J. Richard, *Thin Solid Films*, **160**, 53 (1988).

227. J. Richard, M. Vandevyver, A. Barraud, and P. Delhaes, *Thin Solid Films* (to appear).

228. T. Nakamura, M. Matsumoto, F. Takei, M. Tanaka, T. Sekiguchi, E. Manda, and Y. Kawabata, *Chem. Lett.*, 709 (1986).

229. Y. Kawabata, T. Nakamura, M. Matsumoto, M. Tanaka, T. Sekiguchi, Ho Komizu, and E. Manda, *Synth. Metals*, **19**, 663 (1987).

230. M. Matsumoto, T. Nakamura, F. Takei, M. Tanaka, T. Sekiguchi, M. Mizuno, E. Manda, and Y. Kawabata, *Synth. Metals*, **19**, 675 (1987).

231. T. Shimidzu, T. Iyoda, M. Ando, A. Ohtani, T. Kaneko, and K. Honda, *Thin Solid Films*, **160**, 67 (1988).

232. K. Hong and M. F. Rubner, *Thin Solid Films*, **160**, 187 (1988).

233. H. Nakahara, K. Fukuda, and M. Sato, *Thin Solid Films*, **133**, 1 (1985).

234. G. Marc and J. Messier, *J. Appl. Phys.*, **45**, 2832 (1974).

235. H. M. Millany and A. K. Jonscher, *Thin Solid Films*, **68**, 257 (1980).

236. G. Marc and J. Messier, *Thin Solid Films*, **68**, 275 (1980).

237. A. Barraud, A. Rosilio, and J. Messier, German Patent No. 2,702,487 (1979).

238. A. K. Engel, T. Yoden, K. Sanui, and N. Ogata, *J. Am. Chem. Soc.*, **107**, 8308 (1985).

239. A. K. Engel, N. Ogata, M. Fowler, and M. Suzuki, Proc. 3rd Int. Symp. on Molecular Electronic Devices, North-Holland, Amsterdam (1986).

240. M. Kamimuto, M. Suzuki, T. Konishi, Y. Imai, M. Iwamoto, and T. Hino, *Chem. Lett.*, 823 (1986).

241. E. P. Honig, *Thin Solid Films*, **33**, 231 (1976).

242. A. Leger, J. Klein, M. Belin, and D. Defourneau, *Thin Solid Films*, **8**, R51 (1971).

243. J. Tanguy, *Thin Solid Films*, **13**, 33 (1972).

244. S. K. Gupta, C. M. Singal, and K. Srivastava, *J. Appl. Phys.*, **48**, 2583 (1977).

245. C. M. Singal, S. K. Gupta, A. K. Kapil, and V. K. Srivastava, *J. Appl. Phys.*, **49**, 3042 (1978).

246. A. K. Kapil, C. M. Singal, and V. K. Srivastava, *J. Appl. Phys.*, **50**, 2856 (1979).

247. A. K. Kapil, S. K. Gupta, C. M. Singal, and V. K. Srivastava, *J. Appl. Phys.*, **50**, 2896 (1979).

248. S. K. Gupta, A. K. Kapil, C. M. Singal, and V. K. Srivastava, *J. Appl. Phys.*, **50**, 2852 (1979).

249. S. K. Gupta, C. M. Singal, and V. K. Srivastava, *Electrocomponent Sci. Technol.*, **3**, 119 (1976).

250. M. Sugi, K. Nembach, and D. Möbius, *Thin Solid Films*, **27**, 205 (1975).

251. W. L. Procarione and J. W. Kaufman, *Chem. Phys. Lipids*, **12**, 251 (1974).

252. K. H. Gundlach and J. Kadlec, *Phys. Status Solidi A*, **10**, 371 (1972).

253. K. H. Gundlach and J. Kadlec, *Thin Solid Films*, **13**, 225 (1972).

254. J. L. Miles and H. O. McMahon, *J. Appl. Phys.*, **32**, 1176 (1961).

255. G. L. Larkins, E. D. Thompson, E. Oritz, C. W. Burkhart, and J. B. Lando, *Thin Solid Films*, **99**, 277 (1983).

256. C. Bubeck, B. Tieke, and G. Wegner, *Ber. Bunsenges. Phys. Chem.*, **86**, 499 (1982).

257. M. Sugi, *Thin Solid Films*, **152**, 805 (1987).

258. T. Fukui, M. Sugi, and S. Iizima, *Phys. Rev. B*, **22**, 4898 (1980).
259. M. Sugi and S. Iizima, *Thin Solid Films*, **68**, 199 (1980).
260. M. Sugi, T. Fukui, S. Iizima, and K. Iriyama, *Mol. Cryst. Liq. Cryst.*, **62**, 165 (1980).
261. M. Sugi, M. Saito, T. Fukui, and S. Iizima, *Thin Solid Films*, **99**, 17 (1983).
262. T. Penner, *Thin Solid Films*, **160**, 241 (1988).
263. M. Saito, M. Sugi, T. Fukui, and S. Iizima, *Thin Solid Films*, **100**, 117 (1983).
264. M. Saito, M. Sugi, and S. Iizima, *Jpn. J. Appl. Phys.*, **24**, 379 (1985).
265. K. Sakai, M. Saito, M. Sugi, and S. Iizima, *Jpn. J. Appl. Phys.*, **24**, 865 (1985).
266. M. Sugi, K. Sakai, M. Saito, Y. Kawabata, and S. Iizima, *Thin Solid Films*, **132**, 69 (1985).
267. M. Saito, M. Yoneyama, M. Scuto, K. Ikegami, M. Sugi, T. Nakamura, M. Matsumoto, and Y. Kawabata, *Thin Solid Films*, **160**, 133 (1988).
268. Y. L. Hua, M. C. Petty, and G. G. Roberts, *Thin Solid Films*, **149**, 163 (1987).
269. K. Kudo and T. Moriizumi, *Jpn. J. Appl. Phys.*, **20**, L553 (1981).
270. G. A. Chamberlain and P. J. Cooney, *Nature*, **289**, 45 (1981).
271. R. O. Loutfy, J. H. Sharp, C. K. Hsiao, and R. Ho, *J. Appl. Phys.*, **52**, 5218 (1981).
272. G. G. Roberts, M. McGinnity, W. A. Barlow, and P. S. Vincett, *Thin Solid Films*, **68**, 223 (1980).
273. P. S. Vincett, W. A. Barlow, R. A. Hann, and G. G. Roberts, *Thin Solid Films*, **94**, 171 (1982).
274. H. Kuhn, *J. Photochem.*, **10**, 111 (1979).
275. H. Kuhn, *Pure Appl. Chem.*, **51**, 341 (1979).
276. E. E. Polymeropoulos, D. Möbius, and H. Kuhn, *J. Chem. Phys.*, **68**, 3918 (1978).
277. U. Schoeler, K. H. Teus, and H. Kuhn, *J. Chem. Phys.*, **61**, 5009 (1974).
278. S. J. Valenty, *Macromolecules*, **11**, 1221 (1978).
279. V. L. Ginzburg, *J. Polym. Sci.*, **C29**, 3 (1970).
280. V. L. Ginzburg, *Contemp. Phys.*, **9**, 355 (1968).
281. W. A. Little, *Phys. Rev. A*, **134**, 1416 (1964).
282. W. A. Little, in: *Low Dimensional Cooperative Phenomena* (H. J. Keller, ed.), p. 35, Plenum Press, New York (1975).
283. F. L. Carter (ed.), *Molecular Electronic Devices*, Vols. 1 and 2, Dekker, New York (1982 and 1987).
284. A. Aviram and P. Seiden, U.S. Patent 3,833,894 (1974).
285. A. Aviram, M. J. Freiser, P. E. Seiden, and W. R. Young, U.S. Patent 3,953,874 (1976).
286. R. M. Metzger and C. A. Panetta, in: *Proceedings of NATO Meeting on Low Dimensional Materials* (P. Delhaes, ed.), Plenum Press, New York (1987).
287. R. M. Metzger, C. A. Panetta, N. E. Heimer, A. M. Bhatti, E. Torres, G. F. Blackburn, S. Tripathy, and L. A. Samuelson, *J. Mol. Electron.*, **2**, 119 (1986).
288. A. Aviram, C. Joachim, and M. Pomerantz, *Chem. Phys. Lett.*, **146**, 490 (1988).
289. R. C. Haddon and A. A. Lamola, *Proc. Natl. Acad. Sci. U.S.A.* **82**, 1874 (1985).
290. A. F. Lawrence and R. R. Birge, in: *Nonlinear Electrodynamics in Biological Systems* (W. R. Adey and A. F. Lawrence, eds.), p. 207, Plenum Press, New York (1984).
291. G. Biczo and P. Rajczy, Proc. Symposium on Molecular Electronics and Biocomputers, *J. Mol. Electron.*, **4**, (1988).
292. J. Tanguy, *Thin Solid Films*, **13**, 33 (1972).
293. I. Lundström and D. McQueen, *Chem. Phys. Lipids*, **10**, 181 (1973).
294. I. Lundström and M. Stenberg, *Chem. Phys. Lipids*, **12**, 287 (1974).
295. I. Lundström, *Phys. Scr.*, **18**, 424 (1978).
296. N. J. Evans, M. C. Petty, and G. G. Roberts, *Thin Solid Films*, **160**, 177 (1988).
297. M. C. Petty and G. G. Roberts, *Electron. Lett.*, **16**, 201 (1980).
298. M. C. Petty and G. G. Roberts, *Inst. Phys. Conf. Ser.* **50**, 186 (1970).
299. A. Ashby, Ph.D. Thesis, New University of Ulster (1976).
300. M. Tabib-Azar, A. S. Dewa, and W. H. Ko, *Appl. Phys. Lett.*, **52**, 206–208 (1988).
301. H. H. Wieder, *Inst. Phys. Conf. Ser.*, **50**, 234 (1980).
302. D. T. Clark, T. Fok, G. G. Roberts, and R. Sykes, *Thin Solid Films*, **68**, 274 (1980).

303. G. G. Roberts, K. P. Pande, and W. A. Barlow, *Electron. Lett.*, **13**, 581 (1977).
304. R. W. Sykes, G. G. Roberts, T. Fok, and D. T. Clark, *Proc. IEEE, Part 1*, **127**, 137 (1980).
305. K. K. Kan, M. C. Petty, and G. G. Roberts, Proc. Raleigh Conf. on the Physics of MOS Insulators, p. 344, Pergamon Press, New York (1980).
306. N. J. Thomas Ph.D. Thesis, University of Durham (1986).
307. M. Maissoneve, Y. Segui, and A. Bui Ai, *Thin Solid Films*, **44**, 209 (1977).
308. T. Yamamoto and M. Morimoto, *Appl. Phys. Lett.*, **20**, 269 (1972).
309. J. G. Simmons and A. El-Badry, *Solid-State Electron.*, **20**, 955 (1977).
310. N. J. Thomas, M. C. Petty, G. G. Roberts, and H. Y. Hall, *Electron. Lett.* **20**, 838 (1984).
311. N. J. Thomas, G. G. Roberts, and M. C. Petty, in: *Insulating Films on Semiconductors* (J. Simonne, ed.), p. 71, Elsevier, North Holland, Amsterdam (1986).
312. D. L. Pulfrey, IEEE Trans. *Electron Devices*, **ED25**, 1308 (1978).
313. H. C. Card and E. H. Rhoderick, *Solid-State Electron.*, **16**, 365 (1973).
314. R. J. Stirn and Y. C. M. Yeh, *Appl. Phys. Lett.* **27**, 95 (1975).
315. R. L. van Meirhaeghe, F. Cardon, and W. P. Gomes, *Phys. Status Solidi*, **59**, 477 (1980).
316. R. L. van Meirhaeghe, W. H. C. Laflère, and F. Cardon, *Solid-State Electron.*, **26**, 1180 (1983).
317. H. C. Card, *Solid-State Electron.*, **20**, 971 (1977).
318. I. M. Dharmadasa, G. G. Roberts, and M. C. Petty, Electron. Lett., **16**, 201 (1980).
319. G. G. Roberts, M. C. Petty, and I. M. Dharmadasa, *Proc. IEEE Part 1*, **128**, 197 (1981).
320. R. H. Tredgold and R. Jones, *Proc. IEEE, Part 1*, **128**, 202 (1981).
321. C. S. Winter and R. H. Tredgold, *Proc. IEEE, Part 1*, **130**, 256 (1983).
322. C. S. Winter and R. H. Tredgold, *J. Phys. D*, **17**, L123 (1984).
323. C. S. Winter, R. H. Tredgold, P. Hodge, and E. Khoshdel, *Proc. IEEE, Part 1*, **131**, 125 (1984).
324. R. H. Tredgold and Z. I. El-Badawy, *J. Phys. D*, **18**, 103 (1985).
325. R. H. Tredgold and Z. I. El-Badawy, *J. Phys. D*, **18**, 2483 (1985).
326. R. H. Tredgold, *Thin Solid Films*, **152**, 223 (1987).
327. H. C. Card and B. L. Smith, *J. Appl. Phys.*, **42**, 5863 (1971).
328. J. Batey, G. G. Roberts, and M. C. Petty, *Thin Solid Films*, **99**, 283 (1983).
329. J. Batey, M. C. Petty, and G. G. Roberts, in: *Insulating Films on Semiconductors* (J. F. Verwey and D. R. Wolters, eds.), p. 141, Elsevier, North-Holland, Amsterdam (1983).
330. J. Batey, M. C. Petty, and G. G. Roberts, *Electron. Lett.*, **20**, 489 (1984).
331. M. C. Petty, J. Batey, and G. G. Roberts, *Proc. IEEE, Part 1*, **132**, 133 (1985).
332. M. T. Fowler, M. C. Petty, G. G. Roberts, P. J. Wright, and B. Cockayne, *J. Mol. Electron.*, **1**, 93 (1985).
333. Y. L. Hua, M. C. Petty, G. G. Roberts, M. M. Ahmad, H. M. Yates, N. Maung, and J. O. Williams, *Electron. Lett.*, **23**, 231 (1987).
334. Y. L. Hua, M. C. Petty, G. G. Roberts, M. M. Ahmad, H. M. Yates, and J. O. Williams, *J. Lumin.*, **40**, 861 (1988).
335. W. A. Barlew, E. Owen, G. G. Roberts, and P. S. Vincett, UK Patent No. 1,572,181 (1980).
336. C. L. Cheng, S. R. Forrest, M. L. Kaplan, P. H. Schmidt, and B. Tell, *Appl. Phys. Lett.*, **47**, 1217 (1985).
337. M. T. Fowler, M. Suzuki, A. K. Engel, K. Asano, and T. Itoh, *J. Appl. Phys.*, **62**, 3427 (1987).
338. W. K. Chan, H. M. Cox, J. H. Abeles, and S. P. Kelty, *Electron. Lett.* (to appear).
339. L. G. Meiners, A. T. Clawson, and R. Nguyen, *Appl. Phys. Lett.*, **49**, 340 (1986).
340. C. D. Fung and G. L. Larkins, *Thin Solid Films*, **132**, 33 (1985).
341. J. P Lloyd, M. C. Petty, G. G. Roberts, P. G. LeComber, and W. E. Spear, *Thin Solid Films*, **89**, 4 (1982).
342. J. P. Lloyd, M. C. Petty, G. G. Roberts, P. G. LeComber, and W. E. Spear, *Thin Solid Films*, **99**, 297 (1982).
343. C. T. Honeybourne, *J. Phys. Chem. Solids*, **48**, 109 (1987).
344. T. Moriizumi, *Thin Solid Films*, **160**, 413 (1988).
345. W. M. Reichert, C. J. Bruckner, and T. Joseph, *Thin Solid Films*, **152**, 345 (1987).

346. S. Baker, M. C. Petty, G. G. Roberts, and M. Twigg, *Thin Solid Films,* **99,** 53 (1982).
347. Y. L. Hua, G. G. Roberts, M. M. Ahmad, M. C Petty, M. Hanack, and M. Rein, *Philos. Mag.* **B53,** 105 (1986).
348. M. J. Cook, A. J. Dunn, M. F. Daniel, R. C. O. Hart, R. M. Richardson, and S. J. Roper, *Thin Solid Films,* **159,** 396 (1988).
349. N. B. McKeown, M. J. Cook, A. J. Thomson, K. J. Harrison, M. F. Daniel, R. M. Richardson, and S. J. Roser, *Thin Solid Films,* **159,** 469 (1988).
350. A. Snow and N. L. Jarvis, *J. Am. Chem. Soc.,* **106,** 4706 (1984).
351. D. W. Kalina and S. W. Crane, *Thin Solid FIlms,* **134,** 109 (1985).
352. G. V. Kovacs, P. S. Vincett, and J. H. Sharp. *Can. J. Phys.,* **63,** 346 (1985).
353. S. Baker, G. G. Roberts, and M. C. Petty, *Proc. IEEE, Part 1,* **1p12130,** 260 (1983).
354. H. Woltjen, W. R. Barger, A. W. Snow, and N. L. Jarvis, *IEEE Trans. Electron Devices,* **32,** 1170 (1985).
355. R. H. Tredgold, M. C. J. Young, P. Hodge, and A. Hoorfar, *Proc. IEEE, Part 1,* **132,** 151 (1985).
356. L. Henrjon, G. Derost, A. Ruaudel-Teixier, and A. Barraud, Proc. 2nd Int. Conf. on Chemical Sensors, Bordeaux (1986).
357. I. Lundstrom, *Sensors and Actuators,* **1,** 403 (1981).
358. N. J. Evans, G. G. Roberts, and M. C. Petty, *Sensors and Actuators* (to appear).
359. G. G. Roberts, M. C. Petty, S. Baker, M. T. Fowler, and N. J. Thomas, *Thin Solid Films,* **132,** 113 (1985).
361. P. M. Burr, P. D. Jeffrey, J. D. Benjamin, and M. J. Uren, *Thin Solid Films,* **151,** L111 (1987).
362. B. Liedberg, C. Nylander, and I. Lundstrom, *Sensors and Actuators,* **4,** 299–304 (1983).
363. C. Nylander, B. Liedberg, and T. Lind, *Sensors and Actuators, 3,* 79 (1982).
364. I. Pockrand, J. D. Swalen, R. Santo, A. Brillante, and M. R. Philpott, *J. Chem. Phys.,* **69,** 4001 (1978).
365. W. Knoll, M. R. Philpott, J. D. Swalen, and A. Girlando, *J. Chem. Phys.,* **75,** 4795 (1981).
366. I. Pockrand, J. D. Swalen, J. G. Gordon, and M. R. Philpott, *Surf. Sci.,* **74,** 237 (1977).
367. C. A. Brown, F. C. Burns, W. Knoll, J. D. Swalen, and A. Fischer, *J. Phys. Chem.,* **87,** 3616 (1983).
368. H. Raether, *Phys. Thin Film, 9,* 145 (1977).
369. J. P. Lloyd, C. Pearson, and M. C. Petty, *Thin Solid Films,* **160,** 431 (1988).
370. A. W. Warner and C. D. Stockbridge, *Vacuum Microbalance Techniques,* p. 55, Plenum Press, New York (1963).
371. W. H. King, *Anal. Chem.,* **36,** 1735 (1964).
372. C. Lu and A. W. Czanderna, *Applications of Piezoelectric Quartz Crystal Microbalances,* Elsevier, New York (1984).
373. J. D. Maines and E. G. S. Paige, *IEE Rev.,* **120,** 1078 (1973).
374. J. N. Zemel, in: *Solid State Chemical Sensors* (T. Janata and R. J. Huber, eds.), p. 163, Academic Press, New York (1985).
375. A. D'Amico, A. Palma, and E. Verona, *Sensors and Actuators, 3,* 31 (1982).
376. H. Wohltjen, *Sensors and Actuators, 5,* 307, (1984).
377. J. Ross and G. G. Roberts, Proc. 2nd Int. Conf. on Chemical Sensors, Bordeaux, p. 704 (1986).
378. G. G. Roberts, B. Holcroft, J. Ross, A. Barraud, and J. Richard, *Br. Polym. J.,* **19,** 401 (1987).
379. B. Holcroft, G. G. Roberts, A. Barraud, and J. Richard, *Electron. Lett.,* **23,** 446 (1987).
380. B. Holcroft and G. G. Roberts, *Thin Solid Films,* **160,** 445 (1988).
381. G. G. Roberts, B. Holcroft, A. Barraud, and J. Richard, *Thin Solid Films,* 160, 53 (1988).
382. M. Aizawa, *Analytical Chemistry Series, Chemical Sensors,* **17,** 683 (1983).
383. C. Nylander, *J. Phys. E, 18,* 736 (1985).
384. H. T. Tien, *Bilayer Lipid Membranes,* Dekker, New York (1974).
385. U. Krull and M. Thompson, *IEEE Trans. Electron Devices,* **32,** 1180 (1985).
386. M. Thompson, R. B. Lennox, and R. A. McClelland, *Anal. Chem.,* **54,** 76 (1982).
387. A. Anya, U. G. Krull, M. Thompson, and H. E. Wong, *Anal. Chim. Acta,* **173,** 331 (1985).

388. K. Hongyo, J. Joseph, R. J. Hubar, and J. Janata, Proc. 4th Int. Conf. on Solid State Sensors and Acutators, p. 808, IEE of Japan, Tokyo (1987).
389. K. H. Dambacher and P. Fromhertz, *Biochim. Biophys. Acta,* **861,** 331 (1986).
390. W. L. Porcarione and J. W. Kauffman, *Chem. Phys. Lipids,* **12,** 251 (1974).
391. W. L. Porcarione and J. W. Kauffman, *Chem. Phys. Lipids,* *14,* 49 (1976).
392. A. Arya, U. K. Krull, M. Thompson, and H. E. Wong, *Anal. Chim. Acta,* **173,** 331 (1985).
393. D. C. Taylor and M. G. B. Mahboubian-Jones, *Thin Solid Films,* **99,** 149 (1983).
394. L. R. McLean, A. A. Durrani, M. A. Whittam, D. S. Johnston, and D. Chapman, *Thin Solid Films,* **99,** 127 (1983).
395. H. Bader, K. Dorn, B. Hupfer, and H. Ringsdorf, *Adv. Polym. Sci.,* **64,** 1 (1985).
396. J. H. Fendler, *Membrane Mimetic Chemistry,* Wiley, New York (1982).
397. J. A. Hayward and D. Chapman, *Biomaterials,* **5,** 135 (1985).
398. M. Sriyudthsak, H. Yamagishi, and T. Moriizumi, *Thin Solid Films,* **160,** 463 (1988).
399. E. E. Uzgiriz, *J. Cell. Biochem.* **29,** 239 (1985).
400. A. D. Brown, *Sensors and Actuators,* **6,** 151 (1984).
401. V. A. Howarth, M. C. Petty, G. H. Davies, and J. Yarwood, *Thin Solid Films,* **160,** 483 (1988).
402. H. Bader, R. Van Wagenen, J. D. Andrade, and H. Ringsdorf, *J. Colloid Interface Sci.,* **101,** 246 (1984).
403. M. C. Wilkinson, B. N, Zaba, D. M. Taylor, D. L. Laldman, and T. J. Lewis (to appear).
404. I. Takatu and T. Moriizumi, *Sensors and Actuators,* **11,** 309 (1987).
405. P. Fromhertz, *Biochim. Biophys. Acta,* **225,** 382 (1971).
406. T. Katsube and M. Hara, *Trans 87 Tech. Dig.,* 4th Int. Conf. on Solid State Sensors and Actuators, 1987, IEEE of Japan, Tokyo, p. 816,
407. M. Aizawa and M. Furuki, Proc. 2nd Int. Conference on LB Films, Schenectady (1985).
408. J. T. Ives, W. M. Reichert, P. A. Suci, and J. D. Andrade, *J. Opt. Soc. Am.,* **53,** 13 (1985).
409. U. J. Krull, C. Bloore, and G. Grubs, *Analyst,* **111,** 259 (1986).
410. Y. Kawabata, M. Matsumoto, M. Tanaka, H. Takahasi, Y. Irinatsu, S. Tamura, W. Tagaki, H. Nakahara, and K. Fukuda, *Chem. Lett.,* 1933 (1986).
411. N. Higashi, T. Kunitake, and T. Kajiyama, *Polym. J.,* **19,** 289 (1987).
412. K. Yoshikawa, T. Omochi, I. Ishii, Y. Kuruda, and K. Iriyama, *Biochem. Biophys. Res. Commun.,* **133,** 740 (1985).
413. H. H. G. Jellinek and M. H. Roberts, *J. Sci. Food Agric.,* **2,** 391 (1951).
414. J. D. Swalen, D. I. Allara, J. D. Andrade, E. A. Chandross, S. Garoff, J. Israelachvili, J. J. McCarthy, R. Murray, R. F. Pease, J. F. Rabolt, K. J. Wynne, and H. Yu., *Langmuir,* **3,** 932 (1987).
415. L. Netzer and J. Sagiv, *J. Am. Chem. Soc.,* **105,** 674 (1983).
416. L. Netzer, R. Iscovici, and J. Sagiv, *Thin Solid Films,* **99,** 235 (1983).
417. L. Netzer, R. Iscovici, and J. Sagiv, *Thin Solid Films,* **100,** 67 (1983).
418. R. Maoz and J. Sagiv, *J. Colloid Interface Sci.,* **100,** 465 (1984).
419. J. Gun, R. Iscovici, and J. Sagiv, *J. Colloid Interface Sci.,* **101,** 201 (1984).
420. N. Tillman, A. Ulman, J. S. Schildkraut, and T. L. Penner, *J. Am. Chem. Soc.* (to appear).
421. T. Kunitake, M. Shimomura, T. Kajiyama, A. Harada, K. Okuyama, and M. Takayanagi, *Thin Solid Films,* **121,** L89 (1984).
422. R. H. Tredgold, C. S. Winter, and Z. I. El Badawy, *Electron. Lett.,* **21,** 554 (1985).

Index

Absorption techniques, 53, 226, 227, 230–231, 234, 237, 249, 264, 287, 296
Acoustoelectric sensors, 394–400
Adsorption, 10, 13, 44, 79, 80, 81, 102, 191, 238, 242–245
AES, *see* Auger electron spectroscopy
Air-water interface, *see also* Water surface
  biomolecules at, 273, 276, 277, 278, 283, 284, 287, 288, 289, 290, 293, 295, 297, 303, 304, 306, 307
  defined, 17
  fatty acids and, 33
  fluorescence experiments at, 252–255
  reflection techniques and, 237, 238, 240, 245, 247
Aitken, J., 5–6
Alternate-layer troughs, 122–123, 341
Aluminum substrates, 116, 142, 158, 170, 172, 179, 182, 191, 192, 295
Anthraquinone dyes, 55
Applications, 317–401
  of electrical properties, 368–376
  in mechanical filters, 324
  in molecular electronics, 317–318
  in optics, 322–323, 351–368
  of passive thin films, 332–338
  of piezoelectric and pyroelectric films, 339–351
  in research, 325–331
  in semiconductors, 375–389
  in sensors, 389–400
  in step gauge, 323–324
  in supermolecular electronics, 318–320
  in transducers, 389–400

Aromatic compounds, 30
Aromatic diamines, 71
Aromatic groups, 142
Aromatic materials, 97
Aromatic molecules, 43–58
Aromatic polycyclic hydrocarbons, 67
  derivatives of, 45–52
ATR, *see* Attenuated total reflection
Attenuated total reflection (ATR), 150–151
Attenuated total reflection (ATR) infrared spectroscopy, 297, 298
Auger electron spectroscopy (AES), 157, 158
Azo dyes, 48, 56, 57
  film deposition and, 97
  fluorescence of, 259
  reflection techniques and, 241–242
  transmission techniques and, 232–233

Benzene, 46, 53, 71, 114
  derivatives of, 43–45
Bilayer lipid membranes (BLM), 224, 251, 254
Bilayers, 55, 298, 306, 401
  in biosensors, 398
  defined, 329
  lipids and, 273, 293, 294, 295, 296, 299, 303–304, 307, 398
  lubrication and, 336
  mechanical properties of, 195
  optical microscopy of, 148
  phospholipids and, 277, 278, 287, 288, 289–290, 302
  polymerization and, 74, 77, 97, 139
  proteins and, 283
  Raman spectroscopy of, 155

Bilayers (*cont.*)
  second-order effects and, 361
  sterols and, 277, 278
  supported, 289–290
Biological membranes, 74, 398
  applications of, 329–331, *see also* specific
    applications
  in emission techniques, 251
  properties of, 297–299
  in semiconductors, 376
  spectroscopy of, 224
Biological transducers, 389–400
Biomolecules, 273–307, 329
  Langmuir monolayers of, 276–287
  preparation of supported multilayers and,
    287–295
  spectroscopic studies of, 295–297
  supported molecular layers and, 295–303
Biosensors, 306, 318, 397–400
BLM, *see* Bilayer lipid membranes
Blodgett, Katharine, 11–13, 33, 35, 93, 113,
    134, 135, 163, 167, 276
  patents filed by, 321–325
Bragg techniques, 140, 141, 159
Brillouin scattering spectroscopy, 194, 344

Carbon–carbon double bonds, 53, 77
  polymer formation and, 73–76
Chemical sensors, 306, 318, 389–400
Chemical transducers, 389–400
Chla, *see* Chlorophyll *a*
Chlorophyll, 61, 177, 284, 292, 294, 301,
    397
Chlorophyll *a* (Chla), 59, 235, 252, 284, 285,
    294, 301, 306, 329
Chlorophyll *b*, 285, 306
Cholesterol, 277, 292, 299, 397, 398
Chromatography, 278
Chromium substrates, 116
Chromophores, 239, 325
  reflection techniques and, 236–237, 239, 241,
    242, 243, 249
  second-order effects and, 359
  spectroscopy and, 224
  transmission techniques and, 225, 227–
    234
Circular troughs, 100–102, 283
cis Double bonds, 35, 286, 305
Close-packed monolayers, 80–81
Collapse
  biomolecules and, 278
  factors influencing, 21, 22

fatty acids and, 32
film deposition and, 28, 95
in mixed monolayers, 26
nonaqueous subphases and, 82
pH and, 33
polymers and, 70, 75
porphyrins and, 59
Compact films, 21, 46, 70
Compact monolayers, 152
Complex monolayers, *see also* Mixed
    monolayers
  energy transfer in, 325–326
  spectroscopy of, 223–267
    emission techniques, 223, 251–263
    infrared spectroscopy, 249–250
    optical methods, 223, 263–267
    reflection techniques, 223, 234–249
    transmission techniques, 223, 224–234,
      245, 264
Compression, 19, 20, 21, 22, 44, 224
  biomolecules and, 277, 278, 282, 283, 284,
    285, 297, 304
  film deposition and, 29, 94, 98, 120, 122–
    123
  fluorescence and, 253, 254
  of hydrocarbons, 45
  hydrophilic headgroups and, 39, 40
  in mixed monolayers, 27
  of polymers, 76
  reflection techniques and, 243
  surface pressure and, 106, 108
Compression mechanisms, 100–105
Computer-controlled troughs, 115
Computers, 338
Condensation polymers, 78
Conduction, 181–187
  defect, 178
  electrical, 368–370
  hopping, 176
  ionic, 176
  ohmic, 176
  space-charge limited, 176, 372
  through multilayers, 175–181
Conductivity devices, 390
Constant perimeter barrier troughs, 102–105,
    115, 122
Continuous fabrication troughs, 120–122
Control systems, 110–112
Coordination compounds, 65
Copolymers, 68, 69, 71, 99, 343
Copper ions, 65
Copper substrates, 116, 297

Coupling to surface plasmons, 266–267
Cyanine dyes, 53, 55, 137, 306
  emission techniques and, 251
  fluorescence of, 252–253, 256–257, 258,
    259, 260–261, 262, 326
  optical information storage and, 367
  photoacoustic spectroscopy and, 266
  photoelectric behavior of, 371, 372
  reflection techniques and, 237, 239, 240,
    243–244, 245, 247–248
  second-order effects and, 354
  transmission techniques and, 227

Dark-field microscopy, 292
Defect conduction, 178
Depolarization, fluorescence, 295–296
Depolymerization, 75, 78
Diacetylenes, 76–78, 113, 328, 330, 362
  biomolecules and, 287, 302
  biosensors and, 398
  electron beam microlithography and, 332
  electron diffraction of, 143, 146
  film thickness and, 139
  infrared spectroscopy of, 154
  in piezoelectric and pyroelectric films, 341
  reflection techniques and, 247
Dielectric breakdown, 192–193
Dielectric properties, 181, 295
Differential scanning calorimetry, 162
Dimers
  biomolecules and, 296
  fluorescence and, 254, 262
  infrared spectroscopy and, 151
  reflection techniques and, 237–238, 239, 240,
    241, 243
Diodes
  electroluminescent, 384
  metal-insulator-semiconductor, 376–380, 390
  for semiconductors, 376–380
Dipoles, 224–225, 299, 302, 342, 350
Dipping, 13, 299, 303, 394
  of chlorophyll, 301
  film deposition and, 30, 31, 96, 109, 118
  film thickness and, 135
  metal ions and, 34
  monolayer spreading and, 115
  of phospholipids, 288–292, 296
  polymers and, 76, 82
  reflection techniques and, 244
  semiconductors and, 378
  transmission techniques and, 231
Double bonds, 35, 37, 43, 50, 71

biomolecules and, 286, 305
carbon-carbon, 53, 73–76, 77
cis, 35, 286, 305
terminal, 68
trans, 35, 286
Dyes, 2, 26, 114, 223, 297, *see also* Pigments;
    specific types
  anthraquinone, 55
  azo, *see* Azo dyes
  conductivity of, 179
  coupling to surface plasmons and, 266–267
  cyanine, *see* Cyanine dyes
  electron nuclear double resonance spectra of,
    161
  emission techniques and, 251–263
  film deposition and, 30, 97, 122
  fluorescence of, 252–263, 279, 282, 326
  hemicyanine, 342, 359, 360
  heterocyclic, 53–58
  holographic techniques and, 264–265
  infrared spectroscopy of, 154
  in liquid crystal alignment, 338
  merocyanine, *see* Merocyanine dyes
  molecular pumping and, 374
  optics and, 148, 165, 166, 367
  oxonol, 371
  photoacoustic spectroscopy and, 265–266
  photoelectric behavior of, 371–372
  piezoelectric and pyroelectric films and, 342
  Raman spectroscopy of, 157
  reflection techniques and, 234–249
  second-order effects and, 354, 359, 362
  semiconducting properties of, 179
  stilbazolium, 363
  superconductivity and, 374
  third-harmonic effects and, 363–364
  transmission techniques and, 224, 225, 226,
    227–234
  triphenylmethane, 55, 56

EFISH, *see* Electric field induced second
    harmonic
Electrical characterization, 167–193
  conductance and, 181–187
  dielectric breakdown and, 192–193
  by infrared spectroscopy, 175
  permanent polarization and, 190–192
  permittivity and, 187–189
  quantum mechanical tunneling in, 169–175
  by Raman spectroscopy, 175
  specimen preparation in, 167–169
Electrical conductivity, 368–370

Electrical measurements, 134, 141
Electrical properties
  of biomolecules, 301–302
  of monolayers and multilayers, 368–376
Electric-field-induced second-harmonic (EFISH)
  experiments, 165
Electric-field-induced second harmonic (EFISH)
  generation, 355
Electroluminescent diodes, 384
Electron beam microlithography, 332–333
Electron diffraction, 134, 325
  of biomolecules, 278, 282, 284, 296, 297,
    306
  of film structure, 141, 142–148
  of polymerizable materials, 77
Electron energy ranges, 328–329
Electronic spectroscopy, 301
Electron microscopy, *see* Electron diffraction
Electron nuclear double resonance (ENDOR),
  161–162
Electron paramagnetic resonance, 264
Electron spin resonance (ESR), 160–161, 162,
  264, 376
Electrooptics, 56, 351, 362
Ellipsometry
  of complex monolayers, 224
  film thickness evaluation by, 136–137, 138
Emission techniques, 223, 251–263, *see also*
  specific techniques
ENDOR, *see* Electron nuclear double resonance
Energy transfer, 325–326
ENFET, *see* Enzyme field-effect transistors
Enhanced device processing, 336–338
Enhanced reflection, 236–237
Enzyme field-effect transistors (ENFET), 399
Enzymes, 329
  biomolecules and, 282, 283, 299–300, 306
  in biosensors, 306, 398, 399
Epitaxial deposition, 98, 118, 148
Equilibrium spreading pressure (ESP), 21–22,
  45
ESP, *see* Equilibrium spreading pressure
ESR, *see* Electron spin resonanace
EXAFS, *see* X-ray absorption fine structure

Faraday cage, 191
Fatty acid derivatives, 32–42, 45
Fatty acids, 6, 11, 32–42, 48, 60, 80, 133, 223,
  325, 352
  biomolecules and, 278, 283, 285–287, 302,
    305

in biosensors, 398
carbon-carbon double bonds in, 75
chemical structure of, 32–33
conductance of, 186, 187
dielectric breakdown for, 192, 193
direct conduction through, 175, 177
electron diffraction of, 146
electron spin resonance and, 161
film deposition and, 35, 36, 38, 41, 96, 97,
  98, 99, 113, 119–120, 122
fluorescence and, 258
incorporation of metal ions into, 33–35
infrared spectroscopy of, 151, 153
insulation and, 370
mechanical properties of, 193, 194
in mixed monolayers, 67
optical microscopy of, 148, 149
permanent polarization in, 191
permeability studies of, 195
permittivity of, 189
photoelectric behavior of, 372
pyroelectric films and, 349
quantum mechanical tunneling through, 170,
  171, 173
reflection techniques and, 241
second-order effects and, 359
semiconductors and, 376, 388
structural evaluation of, 142
substrate preparation and, 115, 116, 118
surface analytical evaluation of, 160
thickness evaluation of, 135, 136, 137, 139,
  140–141
Fatty acid salts, 186, 187
  direct conduction through, 175
  electron beam microlithography and, 333
  electron energy and, 328
  infrared spectroscopy of, 151, 152
  magnetic monolayers and, 326
  quantum mechanical tunneling through, 170
  Raman spectroscopy of, 154–155
  refractive indices of, 163
  structure evaluation of, 142
  surface analytical evaluation of, 160
  thickness evaluation of, 136, 139, 141
Fermi level, 185, 193
Fiber optics, 235, 252
Field-effect devices, 384–389, 390–391
Film deposition, 25, 27–32, 53, 55, 56, 82, 83,
  93–123, 324
  biomolecules and, 288
  compression mechanisms and, 100–105
  experimental techniques in, 113–120

fatty acids and, 35, 36, 38, 41, 96, 97, 98, 99, 113, 119–120, 122
film structure and, 143–144
horizontal technique of, 55, 93
insulation and, 370
mechanical properties and, 193
metal ions and, 33
mixed, 94
molecular structure and properties and, 27–32
of non-Langmuir-Blodgett films, 80
of peptides, 70, 71
permanent polarization and, 190, 192
of polymers, 68
of polypeptides, 70–71
principles of, 93–98
semiconductors and, 376
subphase and, 29, 32, 93, 94, 96, 97, 98–101, 102, 106, 110, 111, 112, 115, 116, 118, 120, 122, 123
substrate and, 29, 30, 93, 94, 95, 96, 97, 110, 112, 115–118, 119, 120, 122, 123
troughs and, 99, 100–113, 120–123
X type, *see* X type deposition
Y type, *see* Y type deposition
Z type, *see* Z type deposition
Film rigidity, 27, 28, 73
Films, *see* specific types
Film stability, 45, 46, 68, 70, 81, 82, 284
Film structure, 141–162
electron diffraction of, 141, 142–148
infrared spectroscopy of, 141, 146, 149–154
optical evaluation of, 141, 148–149
Raman spectroscopy of, 141, 154–157
surface analytical techniques and, 157–160
X-ray diffraction of, 141, 146
Film thickness, 6, 9, 176
biomolecules and, 284, 292, 293, 297
conduction and, 184–185, 186, 368
early estimates of, 10
electrical measurements of, 134, 141
electron beam microlithography and, 332
evaluation of, 134–141
fluorescence and, 257, 258
mechanical properties and, 194
optical devices and, 322–323
permeability studies of, 195
permittivity and, 189
pyroelectric properties and, 348, 350, 351
quantum mechanical tunneling and, 170, 171, 172
semiconductors and, 377
step gauge and, 324

surface acoustic waves and, 337, 397
surface plasmons and, 267
third-harmonic effects and, 363–364
transmission techniques for, 224
waveguiding and, 352
Floating monolayers, 11, 133
applications for, 321
deposition of, 93, 97, 99, 100, 105, 112, 113, 118, 123
electron diffraction of, 142
nonlinear optical effects of, 165
permanent polarization in, 190, 191
permeability studies of, 195
piezoelectric properties of, 344
Raman spectroscopy of, 157
second-order effects in, 358
spectroscopy and, 224
thickness evaluation of, 137
Fluorescence, 51, 53, 251, 279, 282, 296, 325, 326
steady-state, 252–263
Fluorescence decay, 259–263
Fluorescence depolarization, 295–296
Fluorescence experiments
at air-water interface, 252–255
on solid substrates, 255–259
Fluorescence microscopy, 149, 304, 305–306
Fluorescence quenching, 122, 255, 256, 258, 265, 266, 307
Fluorescent bleaching, 303
Foster relationship, 263
Fourier transform studies, 150, 151, 249, 297, 303
Franklin, Benjamin, 2, 3–4, 6, 9
Free energy, 18
Frequency doubling, 56, 165
Fresnel equation, 225, 235

Gas chromatography, 278
Gas sensors, 390, 394
Germanium substrates, 150–151, 290
Gibbs free energy, 18
Glass substrates, 27, 95, 108, 118, 141, 142, 191, 195, 226, 250, 288, 289, 291, 294, 298, 300, 307, 330
Gold substrates, 30, 116, 155, 158, 159, 174, 179, 294, 297, 372, 375, 378, 381

Hall effect, 369
Headgroups, 50, 59, 80, 82, 83, 119
of heterocyclic compounds, 56–57
hydrophilic, *see* Hydrophilic headgroups

Headgroups (*cont.*)
  of lipids, 291, 297, 298
  of porphyrins, 59
  pyroelectric films and, 347
Hemicyanine dyes, 342, 359, 360
Heterocyclic compounds and dyes, 53–58
Highest occupied molecular orbital (HOMO), 256
Holography, 264–265, 357
HOMO, *see* Highest occupied molecular orbital
Hopping conduction, 176
Horizontal dipping, 289–290, 293
Horizontal film deposition, 55, 93
Hydrophilic chains, 71
Hydrophilic films, 97
Hydrophilic groups, 24, 27, 43, 45, 59, 61, 64, 77, 82
  fluorescence and, 257–258
  of polymerizable materials, 76
  of polymers, 68
Hydrophilic headgroups, 18–19, 26, 29, 39–42, 69, 74, 239
  film deposition and, 97
  reflection techniques and, 239
Hydrophilic materials, 26
Hydrophilic molecules, 32, 53, 60, 64
Hydrophilic substrates, 27–28, 94, 116, 146, 288, 291, 294
Hydrophobic chains, 20, 38
Hydrophobic groups, 50, 64
Hydrophobic materials, 26
  mixed monolayers of, 67
Hydrophobic molecules, 32, 53, 60, 64
Hydrophobic rings, 43
Hydrophobic substrates, 29, 30, 94, 116, 258, 288, 289, 291
Hydrophobic tailgroups, 18–19, 20, 40, 82
  of fatty acids, 35–39
  of polymers, 69

IET, *see* Inelastic tunneling spectrum
Image reproduction, 321
IMFET, *see* Immunosensitive field-effect transistors
Immunosensitive field-effect transistors (IM-FET), 399
Immunosensors, 306
Impurity conduction, 178
Indirect optical methods, 263–267
Induced bifringence electrooptic modulators, 357
Inelastic tunneling spectrum (IET), 174
Information storage systems, 338

Infrared (IR) spectroscopy, 135, 141, 146, 149–154, 175, 225, 249–250, 297, 298, 303
  attenuated total reflection, 152, 297, 298
Inorganic films, 320, 358, 369, 378, *see also* specific types
Insulated-gate field-effect transistors, 379
Insulators, 370–371, 380, 381, 382, 385, 386, 387, 388, 396
Interferometry, 134–136, 138
Intermolecular attractive forces, 10
Ionic conductivity, 176
Ionization, 33
Ion-sensitive field-effect transistors (ISFETS), 399
Iron, 60
  magnetic monolayers and, 328
IR spectroscopy, *see* Infrared spectroscopy
ISFETS, *see* Ion-sensitive field-effect transistors
Isotherms, 25, 33, 48, 103, 113, *see also* Surface pressure/area isotherms
  control of, 110, 112, 115
  of fatty acids, 35, 37, 98, 99
  film deposition and, 29
  fluorescence and, 253
  of lipids, 289
  of phospholipids, 279
  of phthalocyanines, 115
  of polymers, 70, 71
  of porphyrins, 59

Joyce-Loeble trough, 109

Kelvin probe technique, 190
Kerr effect, 165, 364
  defined, 356
Kretscmann configuration, 164

Langmuir, Irving, 2–3, 9–11, 13, 33, 35, 93, 94, 100, 113, 118, 134, 135, 163, 190, 276, 289, 320, 329, 334, 401
  patents filed by, 321–325
Langmuir balance, 105, 106
Langmuir-Blodgett (LB) films
  defined, 11
Langmuir-Blodgett trough system, 113
Langmuir films, 11, 25, 302
Langmuir/Schaefer method, 343–344
Laser-induced phonon spectroscopy, 297, 303
LB, *see* Langmuir-Blodgett
Lead substrates, 116, 169
Light
  biomolecules and, 273

emission of, *see* Emission techniques
fiber optic, 235
monochromatic, 235
oblique incidence, 241–242
polarized, 227–234, 240–241
reflection of, *see* Reflection techniques
transmission of, *see* Transmission techniques
Linear electrochromic effects, 166
Linear electrooptic effect, *see* Pockels effect
Linear troughs, 100
Lipids, 42, 51, 330
    biomolecules and, 274, 278, 283, 284, 292, 293, 294, 296, 297, 298, 299, 302, 305, 306
    diacetylenes and, 78
    film deposition and, 119
    importance of, 40
    polymerization and, 74
Liquid-condensed films, 22
Liquid crystal alignment, 149, 337–338
Liquid crystals, 318, 320, 321, 369, 388
Liquid-expanding films, 22
Lithium niobate
    in molecular electronics, 317
    in optics, 353, 357
    second-order effects and, 355, 361
    surface acoustic waves and, 336, 395, 396
Lithography, 336
Lowest unoccupied molecular orbital (LUMO), 256
Lubrication, 13, 334
LUMO, *see* Lowest unoccupied molecular orbital
LVDT, *see* Linear variable transformer

Magnetic ions, 161
Magnetic monolayers, 326–328
Maxwell-Wagner interfacial polarization, 182, 187
MBE, *see* Molecular beam epitaxy
Mechanical filters, 324
Mechanical properties, 193–195
Membranes, 274, 294
    of biomolecules, 273, 285, 295, 300, 303
    permeable, 329–331
    purple, 294, 300, 301
Mercury subphases, 81, 82, 98
Mercury substrates, 136, 168, 182
Merocyanine dyes, 55, 56–57
    conductivity of, 179
    electron nuclear double resonance spectra for, 161

infrared spectroscopy of, 154
in liquid crystal alignment, 338
nonlinear optical effects in, 166
photoelectric behavior of, 371
second-order effects and, 354
third-harmonic effects and, 363
Metal cations, 33
Metal-insulator-semiconductor (MIS) diodes, 376–380, 390
Metal-insulator-semiconductor (MIS) solar cells, 380, 381, 384
Metal-insulator-semiconductor (MIS) structures, 177, 192, 193, 386, *see also* specific structures
Metal ions, 33–35, 59, 119, 158, 326, 330, *see also* specific metals
    in electron beam microlithography, 333
    film deposition and, 119
    magnetic monolayers and, 326
    in phthalocyanines, 63, 65
    in polymers, 76
    in porphyrins, 60
Metal-organic chemical vapor deposition (MOCVD), 378
Metals, *see also* Metal ions; Metal substrates
    adhesion between polymers and, 332
    film deposition and, 30
    magnetic monolayers and, 326, 328
Metal substrates, 30, 108, 115, 116, 135–136, 142, 155, 168–169, 187–188, 191, 192, 250, 295, 297, 378, 381, 388, *see also* specific metals
    biomolecules and, 290, 291, 294, 297, 299, 306, 307
    conduction and, 179, 182
    infrared spectroscopy of, 152
    lubrication and, 334
    mechanical properties and, 193
    optical microscopy and, 150–151
    organic rectification and, 375
    photoelectric behavior of films on, 371–372
    quantum mechanical tunneling and, 170, 172, 174
    Raman spectroscopy of, 155, 157
    refractive indices and, 164
    second-order effects and, 358
    surface analytical evaluation of, 157–159
Mica substrates, 108, 193, 299, 306, 334
Microgravimetric sensors, 336
Microlithography, 332–333
Microscopy
    fluorescence, 149, 304, 305–306

Microscopy (*cont.*)
  optical, 77, 148–149, 293
  polarized, 292
  scanning tunnelling, 162
Miller-Abrahams-type expression, 185
MIS, *see* Metal-insulator-semiconductor
Mixed film deposition, 94
Mixed monolayers, 26–27, *see also* Complex
    monolayers
  biomolecules and, 277, 284, 285, 286, 287,
      293, 294, 298, 306, 307
  hydrophobic, 67
  photoelectric behavior of, 372
  phthalocyanine in, 65
  of polymers, 68
  porphyrins in, 60
Mixed multilayers, 53
MOCVD, *see* Metal-organic chemical vapor
    deposition
Model membrane systems, 93
Modulation spectroscopy, 166
Molecular beam epitaxy (MBE), 319, 320, 378
Molecular cooperation, 223, 373–375
Molecular electronics, 317–318, 337, 375
Molecular layers
  supported, 295–303
Molecular orientation, 241–242
Molecular pumping, 373–374
Molecular rectification, 373–375
Molecular rings, 43–58
Molecular structure and properties, 17–83
  of fatty acids, 32–42
  film deposition and, 27–32
  modeling of, 24–25
  monolayer formation and stability and, 18–19
  of phthalocyanines, 59–67
  of polymers, 68–79
  of porphyrins, 59–67
  rings and, 43–58
Monochromatic light, 235
Monochromator, 239
Monolayer formation, 18–19
Monolayers, *see* specific types
Monolayer spreading, 26, 45, 94, 114–115
  biomolecules and, 289
  film deposition and, 118
  fluorescence and, 252, 254
  reflection techniques and, 243
  transmission techniques and, 225
  troughs and, 102
Monolayer stability, 17, 18–19, 38, 44, 50, 51,
    55, 68, 75, 154, 169, 284

  biomolecules and, 278, 289, 293
  polymers and, 68, 75
Monolayer transfer, 11, 98, 118–120, 135, 138,
    144, 151–152, 238, 247, 306, 317

Nanolithography, 333
Near edge X-ray absorption fine structure (NEX-
    AFS), 160
Nernst diffusion layer, 239
Neutron diffraction and reflection, 140–141, 299
NEXAFS, *see* Near edge X-ray absorption fine
    structure
Nonaqueous subphases, 81–82, *see also* specific
    types
Non-Langmuir-Blodgett multilayers, 79–81
Nonlinear coherent directional couplers, 357
Nonlinear optical effects, 134, 165–166, 352,
    353–357, 358–368
  second-order, 56, 354–356, 358–362
  third-order, 354, 356–357

Oblique incidence light, 241–242
Ohmic conductivity, 176
Ohm's law, 185
Oil-water interface, 2, 3–9, 13, 33, 278
  biomolecules at, 282, 283, 304
Oil-water troughs, 33
Oleophilic tailgroups, 18
Optical devices, 322–323, *see also* specific
    devices
Optical disks, 336
Optical fibers, 392
Optical information storage, 365–368
Optical lithography, 332
Optical methods, 44, 53, 60, 134, 141, 223,
      263–267, 292, 351–368, 399–400, *see
      also* Nonlinear optical effects; specific
      methods
  indirect, 263–267
  organic films and, 357–358
Optical microscopy, 77, 148–149, 293
Optical sensors, 392–394, 400
Optical storage, 318
Optical thickness, 135
Optical waveguiding, 70, 141, 352–353, 389,
    392
Optoelectronics, 317, 353
Organic rectification, 374–375
Oxirans, 36, 78, 79, 173
Oxonol dyes, 371

PAS, *see* Photoacoustic spectroscopy
Passive thin films, 332–338
Patents, 321–325
Permanent polarization, 190–192
Permeability studies, 195
Permeable membranes, 329–331
Permittivity, 187–189, 370, 377
pH, 46, 48, 323
  biomolecules and, 273, 297, 300, 304
  in control systems, 111
  fatty acids and, 32, 33, 40, 41
  film deposition and, 32, 96, 97, 99, 119, 120
  magnetic monolayers and, 326
  metal ions and, 33, 34, 35
  waveguiding and, 352
Phase transition, 20
Phase transitions, 239
  biomolecules and, 277, 278, 282, 283, 286,
    296, 297, 302
  of complex monolayers, 239–240
  evaluation techniques for, 161
  fluorescence and, 252, 253
  of mixed monolayers, 26
  of polymers, 68
  reflection techniques and, 239–240
Phospholipids, 329
  anlogues of, 287
  biomolecules and, 273, 277–282, 283, 284,
    285–287, 295, 296–297, 300, 302, 304,
    305
  in biosensors, 398
  conductance of, 185
  dipping of, 288–292
  in electron beam microlithography, 332
  film deposition and, 97
  fluorescence and, 253, 254
  permittivity of, 189
  polymerizable, 285–287, 295
  polymerizable derivatives of, 291–292
Phosphorescence, 251, 255, 259, 263
Photoacoustic spectroscopy (PAS), 162, 265–
  266, 301
Photoconducting polymers, 318
Photoelectric behavior, 371–372
Photosynthesis, 59, 294, 295, 301, 329
Photovoltaic cells, 380–384
Phthalocyanines, 59–67, 82, 384
  conductivity of, 179, 390
  electron diffraction of, 147–148
  film deposition and, 97
  infrared spectroscopy of, 154
  monolayer spreading and, 115

optics and, 366, 392
permittivity of, 189
in piezoelectric and pyroelectric films, 339
quartz oscillators, 394
in semiconductors, 384
surface acoustic wave oscillators and, 396
Picosecond transient digitizers, 357
Piezoelectric crystals, 394
Piezoelectric films, 192, 339–344, 395, 397
Piezoelectric polymers, 318
Pigments, 284–285, 290, 293–295, 297, 301,
  306, *see also* Dyes; specific types
Platinum substrates, 108, 116, 301, 306
Pockels, Agnes, 8–9, 13
Pockels effect, 165, 166, 356, 362
Point-dipole model, 224–225
Polarization, permanent, 190–192
Polarized light, 227–234, 240–241
Polarized microscopy, 292
Polymeric materials, 346, 351
Polymerizable materials, 35, 42, 68–79, 332
  in insulation, 370, 371
  in piezoelectric and pyroelectric films, 341
Polymerizable monolayers, 17
Polymerizable phospholipids, 285–287, 295,
  303
  derivatives of, 291–292
Polymerization, 28, 36, 37
  biomolecules and, 300, 302–303, 304
  carbon-carbon double bonds and, 73, 74, 75
  of diacetylenes, 76–77
  film deposition and, 97
  film thickness and, 139
  of oxirans, 78
  second-order effects and, 359
  semiconductors and, 387
  trough environment and, 113
Polymers, 17, 68–79, 83, 330, 331, 346, 351,
  see also Polymerizable; Polymerization
  adhesion between metals and, 332
  in biosensors, 398
  carbon-carbon double bonds and, 73–76
  compared with phthalocyanines, 64
  condensation, 78
  conductivity of, 181, 369
  early use of, 9
  in electron beam microlithography, 332
  electron diffraction of, 142, 146
  electron energy and, 328
  film deposition and, 30, 97
  lubrication and, 335
  optical information storage and, 367

Polymers (*cont.*)
  in optical sensors, 393
  photoconducting, 318
  piezoelectric, 318
  piezoelectric properties of, 344
  preformed, *see* Preformed polymers
  pyroelectric properties of, 346
  in semiconductors, 382
  as subphase, 82
  substrate preparation and, 116
  in thermal imagers, 346
  third-harmonic effects and, 364
  water-soluble, 35
Polypeptides, 70–71
  in biomolecules, 274, 300, 307
  permittivity of, 189
  synthetic, 293
Poole-Frenkel theories, 176, 177, 178, 369
Porphyrins, 59–67, 61, 231, 236, 239
  biomolecules and, 294, 297
  conductance of, 187, 390
  film deposition and, 97
  fluorescence and, 252
  reflection techniques and, 237, 240, 241, 242,
    243, 250
Preformed polymers, 17, 68–73, 78, 82, 137,
  353
Pressure/area isotherms, 8, *see also* Surface
    pressure/area isotherms
  of biomolecules, 291, 302, 304
  biomolecules and, 278
  solvents and, 114
Proteins, 9, 13
  in biomolecules, 273–274, 275, 282–284,
    293–295, 299–301, 303–304, 305, 306,
    307
  in biosensors, 398
  in electron beam microlithography, 333
  monolayer spreading and, 115
  as subphase, 82
Protein synthesis, 79
Purple membranes, 294, 300, 301
Pyroelectric detectors, 325
Pyroelectric films, 339–343, 344–351, 361, 365

Quadratic electrochromic effects, 166
Quadratic electrooptic effect, *see* Kerr effect
Quantum detectors, 344
Quantum mechanical tunneling, 169–175, 370
Quantum well semiconductors, 318
Quartz oscillators, 394–395
Quartz substrates, 108, 118, 226, 289, 291, 307

Radioactive films, 325, 328
Raman spectroscopy, 250, 292, 297, 299, 303
  electrical characterization by, 175
  of film structure, 141, 154–157
Rayleigh, Lord, 1, 3, 6–9, 10, 13, 334
Reflection high-energy electron diffraction
    (RHEED), 142, 144, 146
Reflection techniques, 164, 223, 234–249, *see
    also* specific techniques
Refractive index, 162–165
RHEED, *see* Reflection high-energy electron
    diffraction
Rhodopsin, 285, 290, 300

Sagiv-type multilayers, 80
SAW, *see* Surface acoustic waves
Scanning tunnelling microscopy (STM), 162
Schottky barrier, 174, 369
Schottky barrier modification, 380–389
Schottky emission, 169, 176, 177, 178
Schottky-like process, 376
Secondary-ion mass spectrometry (SIMS), 157,
    264
Second-harmonic generation (SHG), 165, 356,
    358, 359, 361, 362
Second-order nonlinear optical effects, 56, 354–
    356, 358–362
Semiconductors, 115, 179, 325, 333, 336, 344,
    368, 371, 375–389, 390–391
  dielectric breakdown and, 193
  diodes for, 376–380
  fluorescence and, 258
  group III-V, 386–388
  organic rectification and, 375
  permittivity and, 188
  quantum well, 318
  Schottky barrier modification in, 380–389
  substrate preparation and, 116–118
  surface acoustic wave oscillators and, 396
  surface plasmons and, 266
  third-order effects in, 356
Semiconductor surfaces, 177
SHG, *see* Second-harmonic generation
Shields, John, 4–5
Silicon, 137, 321, 336
  in electron beam microlithography, 332
  lubrication and, 336
  in molecular electronics, 317
  in nonlinear optics, 353
  semiconductors and, 376, 379, 387, 388–389
  in supermolecular electronics, 318, 319
  in thermal imagers, 346, 347
  third-harmonic effects and, 364

Silicon logic circuits, 374
Silicon substrates, 146, 148, 151, 289, 307, 388
Silicon transistors, 385
Silver ions, 142, 333
Silver substrates, 135–136, 152, 155, 157, 158, 159, 164, 174, 294, 295, 297, 307, 358
SIMS, *see* Secondary-ion mass spectrometry
Single barrier troughs, 100–101, 113–114
Skeletonization, 120, 163, *see also* Skeletonized films
    defined, 35
Skeletonized films, 13, 321, 322, 323–324
Snell's law, 134–135
Solid multilayers, 17
Soluble films, 41
Solvents, 19, 81
    biomolecules and, 277, 278
    in film deposition, 29, 99, 118, 120
    monolayer spreading and, 114–115
Space-charge limited conductivity, 176
Spectroscopy, 223–267, 290, 295–297, *see also* specific types
    Brillouin scattering, 194, 344
    electronic, 301
    emmission techniques, *see* Emission techniques
    infrared, *see* Infrared spectroscopy
    laser-induced phonon, 297, 303
    optical methods, *see* Optical methods
    photoacoustic, 301
    Raman, *see* Raman spectroscopy
    transmission techniques, *see* Transmission techniques
    ultraviolet, 154
    visible, 149–154
Spiropyrans, 55
SPR, *see* Surface plasmon resonance
Stark effect, 166
Static electricity, 6
Static friction, 334
Steady-state fluoresence, 252–263
Steel substrates, 291
Step gauge, 323–324
Sterols, 273, 277–282
Stilbazolium dyes, 363
STM, *see* Scanning tunnelling microscopy
Strutt, John William, *see* Rayleigh, Lord
Subphases, 17, 21, 46, 49, 50, 133, 144
    biomolecules and, 278, 283, 284, 288, 290, 291, 293, 299, 304, 306
    containment and, 98–100
    fatty acids and, 32, 33, 39, 40, 41

film deposition and, 29, 32, 93, 94, 96, 97, 98–101, 102, 106, 110, 111, 112, 115, 116, 118, 120, 122, 123
    fluorescence and, 252
    magnetic monolayers and, 326
    metal ions in, 60
    monolayer spreading and, 115
    nonaqueous, 81–82
    optical microscopy of, 148, 149
    permanent polarization and, 191
    phthalocyanines and, 62, 63
    polymerization and, 139
    polymers and, 71, 72, 76, 78
    reflection techniques and, 238, 239, 245, 250
    surface pressure and, 106
    transmission techniques and, 224, 225
    troughs and, 100–101, 102, 120
Subphase surface cleaning, 113, 114
Superconductivity, 374
Supermolecular electronics, 318–320
Supermolecular structures, 122
Supported bilayers, 289–290
Supported molecular layers, 295–303
Supported monolayers, 302
Supported multilayers, 287–295
Surface acoustic wave (SAW) devices, 336–337, 394, 395–397
Surface analytical techniques, 157–160, *see also* specific types
Surface area, 18, 21
Surface area measurement, 108
Surface phase changes, 247–249
Surface plasmon resonance (SPR) technique, 141, 164–165, 392, 393–394
Surface plasmons, 156, 157, 164–165, 265
    coupling to, 266–267
    optical sensors and, 392
    second-order effects in, 358
    waveguiding and, 352
Surface potential, 285, 341
Surface pressure, 55, 144–146, 284
    biomolecules and, 278, 283, 284, 285, 288, 297, 299, 305
    control of, 110, 112
    defined, 19
    film deposition and, 29, 96, 103, 114, 115, 120, 121
    fluorescence and, 253, 255
    monolayer spreading and, 115
    of nonaqueous subphases, 82
    reflection techniques and, 247, 248, 249
    troughs and, 103, 114

Surface pressure/area isotherms, 19–23, 46, *see
    also* Pressure/area isotherms
  of anthracene derivatives, 49
  of polymers, 72
  reflection techniques and, 241
  troughs for, 100
Surface pressure measurements, 105–108, 243
Surface tension, 6, 8–9
  film deposition and, 31, 98
  monolayer formation and stability and, 18, 19
  of nonaqueous subphases, 81
  surface pressure and, 106, 107
Synchrotron radiation studies, 159

Tailgroups, 31, 50, 59, 80
  in film deposition, 29, 31
  hydrophobic, *see* Hydrophobic tailgroups
  interaction between water and, 20
  oleophilic, 18
TED, *see* Transmission electron diffraction
Temperature, 49, 64
  biomolecules and, 278, 283, 291, 292, 296,
    297, 305
  conductance and, 180–181, 183–184
  in control systems, 111
  film deposition and, 96
  film thickness and, 138
  fluorescence and, 255
  insulation and, 371
  monolayer stability and formation and, 18
  optical information storage and, 366
  polymers and, 68
  pyroelectric films and, 347, 350, 351
  quantum mechanical tunneling and, 172–173
  surface acoustic wave oscillators and, 396
  surface pressure/area isotherms and, 19, 22,
    23
Terminal double bonds, 68
Thermal imaging, 345–347, 350–351, 371
Thermally stimulated current (TSC) technique,
    192
Thickness, *see* Film thickness
Thin-layer chromatography, 278
Third-harmonic effects, 362–364
Third-order effects, 354, 356–357
Tin substrates, 157, 169
trans double bonds, 35, 286
Transmission electron diffraction (TED), 142,
    143, 147
Transmission techniques, 223, 224–234, 245,
    264, *see also* specific techniques
Triphenylmethane dyes, 55, 56

Triple bonds, 37, 287
Trough environment, 112–113
Troughs, 82–83, 100–105, 114, 120–123, 235,
    288, 293, 321, 324, *see also* specific
    types
  alternate-layer, 122–123, 341
  alternatives to, 79, 81
  ancillary equipment for, 105–112
  Atemeta, 114
  circular, 100–102, 283
  constant perimeter barrier, 102–105, 115, 122
  continuous fabrication, 120–122
  Joyce–Loebl, 109
  Langmuir, *see* Langmuir trough
  linear, 100
  oil–water, 33
  single barrier, 100–101, 113–114
  surface cleaning of, 113–114
TSC, *see* Thermally stimulated current

Ultraviolet (UV) radiation, 97, 113, 247, 259,
    292, 302, 325
Ultraviolet (UV) spectroscopy, 154
Unstable films, 50
Unstable monolayers, 38
UV, *see* Ultraviolet

Van der Waals forces, 153, 299
Vapor sensors, 390
Vectorial effects, 56
Vertical dipping, 293
Vertical support, 288
Viscosity, 25, 29, 30, 105, 194, 232, 305
Visible spectroscopy, 149–154

Water surface, 11, 13, 43, 44, 45, 82, 397, *see
    also* Air-water interface
  advantages of, 17
  fatty acids and, 33, 40, 41, 67
  film deposition and, 27–28, 29, 31, 94
  molecular structure modeling and, 24–25
  monolayer formation and stability on, 18–19
  optical microscopy of, 149
  phthalocyanines on, 63, 64, 65
  piezoelectric films and, 344
  polymerization on, 73, 74, 75
  polymers and, 69, 71
  porphyrins on, 61, 62
  reflection techniques and, 235, 239, 241, 242,
    249
  transmission techniques and, 225
Wavefront conjugation, 357

Wilhelmy balance, 106, 107
Wilhelmy plate, 29, 105, 107, 110, 121, 278, 288

XANES, *see* X-ray absorption near edge structure
XPS, *see* X-ray photoelectron spectroscopy; X-ray photoemission
X-ray absorption fine structure (EXAFS), 160
X-ray absorption near edge structure (XANES), 160
X-ray diffraction, 11, 134
  of biomolecules, 282, 292, 293, 296, 299, 300, 304, 306
  film deposition and, 97
  of film structure, 141, 146
  film thickness evaluation by, 135, 137–139
  of polymers, 73
X-ray lithography, 374
X-ray photoelectron spectroscopy (XPS), 157, 158, 264
X-ray photoemission (XPS), 328
X type deposition
  applications of, 325
  biomolecules and, 291, 306
  defined, 30
  film thickness and, 137, 138
  nonlinear optical effects and, 166
  permanent polarization and, 190, 191, 192
  in piezoelectric films, 339, 341, 344
  of polymers, 71, 73

principles of, 94, 95, 96–97
in pyroelectric films, 339, 341, 347
second-order effects and, 358
XY type deposition, 94

Y type deposition, 32, 44, 55, 56
  biomolecules and, 291, 293, 300, 306
  defined, 30
  equipment for, 109
  film thickness and, 137, 139
  nonlinear optical effects and, 165, 166
  permanent polarization and, 190, 191
  in piezoelectric and films, 339, 341
  of polymerizable materials, 77
  of polymers, 70
  porphyrins and, 61
  principles of, 94, 95, 96, 97
  in pyroelectric films, 339, 341, 348

Z type deposition, 44, 56
  biomolecules and, 293
  defined, 30
  nonlinear optical effects and, 165, 166
  permanent polarization and, 192
  in piezoelectric films, 339, 340, 341
  of polymers, 70, 73
  porphyrins and, 61
  principles of, 94, 95, 97, 98
  in pyroelectric films, 339, 340, 341, 348, 349
  second-order effects and, 358